NDT in Canada 2016 & 6th International CANDU In-Service Inspection Workshop 2016

Burlington, Canada
15 - 17 November 2016

ISBN: 978-1-5108-3576-4

Printed from e-media with permission by:

Curran Associates, Inc.
57 Morehouse Lane
Red Hook, NY 12571

Printed by Curran Associates, Inc. (2017)

NDT.net
Rheintalstr 3
53498 Bad Breisig
Germany

Phone: + 49 2633 4756670
Fax: + 49 3212-1030513

info@ndt.net

Additional copies of this publication are available from:

Curran Associates, Inc.
57 Morehouse Lane
Red Hook, NY 12571 USA
Phone: 845-758-0400
Fax: 845-758-2633
Email: curran@proceedings.com
Web: www.proceedings.com

TABLE OF CONTENTS

TQNPP FEEDER PIPE DESIGN MODIFICATION AND VALIDATION DURING OPERATION 1
J. Shang

THIRD QINSHAN NUCLEAR POWER PLANT FEEDER PIPE DESIGN MODIFICATION AND VALIDATION DURING OPERATION 11
J. Shang

IN-SERVICE INSPECTION FOR MULTIPLE TYPE NPPS IN CHINA 20
H. Ouyan

APPLICATION OF FLEXIBLE PAUT PROBES FOR SMALL DIAMETER FLOW ASSISTED CORROSION ELBOW INSPECTION II 39
V. Jayasinghe, C. Eaton, B. Millejours

FIELD IMPLEMENTATION OF A TRANSIENT EDDY CURRENT SYSTEM FOR CARBON STEEL PIPE THICKNESS MEASUREMENTS 47
J.A. Buck, C. Kramer, J. Lei, B.A. Lepine

DEPLOYMENT OF PULSED EDDY CURRENT AS AN INSPECTION TOOL FOR NUCLEAR POWER PLANTS 55
V. Demers-Carpentier, M. Rochette, M. Grenier, C. Tremblay, M.M. Sisto, F. Hardy, M. Turgeon

BURIED PIPE NDE OVERVIEW 62
C. Latiolais, S. Kenefick

IMPACT OF ID MAGNETITE ON EDDY CURRENT SIGNALS IN DARLINGTON NGS STEAM GENERATOR TUBES 88
S. Sullivan

APPLICATION OF BAYESIAN INVERSION TECHNIQUES ON MULTI FREQUENCY EDDY CURRENT DATA FOR SUB SURFACE CRACK PROFILING IN STEAM GENERATOR TUBING 103
S.A. Warraich, T.M.R. Khan, M.U. Hassan

FINITE ELEMENT SIMULATION OF LASER-BASED NON-CONTACT NONDESTRUCTIVE EVALUATION 112
S.M. Mirsadeghi

EDDY CURRENT DETECTION OF PARTIALLY CONDUCTING FLAWS IN PICKERING NGS STEAM GENERATOR TUBES 124
S. Sullivan

QUALIFICATION OF ULTRASONIC TESTING OF CAT® RIDGED DUMP TRUCKS BALL STUDS FROM THE INFERIOR POSITION OF THE STUD 132
N.K. Young

HOW TO IMPLEMENT A POD INTO A HIGHLY EFFECTIVE INSPECTION STRATEGY 146
M. Wright

CANDU INSPECTION QUALIFICATION BUREAU (CIQB) THE APPLICATION OF THE EUROPEAN QUALIFICATION METHOD TO INSPECTION OF CANDU NUCLEAR COMPONENTS 154
J. Weed, K. Jang

20 YEARS OF PERFORMANCE DEMONSTRATION ACTIVITIES IN UNITED STATES 164
C. Latiolais

NDE RELIABILITY UPDATE: A GUIDE FOR THE REVIEW AND IMPLEMENTATION OF RECENT U.S. NUCLEAR POWER INDUSTRY NDE RELIABILITY GUIDELINES AND REPORTS 173
C. Latiolais

THICK PLATE DEMONSTRATION PIECES WITH CRACKS ON DEMAND FOR PAUT SCAN TO CSA W59-13 REQUIREMENTS 179
V. Vaidya, T. Ouellet, J. Carignan, J. Magny, P. Holloway

A MODEL-BASED STUDY OF THE EFFECT OF SEMI-ELLIPTICAL SURFACE NOTCH GEOMETRY ON THE SIGNAL OF A SPLIT-D EDDY CURRENT PROBE 189
E. Mohseni, M. Viens, D.R. Franca

ALTERNATIVE EXAMINATION COVERAGE: ASME CODE CASE N711 199
P. O'Regan

DEVELOPMENT OF TARGET FLAW SIZES FOR CANDU PRESSURE VESSEL WELDS FOR USE IN INSPECTION QUALIFICATION 207
S. Yang, R. Yee, K. Jang, J. Baron, J. Weed

ANDE-1 COMPARISON WITH OTHER NDE PQ&C PROCESSES 223
M.L. Turnbow

TRUE ADVANCEMENTS FOR LONGITUDINAL WELD PIPE INSPECTION IN PA 233
P. Rioux, J. Turcotte, P. Cyr

FLAW DETECTION AND CHARACTERIZATION USING ULTRASOUND 241
A. Karpelson

PULSED EDDY CURRENT MEASUREMENTS OF BULK STEEL PIPE WALL THICKNESS 280
K. Faurschou, J. Morelli, T.W. Krause, P.R. Underhill

CINDE IN CONJUNCTION WITH CANDU: ULTRASONIC CREEP WAVE TECHNIQUE 293
N. Marx

A METHOD FOR QUANTITATIVE ANALYSIS OF TRANSIENT EDDY CURRENT TESTING 310
D. Horn, B.A. Lepine, J. Lei

LOW FREQUENCY EDDY CURRENT FOR SUB-SURFACE CRACK SIZING 318
P.R. Underhill, T.W. Krause

LOW TEMPERATURE COMPLIANCE TESTING OF WIND TURBINE APPLICATIONS FOR THE COLD CLIMATE MARKET 328
N/A

PULSED EDDY CURRENT DETECTION OF LOOSE PARTS IN NUCLEAR REACTOR STEAM GENERATORS 339
T.W. Krause, D.P. Johnston, J.A. Buck, P.R. Underhill, J. Morelli

NEW METHODOLOGY FOR NDT ASSESMENT OF IN-SERVICE WOOD POLES 350
M.J. Rodriguez-Roblero, F.J. Tallavo, G. Cascante, M.D. Pandey

THREE-DIMENSIONAL NONDESTRUCTIVE EVALUATION OF CYLINDRICAL OBJECTS (PIPE) USING AN INFRARED CAMERA COUPLED TO A 3D SCANNER 358
F.B.D. Dizeu, S. Hesabi, D. Laurendeau, A. Bendada

APPLICATION OF A FMC/TFM ULTRASONIC SYSTEM TO INSPECTION OF AUSTENITIC WELDS 369
R.T. Grotenhuis, A. Chen, A. Hong, Y. Verma

NON-DESTRUCTIVE MEASUREMENT OF RESIDUAL STRESSES BY ULTRASONIC METHOD 378
Y. Kudryavtsev, J. Kleiman

FINITE ELEMENT MODEL OF EDDY CURRENT MEASUREMENT OF PRESSURE TUBE TO CALANDRIA TUBE GAP 395
G. Klein, M. Luloff, T.W. Krause, J. Morelli

CRACK DETECTION WITH SELF-POWERED WIRELESS SMART COATING SENSOR 409
N. Wu, S. Osho, F. Faisai, S. Zhao

STEAM GENERATOR TUBE INSPECTION AND FOREIGN OBJECT RETRIEVAL IN WOLSONG NPP 422
J.U. Lee

Author Index

中核核电运行管理有限公司
CNNP Nuclear Power Operations Management Co., Ltd.

TQNPP Feeder Pipe Design Modification and Validation during Operation

Third Qinshan Nuclear Power Plant

November,2016

中核核电运行管理有限公司
CNNP Nuclear Power Operations Management Co., Ltd.

Contents

1. Background
2. Wall thinning mechanism
3. Design modification
4. Operation control and monitoring
5. Conclusion

1. Background

- Dec. 1996 , TQNPP was approved by China and Canada governments.
- Media reported PLG& G-2 Feeder bend thinning rate exceed design.

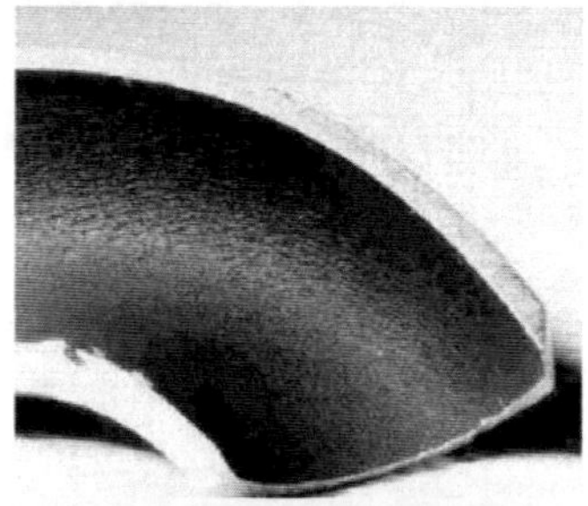

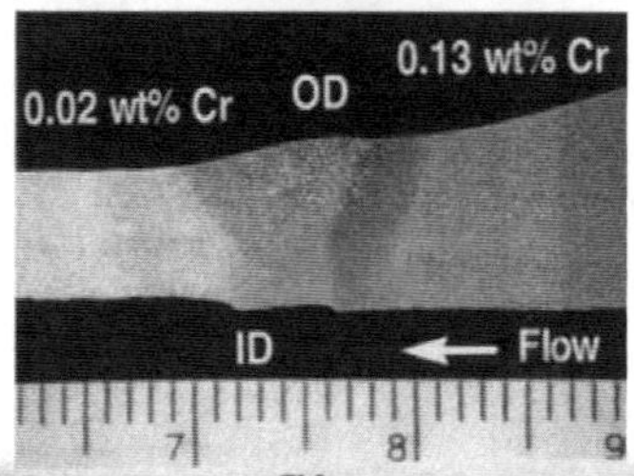

中核核电运行管理有限公司
CNNP Nuclear Power Operations Management Co., Ltd.

2. Wall thinning mechanism

- Mechanism of FAC of Carbon Steel

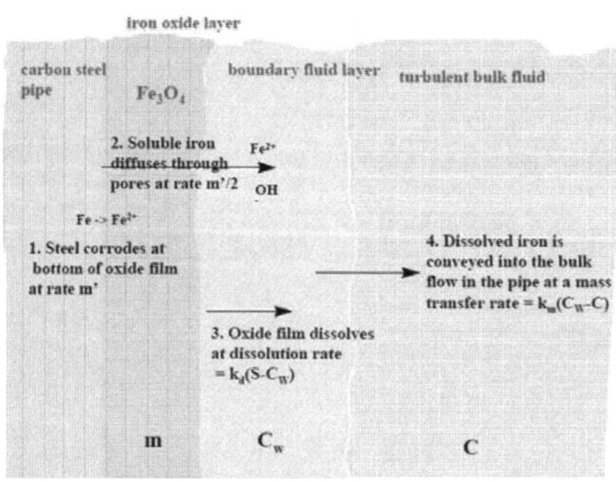

2. Wall thinning mechanism

- **Iron transport around the CANDU PHTS**

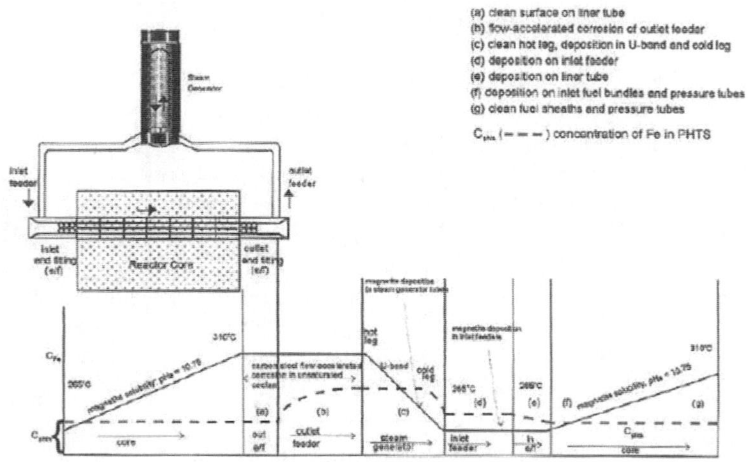

中核核电运行管理有限公司
CNNP Nuclear Power Operations Management Co., Ltd.

3、Design modification

- **Increase Cr content：** NRU reactor U-2 loop test
- **Adjust pHa in PHTS：** Autoclave circuit test

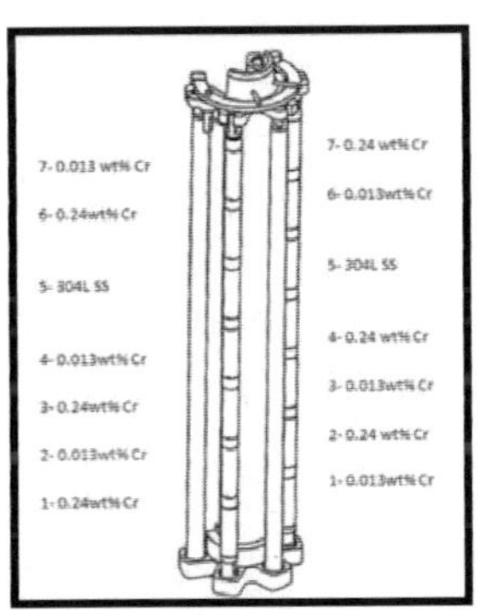

NRU reactor U-2 loop test

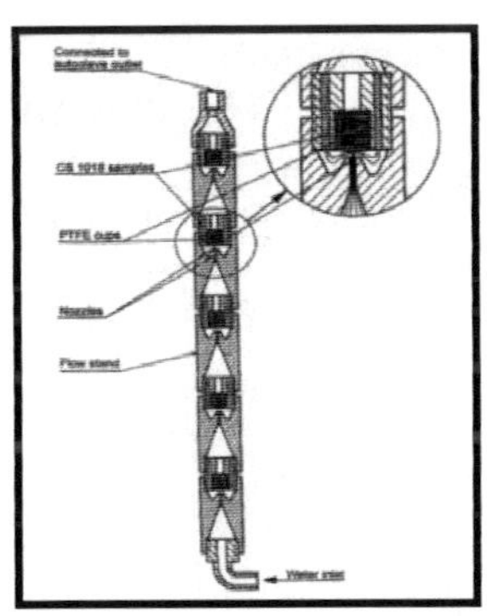

Autoclave circuit test

中核核电运行管理有限公司
CNNP Nuclear Power Operations Management Co., Ltd.

3、Design modification

NRU reactor U-2 loop test

- FAC rate during a test of 611 days

position	FAC rate 0.013%Cr (μm/EFPY)	FAC rate 0.24%Cr (μm/EFPY)	FAC reduction (%)
1	58	19	68
2	49	18	64
3	44	18	59
4	39	18	55
6	36	18	49
7	33	17	49

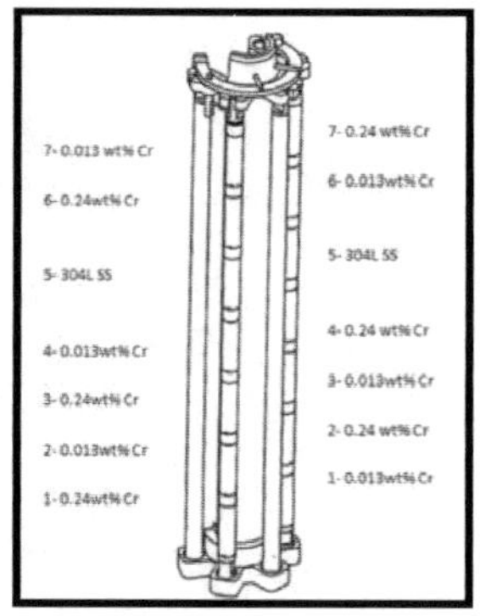

中核核电运行管理有限公司
CNNP Nuclear Power Operations Management Co., Ltd.

3、Design modification

Autoclave circuit test

- pHa decrease from 10.8 to 10.2, the reduction of FAC corrosion rate is about 25～45%.

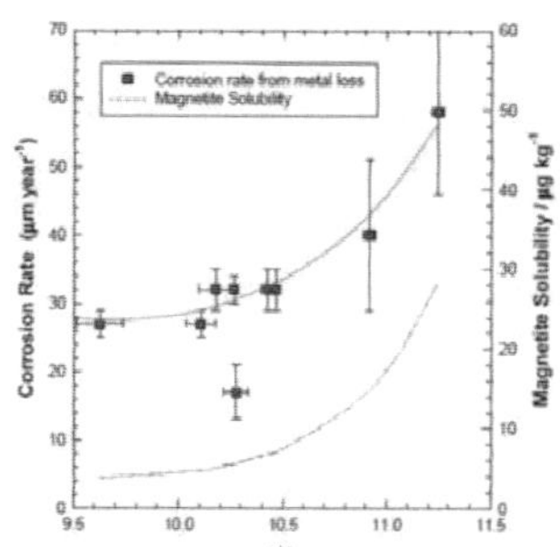

中核核电运行管理有限公司
CNNP Nuclear Power Operations Management Co., Ltd.

3、Design modification

Evaluation of design life

- Point Lepreau: 2.5″- 120μm/EFPY, 2″-110μm/EFPY
- TQNPP: Calculate with reduction of 60%

TQNPP	2 inch	2.5 inch
Nominal wall thickness (mm)	5.54	7.01
FAC corrosion allowance (mm)	2.12	2.88
FAC Rate (mm/EFPY)	0.044	0.048
Predicted life (EFPY)	48	60

中核核电运行管理有限公司
CNNP Nuclear Power Operations Management Co., Ltd.

4、Operation control and monitoring

- **PHTS Chemistry Control： pHa 10.2~10.4**

	Unit1 pHa		Unit2 pHa	
	Average	STDEV	Average	STDEV
2003	10.31	0.11	10.42	0.17
2004	10.31	0.09	10.30	0.08
2005	10.34	0.09	10.36	0.09
2006	10.34	0.08	10.36	0.09
2007	10.31	0.07	10.31	0.08
2008	10.31	0.05	10.32	0.08
2009	10.33	0.06	10.32	0.07
2010	10.32	0.06	10.31	0.06
2011	10.35	0.05	10.33	0.07
2012	10.37	0.08	10.35	0.07
2013	10.34	0.07	10.34	0.06
2014	10.32	0.05	10.34	0.08
2015	10.34	0.07	10.30	0.06

4、Operation control and monitoring

➢ **2015, 2 units PHTS pHa values**

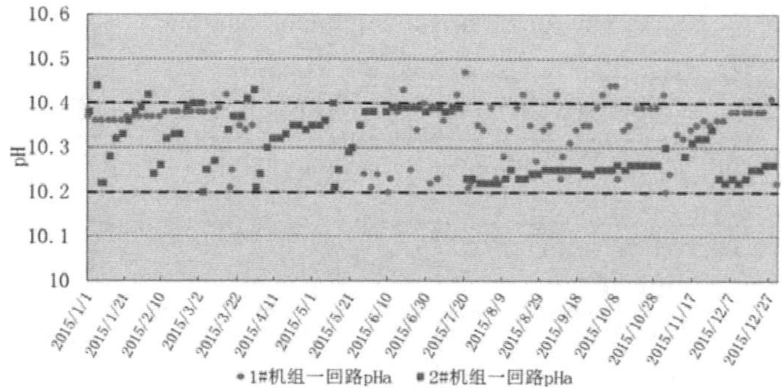

中核核电运行管理有限公司
CNNP Nuclear Power Operations Management Co., Ltd.

4、Operation control and monitoring

Manual and Semi-auto UT thickness measurement

➢ Accuracy±0.03mm ,roughness <1.6um

Manual **Semi-auto**

4、Operation control and monitoring

Inspection history

- measured all Feeder pipe thickness in factory
- 1ST ISI after operated 5 years
- 2nd ISI after operated 10 years

	1st ISI OT103	2nd ISI OT106/107	1st ISI OT204	2nd ISI OT206/207
Outlet Feeder Number	33	25	34	31
Valid Feeder Number	24	23	16	21

中核核电运行管理有限公司
CNNP Nuclear Power Operations Management Co., Ltd.

4、Operation control and monitoring

Analysis methods:

1. Linear regression analysis for FAC rate using (factory/ first ISI/ second ISI) data
2. Linear regression analysis for FAC rate versus flow velocity
3. Not distinguish 2" and 2.5"

4、Operation control and monitoring

- **1ST ISI , 5 years, no relation between FAC rate and velocity**

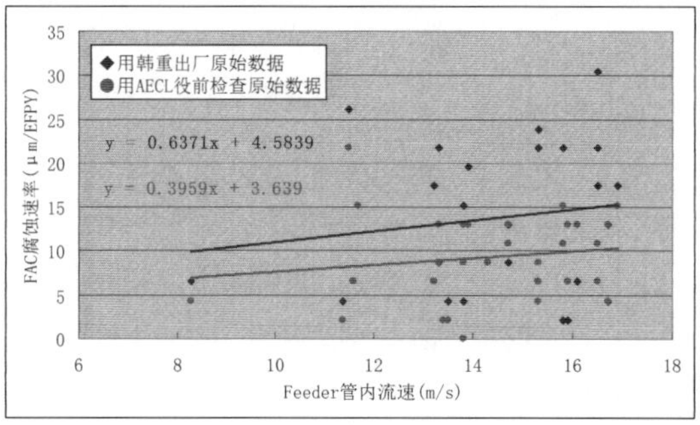

中核核电运行管理有限公司
CNNP Nuclear Power Operations Management Co., Ltd.

4、Operation control and monitoring

2nd ISI, 10 years, obvious relation between FAC rate and velocity

- Linear regression formula , maximum rate : 19 um/EFPY
- plus 2 × STDEV , maximum rate: 32 um/EFPY

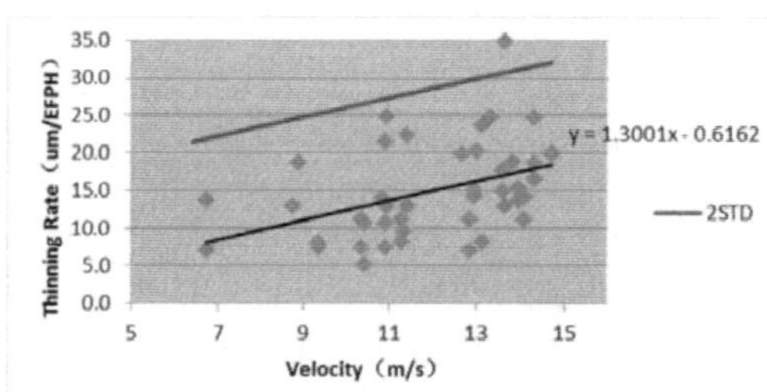

中核核电运行管理有限公司
CNNP Nuclear Power Operations Management Co., Ltd.

4、Operation control and monitoring

- **Predicted life**

TQNPP	2 inch	2.5 inch
Nominal wall thickness (mm)	5.54	7.01
FAC corrosion allowance (mm)	2.12	2.88
FAC Rate (mm/EFPY)	0.032	
Predicted life (EFPY)	66	90

中核核电运行管理有限公司
CNNP Nuclear Power Operations Management Co., Ltd.

5. Conclusion

- The modifications to increase the Cr content to 0.20~0.40wt%, and control the pHa value within the range of 10.2~10.4. The engineering verification test over four years and the operation of TQNPP over ten years have proved the design modifications are successful.
- TQNPP Feeder pipe Cr content was controlled at 0.32~0.33wt%. After over ten years of operation, the in-service inspection results showed that the FAC thinning rate was less than 32μm/EFPY, TQNPP Feeder pipe can be used safely in the design life of 34 EFPY, and the life can be extend up to 66 EFPY.

中核核电运行管理有限公司
CNNP Nuclear Power Operations Management Co., Ltd.

Question?

Third Qinshan Nuclear Power Plant Feeder Pipe design modification and validation during operation

SHANG.Junmin

China Nuclear Power Operations Management Co.,Ltd. Haiyan, Zhejiang, China
Phone: +86-573-86652281, e-mail:shangjm@cnnp.com.cn

Abstract: Flow accelerated corrosion (FAC) is the main cause to the Feeder wall thinning in CANDU PHWR reactor nuclear power plant, AECL made design modification to the Feeder pipes in TQNPP, by increasing the chrome content to 0.33wt% the first time. In this article, the mathematical statistical analysis method is used to analyze the wall thickness measured data of the Feeder pipes of TQNPP. The analysis shows that the Feeder pipes wall thinning complies with the FAC theory, which proves that the AECL design modification is effective, and the Feeder pipes of TQNPP can meet the design life and even extend the service life.

Key words: Feeder, design modification, FAC, thinning,

1. General

In Dec. 1996, the governments of China and Canada approved the construction of two CANDU 6 PHWRs at Third Qinshan Nuclear Power Plant (TQNPP). Just at that time, serious flow accelerated corrosion (FAC) problem was found at the outlet Feeder pipe of the Point Lepreau and G-2 CANDU 6 nuclear power plants in Canada: the wall thinning rate of the outlet Feeder pipe bend was much higher than the design value, which meant that the requirements for 30-year design life could not be met [1]. TQNPP and AECL (Atomic Energy Canada Ltd.) conducted special emergency meeting on this issue, AECL analyzed the FAC corrosion mechanism of the outlet Feeder pipes and proposed design modification and engineering verification plan. After over four year's laboratory tests, it was verified that the design modifications were appropriate. The Feeder pipes made for TQNPP according to the design modifications have been operated for over 10 years. The two in-service inspections verified that the wall thinning rate of the outlet Feeder pipes first bend was lower than the expected design value, so the 40 years with 85% capacity design life can be guaranteed.

2. Feeder pipe wall thinning mechanism and design modifications

2.1 Causes of Feeder pipe wall thinning [1] [2]

For the primary heat transfer system of CANDU 6 reactor, the coolant inlet temperature is 266℃ and outlet temperature 310℃, with a temperature difference of 44℃; the maximum coolant flow rate is 17.6m/s; the pHa range of the coolant is 10.2~10.8; the Feeder pipes are made of SA-106B, and in early Feeder pipes the Cr content was 0.02wt%, these factors have provided necessary conditions required for FAC.

Fig. 1 shows the dissolved iron transport process in the primary heat transfer system of the CANDU 6 reactor: after the coolant enters the steam generator at 310℃, its temperature starts to decrease, after the temperature has decreased to a certain level, the iron dissolved in the coolant was saturated and deposited, the iron dissolved in the coolant deposits inside the steam generator heat transfer tubes, the cold leg of main pipes and inlet Feeder pipes and the dissolved iron concentration in the coolant thus decreases. The coolant is then heated up to 310℃ in the core, and the dissolved iron in the coolant becomes unsaturated again. At the outlet Feeder pipes, outlet header and main pipe hot leg, corrosion occurs due to the FAC mechanism. The most serious corrosion occurs at the outlet Feeder pipes first bend. At this position, the difference between the actual value of coolant iron concentration and the saturated solubility is the highest, and the water flow is in an unstable turbulent state, favorable to the dissolution, diffusion and transport acceleration of iron.

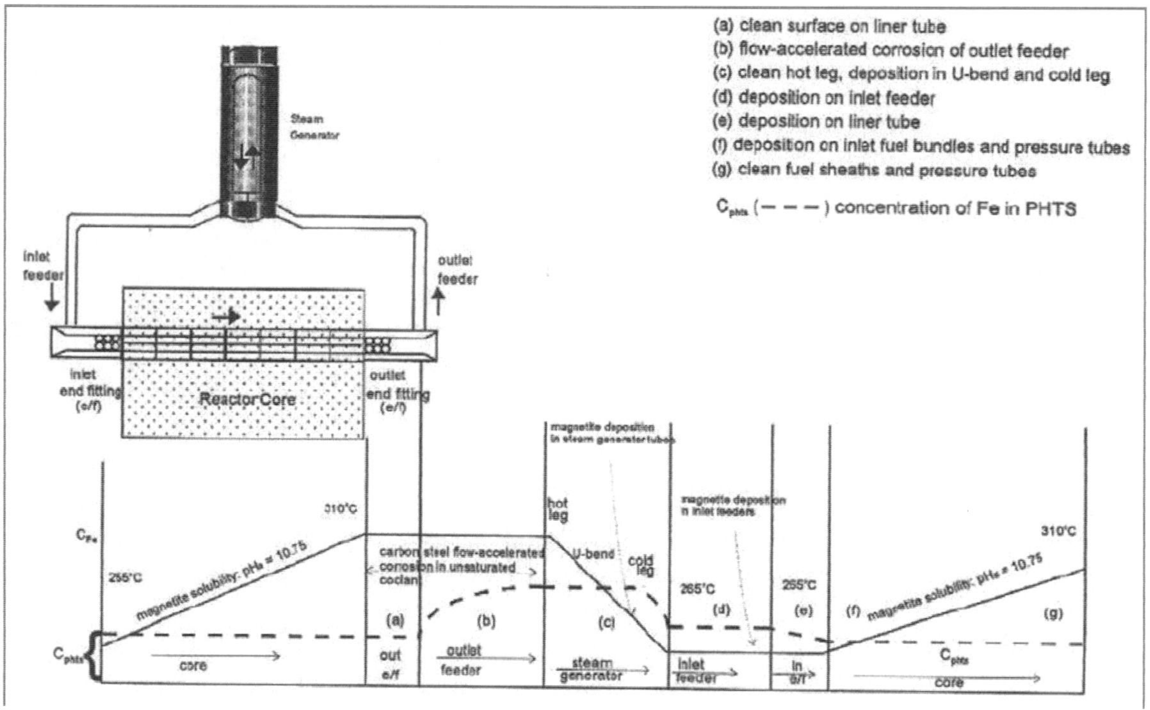

Fig. 1 Iron transport in the primary heat transfer circuit of CANDU

2.2 Design modifications and engineering verification tests

When the construction of TQNPP was planned, CANDU 6 reactor was already a mature commercial operating reactor, so the coolant temperature and flow rate of the primary heat transfer system could not be easily changed. According to the research of the FAC mechanism in the industry, the Feeder pipes FAC rate could only be reduced in practice by adjusting the Cr content and the pHa value of coolant.

2.2.1 Increase Cr content

The Feeder pipes are made of SA-106B carbon steel, which has a permitted Cr content range of 0~0.4wt%. However, in the Feeder pipes of early CANDU units, the Cr content was only about 0.02wt%. According to research, the FAC rate would be reduced substantially when the Cr content in carbon steel exceeds 0.04wt%. TQNPP agreed to increase the Cr content of the Feeder pipes to 0.20 ~ 0.40wt%, and performed a number of engineering verification tests to verify the modification, the successful test was performed on the NRU reactor U-2 loop.

In the U-2 loop at the Chalk River Laboratory in Canada, FAC comparison tests for SA-106B carbon steel with two different Cr contents were conducted. The test sample rack is as shown in Fig. 2. On the rack, two strings of test samples were arranged, and the samples of 0.013wt%Cr and 0.24wt%Cr were arranged alternatively. In the test loop, the PHa was ~10.3 (equivalent to PHa10.7 in the heavy water in the primary circuit), water temperature ~307℃, and flow rate 10m/s, the sample rack was designed to make the water flow in a serious turbulent state, so that all other conditions were similar to those in the outlet Feeder pipes of CANDU 6 reactor, except for the flow rate which was a bit lower. Under such conditions, FAC corrosion could occur on samples.

The samples were analyzed for thinning at the time of reactor shutdown maintenance. During the test period, sample analyses were made 14 times.

After 611 days tests, the statistic values of the FAC thinning rate reduction with high Cr content and low Cr content are presented in table 1. It can be seen that, at different locations of the rack, the FAC reduction percentage between samples of 0.24wt%Cr and 0.013wt%Cr are different, with an average value of approximately 50%. The reduction is as high as 68% at position 1 where the flow is highly turbulent. The FAC corrosion rate of the sample with 0.013wt%Cr measured in the test was 58μm/EFPY, at Point Lepreau, the measured maximum FAC corrosion rate at the 2.5″ outlet Feeder pipes first bend was 120μm/EFPY, 2.0″ outlet Feeder pipes first bend was 110μm/EFPY. The test value was about the median value in the measured distribution, as the maximum flow rate of the Feeder pipes in Point Lepreau is 17.6m/s, and the flow rate in the test was lower at 10m/s, therefore this result was reasonable.

Table 1 FAC rate of samples with high and low Cr contents during a test of 611 days

Sample position	**Corrosion rate of samples with 0.013wt%Cr (μm/EFPY)**	**Corrosion rate of samples with 0.024wt%Cr (μm/EFPY)**	**FAC reduction (%)**
1	58	19	68
2	49	18	64
3	44	18	59
4	39	18	55
6	36	18	49
7	33	17	49

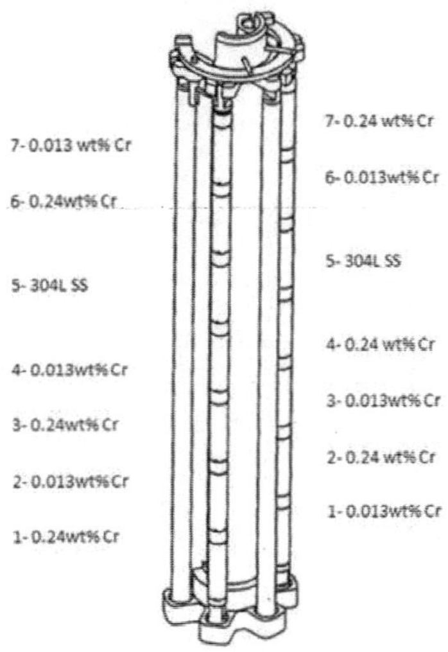

Fig. 2 FAC test sample rack in U2 loop

2.2.2 *Adjust pHa value*[2]

According to technical requirements, in the primary heat transfer system of the CANDU 6 unit, the coolant pHa value should be 10.2~10.8, and outlet temperature 310℃. The research results show that at 310℃ and in the range of pHa>10.0, the higher pHa, the higher solubility of Fe_3O_4 in the coolant. Therefore within the scope of the original technical requirements, the coolant pHa value was adjusted to the lower limit as appropriate. In the operation, pHa was controlled at 10.2~10.4 whenever possible. This practice can reduce the solubility of Fe_3O_4 in the coolant, thus mitigating the FAC rate.

To verify the effect of pHa value on the FAC rate, the autoclave jet impingement test was designed, and the test apparatus is as shown in Fig. 3.

In the autoclave circuit, the water temperature is 310℃ and water flow 1.2m/s, at the high temperature of 310℃, the pHa value was controlled at 9.2~10.8 (equivalent to pHa of 9.6~11.2 in heavy water), and the test lasted 14 days for each pH value. The test results are as shown in Fig. 4. It can be seen that, with the increase of pH value, the solubility of Fe_3O_4 in the solution increases, and the homogeneous corrosion rate on the carbon steel surface also increases. As converted, it is corresponding to reduction of pHa value of heavy water from 10.8 to 10.2, and the corrosion rate is reduced to 25~45% of the original.

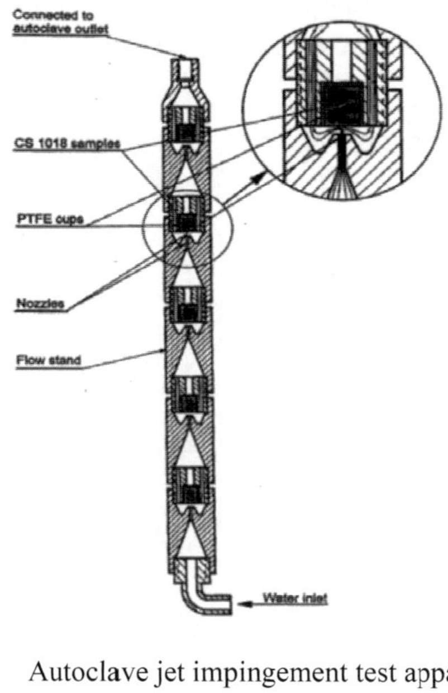

Fig. 3 Autoclave jet impingement test apparatus

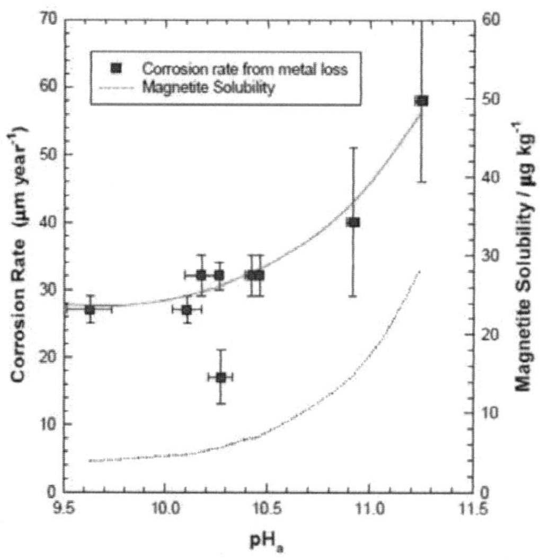

Fig. 4 Effect of pHa on corrosion and release of dissolved iron

2.3 Conclusions of engineering verification tests

1) It was verified in the U2 loop test that, after 611 days test, the FAC rate is reduced by about 50% with carbon steel of 0.20wt%Cr as compared with carbon steel of 0.013wt%Cr.

2) It was verified in the autoclave jet impingement test that: under the water chemistry conditions of the Feeder pipes of CANDU unit (pHa: 10.2~10.8), the pHa decrease from 10.8 to 10.2, the reduction of FAC corrosion rate is about 25～45%.

2.4 Evaluation of design life

For Feeder pipes of 2″ and 2.5″, the minimum permitted wall thickness and the corrosion allowance are as shown in Table 2.

Table 2 Design data of Feeder pipes

	2 inch	2.5 inch
Nominal wall thickness (mm)	5.54	7.01
FAC corrosion allowance (mm)	2.12	2.88
Predicted life（EFPY）	48	60

At Point Lepreau, the FAC maximum corrosion rate of 2.5" outlet Feeder pipes was 120μm/EFPY and 110μm/EFPY of 2" pipes, according to the evaluation of FAC rate reduction by 60% as shown in the test results, after the design modifications, the FAC rate of the Feeder pipes at TQNPP is respectively 48μm/EFPY and 44μm/EFPY of 2.5" and 2" pipes, and the service life will extend to 48 EFPY for 2" pipes and 60 EFPY for 2.5″ pipes.

3. Operation control and monitoring

3.1 Chemical control[3]

In TQNPP chemical control procedure, it is specified that the pHa value of heavy water in the primary heat transfer system should be controlled at 10.2~10.8, with the desired value of 10.2~10.4, and corrective actions shall be taken if this scope is exceeded.

To measure the pHa value of primary circuit, 2 samples are taken every week, and the statistics of the two TQNPP units for years are presented in Table 3. It can be seen that the pHa value has been strictly controlled within the desired value range of 10.2~10.4. If this scope is exceeded, proper actions would be taken in order to control the FAC rate of the Feeder pipes.

Table 3 PHa value of primary circuit in the CANDU 6 units of Qinshan

	pHa of Unit 1#		pHa of Unit 2#	
	Average value	Standard deviation	Average value	Standard deviation
2003	10.31	0.11	10.42	0.17
2004	10.31	0.09	10.30	0.08
2005	10.34	0.09	10.36	0.09

	pHa of Unit 1#		pHa of Unit 2#	
	Average value	Standard deviation	Average value	Standard deviation
2006	10.34	0.08	10.36	0.09
2007	10.31	0.07	10.31	0.08
2008	10.31	0.05	10.32	0.08
2009	10.33	0.06	10.32	0.07
2010	10.32	0.06	10.31	0.06
2011	10.35	0.05	10.33	0.07
2012	10.37	0.08	10.35	0.07
2013	10.34	0.07	10.34	0.06
2014	10.32	0.05	10.34	0.08
2015	10.34	0.07	10.30	0.06

3.2 In-service inspection

3.2.1 Error in ultrasonic thickness measurement

At TQNPP, manual and semi-automatic ultrasonic thickness measurement methods have been adopted to measure the thickness at the outlet Feeder pipes first bend. The instruments and positioning clamps used in the two methods are as shown in Figs. 5 and 6.

There is system error in the ultrasonic gauge, probe and analysis software themselves, with the two methods, the measured accuracy on a test piece with surface roughness <1.6um is ±0.03mm, the same as the devices used in other CANDU plants.

With the extension of operation time, and under the effect of FAC, the inner surface of Feeder pipes was scoured into scallops shape. Fig. 7 shows the SEM image of the inner surface of S08 channel Feeder pipes at Point Lepreau after operation for 12.4 EFPY, the size of individual scallops ranges from 0.1mm to 0.9mm, and the roughness peak value is about 0.1mm. The inner surface roughness of Feeder pipes affects the accuracy of ultrasonic thickness measurement, and the measured wall thickness data will change with any slight change of the position and angle of the probe.

Another difficult point of ultrasonic thickness measurement is the very small gap between the Feeder pipes, there is very small space to mount the measuring device at the bend and the operation is difficult. Therefore, the in-service wall thickness measurement of the Feeder pipes of the CANDU units is quite difficult.

Fig. 5 Manual ultrasonic thickness measuring instrument and positioning device

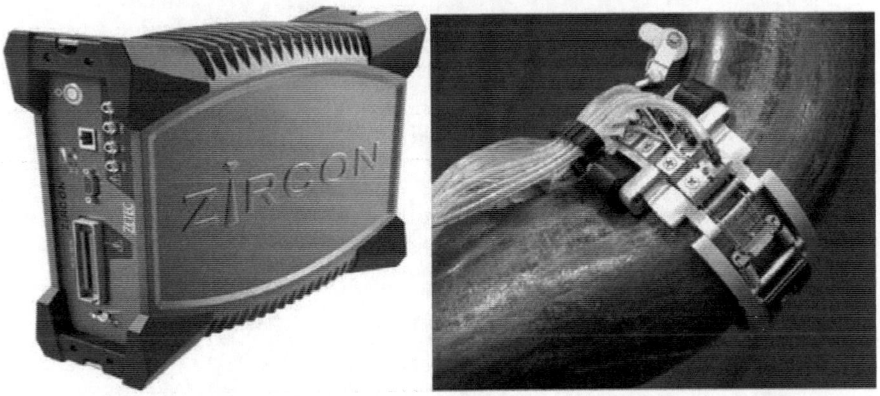

Fig. 6 Semi-automatic ultrasonic thickness measuring instrument and positioning device

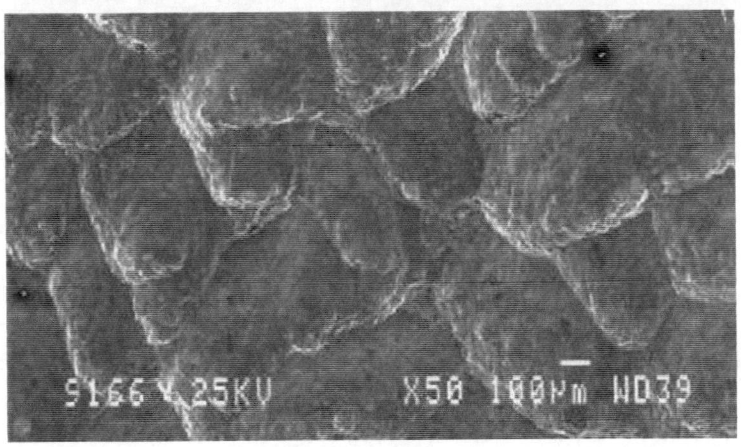

Fig. 7 SEM image of Feeder inner surface

3.2.2 In-service wall thickness measurement[4][5]

Due to the error of the system itself and the roughness of Feeder inner surface, it is quite difficult to find the thinnest point with ultrasonic measurement. In processing the thickness measurement data, those greater than the original wall thickness and obviously big values were taken as invalid data, which is in line with the conventional analysis method, and the analysis of thinning trend after discarding such invalid data is more realistic.

For the two units at TQNPP, the second in-service inspection was performed at 8 ~10EFPY. After operation for about 10 years, the feeder wall was thinned obviously. For unit 1, 25 outlet Feeder pipes were measured in the OT106 in 2012 and OT107 in 2013, data for 23 pipes were valid; for unit 2, 31 outlet Feeder pipes were measured in the OT206 in 2012 and OT207 in 2014, data for 21 pipes were valid.

The wall thickness thinning rate of each Feeder pipe was analyzed based on the manufacturer and in-service thickness measuring data using the linear regression method. Mathematical statistical analysis was performed for the thinning rate and flow rate of Feeder pipes, and a scatter diagram of these data was plotted (Fig. 8). Linear regression analysis was made for the relationship between the thinning rate and flow rate, and the linear fitting formula in the figure is:

$y = 1.3001x - 0.6162$ (μm/EFPY)

As estimated using this formula, the FAC thinning rate of the Feeder with the maximum flow rate (14.7 m/s) is 19μm/EFPY.

In a conservative estimation, by applying the fitting formula in Fig.8 plus two times the positive standard deviation, the FAC rate of Feeder pipes at the maximum flow of 14.7m/s is 32μm/EFPY. This is the maximum possible corrosion rate for the Feeder pipes in TQNPP, and it is sufficiently conservative to estimate the service life using this corrosion rate.

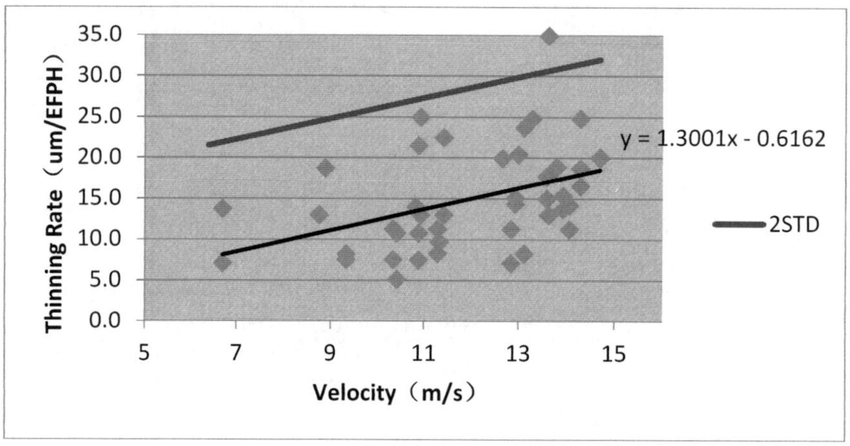

Fig. 8 Thinning trend of Feeder s at TQNPP

3.2.3 Evaluation of service life

To estimate the service life using the FAC corrosion allowance in Table 2, it is 66 EFPY for 2″Feeder pipes and 90 EFPY for 2.5″ Feeder pipes.

In the engineering verification test of AECL, the Cr content was 0.24wt%, and it was estimated that the service life would be 48 EFPY for 2″ Feeder pipes and 60 EFPY for 2.5″ Feeder pipes. The actual Cr content of Feeder pipes at TQNPP is 0.32%, and the measured FAC corrosion rate is lower than that measured in the verification test, therefore it is reasonable to get longer service life than estimated.

4. Conclusion

The modifications adopted to solve the problem of FAC thinning of Feeder pipes are to increase the Cr content to 0.20~0.40wt%, and at the same time control the pHa value within the range of 10.2~10.4. The engineering verification test over four years and the operation of TQNPP over ten years have proved that the design modifications are successful.

In the manufacturing of the Feeder pipes of TQNPP, the Cr content was controlled at 0.32~0.33wt%. After over ten years of operation, the in-service inspection results showed that the FAC thinning rate was less than 32μm/EFPY, TQNPP Feeder pipe can be used safely in the design life of 34 EFPY, and the life can be extend up to 66 EFPY.

5. Reference

[1] AECL (1997) "Report on Feeder Design and Performance" 98-33128-200-001 Rev1

[2] AECL (2001) "Assessment on CANDU6 Feeder Design and Performance" 98-33128-ASD-001 Rev1

[3] "Plant chemical control" of TQNPP 98-93400-OM-001

[4] "Wall thickness measuring report for Feeder bend and Grayloc area in outage 206" 9802-QY-33126-NDE-12002

[5] "Wall thickness measuring report for Feeder bend and Grayloc area in outage 207" 9802-QY-33126-NDE-14001

中国核工业集团公司
China National Nuclear Corporation

In-Service Inspection for Multiple Type NPPs in China

Honghao Ouyang

Director of International Business

China Nuclear Power Operation Technology Corporation, LTD. (CNPO)

6th International CANDU ISI Workshop/NDT in Canada 2016 Conference

Content

1. **Overview for NPP in China mainland**
2. **Requirement and strategy for ISI**
3. **General ISI technologies for NPPs**
4. **Specific ISI technologies for NPPs**
5. **ISI Experience for Multiple Type NPP**
6. **Conclusion**

1. Overview for NPP in China mainland

Units	Province	Net capacity (each)	Operator	Commercial operation
Daya Bay 1&2	Guangdong	944 MWe	CGN	1994
Qinshan Phase I	Zhejiang	298 MWe, 1020 MWe	CNNC	April 1994
Qinshan Phase II, 1,2,3,4	Zhejiang	610 MWe	CNNC	2002, 2004,2010, 2012
Qinshan Phase III, 1&2	Zhejiang	678 MWe	CNNC	2002, 2003
Fangjiashan 1,2	Zhejiang	1020 MWe	CNNC	Dec 2014, (Feb 2015)
Ling Ao Phase I, 1&2	Guangdong	938 MWe	CGN	2002, 2003
Ling Ao Phase II, 1&2	Guangdong	1026 MWe	CGN	Sept 2010, Aug 2011
Tianwan 1&2	Jiangsu	990 MWe	CNNC	2007, 2007
Ningde 1,2,3,4	Fujian	1020 MWe	CGN & Datang	April 2013, May 2014,
Hongyanhe 1,2,3	Liaoning	1024 MWe	CGN & CPI	June 2013, May 2014,
Yangjiang 1,2,3	Guangdong	1021 MWe	CGN	March 2014,
Fuqing 1,2,3	Fujian	1020 MWe	CNNC & Huadian	Nov 2014
Total			30 units	

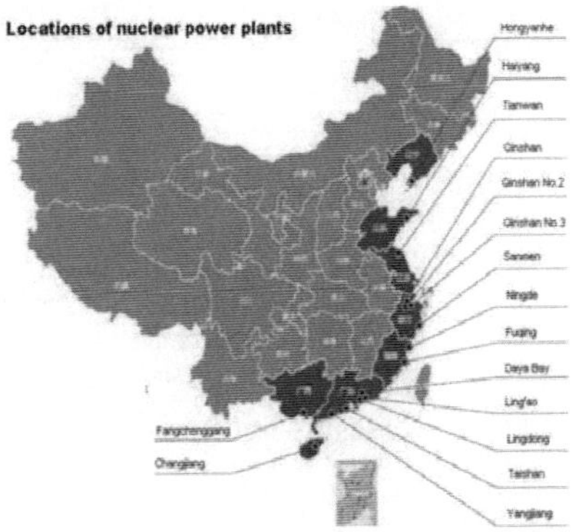

Up to Feb. 2016, 30 units in operation, 24 units Under construction, and about 40 units Planned.

□ Application codes/standards and characterization for NPPs

Units	ISI Code and standard	Main characterizations of NPPs
Daya Bay 1&2	RSE-M 1997	Examination system of RPV and SG is very similar as the PWR NPPs applying with ASME. Only NDT tech. are little different, and manipulators will be modified according to the different NPPs. Specially, for example, examination system and technology for RPV BMI and the seal welds of CRDM.
Qinshan Phase II, 1&2		
Qinshan Phase II, 3&4		
Fangjiashan 1&2		
Ling Ao Phase I, 1&2		
Ling Ao Phase II, 1&2		
Ningde 1,2,3,4,5,6		
Hongyanhe 1,2,3,4,5,6		
Yangjiang 1,2,3,4,5,6		
Fuqing 1,2,3,4,5,6		
Qinshan Phase III, 1&2	CAN/CAS	The pressure Tube is very different from the RPV of PWR NPPs. Specially, thickness Measurement for CANDU feeder pipe and Video Examination for Support and Hanger of Feeder Pipe
Tianwan 1,2,3,4	Russian code	For horizontal Steam Generator, the special SG ET system has been used during ISI
Qinshan Phase I	ASME 1998	For AP1000, Specially, examination system and technology for the weld of RCP to SG nozzle of AP1000 and RPV DVI
Sanmen 1,2		
Haiyang 1,2		

2. Requirement and Strategy for ISI

The different type NPPs comply with different PSI/ISI code

Codes and standards used for NPP:

- French RSE-M(RCC-M)
- USA ASME
- Russia standards
- Canadian standards

For example, different examination scope, periodic, sampling and NDT technology/equipment for the different type NPPs.

2.Requirement and Strategy for ISI

- The strategy of ISI for multiple type reactor in China:
  - To analyze the structure characteristic, application codes and requirement of ISI for all type NPP
  - To R & D the different NDT equipment and to use the specific NDT technologies to satisfied ISI requirement for different type NPPs
  - To keep the NDT data acquisition system is same or similar for different type reactor, i.e. UT and ET instrument, and software

2.Requirement and Strategy for ISI

- To keep some NDT equipments (manipulators), main parts are same or similar
- To manufacture the different mockup NPP for function testing and training
- To prepare different ISI program for different NPPs

3.General ISI Technologies for NPPs

3.General ISI Technology for NPP

3.1 Examination system and technology for Reactor Pressure Vessel

a) Scope of examination: Shell weld, nozzle weld and safe-end weld
b) Technology: Contact ultrasonic, RT and remote Video Exam
c) Manipulator: CRUS/SUPREEM

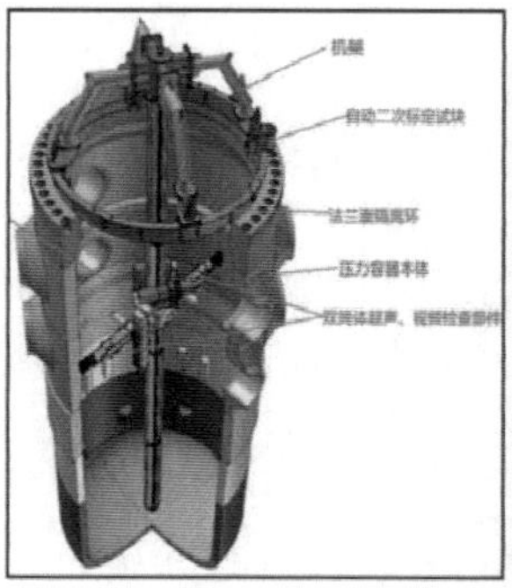

CRUS examination system for RPV

3.General ISI Technology for NPP

1) For shell and nozzle welds: detection by 45T2, 60T2,70TRL probes, sizing by high frequency longitudinal wave probe(i.e. 45L4, 45TRL4)

2) For safe-end welds: UT(45TRL and 70TRL) & RT examination

3) For flange thread area: UT(0L2+45T2) examination

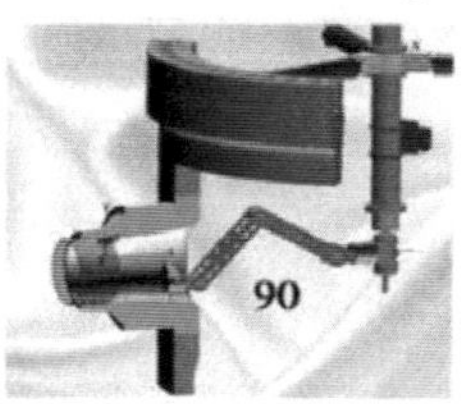

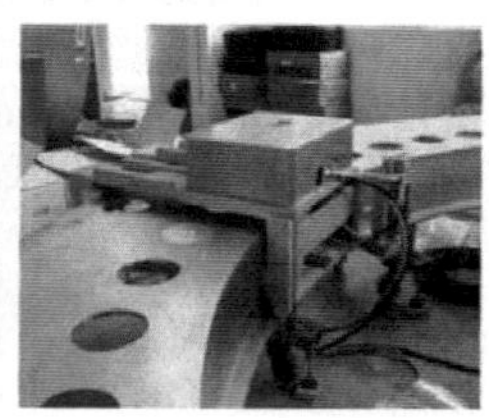

SUPREEM examination system for RPV

3.General ISI Technology for NPP

3.2 Examination system and technology for Steam Generator tubes ET

a) Scope of examination: whole length for 100% tubes
b) Technology: Bobbin, MRPC, Array probe
c) Equipment: "BEEBOT" probe fixture, "CEddy" data acquisition and analysis software system

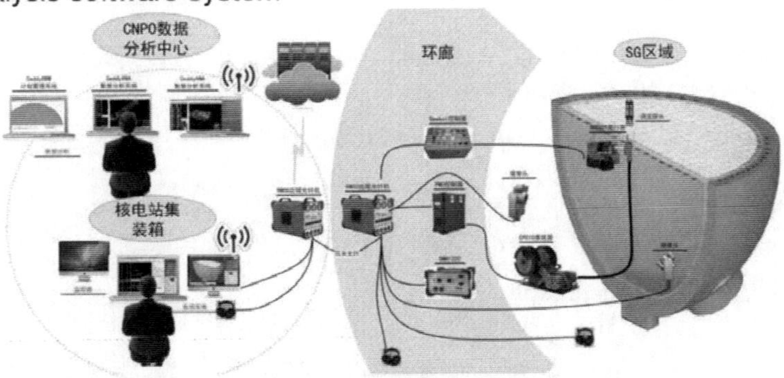

CEddy Examination system for SG tubes

3.General ISI Technology for NPP

- For whole tube: Bobbin probe, MRPC probe for retesting
- For expansion: MRPC or array probe

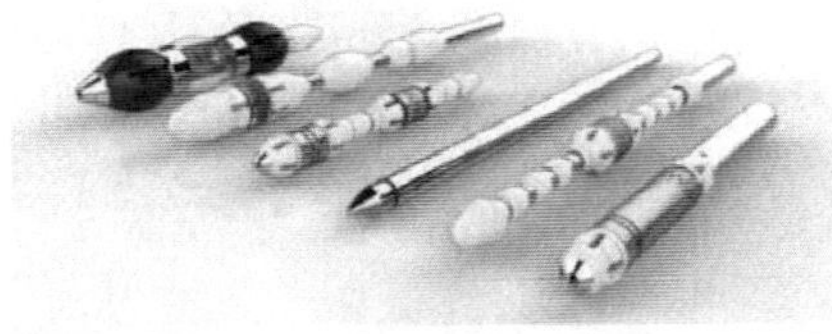

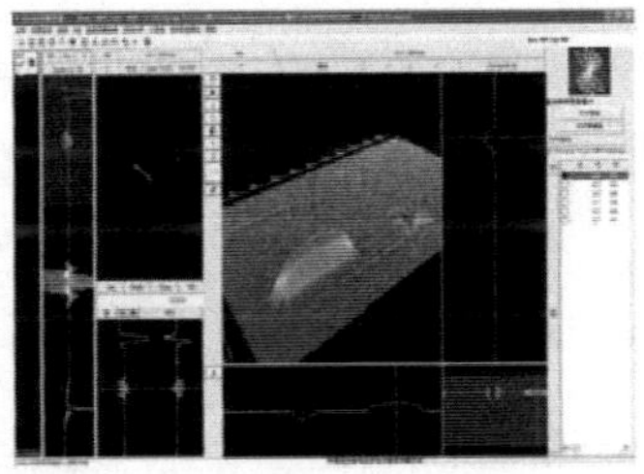

3.General ISI Technology for NPP

3.3 Examination system and technology for Piping welds

a) Scope of examination: weld and head-affected zone

b) Technology:

•For ferrite piping welds: UT examination - detection by transverse wave probe , sizing by longitudinal wave probe.

•For austenite piping welds: UT examination - longitudinal wave probe for welds, transverse wave probe for head-affected zone.

c) Equipment: AUMA manipulator for different pipe size

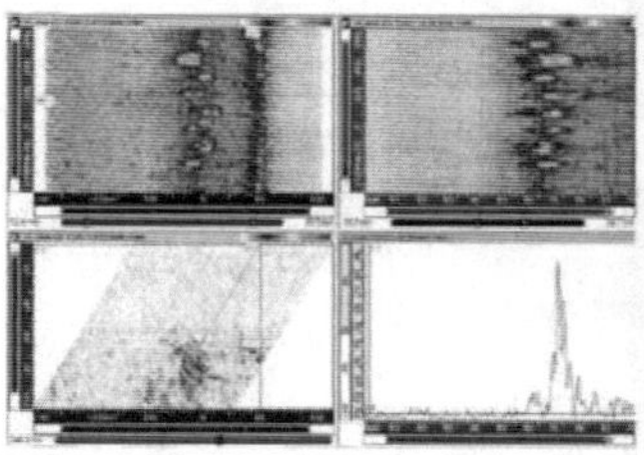

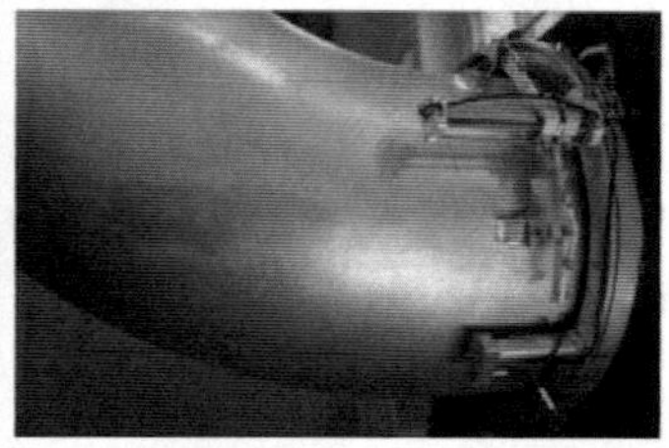

3.General ISI Technology for NPP

3.4 Examination system and technology for Upper Closure Head Penetration of Nuclear Reactor Vessel

a) Scope of examination: the "J" groove welds of CRDM

b) Technology: ET for the surface of welds, UT for the whole volume of welds (TOFD + Pulse Echo)

c) Equipment: CHEAS machine

3.General ISI Technology for NPP

3.5 Examination system and technology for RCCA

a) Scope of examination: defect types including swelling, wear and crack.
b) Technology: immersion focusing UT probe and Eddy current external probe
c) Equipment: RAID machine

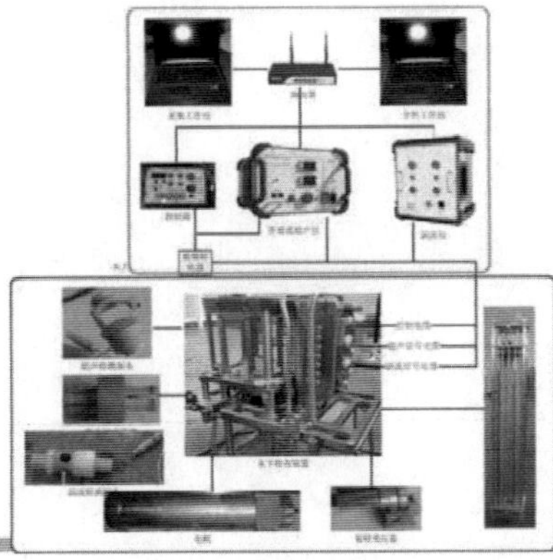

6th International CANDU ISI Workshop/NDT in Canada 2016 Conference 15

3.General ISI Technology for NPP

3.6 Examination system and technology for RPV Stud and Nut

a) Scope of examination: screw region for ET, whole volume for UT
b) Technology: Ultrasonic immersion transducers and eddy current pencil probes
c) Equipment: STAR machine

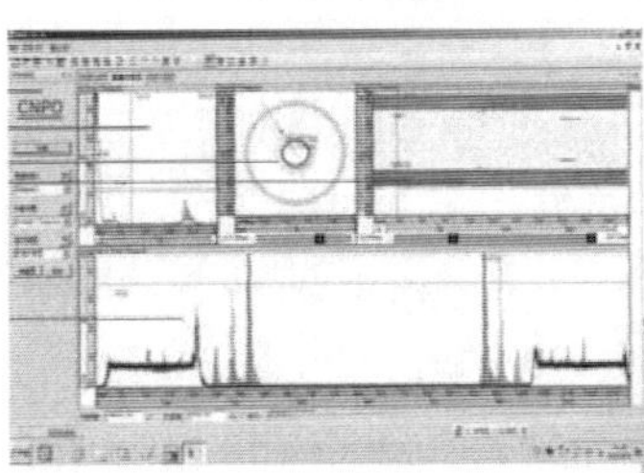

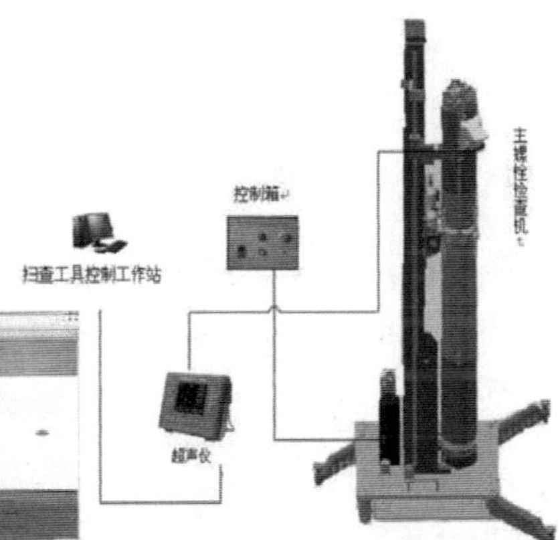

6th International CANDU ISI Workshop/NDT in Canada 2016 Conference 16

4. Specific ISI technology for NPP

4.1 ISI Technology for M310

1) Examination system and technology for RPV BMI

a) Scope of examination: penetration nozzle and heated affected zone.

b) NDT Technology:

- a) Immersion focusing Ultrasonic transducer, Eddy current bobbin and rotating probe
- b) ET is used for the inner surface of penetration nozzle
- c) UT is used for the nozzles and welds between nozzle and vessel.

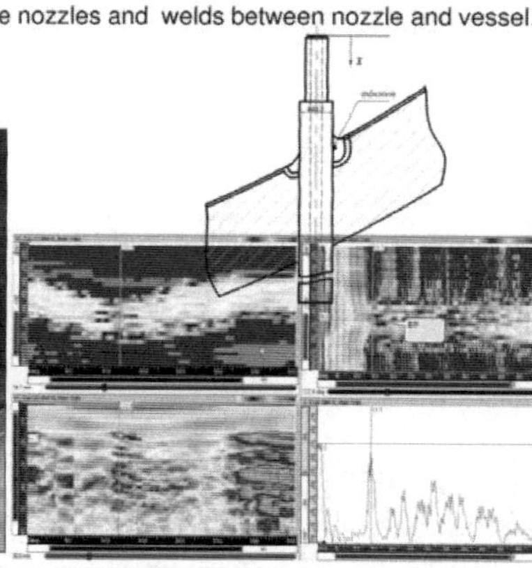

4.1 ISI Technology for M310

UT Technology:

① axial/circ. TOFD
② S45°
③ L21°

ET technology:

① MRPC for detection
② Bobbin for sizing

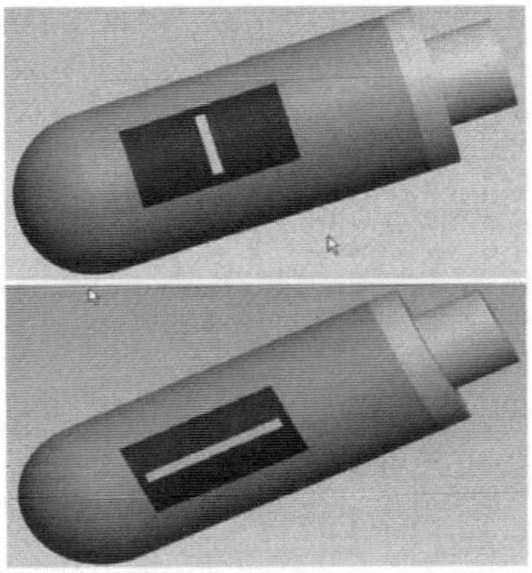

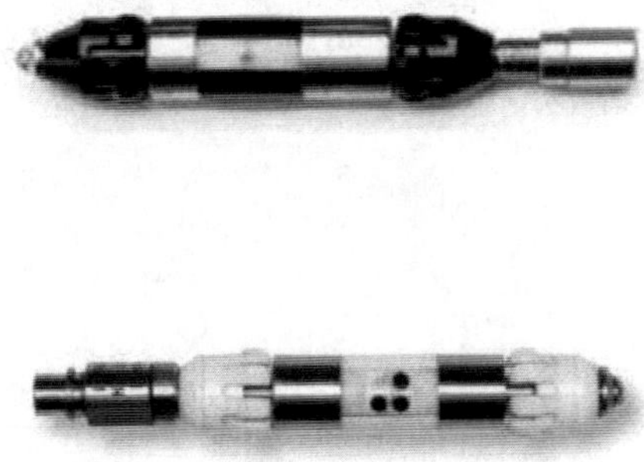

4.1 ISI Technology for M310

2) Examination system and technology for fail fuel rod assemblies
   a) Scope of examination: fuel tube
   b) NDT Technology: the lamb wave UT, Through to checking if there is the water in the tube to confirm the leakage condition of it
   c) Equipment: FUMA manipulator

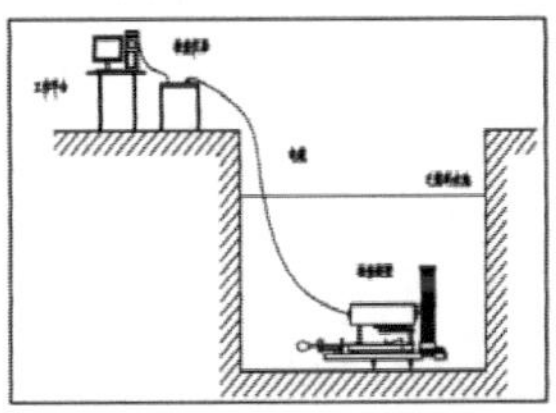

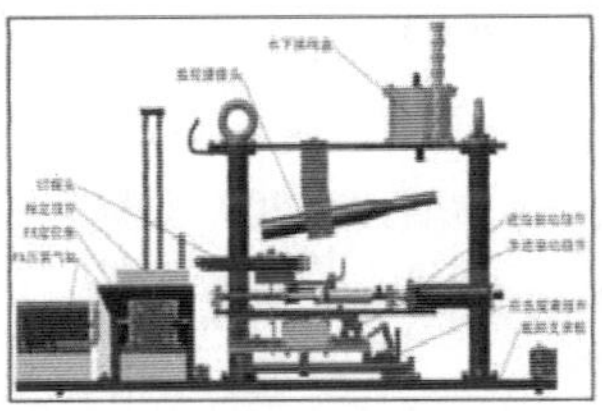

4.1 ISI Technology for M310

3) Examination system and technology for the seal welds of CRDM
   a) Scope of examination: seal weld and base material in upper, middle and down part welds
   b) Technology: water immersion focusing ultrasonic transducer and Eddy current array probe

4.1 ISI Technology for M310

UT technology:

Axial examination:

1) Frequency:>10MHz, shear wave, Angle:35~40°
2) T-R transducer : Frequency:>10MHz, longitudinal wave, Angle: 44° .
3) TOFD transducers

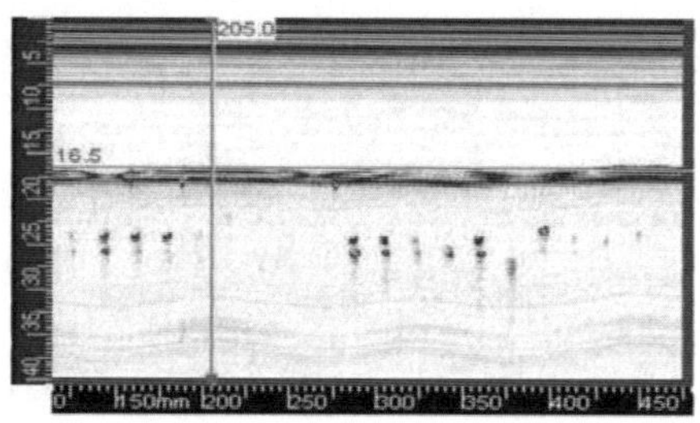

Circumferential examination:

1) T-R transducer: Frequency:>10MHz, shear wave, Angle: 45°

ET Examination :

Two types of ET Probe are used to the examination

1) Cross Winded (CW) probe is for detection
2) Parallel Coil (PC) probe is for depth measurement

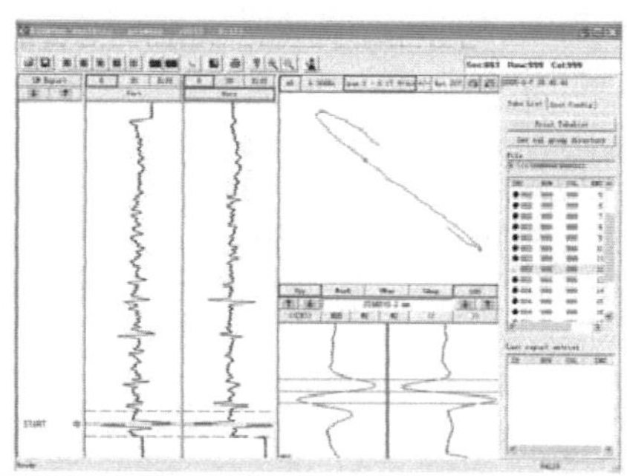

4.2 ISI Technology for AP1000

1) Examination system and technology for the weld of RCP to SG nozzle of AP1000
   a) Scope of examination:1/3 inner thickness of welds
   b) Technology: scanning from ID, the dual element longitudinal wave probes(TRL) are used for UT, due to the casting austenite material of pipe.

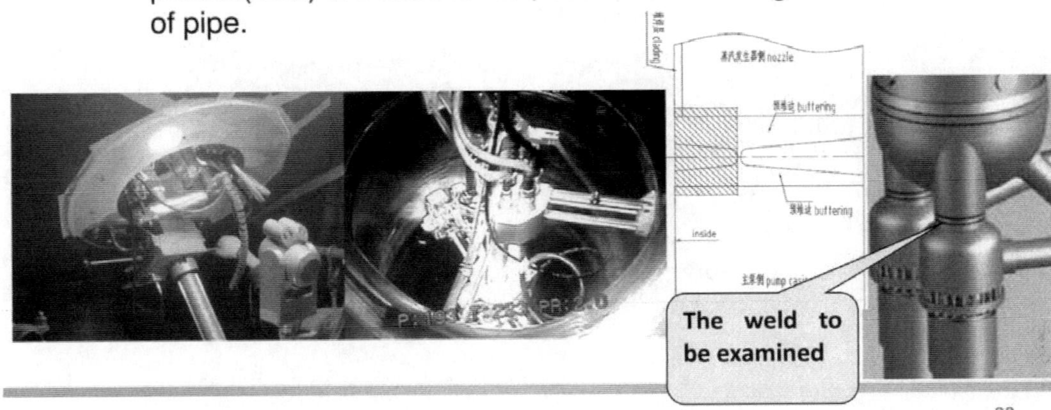

4.2 ISI Technology for AP1000

2) Ultrasonic examination system and technology of AP1000 RPV DVI
   a) Scope of examination: safe-end welds and heated affected zone
   b) NDT Technology: the water immersion transducers are used for UT

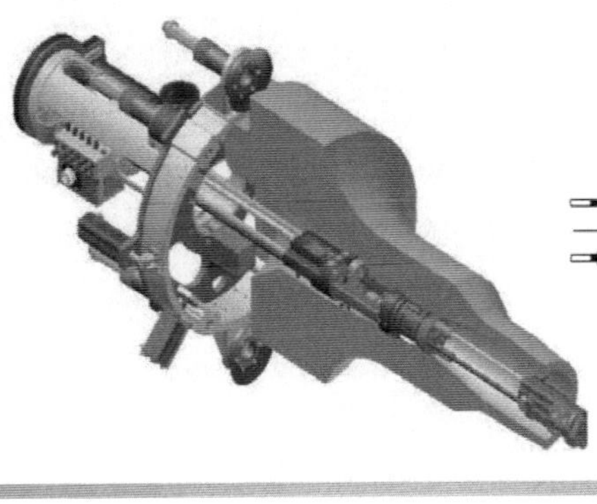

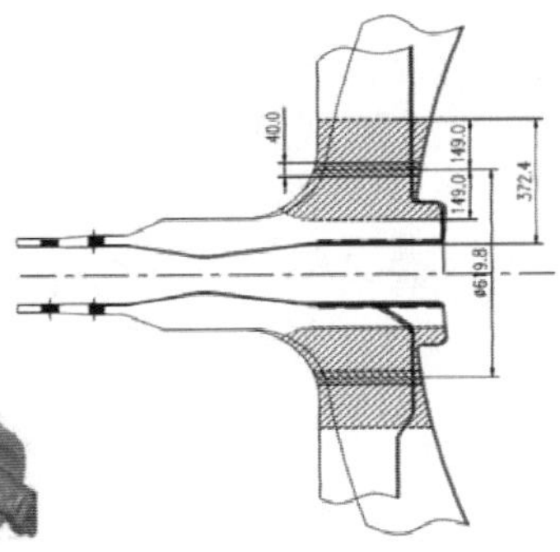

4.3 ISI Technology for WWER1000

1) Examination system and technology for WWER1000 SG tube ET
   a) Scope of examination: whole length tubes and collector ligaments
   b) Technology: Bobbin, MRPC probe
   c) Equipment: CEMA manipulator

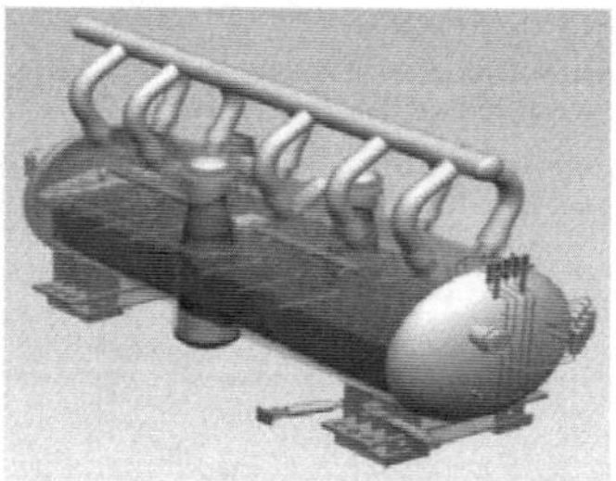

4.4 ISI Technology for CANDU600

1) Thickness Measurement for CANDU feeder pipe
   a) Scope of examination: bend area of feeder pipe
   b) Technology: to use the high frequency ultrasonic (phased array imaging tech.)

CANDU heat transfer feeder pipe position

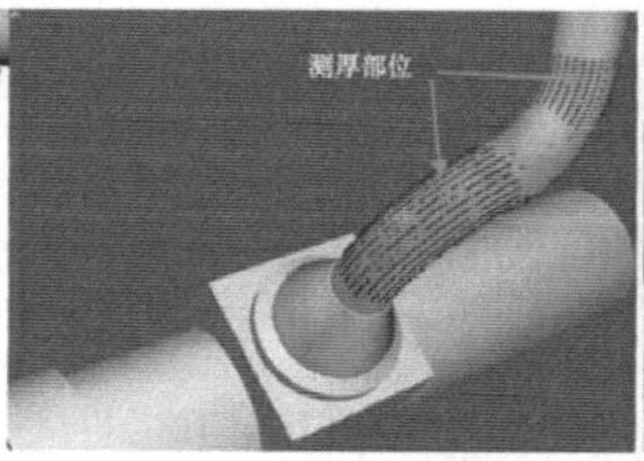

The inspection scope of Feeder pipe

The inspection condition of Feeder pipe

4.4 ISI Technology for CANDU600

2) Video Examination for Support and Hanger of Feeder Pipe
   a) Scope of examination: integrity for pipe support and hanger
   b) Technology: remote video examination
   c) Equipment: FVIS machine

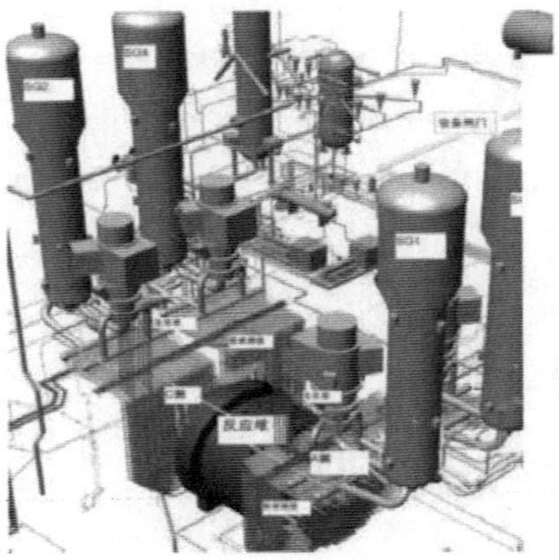

4.5 ISI Technology for HTR

1) UT examination for RPV
   a) Scope of examination: weld of vessel
   b) Technology: UT examination from outside, detection by transverse wave transducers, and sizing by higher frequency longitudinal wave transducers

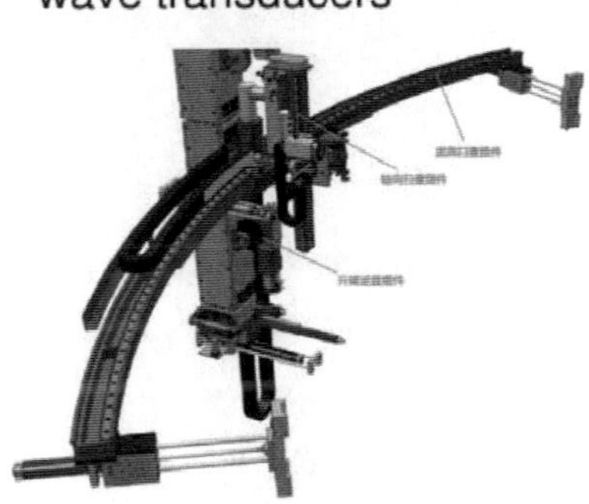

4.6 ISI Technology for CEFR

1) Remote field ET for SG tubes
   a) Scope of examination: 100% tubes
   b) Technology: remote field eddy current testing

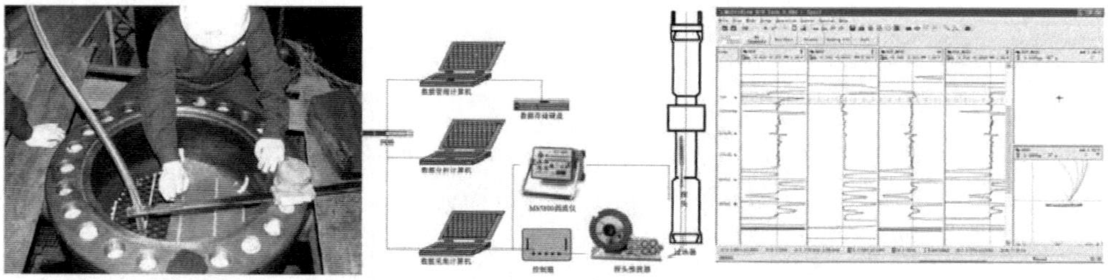

5. ISI Experience for Multiple Type NPP

- For RPV examination
  - Requirement and characterization of RPV examination:
    - Different code and standard
    - ExaminationTime limitation
    - different acceptance criteria
  - Recent application and resolving method
    - Little different of manipulator used for different RPV, i.e. different support legs
    - Control unit and software are same
    - Same UT instrument and data treatment software
    - Similar UT transducers, for defect detection and sizing
    - Different recording level, i.e. diameter 2mm SDH for RSE-M and ASME, diameter 3.1mm FBH for Russia Code

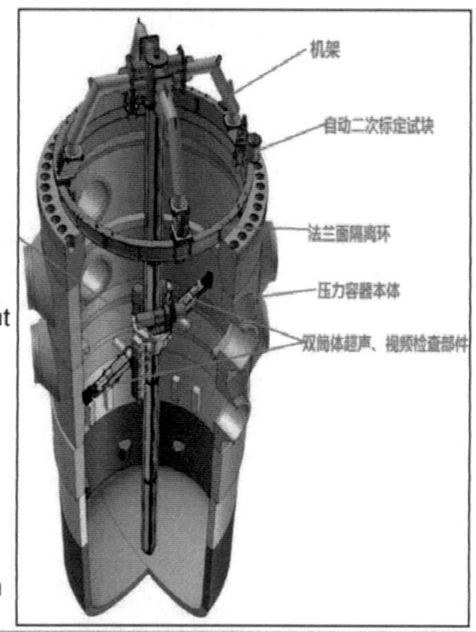

5. ISI Experience for Multiple Type NPP

- Multiple scanning platform, i.e. scanning for different welds in same time
- In the future, pulse echo UT + Phased array UT

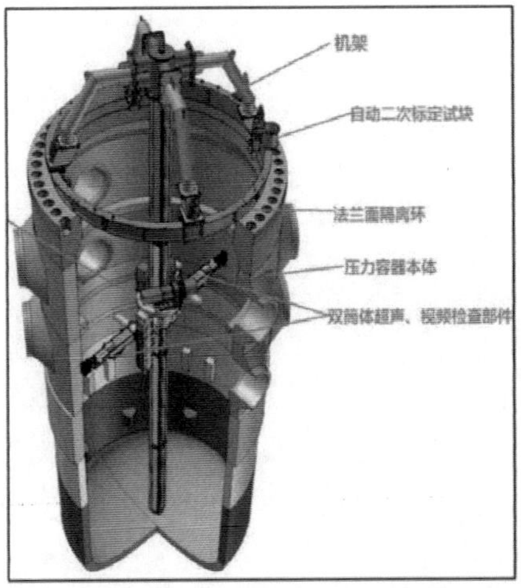

5. ISI Experience for Multiple Type NPP

- For SG tubing ET
  - Requirement and characterization of SG tubing ET:
    - Different code and standard
    - Different structure of SG, i.e. PWR and WWER
  - Recent application and resolving method
    - BEEBOT/Pegasys for PWR, CEMA for WWER

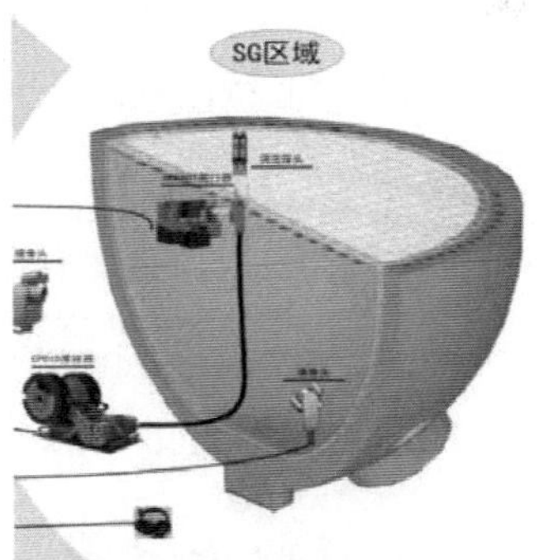

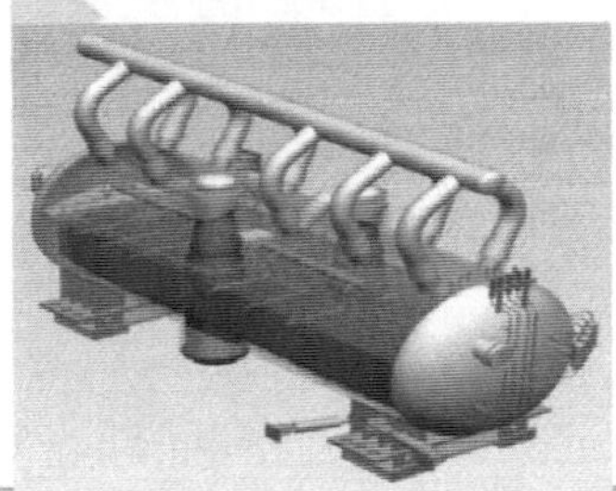

5. ISI Experience for Multiple Type NPP

- BOBBIN + MRPC Eddy current probe
- Different calibration tubes, i.e. ASME calibration tubes or RSE-M calibration tubes for relative codes
- Automatic Data Acquisition, and remote data analysis(In Wuhan, out of the NPP sites)

- In the future, array probe will be used roundly
- And also, Automatic data analysis technology will apply to the SG tubing Eddy current examination

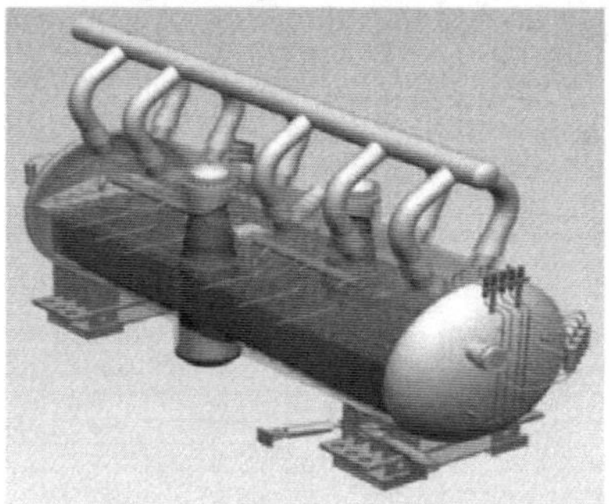

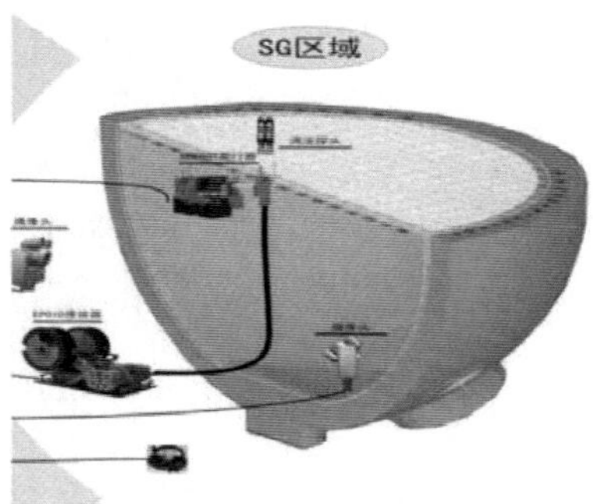

5. ISI Experience for Multiple Type NPP

- For Stainless Steel Pipe Welds UT
  - Requirement and characterization of pipe weld UT:
    - NDT technology shall be qualified
    - A-scan UT is difficult for sizing of defects during qualification
    - High reliability UT technology is required
  - Recent application and resolving method
    - Automatic pulse echo UT
      - Use the auto manipulator (AUMA)
      - Multiple channel ultrasonic instrument(pulse echo UT)
    - Manually Phased Array ultrasonic examination technology for SS pipe weld UT

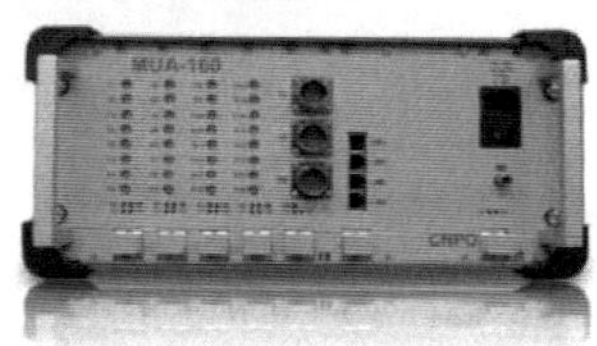

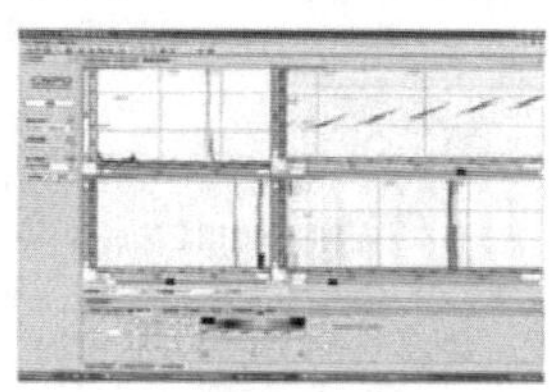

5. ISI Experience for Multiple Type NPP

- For ISI NDT qualification of multiple type NPP by different code
  - Basically, RSE-M methodology has been used for NDT qualification in China
  - ISI applications are divided to three type qualification, i.e. general, conventional, and specific qualifications
  - To include technical justification(TJ) and experiment trial
  - For ASME code NPP, specific qualification shall be according to the Appendix VIII of ASME XI

6. Conclusion

- For the different NPPs, the some new special tools and technology will be used for specific component or parts. For example:

Reactor TYPE	Typical difference technology for ISI
M310	Some special NDT technologies and equipment such as BMI
AP1000	Special tool applying to RCP for AP1000
WWER1000	Special tools for horizontal SG tubes ET.
CANDU	Remote Video Examination for Support and Hanger of Feeder Pipe
HTR	RPV inspection from outside of vessel because of unreachable from inside

6. Conclusion

- The most of general ISI technologies can be used for different type NPPs, and just only adjust some NDT parameters and relative tools.

Thanks for your attention!

Application of Flexible PAUT probes for small diameter Flow Assisted Corrosion elbow inspection II

Vajira JAYASINGHE[1], Chris EATON[1], Brian MILLEJOURS[1]

[1] IMS Development Lab, 1910 Clements Rd, Pickering ON
Phone: 905-427-4010; E-mail: vajira.jayasinghe@opg.com, chris.eaton@opg.com, brian.millejours@opg.com

Abstract
This paper presents an update to the application of newer flexible Phased Array UT (PAUT) probes that can be contoured to the surface of the component to be inspected. Development was based on introducing PAUT in lieu of Radiography (RT) as part of the FAC inspection program.

Currently for piping elbows of NPS 4" and less, there are limited PAUT solutions available. Most scanners cannot traverse elbows, or are not configured for smaller pipe diameters at all. The goal with application of a flexible array probe was to eliminate the need for multiple probes/wedge combinations and utilize a single design over a range of elbow diameters to determine wall loss.

After successful field trials, use of the flexible probes was expanded to actively replace RT in the plant FAC inspection program for butt-welded joints. Capabilities with C-scan mapping and post-processing to allow PAUT data mapping on 3D components were explored, with goal to provide visualization of the identified defects to facilitate replacement

Keywords: PAUT, FAC, pipe elbows, ultrasonics, flexible probe, Phased Array

1. Introduction

Ontario Power Generation's (OPG's) plants are in the process of replacing radiography testing (RT) with Phased Array Ultrasonic Testing (PAUT) on smaller diameter piping as part of the Flow Assisted Corrosion (FAC) program. It is preferable to use phased array ultrasonic testing in lieu of radiography for inspection accuracy, reducing manpower, and minimizing safety risk. However a significant drawback still remains with regard to the visual representation of the data.

This paper continues from where the previous entry left off. That paper summarized the creation of applicable test pieces which incorporated typical Flow Assisted Corrosion patches, followed by applying several different flexible probe sets to determine limitations.[1] Linear and sectorial scan parameters were manipulated for each probe on the test blocks with an end goal of determining measurement accuracies, elements in contact, and optimal PAUT settings. Since that point in time, further development has continued with a focus on the potential to introduce encoded data scanners, mapping the acquired PAUT data onto a representative 3D model, and streamlining the process for simplified field application.

2. Background

The previous paper presented application of newer flexible PAUT probes which can be contoured to the surface of the component being inspected. The use of this type of probe was predicated by a need to inspect piping elbows of nominal pipe size (NPS) 4" and less for which there are limited PAUT scanner solutions available. Most scanners cannot traverse elbows, or are not configured for smaller pipe diameters. The application of a flexible array probe over more traditional manual PAUT probes was undertaken to eliminate the need for multiple probe/wedge combinations and simplifying the inspection by utilizing a single design over a range of elbow diameters.

Following successful lab testing, the usage of flexible array probes as a potential replacement for RT was field tested at an OPG power plant in 2015. It has been introduced this year to replace radiography in several feed-water heater and boiler line locations. The implementation of this program is preferable due to several factors including lower manpower requirements, minimizing hazards/dose, and minimizing impact to other work groups. The primary drawback at this time is in the ultrasonic testing (UT) data presentation to the engineer.

Currently, this inspection is taking place at one of our nuclear facilities with all of the special provisions working in such an environment entails. The procedure involves a relatively simple manual inspection using a 0° electronic scanning (E-scan) setup of a grid-mapped component. Encoded solutions are employed for larger diameter piping (>3") inspections as necessary.

2.1 Test Pieces

Several piping elbow test pieces were developed based on FAC samples taken from the field. Based on a summary of affected components in the FAC program, both 90° and 45° elbows were created to cover a range of piping diameters (NPS ¾", 1", 1½", 2", 2½", 3", 4"). Three defects patches were machined into the components based on relevant FAC degradation mechanisms as outlined by Figure 1 [1].

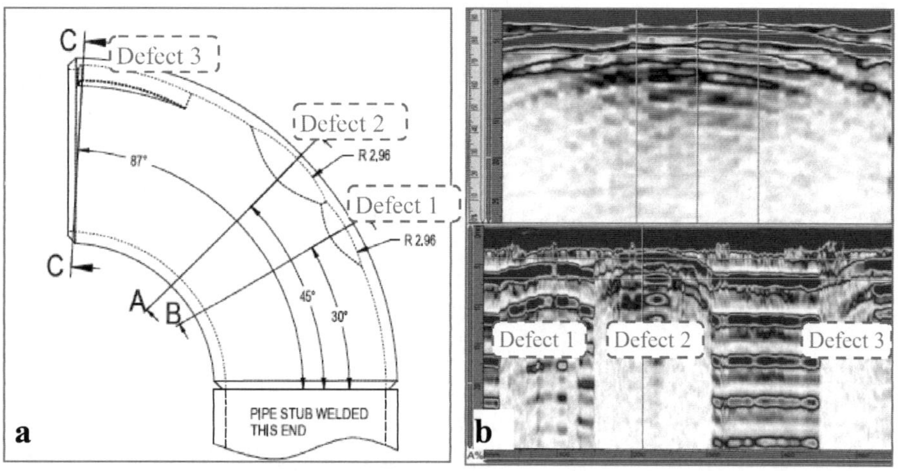

Figure 1: Schematic of piping elbow with FAC defects (a), with representative PAUT of sample (b).

These test pieces were used to develop the inspection and as such were designed to test the full capabilities of the probes under evaluation. Three FAC areas were machined into the test pieces using specially formed Electrical Discharge Machining (EDM) electrodes. Two patches mimic FAC at the midpoint of the elbow with 25% wall-loss and 50% wall-loss, and a third patch mimicking 40% loss following a weld. It is useful to note that the FAC pattern utilized is considerably more pronounced than typical FAC patches encountered in the field; this was done purposely in order to test the potential worst case scenario of a reflector with inadequate parallel surface, creating a loss of signal [1].

3. Scanner Evaluation

One of the main improvements identified during the replacement of RT was to provide the engineer a better visualization of any identified defects. This immediately lent itself to an encoded scan that could be mapped either to CAD models or utilized in a .csv file for export. The primary purpose for the scanner was to enhance accuracy over a time-based scan mapped over a grid system and also to facilitate mapping of the PAUT data onto 3D components.

Different scanner options were evaluated on the test pieces mimicking FAC corrosion patches at different elbow diameters. The preferred solution for both cost and convenience was a ready-made off the shelf item. To that end, both the JIREH based low profile encoded scanners (Cobra and Circ-it) were looked at initially, but their design did not lend itself well to traversing an elbow. A wheel probe design was considered due to the inherent flexibility of the membrane, but unfortunately was not compatible once the piping diameter was less than 3". Additional off-the-shelf scanners for small diameter piping were initially evaluated, and when it was evident that a suitable scanner could not be found the evaluation was expanded to custom solutions.

Ultimately several promising options were further assessed: a custom encoder housing for a flexible array probe, a repurposed flexible housing system, and a manual PAUT probe XY scanner.

3.1 Custom Flexible Probe Housing (e2sense probe)

A custom flexible array probe (10 MHz, 32 element 1.0 mm pitch) from e2sense was designed with a detachable housing mount which used different 'pressure pads' to control the flexibility of the probe around different diameter elbows (see Figure 2a). The pads would fit in between the housing shell and the probe to distribute the appropriate pressure to the probe to maintain coupling (see Figure 2b). This foam could be readily adjusted or switched out entirely for a different material. While the system worked well in principle there were difficulties maintaining uniform contact with the part surface. A gap tended to develop along the profile of the elbow creating contact issues as shown in Figure 2d).

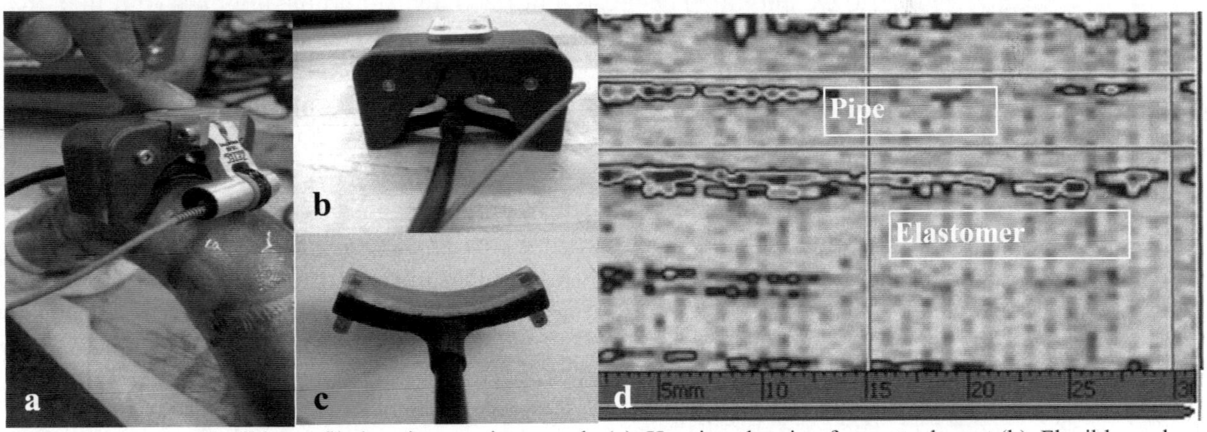

Figure 2: Custom Flexible profile housing on pipe sample (a), Housing showing foam attachment (b), Flexible probe with attached elastomer (c), and sample data E-scan file from acquisition (d).

3.2 Repurposed ECA Scanner (Olympus probe)

An Olympus designed 5 MHz, 64 element, 1.0 mm pitch flexible array probe did not come with any OEM encoding options for the probe. An updated design of the probe housing offered a potential option to attach the mini-wheel encoder, however this proved incompatible during actual testing on pipe elbows. An alternative option was proposed when observing the standard diameters covered in the Eddy-Current-Array (ECA) kit.

The ECA probe housing and encoder was removed and used in concert with the flexible array probe (see Figure 3). The Olympus PAUT probe is not as flexible as the ECA, so some care had to be used in how the probe was contoured to the scanner geometry. Also, the shape of the scanner created a cantilever on the elbow, affecting either probe contact or encoding quality as the scanner traversed the part. A sample set of results is shown in Figure 3c.

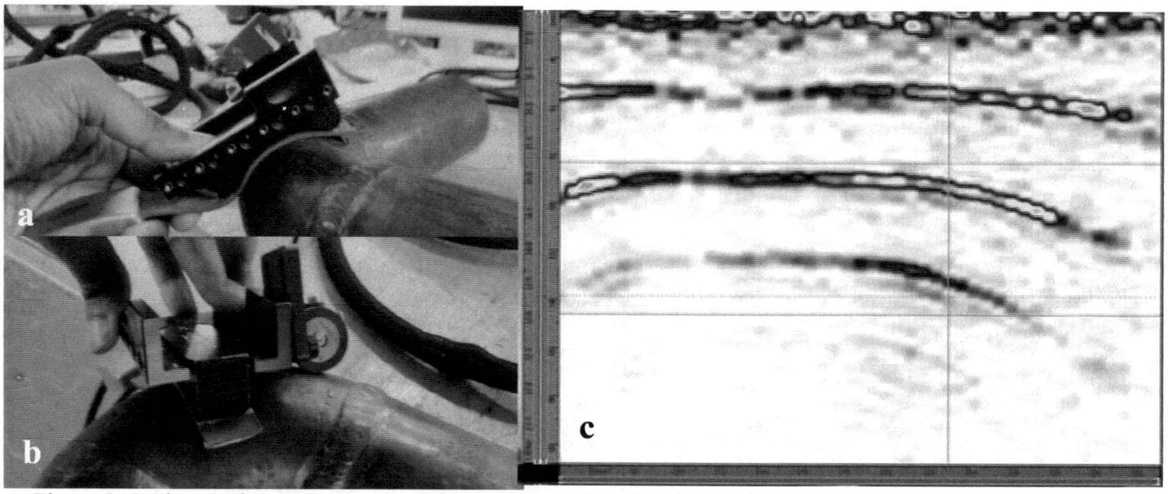

Figure 3: Probe curving on NPS 2.5" pipe (a), side profile of probe combination (b), and sample E-scan file from acquisition (c).

3.3 XY scanner (Imasonic probe)

The final scanner tested is a more traditional small footprint model, which was used in order to ensure a solution was available for deployment if an encoded solution was required for field execution. This design utilized a custom housing built to utilize manual PAUT probes with an encoder for the x-direction and indexer box for the y-direction (as shown in Figure 4). The encoder unit can be separated from the probe through a magnetic attachment, allowing a manual scan to be conducted and followed up with an encoded scan when recordable indications are observed. This results in a far more adaptable design, as a single line scanner can be prone to errors when axially scanning elbows. While the probe footprint is adequate to maintain contact from the elbow left cheek to right cheek, it cannot effectively be used on smaller elbow intrados. Additionally there is no practical method to avoid encoder slip during an axial scan, and a circumferential scan is impossible at the bend. Since it covers a distinctly smaller area, a gridding system is necessary on the pipe to ensure adequate coverage and that proper probe overlap is maintained during the scan.

Use of a manual PAUT probe allows more versatility in the system, which enables it to be deployed as a complementary technique should additional degradations mechanisms be discovered. The scanner's usage is currently limited to piping elbows 3" and above. The probe holder is designed to accommodate the shoe for either axial or perpendicular scanning depending on the expected orientation of the defects.

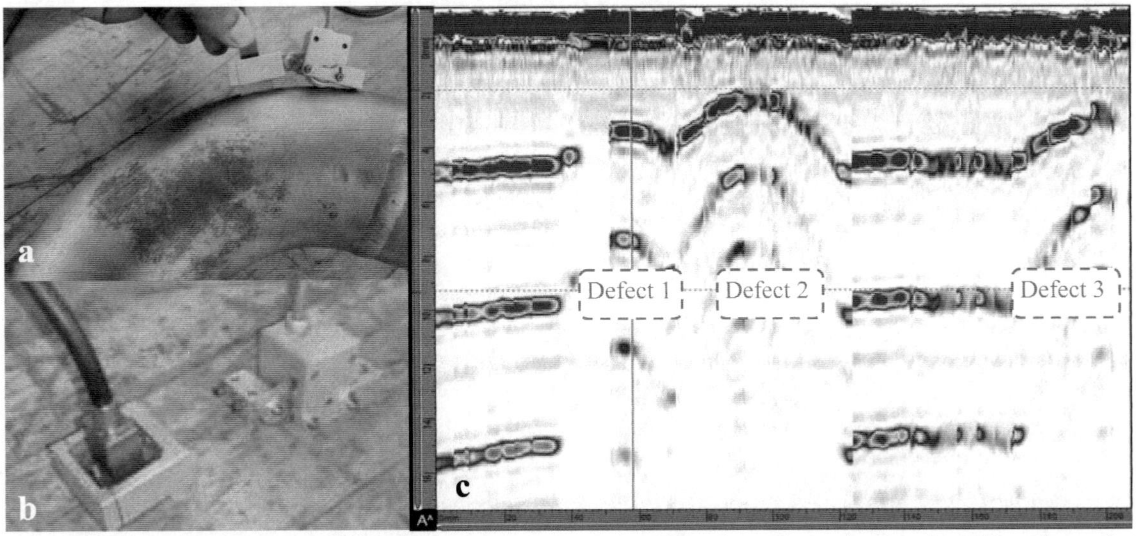

Figure 4: Scanner system traversing a 3" elbow (a), Scanner with probe and encoder detached (b), and a typical B-scan response (c).

4. 3D Visualization of Data

The primary goal of incorporated 3D visualization to inspection results is to simplify presentation of the inspection data. A clearly outlined area can greatly enhance the characterization of a defect, provide additional information when searching for potential causes and aid in prioritizing repair/replacement efforts. A key caveat to these benefits is that it needs to be balanced by the practicality of the inspection. The search for a perfect picture should not significantly increase the complexity of the inspection. For that reason, more advanced methods of data recording were not pursued. We want to create a streamlined system that a technician can use to help the engineer/evaluator understand the significance of the findings. This ideally is accomplished in a procedural manner, with minimal training required, and without turning it into an exercise in graphical manipulation.

These goals are aided somewhat by the simpler interpretation of UT thickness measurements as opposed to weld inspection which would require accuracies of the weld bevel prep to match up with the PAUT data, or turbine blades which have non-uniformities across batches. However, an acknowledged challenge in the approach is manufacturing tolerances with piping elbows and fittings which could still cause larger errors in mapping.

The process required use of either an MX-2 or Topaz, a 3D license of UltraVision 3.7 and a CAD model of the component inspected. An overview of the step by step process was as follows:

1. Importing the data-file from either an MX-2 or a Topaz into UltraVision 3.7.
2. Import custom geometry as the specimen (in this case a pipe elbow).
3. Map the scan path of the probe over the same area inspected onto the CAD model.
4. Set up contours in UltraVision to adjust the volumetric merge parameters by boxing in the relevant data (ie. eliminate multiple backwalls, extraneous scan data).
5. Perform a volumetric merge.
6. Evaluate the alignment between D-scan, B-scan and the projection for accuracy and adjust the process from 3. as necessary.

For testing on a block with known defects, modelling was pursued with the actual CAD model of the reference block to enhance the data visualization. As shown in Figure 5, a single line scan over the critical orientation of the defects can be merged with the CAD model to enhance them. These defects line up with the presentation in Figure 1, showing clear identification of each indication when compared to Figure 5d. The final presentation of the CAD model overlaid with the image is consistent with these identified thickness changes (see Figure 5a).

Subsequently, modelling using a standard CAD model was used as a comparison with a scan from a field sample (see Figure 6). In both cases, while the merge is consistent with the results, the overall presentation does not greatly aid in distinguishing thin areas.

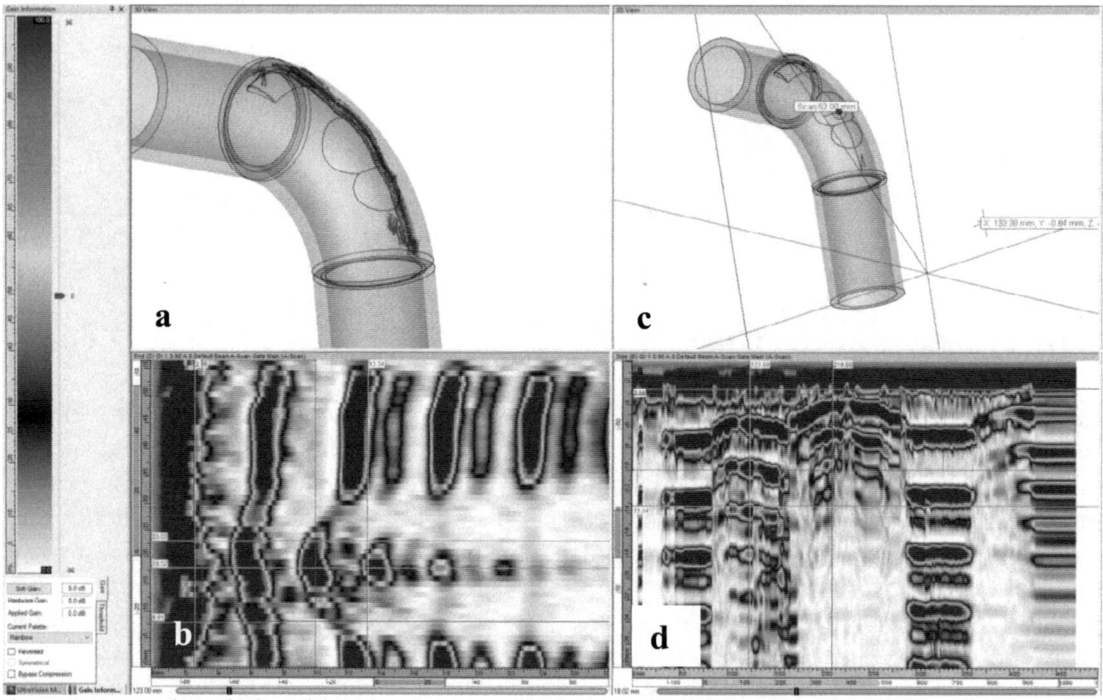

Figure 5: Screenshot of UltraVision showing scan of test piece with data overlaid onto CAD model (a), E-scan of the test piece (b), Probe scan path on part (c), B-scan of the test piece (d).

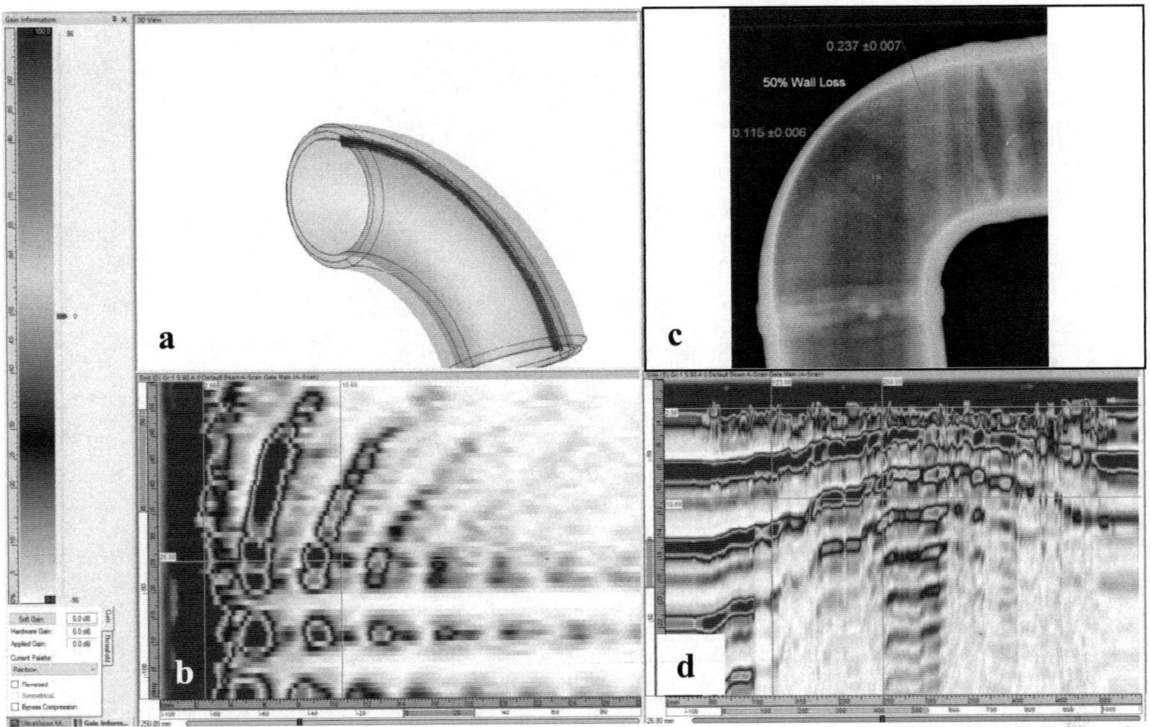

Figure 6: Scan of a field sample, with data overlaid onto a generic CAD model of same dimension pipe elbow Screenshot of UltraVision showing scan of a field sample, with data overlaid onto a generic CAD model of the same pipe dimension (a), E-scan of the test piece (b), RT shot of the same elbow (c), B-scan of the test piece (d).

4.1 3D laser mapping component

Creating a CAD model by laser mapping the inspected part either prior to or following the examination was considered as an additional step that could be undertaken for greater accuracy. A lab trial was conducted to determine the viability of such a solution in the field. Using a Creaform laser scanner, a field sample was dotted, scanned, and converted to a compatible CAD model (as shown in Figure 7). The subsequent model was imported into UltraVision and the same data mapping process as outlined before was repeated.

It may also be possible to approach this process inversely through utilizing 3D laser mapping pipe-check software with the import of UT data as a CSV file. However this would directly result in a degradation of the data and it is unclear whether the elbow transition point can be accurately mapped in the software. This option was not explored in detail.

Overall, for the purposes of scanning piping elbows this process is somewhat laborious and the specific benefits are not necessarily realized for an FAC program. However, the process provides an ideal testing ground for the potential application to more complex surfaces such as nozzles or weldolets which may differ from patch to patch and possess significant variability.

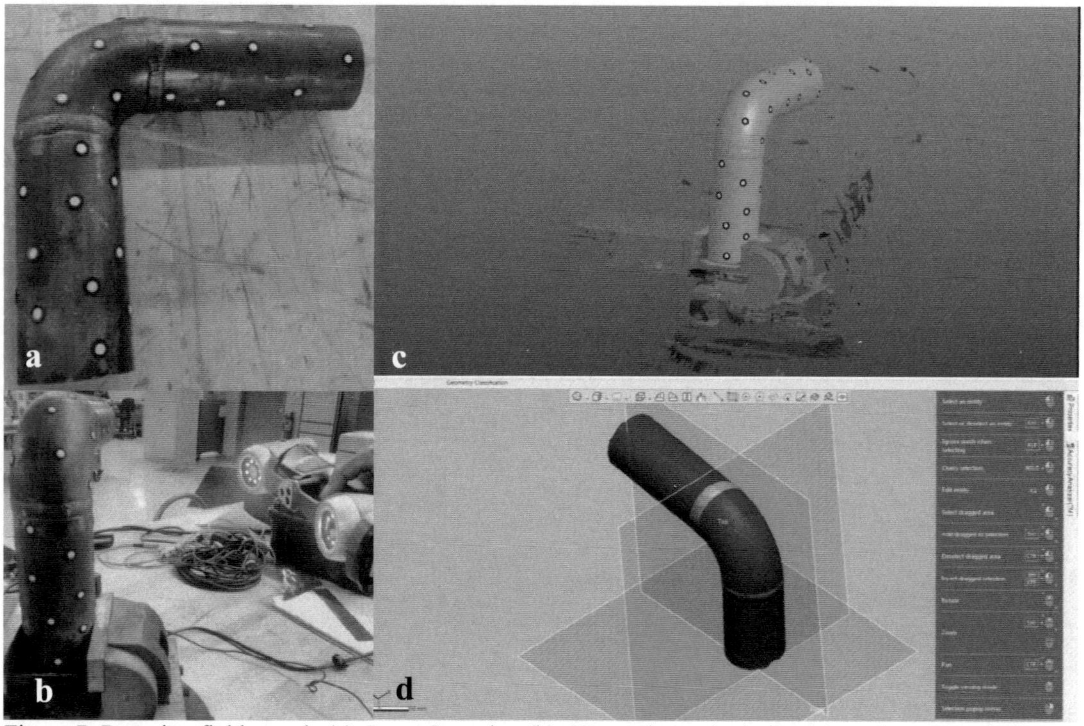

Figure 7: Prepping field sample (a), Laser Scanning (b), 3D mesh manipulation (c), and CAD modelling of a field sample prior to import (d).

5. Conclusion

An ideal encoding solution for this application has yet to be found. While several solutions exist to provide an adequate solution for the time being, new options will continue to be explored.

The process of matching a CAD component to a single-line encoder scan is relatively simple, and the key evolution for this would be in the ability to pre-configure setup files with CAD models and a scan plan uploaded. This capability would allow for the post-processing to be streamlined into a step by step boxing of the area of interest, followed by processing the merge which would not add significantly to off-line data analysis already used in such a situation.

6. References

1. V Jayasinghe 'Application of Flexible PAUT Probes for Small Diameter Flow Assisted Corrosion Elbow Inspection', ASNT Annual Conference 2015, Salt Lake City, ASNT, 2015

2. J. Berlanger et al, Advanced 3D Visualization and Analysis of Phased Array UT Inspection Data, Quebec City, NDT.net, 2012

3. Zetec, Inspection of Complex Parts using CREAFORM 3D Scanner & UltraVision 3, Quebec City, 2011

Field Implementation of a Transient Eddy Current System for Carbon Steel Pipe Thickness Measurements

Jeremy A. BUCK [1], Colin KRAMER [1], Jia LEI [1], and Brian A. LEPINE [1]

[1] Inspection, Monitoring and Dynamics Branch, Canadian Nuclear Laboratories; Chalk River, Canada
Phone: +1 613 584 3311; e-mail: jeremy.buck@cnl.ca, colin.kramer@cnl.ca, jia.lei@cnl.ca, brian.lepine@cnl.ca

Abstract
Safe storage of nuclear waste is a concern faced by the nuclear industry. Proper planning and continuous monitoring are important aspects of successful nuclear waste management programs. Containers and storage cells are often made of carbon steel, which possesses good structural properties and acts as a neutron absorber. Although designs of long-term storage cells are robust, they may necessitate inspections to assure that the aging process has not affected storage container integrity. Tile holes are concrete in-ground containment cylinders lined with a 5 m long 250 mm Schedule 40 carbon steel pipe, with 9.3 mm wall thickness. They are placed in engineered storage arrays, in which sealed containers of fissile waste are stored for decades. An inspection of ten tile holes was performed for general wall-thickness measurements using a custom probe head that combined ultrasonic testing (UT) and transient eddy current (TEC) testing. The focus of this paper is the TEC probe design, data acquisition, and analysis.
Two sets of four coil-pairs with differing coil arrangements, in transmit/receive and reflection modes, were incorporated in the final product and were alternately placed at 45° intervals circumferentially on the probe head. All TEC measurements were non-contact, performed with 12.7 mm nominal lift-off. Three analysis methods were applied to collected waveforms to estimate wall thickness: single-crossing voltage threshold, power law curve fitting, and a quantitative multi-frequency analysis technique adapted to TEC. These methods were performed by three analysts independently. Results were found to be in good agreement internally, and externally with UT thickness measurements. No regions of local thinning were identified, but a helix of minor wall thickness changes, attributed to the pilgering manufacturing process, was observed in the data by all parties. This application demonstrates the capability of TEC to accurately estimate steel wall thickness with moderate probe lift-off.

Keywords: Electromagnetic Testing, Transient Eddy Current, Data Analysis, Probe Development, Steel Degradation

1. Introduction

Safe storage of nuclear waste is a concern faced by nuclear operators world-wide. Proper planning and continuous monitoring are critical components of safe and effective nuclear waste management programs. In many cases, nuclear waste is placed in medium-term temporary storage systems, awaiting licensing and construction of more permanent repositories. Tile holes are one such system used for waste storage. Tile holes are vertical concrete in-ground storage locations, with additional integrity and neutron absorption provided by a lining of Schedule 40 carbon steel pipes, in which sealed waste containers are stored.
Nondestructive evaluation (NDE) is an important aspect of life management for nuclear waste storage systems. Eddy current testing (ET) and ultrasonic testing (UT) are two commonly employed techniques for in-service inspections, relying on electromagnetic induction and mechanical wave reflection, respectively. For ferromagnetic material inspections, conventional ET has very limited depth of penetration due to decreased electromagnetic skin depth [1]. Transient eddy current (TEC), an inspection technique which utilizes a broadband pulse instead of discrete frequencies for eddy current induction, has been shown to outperform sinusoidal ET for estimating ferromagnetic plate thickness [2].
Analysis of TEC signals is a continuously developing area, with no widely accepted method. Current analysis techniques for TEC data vary based on application, with different approaches focusing on different aspects of the data. A lift-off intersection (LOI) point has been identified as a time-domain feature for estimating material thickness largely insensitive to lift-off variations [3]. Thicknesses of ferromagnetic pipes have been estimated in the presence of

lift-off and conductive cladding by analyzing the slope of TEC signal decay [4]. Power-law curve fitting has been applied to estimate trefoil carbon-steel support-structure thinning from within steam-generator tubes [5]. Principal components analysis (PCA) has been applied to TEC responses to measure corrosion of mild steels [6], and estimate material loss of steam generator tube, and carbon-steel drilled-support structures [7], [8]. Power-spectrum densities of Fourier transformed TEC responses have been examined for the detection of thickness variations in stainless steel [9]. TEC spectral responses have also been related to lift-off measurements, and analysis of material conductivity and permeability changes [10].

Ten tile holes were inspected with a series of TEC and UT probes, mounted on a common inspection head. The two techniques were complimentary in this application. While UT is an excellent thickness-measurement tool for thick materials, analysis becomes more difficult for thinner regions. TEC has been shown to be sensitive over a wide range of thicknesses, and would prove more reliable in situations of surface corrosion as surface coupling is not an issue. The focus of this work was the comparison of three TEC analysis methods: voltage threshold, power-law fit, and a patented transient eddy-current multi-frequency analysis (TEC MFA) [11]. Good agreement with UT thickness measurements demonstrates the quality of results obtainable from field applications of TEC.

2. Theory

2.1 Transient Eddy Current

Transient eddy current is an emerging NDE technique that relies on the principle of electromagnetic induction. This phenomenon can be described by Faraday's law for a coil with N turns of wire [12]:

$$\boldsymbol{\varepsilon = -N\frac{d\varphi}{dt}}, \quad \textbf{(1)}$$

where φ is magnetic flux, and ε is the induced electromotive force. By passing a time-varying current through a coil of wire, a change in flux can be produced. This induces an opposing electromotive force, and therefore currents, in nearby conducting materials. These currents are called eddy currents due to their circular nature. By measuring the effects of magnetic fields produced by induced eddy currents, information such as material thickness and the presence of defects can be ascertained. Electromagnetic waves decay with increasing depth in conducting materials. Under steady-state conditions (as in conventional ET), the depth at which the eddy currents diminish to $1/e$ (37%) of their surface value is called the skin depth, and is approximated in mm as [1]:

$$\boldsymbol{\delta = 50\sqrt{\frac{\rho}{\mu_r f}}}, \quad \textbf{(2)}$$

where ρ is the material resistivity in μΩcm, f the frequency of eddy currents in Hz, and μ_r is the relative magnetic permeability. While conventional ET utilizes one or more discrete frequencies for eddy current induction, TEC generally uses a square-wave broadband pulse. This method helps overcome skin depth limitations encountered by conventional ET, and results in eddy currents that decay according to a diffusion process. This becomes particularly important for ferromagnetic materials such as carbon steels, which have μ_r much greater than 1, significantly reducing depth of penetration for conventional ET frequencies. Diffusion of magnetic flux $\vec{B}$ at low frequencies (less than $10^8\ Hz$) can be described by Maxwell's equations as [12]:

$$\nabla^2 \vec{B} = \mu\sigma \frac{\partial \vec{B}}{\partial t}, \tag{3}$$

which has general solutions of the form [13]:

$$\vec{B} = f\left(e^{-t/\tau_D}\right). \tag{4}$$

The characteristic diffusion time, τ_D, governs the rate at which transient eddy currents decay within conducting media, and is approximated as [13]:

$$\tau_D \sim \mu\sigma\ell^2, \tag{5}$$

where σ and μ are material conductivity in S/m and permeability in H/m, respectively, and ℓ is the characteristic length of the specimen in m, such as thickness. Eddy currents take longer to decay in ferromagnetic materials due to this diffusion time. Currently, there is no single accepted analysis method for TEC, comparable to impedance plane analysis commonly used in ET. In the work described herein, three different analysis techniques were applied to the TEC field measurements for wall thickness inspection: voltage threshold, logarithmic fit, and multi-frequency analysis for TEC.

2.2 Voltage Threshold

Signal amplitude measurements at a specific time slice, such as the LOI point, have been used to extract material thickness information from TEC data [3]. Similarly, zero-voltage crossing times have been shown to correlate to material thickness in reference-subtracted TEC data [14]. An analysis technique was proposed, which used a late-time voltage threshold to discriminate carbon steel material thickness. It was confirmed, experimentally, that the later time decay region of TEC signals were relatively insensitive to lift-off variations. By measuring the crossing time of TEC signals for multiple wall thicknesses on a known specimen, a calibration curve could be generated and applied to signals of similar material.

2.3 Power law curve fit

The shape of TEC pickup coil responses can be described by a sum of decaying exponentials. When multiple decay processes are present in a complex system, the series of exponential decays can be approximated by a power law [15]. The power law decay coefficient of TEC signals has been found to be related to material thickness [5]. By windowing decay signals and obtaining power coefficients for known samples at upper and lower bounds for thicknesses of interest, calibration curves can be produced. Given similar acquisition conditions, calibration curves can then be applied to field measurements.

2.4 Transient Eddy Current Multi-frequency Analysis

Multi-frequency analysis (MFA) is a technique employed in ET inspections that uses data from two or more discrete excitation frequencies to improve flaw detection and subsequent characterization. It relies on the fact that characteristic parameters of the signal (such as amplitude and phase) change with excitation frequency, material type, and flaw type. By performing a Fourier transform on TEC signal decay regions, a series of amplitudes and phases can be produced for a specific measurement condition. Using samples with known characteristic parameters, theoretical amplitude and phase for a set of frequencies are calculated to generate a set of constants. Amplitudes and phases of experimental data obtained from a Fourier transform are compared to those obtained from analytic expressions, floating characteristic parameters of interest. A least-square minimization routine is performed to determine what characteristic parameter values correspond to that experimental measurement. This patented analysis technique taps into the understanding of discrete frequency eddy current behaviour, and can be modified for different parameters of interest [11]. This technique is presented in more detail in this conference [16].

3. Experimental Technique

3.1 Tile Holes

Inspection targets for wall thickness measurements were a series of ten tile holes. Tile holes are vertical concrete in-ground containment cylinders placed in engineered storage arrays, in which containers of fissile waste are stored. Each tile hole is lined by a 5 m long, 250 mm Schedule 40 ASTM A53 steel pipe, with 9.3 mm nominal wall thickness.

A 600 mm long, 250 mm Schedule 40 carbon steel pipe sample was procured for laboratory testing. Outer diameter (OD) and inner diameter (ID) concentric grooves, both 100 mm in length axially, were machined on the pipe, providing known thickness variations for calibration purposes. The pipe was annealed after machining to eliminate cold-work effects. For every tile hole inspection, the calibration pipe was placed on top of the tile hole to provide in-situ data for calibration.

3.2 Probe

The inspection head combined both UT and TEC surface probes, allowing them to follow the contour of the inspection surface while remaining within 10 mm of the inner pipe wall. TEC probes were located 130 mm above the inspection head base, and 65 mm above UT probes. A total of 12 coils constituted the eight TEC probes for the inspection. The four transmit coils were placed every 90°, with four receive coils each placed coaxial with a transmit coil (reflection mode, odd numbered), and a further four receive coils 45° offset from a transmit coil (transmit/receive mode, even numbered), shown schematically in Figure 1. Nominal coil parameters are shown in Table 1.

3.3 Data Acquisition

The driving waveform was a 7 V amplitude step function at 8 Hz and 20% duty cycle, controlled by an InspectionWare[1] workspace and output through an NI PCA-5402 function generator. An Eddyfi prototype TEC system amplified the driving signal to 10 V and supplied the necessary current for the transmit coils. Receive coils were connected to a fixed 10X gain pre-amplifier, and the resulting signals were digitized by an NI 6353 USB DAQ at 100 kHz. Receive-coil responses were averaged over 5 cycles by the software. When scanning tile holes, a position-based encoder was used to monitor probe location. The probe head was driven at

[1] InspectionWare is a software platform from UTEX scientific

10 mm/s down the tile hole, pausing every 100 mm to collect data. Figure 2 shows calibration waveforms as collected by the Eddyfi prototype TEC system.

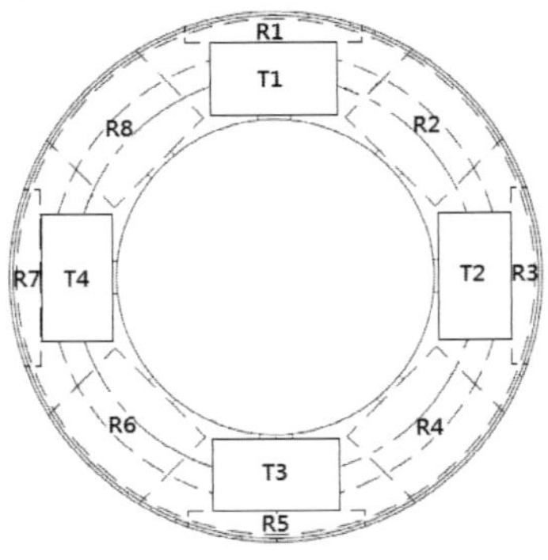

Figure 1: Coil arrangement on inspection head.

Table 1: Nominal coil parameters.

Coils	T1, T2, T3, T4	R1, R3, R5, R7	R2, R4, R6, R8
ID (mm)	38	53	38
OD (mm)	51	76	51
Height (mm)	38	3	38
Wire Gauge	24	33	24
Turns	737	812	737
R_{DC} (Ω)	8.6	109.0	8.6
L (mH)	15.6	60.8	15.6

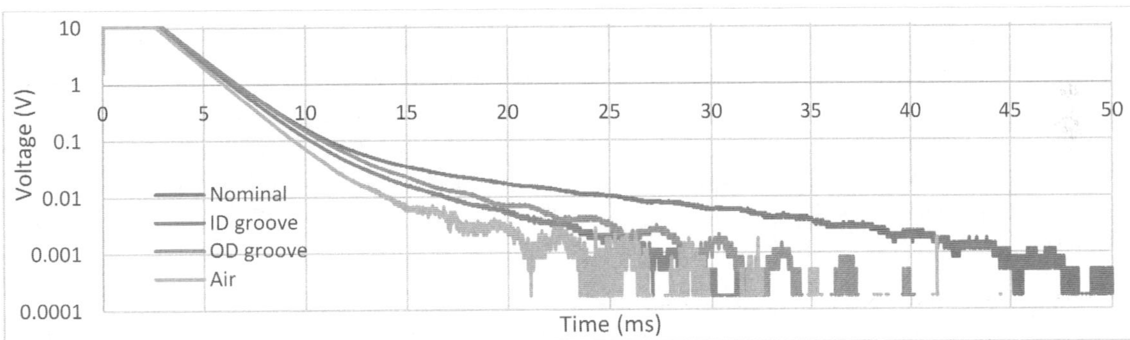

Figure 2: Semi-log plot of sample signals collected by coil R1 for in-situ calibration.

4. Results and discussion

The completed tile hole inspections were targeting regions of local pipe thinning, as well as overall wall loss. To compare the effectiveness of different TEC analysis techniques, the results were directly compared with UT measurements collected during the same inspection campaign. Data was collected every 1 mm with UT probes, as described in Section 3. To compare with TEC results, the reported UT thickness values were averaged over 50 measurements about the TEC collection location, corresponding to the TEC probe footprint. For all analysis methods, responses from each receive coil were calibrated independently due to the slight eccentricity of the calibration-pipe grooves. Thickness results from TEC analyses of the first tile hole are shown in Figure 3(a)-(c). UT results are shown in Figure 3(d) for comparison.

A voltage threshold of 10 mV was used to analyze field data, resulting in crossing times between 15 and 30 ms. The 10 mV threshold was a compromise between sensitivity to thickness and low-voltage noise. Thickness results from voltage threshold analysis of the first tile hole inspected are shown in Figure 3(a). Time-window selection was required for proper fitting of acquired TEC data using the power law method. A window range from 3 ms to 30 ms was used based on laboratory data to produce the best power-law fit. Decay coefficients of power-law fits for in situ calibration pipe measurements were used to generate a 3-point calibration line. Thickness results from power-law analysis of the first tile hole inspected are shown in Figure 3(b). Laboratory measurements were also processed using TEC MFA to fit a number of constant parameters that could then be applied to field data. Thickness results from TEC MFA of the first tile hole inspected are shown in Figure 3(c). The ripple structure seen in Figure 3 has been attributed to a pilger manufacturing process as opposed to pipe degradation. Seven of the ten pipes exhibited this pilger structure to varying degrees. The remaining three tile holes looked considerably smoother. It has been postulated that those pipes were seam-welded during manufacturing. An interpreted thickness increase along one channel in these pipes suggests the location of the weld. Mean values from each TEC analysis technique were calculated for each tile hole inspected, shown in Table 2. Standard deviations calculated from TEC results represent the spread inherent in the data. Percent differences between TEC and UT means are indicative of the extent of over- or under-sizing for each analysis technique. The voltage threshold technique results deviated the most from UT results. Power law and TEC MFA results were comparable to UT results, exhibiting 50% lower standard deviations compared to the voltage threshold.

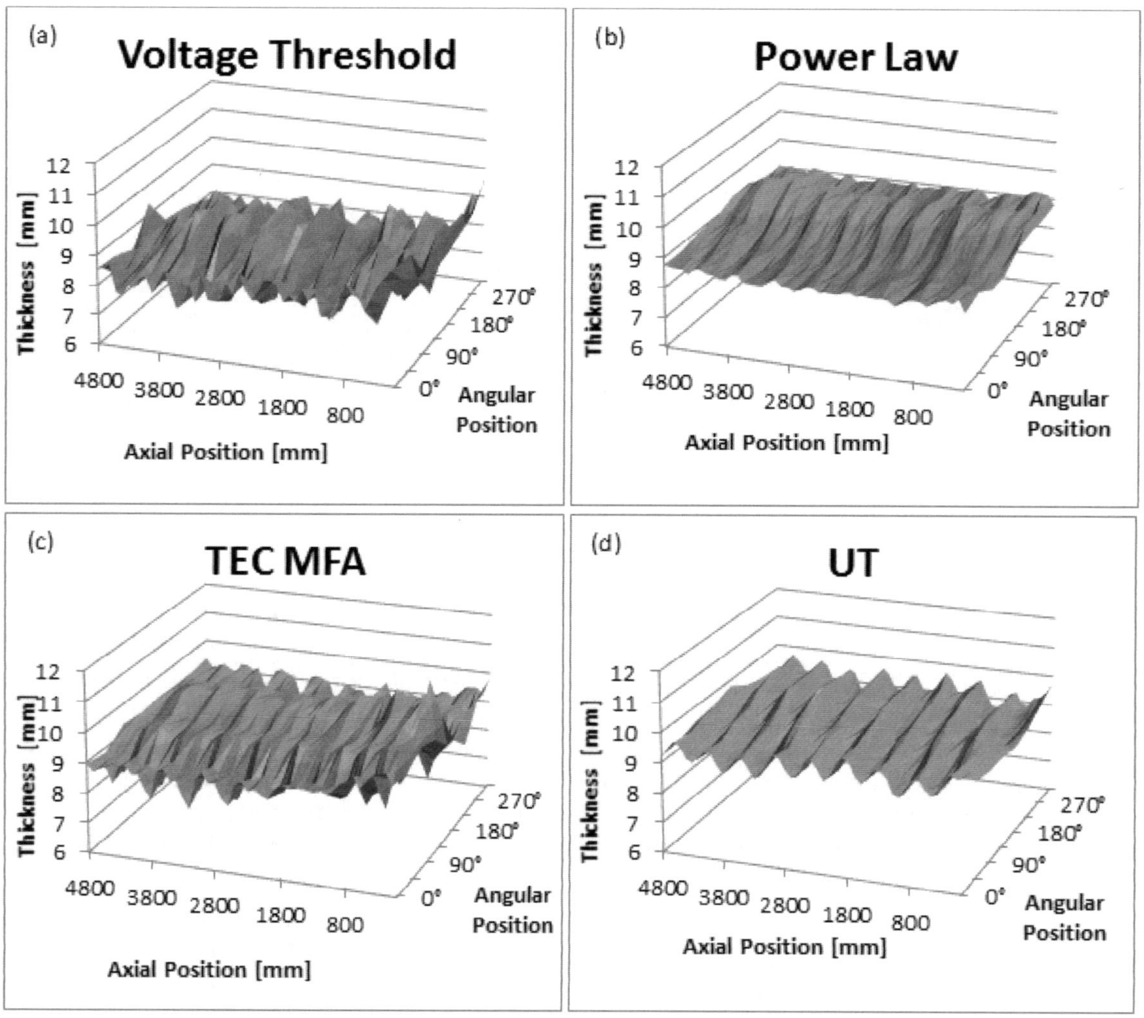

Figure 3: Pipe thickness estimates from (a) UT, (b) voltage threshold analysis, (c) power law analysis, and (d) TEC MFA. Internal ripple structure is apparent in all results.

The root-mean-square error (RMSE) was calculated for each TEC analysis technique for two of the ten tile hole scans: one of each type of pipe (pilger and seam-welded). RMSE calculations were based on the locally averaged (over the TEC footprint) UT thickness values. The results are compared in Table 3 for each analysis method.

Table 2: Overall pipe thickness mean and standard deviation (in mm) as determined by different analysis methods. Tile hole class indicates pilger (P) or seam-welded (S). Percent differences were determined by comparing mean thicknesses of TEC with UT.

Tile Hole		UT	Voltage Threshold			Power Law			TEC MFA		
Number	Class	Mean	Mean	SD	% diff	Mean	SD	% diff	Mean	SD	% diff
1	P	9.3	8.2	0.5	-12	8.8	0.3	-5	9.1	0.4	-2
2	P	9.2	7.4	0.9	-19	8.8	0.4	-4	9.5	0.4	3
3	P	9.8	8.2	0.6	-16	9.2	0.4	-6	9.6	0.4	-2
4	P	9.6	7.6	0.7	-20	9.4	0.4	-3	9.4	0.3	-2
5	P	9.6	8.7	1.1	-10	9.2	0.3	-5	9.3	0.4	-3
6	P	9.5	8.5	1.1	-11	8.7	0.3	-8	9.3	0.4	-2
7	S	9.3	7.8	0.9	-16	9.5	0.2	2	9.7	0.3	5
8	S	9.3	7.7	0.7	-17	9.5	0.4	2	9.7	0.3	4
9	P	9.6	8.4	0.5	-13	9.2	0.3	-5	9.6	0.4	0
10	S	9.4	7.4	0.7	-21	9.3	0.3	-1	9.6	0.2	2
Average		9.5	8.0	0.8	-15	9.2	0.3	-3	9.5	0.4	0

Table 3: RMSE (in mm) for all TEC analysis methods for a pilger pipe (1) and seam-welded pipe (8).

Tile hole	Voltage Threshold	Power Law	TEC MFA
1	1.22	0.65	0.43
8	2.07	0.29	0.26

No regions of local wall thinning were indicated by either UT or any of the TEC analyses. Additionally, wall thicknesses reported by UT, power law, and TEC MFA mean wall thicknesses were within 10% of nominal, indicating no overall degradation.

Voltage threshold analysis was the simplest technique applied to the tile hole TEC data, but was also the least accurate and precise, under-sizing thickness by an average of 15%. Of the three analysis techniques, TEC MFA thickness estimates were most accurate, with an average percent difference of 0%. Quantitatively, TEC MFA mean thickness estimates were on average 3% and 15% more accurate than power law and voltage threshold analyses, respectively.

It is suspected that the two classes of pipes, pilgered and seam-welded, had slightly different electromagnetic properties. According to manufacturing specifications all pipes inspected were of the same steel type, but slightly different shape responses were observed. It should be noted that for seam-welded pipes, power-law precision was similar or better than TEC MFA. Electromagnetic material properties were entered as fixed variables in the TEC MFA algorithm, leading this technique to be slightly more susceptible to calibration errors due to material differences.

5. Conclusion

An inspection campaign was completed in which ten tile holes with 5 m long, 250 mm Schedule 40 liner pipes with 9.3 mm nominal wall thickness were scanned by a combination

UT and TEC inspection head. The maximum lift-off encountered by the probe was 10 mm during inspection. Three TEC analysis methods were employed: voltage threshold, power law analysis, and TEC MFA. In general, there was good agreement between UT and TEC results. No regions of local wall thinning were identified, and reported thickness values indicated no overall degradation. Power law and TEC MFA proved to be more accurate and robust methods for estimating thickness. These two techniques utilized more of the available information, and are more directly linked to underlying electromagnetic principles than the voltage threshold technique, which was more empirical in nature. TEC MFA produced the most accurate and precise wall thickness estimates in almost every case, in comparison with UT measurements. Wall thickness of ferromagnetic steel pipes has been successfully estimated by TEC with moderate lift-off. The field-ready inspection system and analysis methods can be easily modified to accommodate different ferromagnetic materials and structure sizes for future inspections.

References

[1] V. S. Cecco, G. Van Drunen and F. L. Sharp, Eddy CUrrent Manual Volume 1: Test Method, Chalk River: Chalk River Nuclear Laboratories, 1981.

[2] X. Chen and Y. Lei, "Excitation current waveform for eddy current testing on the thickness of ferromagnetic plates," *NDT&E International,* vol. 66, pp. 28-33, April 2014.

[3] B. A. Lepine, J. S. Giguère, D. S. Forsyth, A. Chahbaz and J. M. S. Dubois, "Interpretation of Pulsed Eddy Current Signals for Locating and Quantifying Metal Loss in Thin Skin Lap Splices," in *Review of Progress in Quantitative Nondestructive Evaluation*, Bellingham, WA, 2002.

[4] W. Cheng, "Pulsed Eddy Current Testing of Carbon Steel Pipes' Wall-thinning Through Insulation and Cladding," *Journal of Nondestructive Evaluation,* vol. 31, pp. 215-224, March 2012.

[5] S. G. Mokros, P. R. Underhill, J. Morelli and T. W. Krause, "Pulsed Eddy Current Inspection of Wall Loss in Steam Generator Trefoil Broach Supports," *Submitted IEEE Sensors Journal August 2016.*

[6] M. Alamin, G. Y. Tian, A. Andrews and P. Jackson, "Principal Component Analysis of Pulsed Eddy Current Response from Corrosion in Mild Steel," *IEEE Sensors Journal,* vol. 12, no. 8, pp. 2548-2553, August 2012.

[7] J. A. Buck, P. R. Underhill, S. G. Mokros, J. E. Morelli, V. K. Babbar, B. A. Lepine, J. Renaud and T. W. Krause, "Pulsed Eddy Current Inspection of Support Structures in Steam Generators," *IEEE Sensors Journal,* vol. 15, no. 8, pp. 4305-4312, August 2015.

[8] J. A. Buck, P. R. Underhill, J. E. Morelli and T. W. Krause, "Simultaneous Multiparameter Measurement in Pulsed Eddy Current Steam Generator Data Using Artificial Neural Networks," *IEEE Transactions on Instrumentation and Measurement,* vol. 65, no. 3, pp. 672-679, March 2016.

[9] D. G. Park, C. S. Angani, G. D. Kim, C. G. Kim and Y. M. Cheong, "Evaluation of Pulsed Eddy Current Response and Detection of the Thickness Variation in the Stainless Steel," *IEEE Transactions on Magnetics,* vol. 45, no. 10, pp. 3893-3896, October 2009.

[10] G. Y. Tian, Y. He, I. Adewale and A. Simm, "Research on spectral response of pulsed eddy current and NDE applications," *Sensors and Actuators A: Physical,* vol. 189, pp. 313-320, 2013.

[11] D. Horn, B. A. Lepine and J. Lei, "Method for non-destructive analysis of multiple structural parameters". United States Provisional Patent 62/289,515, 1 February 2016.

[12] D. J. Griffiths, Introduction to Electrodynamics 3rd ed., Upper Saddle River: Prentice Hall, 1999.

[13] H. C. Ohanian, "On the approach to electro- and magneto-static equilibrium," *American Journal of Physics,* vol. 51, no. 11, pp. 1020-1022, 1983.

[14] J. A. Bieber, S. K. Shaligram, J. H. Rose and J. C. Moulder, "Time-Gating of Pulsed Eddy Current Signals for Defect Characterization and Discrimination in Aircraft Lap-Joints," in *Review of Progress in Quantitative Nondestructive Evaluation Vol. 16*, La Jola, CA, 1996.

[15] W. J. Reed and B. D. Hughes, "Power-law distributions from exponential processes: an explanation for the occurence of long-tailed distributions in biology and elsewhere," *Journal of Mathematical Science,* vol. 8, pp. 329-339, March 2003.

[16] D. Horn, B. A. Lepine and J. Lei, "A Method for Quantitative Analysis of Transient Eddy Current Testing," in *NDT in Canada 2016*, Burlington, 2016.

Deployment of Pulsed Eddy Current as an Inspection Tool for Nuclear Power Plants

Vincent DEMERS-CARPENTIER[1], Maxime ROCHETTE[1], Marc GRENIER[1], Charles TREMBLAY[1], Marco Michele SISTO[1], Florian HARDY[1], Martin TURGEON[1]

[1] Eddyfi, 2800 rue Louis-Lumière, bureau 100, Québec (Québec), G1P 0A4, Canada
Phone: (418) 780-1565 ; e-mail: mturgeon@eddyfi.com

Abstract
Pulsed Eddy Current (PEC) has been successfully deployed over the last decades for a variety of corrosion-related applications, most notably for Corrosion Under Insulation (CUI) inspections, Corrosion Under Fireproofing (CUF) and Flow Accelerated Corrosion (FAC). This technology has proven to be an efficient screening tool, allowing for detection of corrosion without having to remove coating or insulating material over typical pipes, tanks and vessels. However, the use of this technique has been largely focused on petrochemical applications. In the spring of 2016, a leading NDT service provider conducted a PEC inspection at a North American power plant using a new-generation instrument. This paper discusses the possibilities and limitations of the technology for addressing NDT needs specific to the nuclear power industry. Field results are presented and discussed, and performances are compared for further insight.

Keywords: Pulsed Eddy Currents, Nuclear Power, Corrosion Under Insulation

1. Introduction

Pulsed Eddy Currents (PEC) is a versatile non-destructive evaluation technique that can measure wall thickness of conductive components at high lift offs[1], [2]. As such, PEC is uniquely suited to detect corrosion under insulation, one of the most important unresolved issues facing the petrochemical industry today[3], [4]. Pulsed Eddy currents has been shown to work through weather jacket (or cladding) and presents major upsides compared to the brute force approach which consists in removing the insulation to perform a visual or ultrasounds inspection[5]. Insulated pipes and vessels are ubiquitous on oil refinery, therefore PEC has been used in a variety of applications[6], especially for inspections where X-rays radiography is inconvenient or down right impossible to use due to component geometry and placement. Recent advancements in PEC technology have contributed to a rapid expansion in the scope of its usage for a wide array of applications[7]. Comparatively, PEC adoption for the nuclear power generation industry has been slow despite the numerous potential benefits. Herein we present experimental data acquired using the PEC instrument developed by Eddyfi (the Lyft™) at a North American power plant and discuss how the technique can provide valuable insight for asset owners worldwide.

PEC is best used as a screening tool owing to its ability to inspect in-service components through insulation and cladding. By allowing to identify corrosion areas outside of shutdown period, it permits the expansion of inspection scope and frequency without increasing its schedule. This broader screening permits a more focused application of complimentary quantitative methods such as radiography and ultrasounds during shutdowns. Lyft is a versatile tool that can inspect insulated pipes, vessels, steam lines, heat exchanger shells, essentially anything made of carbon steel for wall thickness up to 2.5'' and insulation between 0 and 8'' whether they are cladded in aluminum sheets or not.

2. Technical information

A pulsed Eddy Currents measurements occurs in three distinct steps, as shown in Scheme 1. First the probe projects a strong magnetic field which fully penetrates the material and charges it with magnetic potential energy. This step typically lasts between 10 and 500 milliseconds depending on material wall thickness, long enough for the magnetic field to stabilize and the eddy currents to die out in the inspected component. The second step is a very abrupt cut-off of the magnetic field which will jump-start strong eddy currents in the component as they oppose the sudden change in magnetic field. During the third step, the magnetic field in the component slowly die out as they expend the stored magnetic potential energy in the form of eddy currents. The rate of this decay is measured by the probe and contains information on the component's wall thickness that is analyzed by the PEC instrument.

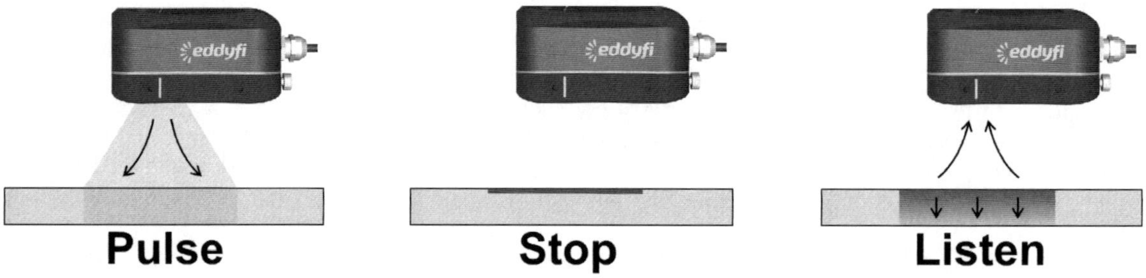

Scheme 1: A typical PEC measurement consists in the three steps of 1- Pulse, 2- Stop and 3- Listen. The signal received by the probe during the third step is analyzed by the instrument to measure the inspected component's wall thickness.

3. Laboratory experimentation

The sizing algorithms developed for the Lyft system analyse a section of the eddy current signal that is earlier in time than what is done by most competitor systems. This provides a number of benefits, chief among them is that it allows the acquisition routine to be performed much faster. First, analysing the signal earlier shortens the receiving phase, which in itself can speed up the acquisition rate by 10% to 30%. Moreover, since we measure the signal earlier in the decay we get a better signal-to-noise ratio, lessening the need to average a large number of pulses. Additionally, the Lyft employs a number of smart filters which further improve the signal-to-noise ratio. These three factors amount to a substantial increase in effective acquisition rate.

A typical use case for the inspection of a CUI component with a current technology PEC tool goes as follow. The inspection crew first traces a grid on the component with a pitch corresponding to a fraction of the probe's footprint. Only then can the inspection truly begin, with the inspector acquiring a data point for each cell in the grid. The first step of tracing the grid is the most time consuming and can amount for more than 50% of the total inspection time.

In contrast, Lyft comes with a fully enabled Dynamic Scanning Mode allowing the users to record the position of every measurement with the rugged encoder wheel embedded into probe. The inspection team therefore only needs to trace parallel lines to serve as guides for the encoded probe measurements. In addition to cutting down in half the setup time of the inspection crew, this dynamic scanning mode allows an increased spatial resolution along the scanning axis. Figure 1 shows the benefits of using the Lyft's dynamic scanning mode by comparing a rough grid mode C-scan with resolution 50.8 mm x 50.8 mm (2 in x 2 in) and a fully encoded, high resolution C-scan recorded with an in-axis resolution of 12.7 mm (0.5 in) and a pitch of 25.4mm (1 in) for the parallel guiding lines. The sample is a 6.35 mm (0.25 in) thick carbon steel plate with a round flaw in the centre. The artificial defect used in the experiment is quite large (101.6mm) to ensure a proper flaw detection with the grid scanning mode but it is not representative of the minimum detectable defect with the Lyft system. Under these experimental conditions and with the averaging set to 4 pulses per data point, the dynamic scanning was achieved at a moving speed of 50.8 mm/s (2 in/s).

Lab mockup sample – Flat bottom hole	
Plate WT	6.35 mm
Insulation height	25.4 mm
Weather Jacket	Aluminum 0.5 mm
Flaw size	101.6 mm diameter Remaining WT 2.1mm
Sample	

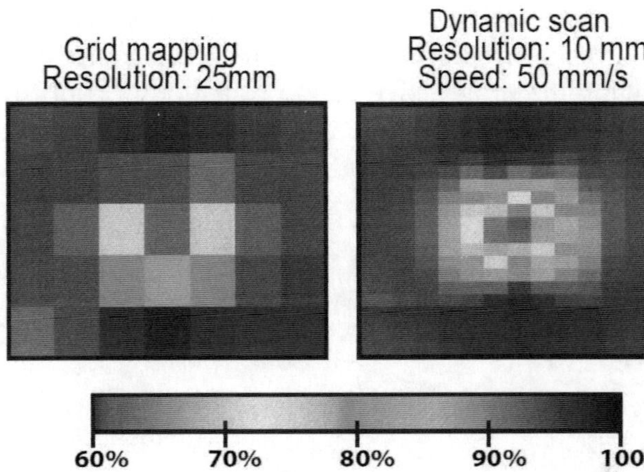

Figure 1. Grid-mode and Dynamic Scanning Mode C-scans of a sample plate with a flat bottom hole flaw.

Sample Pipe #0384	
Pipe Diameter	273 mm
Wall Thickness	9.3 mm
Insulation Thickness	50.8 mm
Weather Jacket	Aluminum 1mm ; Galvanized Steel 0.6 mm
Defect Type	Uniform Steps
Steps Length	304.8 mm
Steps Remaining Wall Thickness	9.3 mm (100%) 8.4 mm (90%) 7 mm (75%) 5.1 mm (55%)
Sample	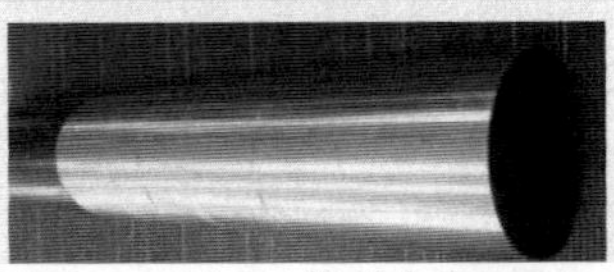

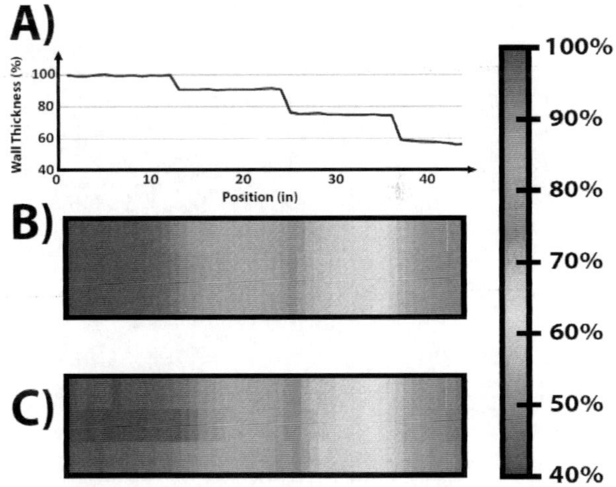

Figure 2. Ultrasound wall thickness measurements (A) compared to *Lyft* grid mode measurements (B, C) taken on a calibration carbon steel pipe wrapped in 50.8 mm (2 in) of insulation with a 1 mm (0.040 in) thick aluminum (B) or a 0.6 mm (0.024 in) thick galvanized steel (C) weather jacket.

It is very common for CUI components to include a layer of conductive material covering the insulation. This weather jacket (or cladding) material can influence the quality and accuracy of PEC measurements. The Lyft's algorithms are specially tailored to minimize the adverse effect of a wide variety of weather jacket materials, including the industry standards of aluminum, stainless steel, and galvanized steel. Figure 2 shows a side-by-side comparison of Lyft measurements on a calibration step pipe wrapped in 50.8 mm (2 in) of insulating material and jacketed in either 1 mm (0.040 in) of aluminum or 0.6 mm (0.024 in) of galvanized steel.

The sizing is mostly unaffected by the presence of the weather jackets as long as the calibration is performed correctly.

It is common for CUI to form in hard-to-reach places such as around a nozzle or pipe support. The challenge then is to differentiate between the mass loss from corrosion and the additional conductive mass present at the junction with the support or nozzle. Figure 3 shows the *Lyft* system's performance in a typical configuration where mass loss occurs near the point of contact between a carbon steel pipe and its saddle support. The flaw, a 20% remaining wall thickness tapered in both the axial and circumferential directions, is mapped in dynamic mode showing the progressive nature of the defect until it gets overridden by the excess mass from the steel saddle support, roughly 50.8 mm (2 in) away from the structure. This is another demonstration of the added insight brought by the increased resolution arising from the use of the *Lyft* system's dynamic mode.

Lab mockup sample – Pipe support	
Pipe Dimension	164 mm OD x 7.11 mm WT
Wrap thickness	50.8 mm
Weather Jacket	Aluminum 1.0 mm
Defect Type	Taper flaw over 180° at pipe support
Scan Type	Dynamic scan
Sample	

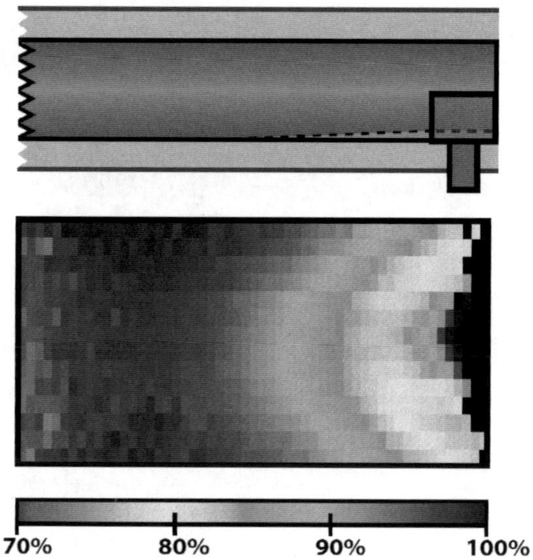

Figure 3. Dynamic Mode C-Scan a carbon steel pipe with a tapered defect near a saddle support.

4. Inspection performed on an active nuclear power plant

In April 2016, Eddyfi assisted in the first ever PEC Lyft™ inspection on an active nuclear power plant on a number of feedwater heaters. Herein we present key results to highlight the possibilities of the PEC technique to provide valuable information to asset owners when it comes to managing corrosion under insulation issues.

Figure 4 shows a prototypical inspection result acquired on a low-pressure feedwater heater with 15mm thick walls, wrapped in 50mm of insulation and 1 mm of aluminum cladding. These vessels being quite large (1.5 m diameter, 9 m long) the inspection is typically divided into a number of scan zones to facilitate the data acquisition and reporting. The C-scan shown is for the upper middle section of the vessel and contains a number of features that demonstrate the capabilities of the instrument. Two sections at the top of the C-scan appear black and correspond to the obstruction of two large nozzles as shown in the schematic representation. On the left hand-side of the C-scan, we observe a thicker region which corresponds to a slight edge effect from the vacuum ring running along the circumference of the vessel. Additionally, running horizontally toward the bottom of the C-scan is a thicker line which is attributed to a steam tracer line located underneath the insulation and cladding. Finally, some mild corrosion, roughly 10% wall loss, can be found between the two nozzles and appear light blue in the C-scan.

Low Pressure FW heater	
Vessel Dimension	9 m long, 1.5 m OD
Insulation Thickness	50 mm
Wall Thickness	15 mm
Cladding	Aluminum Steel 1.0 mm
Defect Type	Real corrosion defect

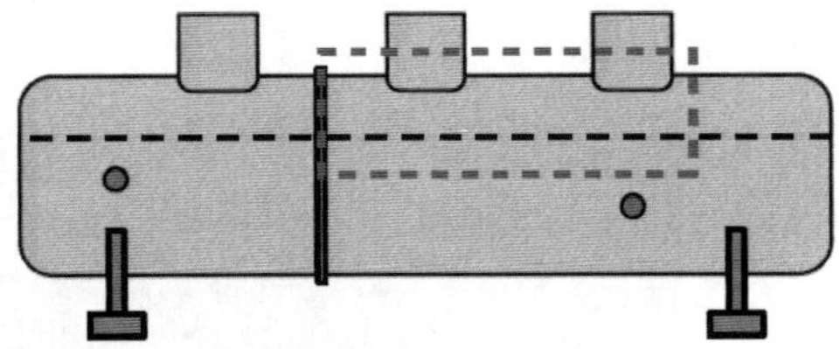

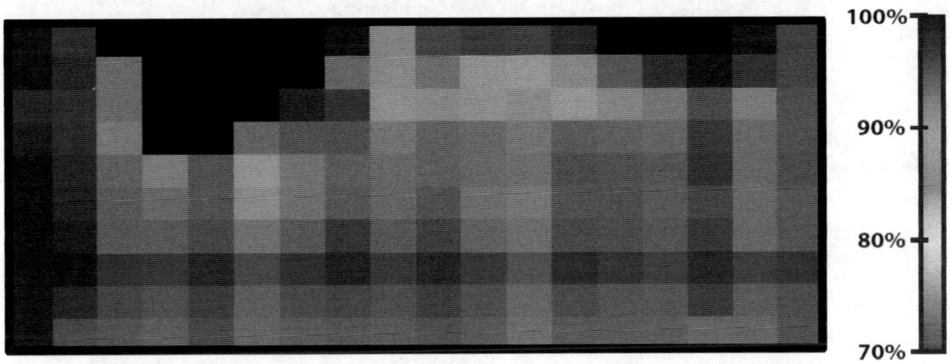

Figure 4: Inspection results for a section located at the top of a low-pressure feedwater heater showing prototypical feature including two nozzles, a vacuum ring, a steam tracer line and some mild corrosion.

Figure 5 shows three distinct scan zones recorded on a high-pressure feedwater heater of similar dimensions and specifications to the previous vessel (wall thickness 15mm, 50mm insulation wrapped in aluminium cladding). The three zones are for the top (A), bottom (B) and nose (C) of the heat exchanger. The first scan zone (A) shows the top part of the component where three large nozzles and two vacuum rings which produced obstruction for the inspection. However, the inspector managed to achieve good coverage of the zone by scanning around these elements and was able to observe mild corrosion (~10% wall thickness loss or 1.5 mm) in some spots, similar to what was reported in figure 4. In contrast, more relatively advanced corrosion was observed at the bottom of the vessel (B). Corrosion forming at the bottom of the vessel is expected given how this is where condensation will typically build up. The wall thickness loss of roughly 25% (or 3.75 mm) is consistent with the asset owner's expectations. The last scan zone (C) shows data acquired at the nose of the heat exchanger. The C-scan reveal some corrosion at the bottom left corner. This wall loss of ~30% is slightly higher than expected although well within the safety margins.

High Pressure FW Heater	
Vessel Dimension	7 m long, 1.3 m OD
Insulation Thickness	50 mm
Wall Thickness	15 mm
Cladding	Aluminum Steel 1.0 mm
Defect Type	Real corrosion defect

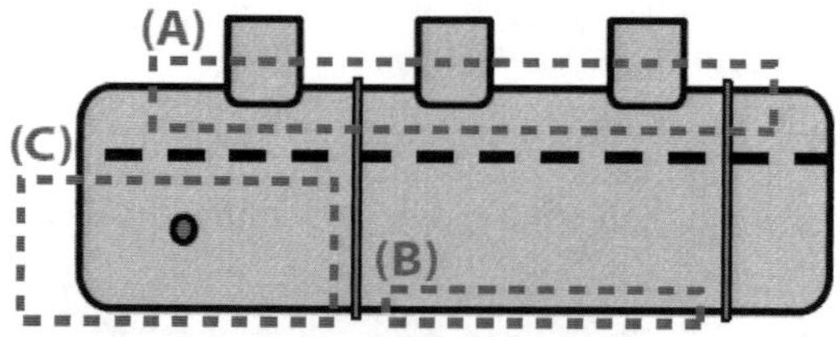

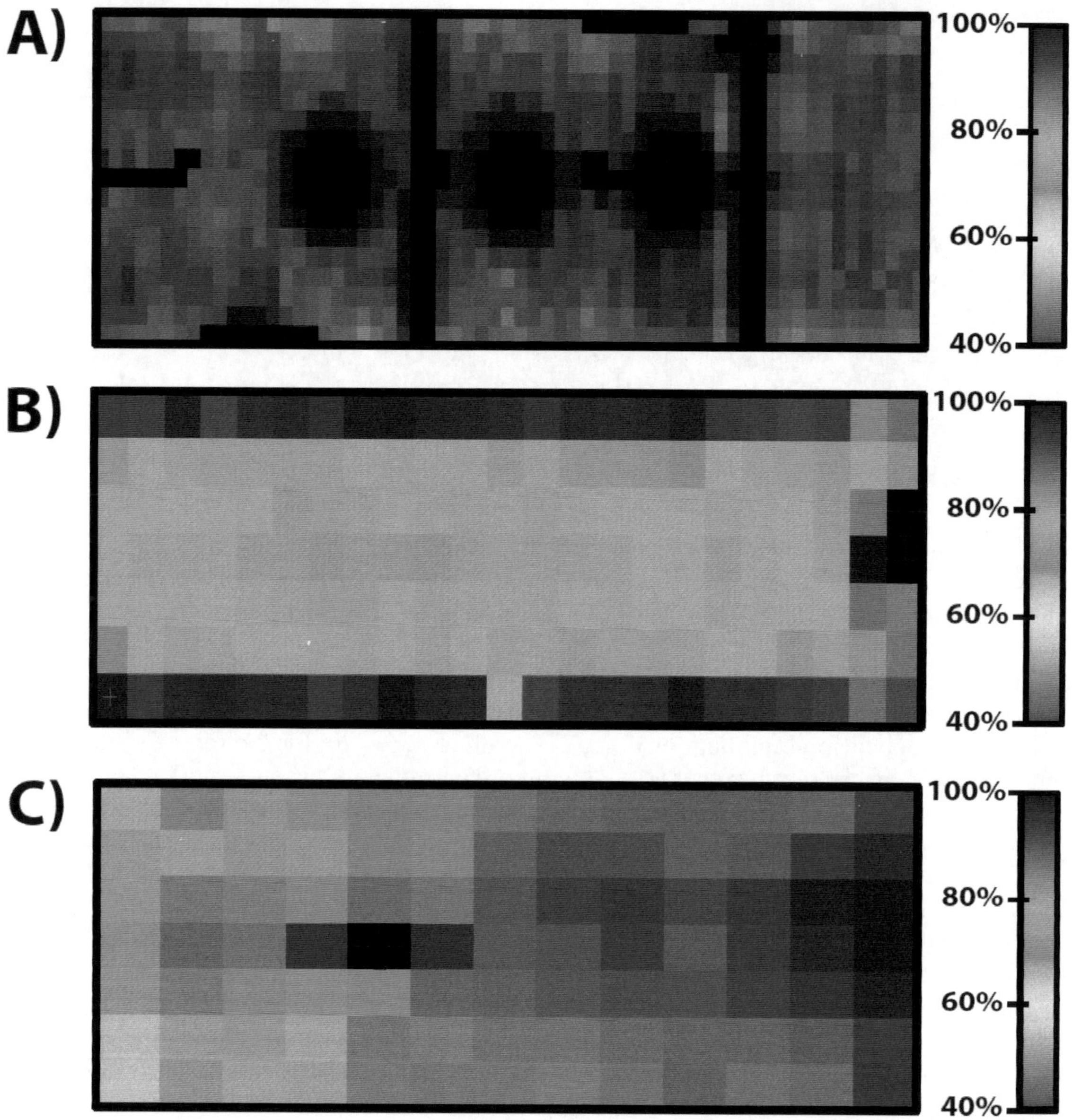

Figure 5: Inspection results for three sections located at the top (A), the bottom (B) and the nose (C) of a high-pressure feedwater heater showing several features including varying degrees of corrosion under insulation.

The feedwater heaters inspected as part of this technical study reveal varying degrees of corrosion. These vessels are scheduled to be unwrapped and quantitatively inspected using ultrasounds in fall 2018. The wall loss evaluation provided by the PEC inspection may incite the NDT coordinator to push ahead or back this schedule in an effort to better focus the short time available during any given shutdown. This is the type of valuable information asset owners need in order to make good decision when it comes to inspection schedule.

5. Conclusion

Pulsed eddy current is a versatile screening technique which is uniquely suited to detect and evaluate corrosion under insulation. In this paper we have shown how Eddyfi's Lyft™ was used to measure CUI on an active nuclear power plant, providing the client valuable information which can be used to better focus the attribution of other NDE techniques in the upcoming shutdowns. The ability to quickly measure wall loss on in-service components through insulation with good accuracy and without having to remove the insulation can be a real asset in the NDE tool belt of the nuclear power generation industry.

References

[1] M. A. Robers and R. Scottini, "PULSED EDDY CURRENT IN CORROSION DETECTION," presented at the 8th ECNDT, Barcelona, 2002.

[2] W. Cheng, "Pulsed Eddy Current Testing of Carbon Steel Pipes' Wall-thinning Through Insulation and Cladding," *J. Nondestruct. Eval.*, vol. 31, no. 3, pp. 215–224, Mar. 2012.

[3] N. Schmoyer, "A Primer on Corrosion Under Insulation (CUI)," Inspectioneering, Jun. 2015.

[4] T. Hanratty, "Corrosion Under Insulation – The Hidden Problem," *Hydrocarb. Asia*, no. APR-JUN, p. 3, 2012.

[5] *API Recommended Practice 583 - Corrosion Under Insulation and Fireproofing*, 1st ed. American Petroleum Institute, 2014.

[6] P. Crouzen and I. Munns, "Pulsed Eddy Current Corrosion Monitoring in Refineries and Oil Production Facilities – Experience at Shell," presented at the ECNDT, 2006, p. 7.

[7] M. Grenier, V. Demers-Carpentier, M. Rochette, and F. Hardy, "Pulsed Eddy Current: New Developments for Corrosion Under Insulation Examinations," presented at the 19th World Conference on Non-Destructive Testing (WCNDT 2016), Munich (GER).

Buried Pipe NDE Overview

6th International CANDU ISI Workshop/NDT in Canada 2016 Conference

Presenter: - Carl Latiolais Senior Program Manger NDE Reliability

Author: Steve Kenefick – Principal Project Manager

Overview

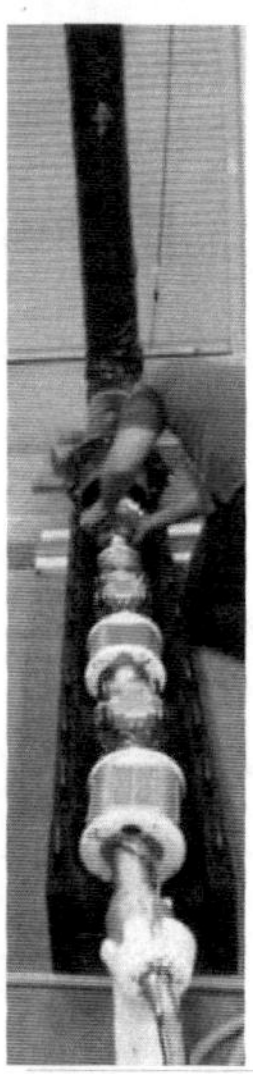

- Operational experience (OE)
- US inspection requirements
- Overview of NDE technologies and methodologies
- EPRI related information and references

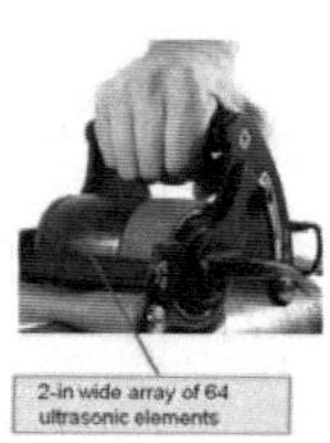

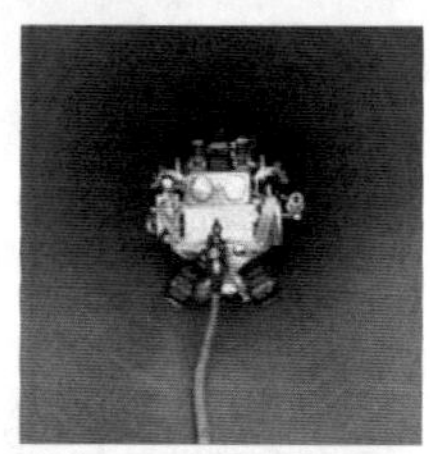

Buried Pipe Operating Experience

- Infrastructure in US is aging
- Buried and underground piping leaks have occurred
  - Low levels of Tritium has been detected on-site
  - Leaks to date have not had significant safety or radiological consequences

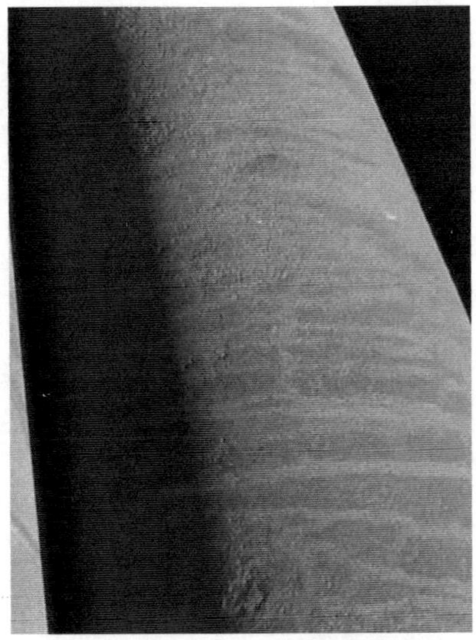

3

NSIAC 09-14 Initiative

- The Nuclear Strategic Issues Advisory Committee (NSIAC) developed an Initiative to address US industry buried pipe leaks
  - NSIAC is composed of US Chief Nuclear Officers (CNO)
  - Industry commitment, not regulation
  - Goal to provide "Reasonable Assurance" of structural and leak integrity of buried pipe with special emphasis on piping that contained radioactive materials
- The NEI Buried Piping Integrity Task Force (BPTIF) was formed to develop direction to meet Initiative
  - Composed of utilities, NEI, INPO, Insurers, and EPRI
  - Developed NEI 09-14 Initiative – Guideline for the Management of Underground Piping and Tank Integrity (Currently revision 4)

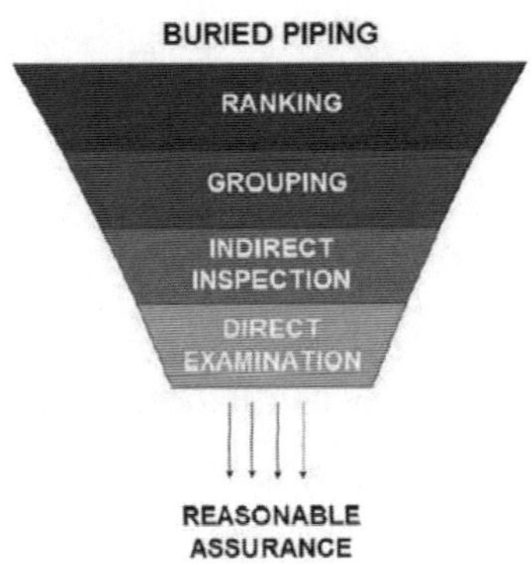

4

Scope of the NSIAC NEI 09-14 Initiative

- All buried and underground piping and tanks that are outside of a building and below grade (whether or not they are in direct contact with the soil) if they:
  - Are safety related,

    OR
  - Contain licensed material or are known to be contaminated with licensed materials,

    OR
  - Contain environmentally sensitive materials

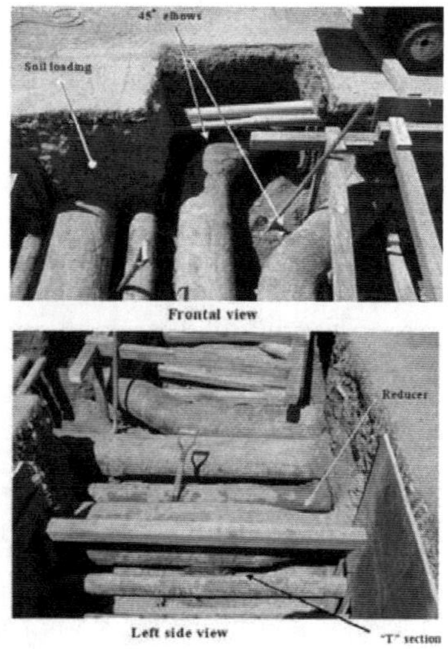

Status of NEI 09-14

- All milestones are complete
  - Corrosion has been found but many systems are in good shape
  - Most corrosion is internal
  - Is now part of an ongoing engineering program
- NRC staff recommended to the Commission that no new regulation is necessary "Commission agreed"
- Ongoing oversight
  - NRC to continue to monitor
    - Objective a reduction in leak trends
  - INPO to continue to monitor

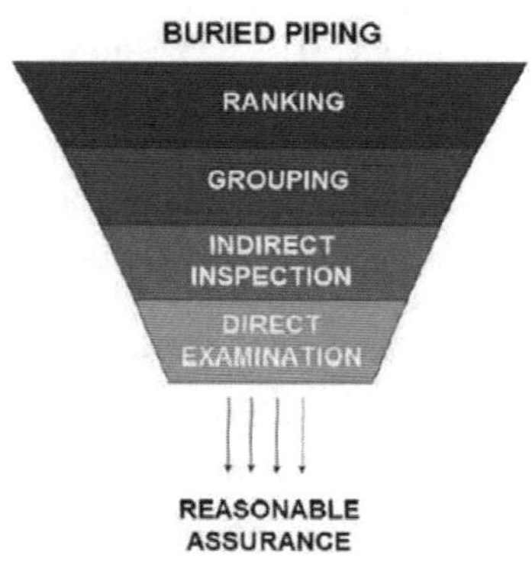

License Renewal

- NRC license renewal addresses underground piping and tanks in NUREG-1801 "*Generic Aging Lessons Learned*" (GALL)
  - Provides aging management programs (AMP) to manage aging effects of systems, structures and components (SSC)
  - Applicant may propose alternatives methods
- Interim Staff Guidance (ISG)
  - LR-ISG-2011-03 – Generic Aging Lessons Learned (GALL) Report Revision 2 AMP XI.M41, "Buried and Underground Piping and Tanks"
- GALL requirements have evolved
  - Rigor of buried pipe requirements has increased
  - Second license renewal (after 60 years) requirements will be more demanding for early implementers

EPRI | ELECTRIC POWER RESEARCH INSTITUTE

Challenges of Examining Underground Pipe

- Many systems are difficult to access
  - Not designed or installed with inspection access in mind
  - Buried deep
  - In layers with other infrastructure
  - Under buildings
- Locations sometimes not well documented
- Some contain radioactive fluids

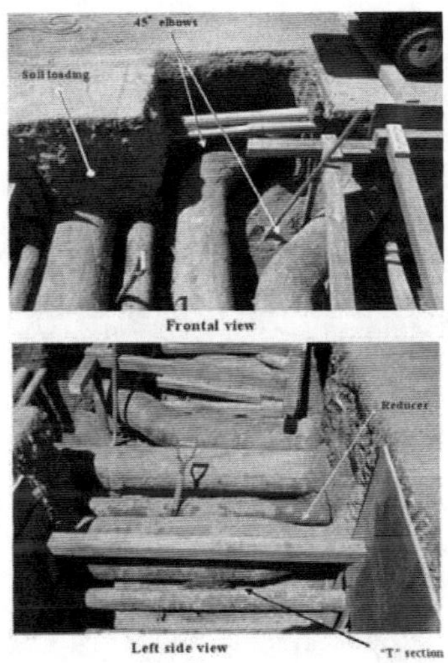

EPRI | ELECTRIC POWER RESEARCH INSTITUTE

Challenges Presented by Underground Pipe

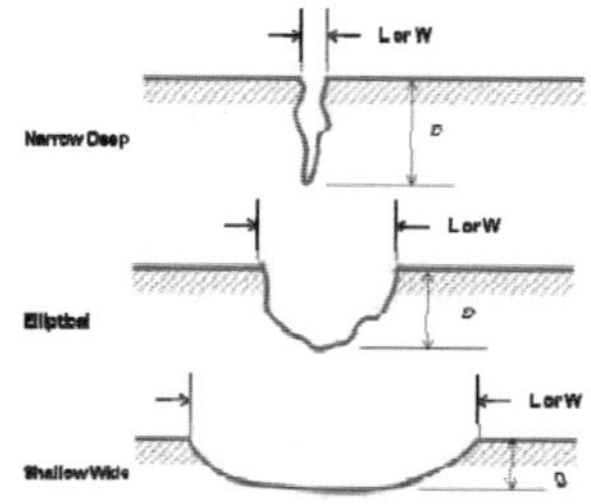

Wide ranging set of variables require careful selection of NDE technology

- Several material types: steel, cast iron, copper alloys, stainless, concrete, asbestos, PVC, polyethylene, FRP, etc.
- Diameters range from small bore to >10-ft
- Multiple joint types: butt-welded, socket welded, flanged, threaded, etc.
- Piping content include raw and treated water, gases, chemicals, and oil
- Coated and lined pipe: cement, high density polyethylene (HDPE), tar, tape wrap
- Varying damage mechanisms and flaw morphologies

EPRI | ELECTRIC POWER RESEARCH INSTITUTE

Buried Pipe NDE Workshop Agenda

- Operational experience (OE)
- Inspection requirements
- Overview of NDE technologies and methodologies
- EPRI references and training

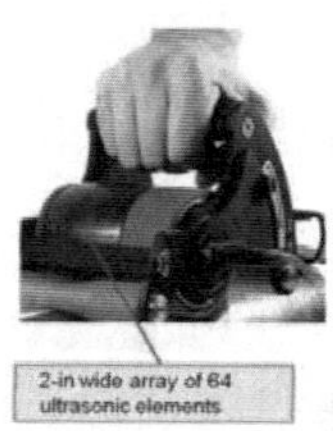

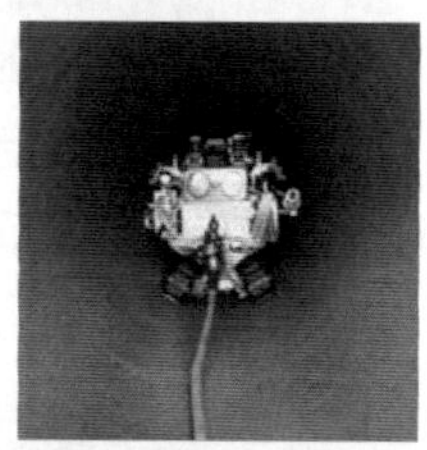

EPRI | ELECTRIC POWER RESEARCH INSTITUTE

Report 3002004395 – Nondestructive Evaluation: Buried Pipe NDE Reference Guide—Revision 3

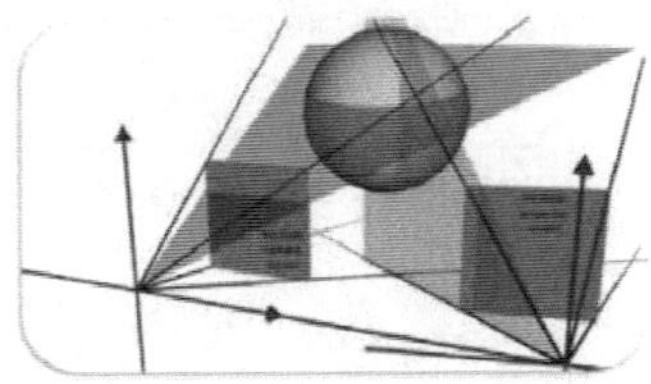

- **Repository of buried pipe NDE information**
  - Ultrasonic Technology
  - Guided Wave
  - Remote Field Testing
  - Magnetic Flux Leakage (MFL)
  - Electromagnetic Technology
  - Radiography
  - Visual Technologies
  - Surface Profilometry
  - Leak Detection
  - Radar and Microwave Inspection

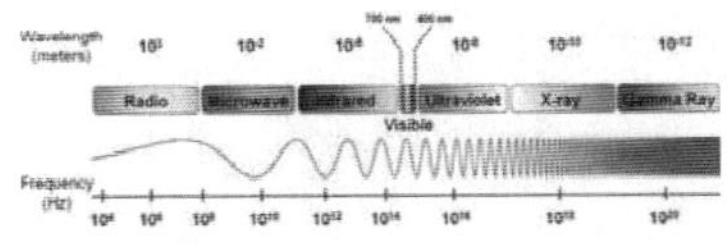

Report 3002004395 – Nondestructive Evaluation: Buried Pipe NDE Reference Guide—Revision 3

- **Repository for buried pipe NDE information**
  - Discontinuity Morphologies
  - Technology overview
  - Sensor technology
  - NDE selection considerations
  - In-line Inspection Delivery Technology

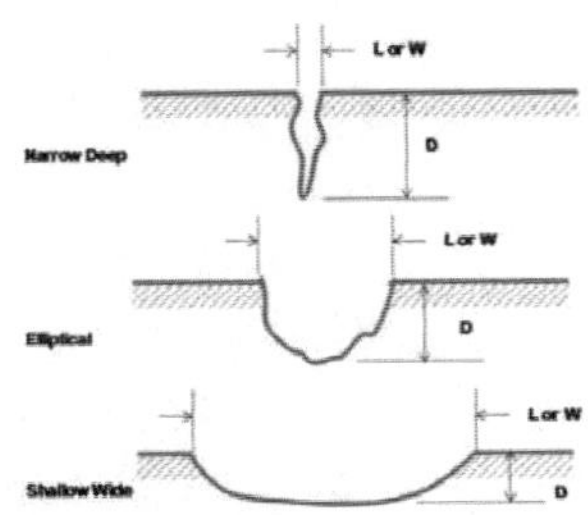

Buried Pipe NDE Methods

In-line	In-line & Outside Pipe	Outside pipe
Remote field testing	Ultrasonics	Guided wave
Magnetic flux leakage	Electromagnetic techniques	Radiography
Saturated Low Frequency Eddy Current	Electromagnetic acoustic transducer	
Remote visual	Laser profilometry	

In-line Delivery Methods
- Robotic Crawlers - Flow through - Pull through

13

EPRI ELECTRIC POWER RESEARCH INSTITUTE

Buried Pipe NDE Technology Capabilities (General)

Method	Volumetric	Measures Thickness	Change in Thickness
Ultrasonics	X	X	X
Guided Wave	X		X
Remote Field Testing (RFT)	X	X	X
Guided Wave	X		X
Magnetic Flux Leakage (MFL)	X		X
Radiography	X		X
Saturated Low Frequency Eddy Current	X		X
Laser Profilometry / Structured White Light			X
Visual			X

14

EPRI ELECTRIC POWER RESEARCH INSTITUTE

Buried Pipe NDE Methods

In-line	In-line & Outside Pipe	Outside pipe
Remote field testing	**Ultrasonics**	Guided wave
Magnetic flux leakage	Electromagnetic techniques	Radiography
Saturated Low Frequency Eddy Current	Electromagnetic acoustic transducer	
Remote visual	Laser profilometry	

In-line Delivery Methods
- Robotic Crawlers - Flow through - Pull through

15

EPRI | ELECTRIC POWER RESEARCH INSTITUTE

Flow Through or Free Swimming In-line Ultrasonic Technology

- Ultrasonic tool inserted into piping system
  - Bi-directional tool – requiring one access point
  - Single direction tool – requires launcher and retriever
- Moved through piping by fluid flow
  - Fluid added behind tool
  - Speed of tool controlled by fluid flow
- Onboard pulser, receiver, digitizer, and data storage
- Can collect data at feet per second
- Axial location of data measured with encoder
- Sophisticated data analysis software

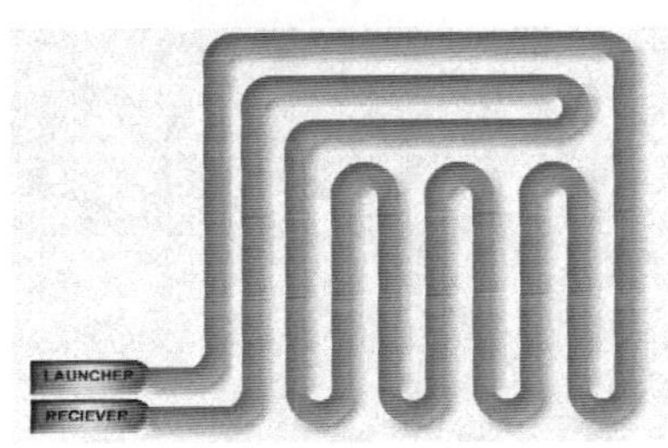

16

EPRI | ELECTRIC POWER RESEARCH INSTITUTE

Flow-through In-line Ultrasonic Array Technology

Transducer array device

- Transducers positioned 360-degrees around tool circumference
  - Number of transducers dependent on diameter and desired coverage
  - Small-diameter tool may have as many as 48 / large-diameter 100s of transducers
- Small to large diameter pipe

17

EPRI ELECTRIC POWER RESEARCH INSTITUTE

Flow Through In-line Rotating 0-degree Ultrasonic Transducer

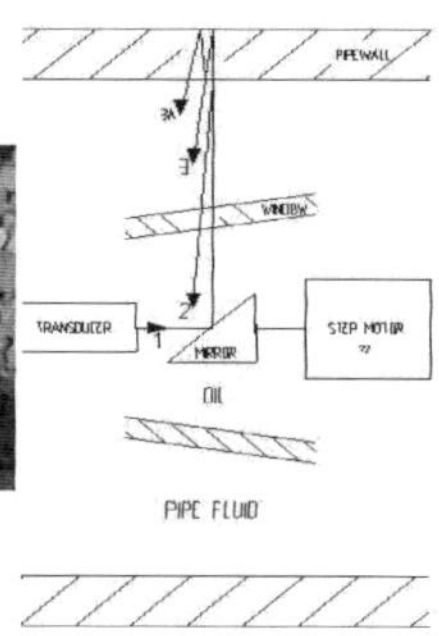

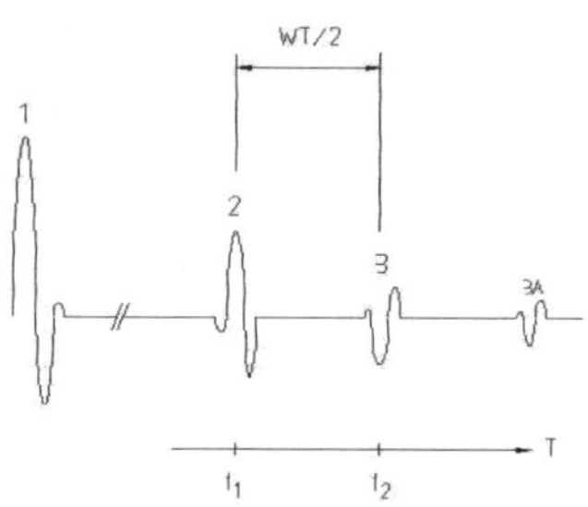

18

EPRI ELECTRIC POWER RESEARCH INSTITUTE

Flow Through In-line Ultrasonic Technology

Technology has been used at multiple plants (one example)

- 2300-feet of piping in ~2.5 hours
  - ~11.5 million measurements
- Ultrasonic Data Acquired
  - 2 degree circumferential increment
  - 10 mm axial increment
- Line contained several elbows
  - Maneuvered back to back elbows without issue

19

Robotically Driven Ultrasonic Rotating Array Technology

- Rotating head with 8 zero-degree transducers
- High resolution cameras

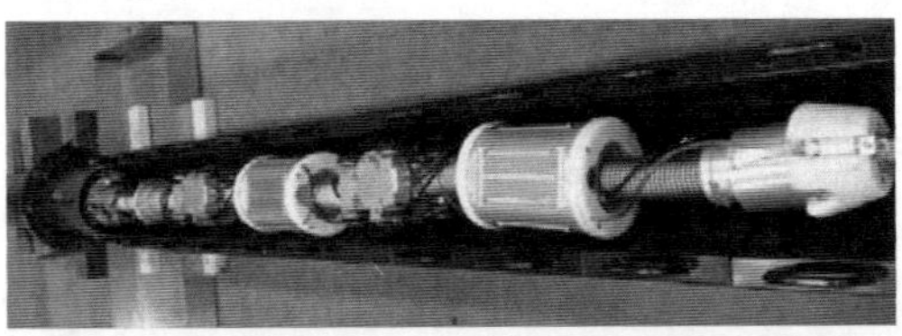

20

NDE Research Results – Phased Array

- Rapid scanning providing 100% coverage with increased sensitivity to corrosion type flaws
- Imaging with permanent data storage capabilities
- Technical basis published in: Buried Pipe NDE Technology Assessment and Development Interim Report (1025219)
- Area scanned in picture is 500-mm by 660-mm
  - ~80K measurements
  - Results can be extracted to a spreadsheet for analysis
  - Results shown in box
- Inspection vendors and utilities are using technology

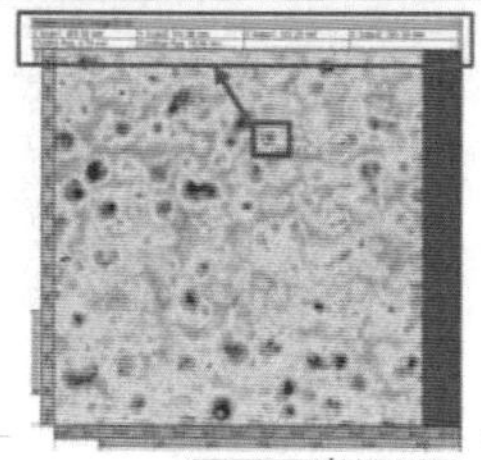

EPRI ELECTRIC POWER RESEARCH INSTITUTE

Electromagnetic Acoustic Transducer (EMATS)

- Self-propelled EMAT and visual crawler
- Probes are placed in contact with inside surface and rotated
- Detects and measures wall thickness
- Examine through coatings and linings
- Reduced surface preparation
- Couplant is not needed

EPRI ELECTRIC POWER RESEARCH INSTITUTE

Buried Pipe NDE Methods

In-line	In-line & Outside Pipe	Outside pipe
Remote field testing	Ultrasonics	**Guided wave**
Magnetic flux leakage	Electromagnetic techniques	Radiography
Saturated Low Frequency Eddy Current	Electromagnetic acoustic transducer	
Remote visual	Laser profilometry	

In-line Delivery Methods
- Robotic Crawlers - Flow through - Pull through

23

EPRI | ELECTRIC POWER RESEARCH INSTITUTE

Guided Wave Concept

- A guided wave is generated by a ring of piezoelectric transducers.
- The guided wave propagates down the pipe
- Changes in the cross-section of the pipe or material properties create reflections.
  - Welds, flaws, corrosion, flanges, valves, wall thickness changes, etc.
- Reflections propagate back to the sensors
- Signal is plotted as a function of distance from the sensors
- Welds are typically used as reference markers and calibration for the test.

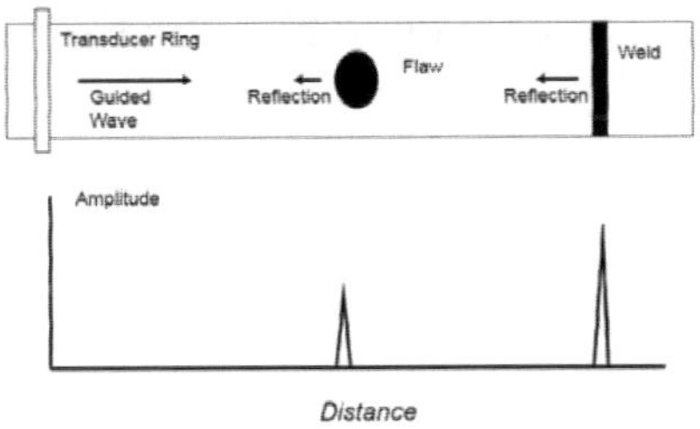

24

EPRI | ELECTRIC POWER RESEARCH INSTITUTE

Guided Wave Data

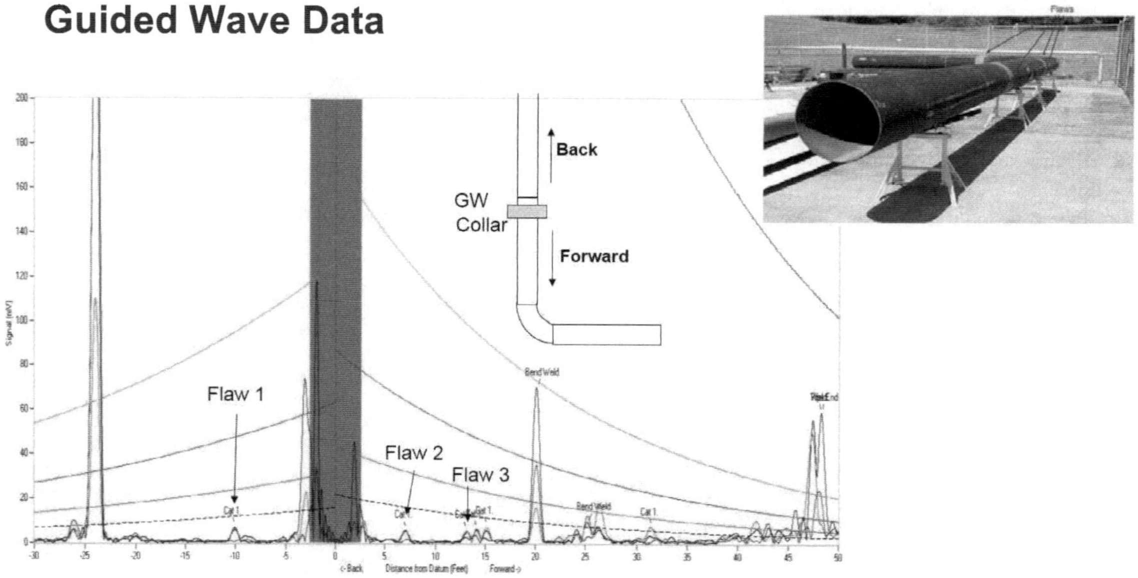

Benefits of Guided Waves

- Efficient 100% volume inspection of a large section
- Can potentially propagate long lengths, although flanges and large valves act as obstructions
- Inaccessible locations including wall penetrations, buried, coated, or obstructed areas
- Limited cleaning and excavation (required at locations where transducer ring is applied), but does require removal of coating
- Does not require access to inside of pipe or cleaning of inside pipe walls
- May inspect while system is operating
- Potential for periodic monitoring via permanently mounting sensors
- Technology available from 2” to 96” diameter

Challenges of Guided Waves

- Thick viscous coatings attenuate signal, thus reducing effective inspection lengths and sensitivity to flaws
- Flanges are barriers for guided wave propagation – inspection beyond a flange is not possible
- Elbows distort the guided wave signal and reduce signal-to-noise making inspection beyond them difficult
- Distinguishing flaws from other nonaxisymmetric reflectors such as welded attachments
- Quantifying size of damage is often not highly accurate

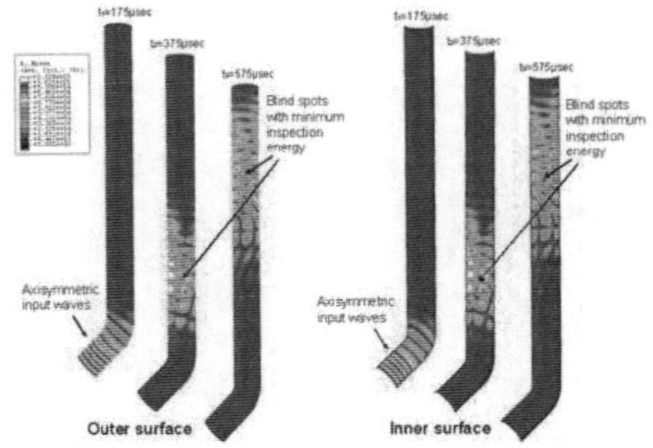

EPRI | ELECTRIC POWER RESEARCH INSTITUTE

Key Guided Wave Resource

- **Buried Pipe Guided Wave Examination Reference Document (1019115)**
  - Guided Wave Basic Theory
  - Data Acquisition
  - Data Analysis
  - Project Management
  - Literature Survey

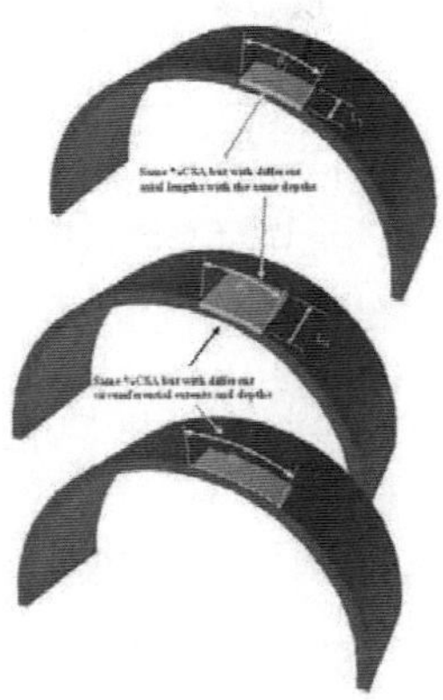

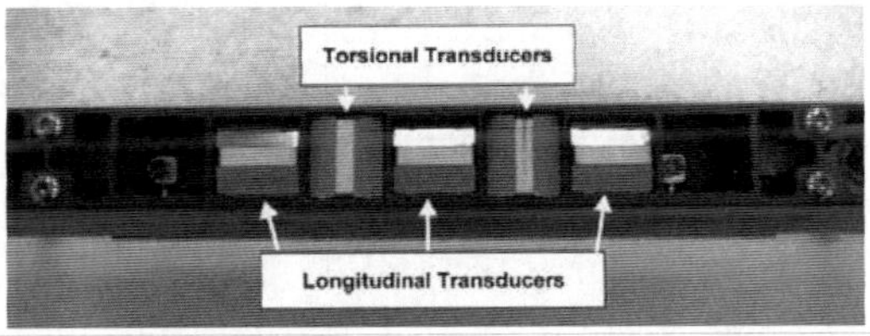

EPRI | ELECTRIC POWER RESEARCH INSTITUTE

Buried Pipe NDE Methods

In-line	In-line & Outside Pipe	Outside pipe
Remote field testing	Ultrasonics	Guided wave
Magnetic flux leakage	Electromagnetic techniques	Radiography
Saturated Low Frequency Eddy Current	Electromagnetic acoustic transducer	
Remote visual	Laser profilometry	

In-line Delivery Methods
- Robotic Crawlers - Flow through - Pull through

RFT Sensor and Energy Flow

- Uses Two coils
  - Exciter (transmitter or send)
  - Detector (receiver or pick-up)
  - Coils typically separated by 2 to 3 pipe diameter
- Measures
  - "Time-of-flight" (phase shift) between coils which indirectly relates to the wall thickness
  - Signal strength (amplitude) between coils

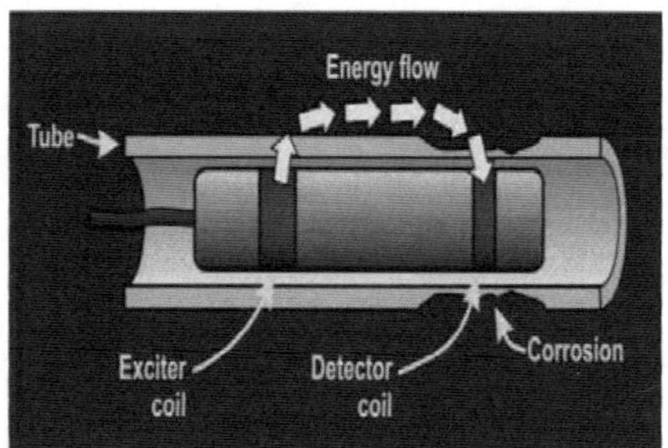

Courtesy of Russell NDE Systems Inc

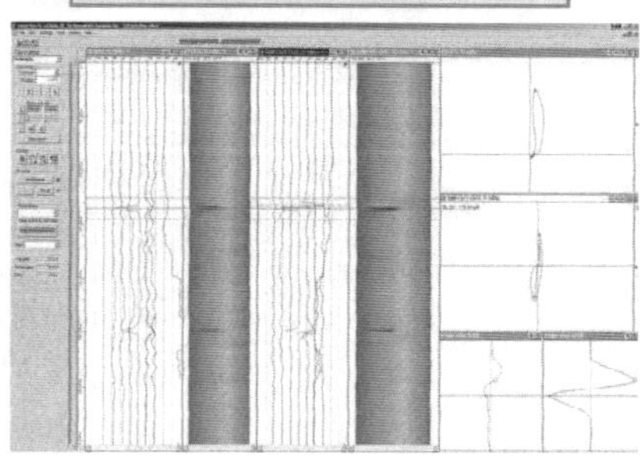

Remote Field Testing (RFT)

- In-line pipe inspections technique
  - Metallic pipe such as carbon steel, cast iron, and ductile iron
  - Pipe diameters ranging from 2-inches to 7 feet
- Can measure wall thickness
- Equal sensitivity to external and internal flaws
- Can be used in fluids such as water and oil

31

EPRI | ELECTRIC POWER RESEARCH INSTITUTE

RFT Application

- Sensors can operate approximately 1-inch from the pipe surface allowing for examination through:
  - Internal scales, tubercles, sand, deposits, and layers of mud
  - Liners such as high density polyethylene
  - Concrete lined piping
  - Coatings such as cement, bitumen, and cold tar wrap
  - Operates both in air or water
- Reduces / eliminates the need for extensive pre-cleaning in most situations

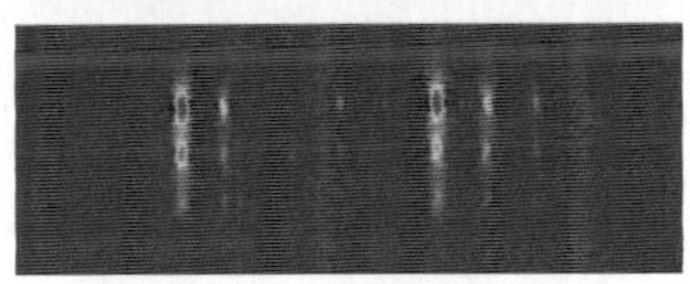

32

EPRI | ELECTRIC POWER RESEARCH INSTITUTE

Buried Pipe NDE Methods

In-line	In-line & Outside Pipe	Outside pipe
Remote field testing	Ultrasonics	Guided wave
Magnetic flux leakage	Electromagnetic techniques	Radiography
Saturated Low Frequency Eddy Current	Electromagnetic acoustic transducer	
Remote visual	Laser profilometry	

In-line Delivery Methods
- Robotic Crawlers - Flow through - Pull through

Saturated Low Frequency Eddy Current (SLOFEC)

Loss-of-material on the far surface alters the magnetic flux, which in turn alters the eddy currents

- Loss-of-material on the near surface directly interrupt the flow of eddy currents.
- These two responses alter the phase angle of the eddy current

Saturated Low Frequency Eddy Current (SLOFEC)

- In-line pipe inspections technique for ferromagnetic materials
  - Detection and resolution of pitting
  - Can differentiate between inside and outside surface discontinuities
  - Can tolerate coatings and liftoff from pipe surface
  - Cannot measure wall thickness

35

EPRI | ELECTRIC POWER RESEARCH INSTITUTE

NDE Technology: Saturated Low Frequency Eddy Current (SLOFEC) Robotics

- Examines through coating and linings
- Self-propelled tethered robot
- Traversed 1.5-diameter bends
- Technology available and used in other industries

36

EPRI | ELECTRIC POWER RESEARCH INSTITUTE

Buried Pipe NDE Methods

In-line	In-line & Outside Pipe	Outside pipe
Remote field testing	Ultrasonics	Guided wave
Magnetic flux leakage	Electromagnetic techniques	Radiography
Saturated Low Frequency Eddy Current	Electromagnetic acoustic transducer	
Remote visual	Laser profilometry	

In-line Delivery Methods
- Robotic Crawlers - Flow through - Pull through

37

EPRI | ELECTRIC POWER RESEARCH INSTITUTE

Magnetic Flux Leakage (MFL)

- Magnetic lines of flux flow between the magnetic poles
  - Areas of corrosion or cracking interrupt this flow
- Flux lines are monitored with various sensor types placed between the magnet poles
- Data is recorded and subsequently analyzed to identify and characterize damage

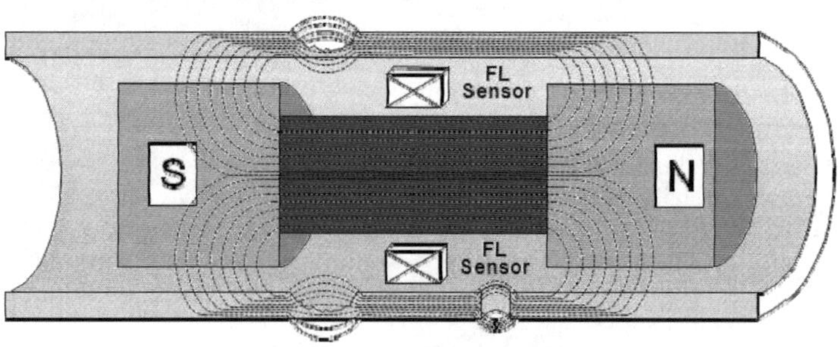

Courtesy of Inline Devices – A Mears Group Company

38

EPRI | ELECTRIC POWER RESEARCH INSTITUTE

Magnetic Flux Leakage (MFL)

In-line pipe inspections of ferromagnetic materials

- Detection and resolution of pitting
- Cannot measure wall thickness but can measure changes in wall thickness
- Cannot differentiate between ID and OD flaws
- Can tolerate coatings
- Fast scan speeds

EPRI | ELECTRIC POWER RESEARCH INSTITUTE

MFL Field Application

Extensively used to examine transmission pipelines

Courtesy of Inline Devices – A Mears Group Company

EPRI | ELECTRIC POWER RESEARCH INSTITUTE

Buried Pipe NDE Methods

In-line	In-line & Outside Pipe	Outside pipe
Remote field testing	Ultrasonics	Guided wave
Magnetic flux leakage	**Electromagnetic techniques**	Radiography
Saturated Low Frequency Eddy Current	Electromagnetic acoustic transducer	
Remote visual	Laser profilometry	

In-line Delivery Methods

- Robotic Crawlers - Flow through - Pull through

Pulsed Eddy Current

- Detection of area corrosion in ferromagnetic materials
  - Non-contact technique
  - Can be used to examine through liners, insulation, and coatings
  - Broad assessment – may miss localized damage
  - Requires minimal cleaning
  - Spot measurement
  - Can be slow

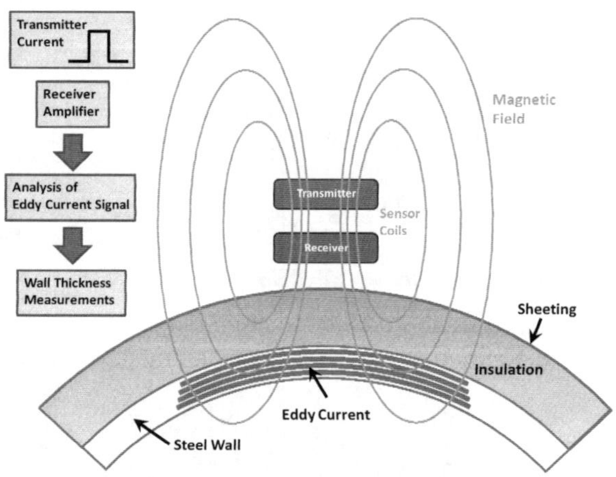

Pulsed Eddy Current

- Probe generates multiple frequencies in the material
  - Measurements are made in the time domain
  - Features near the inspection coils will be seen first and more distant features will be seen later in time

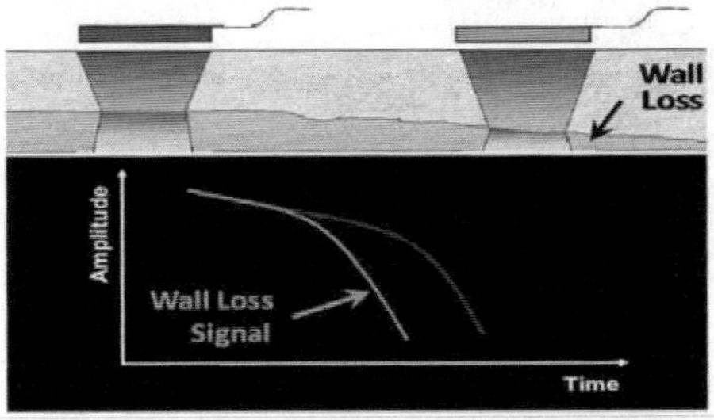

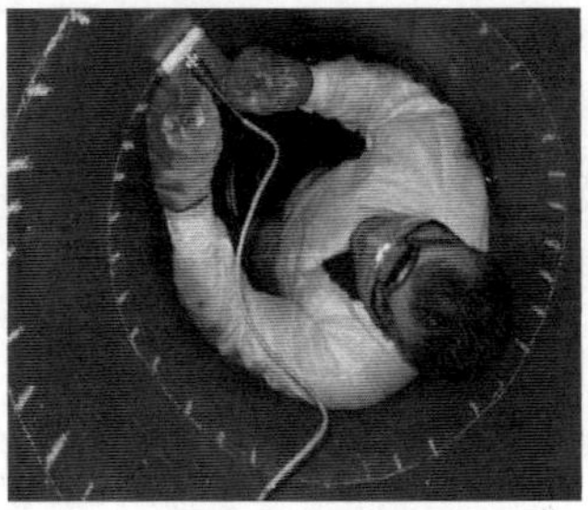

EPRI ELECTRIC POWER RESEARCH INSTITUTE

Buried Pipe NDE Methods

In-line	In-line & Outside Pipe	Outside pipe
Remote field testing	Ultrasonics	Guided wave
Magnetic flux leakage	Electromagnetic techniques	Radiography
Saturated Low Frequency Eddy Current	Electromagnetic acoustic transducer	
Remote visual	**Laser profilometry**	

In-line Delivery Methods
- Robotic Crawlers - Flow through - Pull through

EPRI ELECTRIC POWER RESEARCH INSTITUTE

3D Optical Scanner

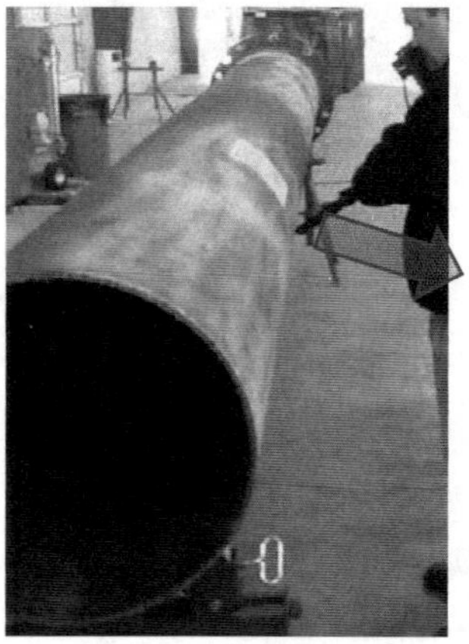

- Development funded by EPRI, Pipeline Research Council International (PRCI), and Chevron
  - Developments demonstrated 2012 Q4
  - Now commercially available
  - Identify region of interest and extracts; length, depth,

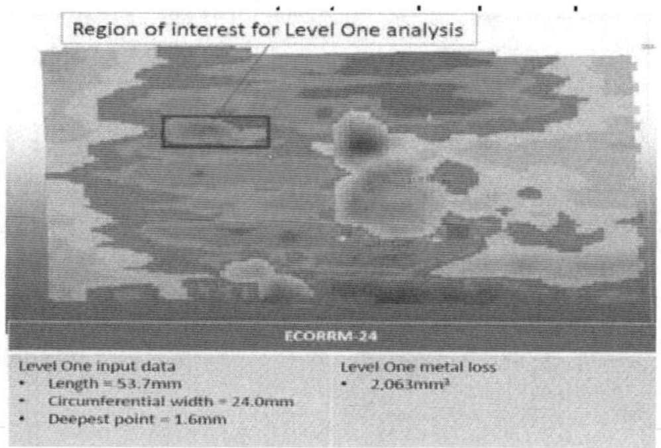

Summary

- The US Industry committed to develop processes to manage buried pipe degradation
  - Goal was to provide “Reasonable Assurance” of structural and leak integrity
- EPRI has worked with the nuclear and petrochemical industry to develop a comprehensive NDE guide that assists nuclear utilities meet this commitment
- US regulators have accepted the industry process

Related Information and References

EPRI Buried Pipe NDE Reports

Report Title	ID Number
Nondestructive Evaluation: Buried Pipe NDE Reference Guide—Revision 3	3002004395
Nondestructive Evaluation: Assessment and Development of Buried Pipe NDE Technology (NDE)	3002000463
Nondestructive Evaluation: Assessment and Development of Buried Pipe NDE Technology, Addendum 1	3002005431
Development of Ultrasonic Phased Array Technology for Corrosion	3002004401
Guidelines for Obtaining Credit for Buried Pipe Guided Wave Examinations	3002000468
Buried Pipe Guided Wave Examination Reference Document (NDE)	1019115
Buried Pipe Direct Examinations Through Coatings	1025228

EPRI Buried Pipe NDE Reports

Report Title	ID Number
Nondestructive Evaluation: Buried Pipe In-Line NDE Depth Sizing Procedure	1025231
Nondestructive Evaluation: Quantification of Real Corrosion in Buried Piping	3002003023
Nondestructive Evaluation: Novel Ultrasonic Sensor Technology for Structure Monitoring	3002003031
Guided Wave Analysis Tools for Buried Pipe	3002000466
Nondestructive Evaluation: Guided Wave Analysis Tools	1025212
Nondestructive Evaluation: Guided Wave Status Report	1022929
Nondestructive Evaluation: Further Developments of Guided Wave Examination Application 2009 Status Report	1019116

EPRI Buried Pipe NDE Reports

Report Title	ID Number
Nondestructive Evaluation: Further Developments of Guided Wave Examination Application	1016675
Nondestructive Evaluation: Buried Pipe Structural Health Monitoring	1025213
Remote Field Technology Assessment for Piping Inspection Including Buried and Limited Access Components	1021153
Catawba Field Trial of EPRI's Large Diameter Buried Pipe Instrumented Vehicle	1016676
Intermediate Diameter Buried Piping Instrumented Vehicle--Evaluation	1022926

EPRI Tank NDE Reports

Report Title	ID Number
Assessment and Evaluation of Nondestructive Evaluation Technologies for Tanks and Containment Liners	3002005448
Nondestructive Evaluation: NDE for Tanks and Containment Liners	3002000462
Inspection Methods for Tanks and Containment Liners	1025215
Inspection Methodologies for Buried Pipes and Tanks	1021561
Nondestructive Evaluation: High-Density Polyethylene NDE Technology	3002000439
Nondestructive Evaluation: Technique Development to Evaluate the Joint Strength of High-Density Polyethylene Butt Fused Pipe Joints	3002003032

51

EPRI | ELECTRIC POWER RESEARCH INSTITUTE

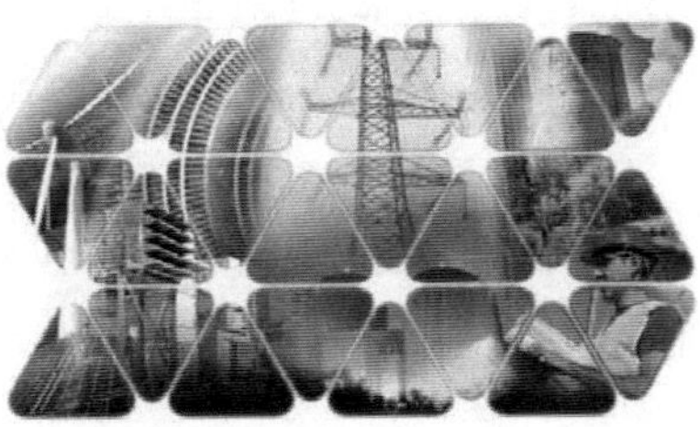

Together…Shaping the Future of Electricity

52

EPRI | ELECTRIC POWER RESEARCH INSTITUTE

Impact of ID Magnetite on Eddy Current Signals in Darlington NGS Steam Generator Tubes

6th International CANDU In-Service Inspection NDE Workshop and NDE in Canada 2016 Conference
November 15 to 17, 2016

Sean Sullivan – Inspection & Maintenance Services

BEHAVIOURS
- Say It, Do It
- Simplify It
- Think Top and Bottom Line
- Integrate and Collaborate
- Tell It As It Is

VALUES • SAFETY • INTEGRITY • EXCELLENCE • PEOPLE & CITIZENSHIP •

PEOPLE POWERING THE FUTURE OPG NUCLEAR

ONTARIO POWER GENERATION

Presentation Overview

- Introduction to the Issue of ID Magnetite in CANDU Steam Generator Tubes
- Summary of ways ID Magnetite Impacts ET Inspections
- Contribution to background noise in ET data from ID Magnetite <u>variations</u>
- Shielding Effects of ID Magnetite on ET Flaw Signals
- Summary/Conclusions

Introduction

- In terms of Steam Generator Tubes at CANDU Nuclear Power Generating Stations the presence of ID Magnetite is common, but not in Steam Generator Tubes at PWR Stations

- It is believed that ID Magnetite Layers in CANDU SG Tubes were generated by Iron Oxide particles that were eroded from Carbon Steel Feeder Pipes used to transport fluid from the Reactor Fuel Channels to the Steam Generators

Observations in Pulled Tube

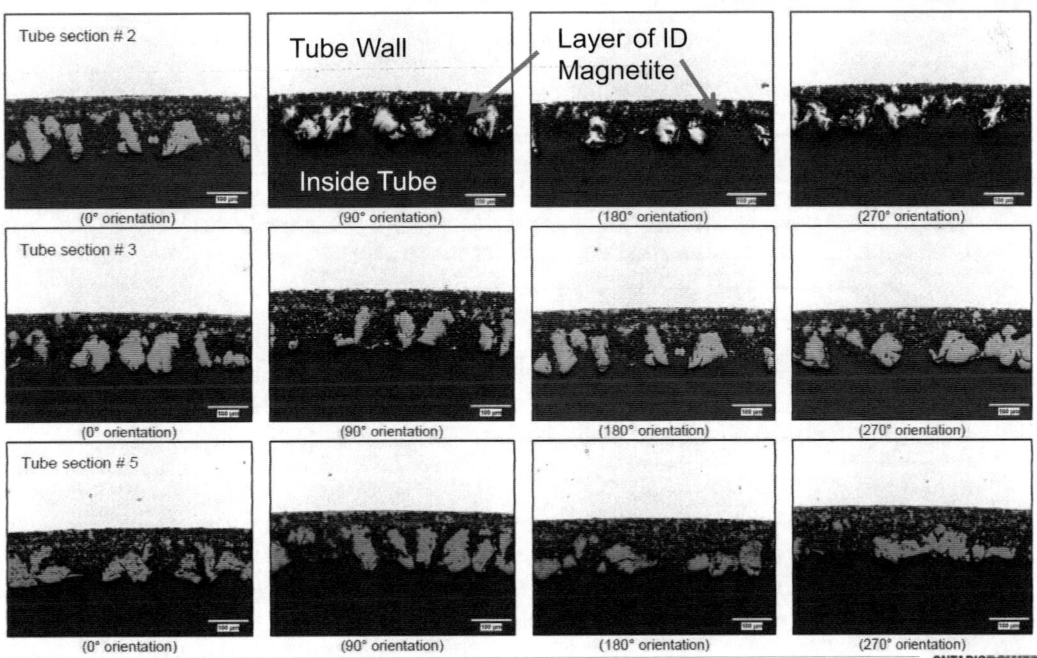

ID Magnetite in Pulled Tube

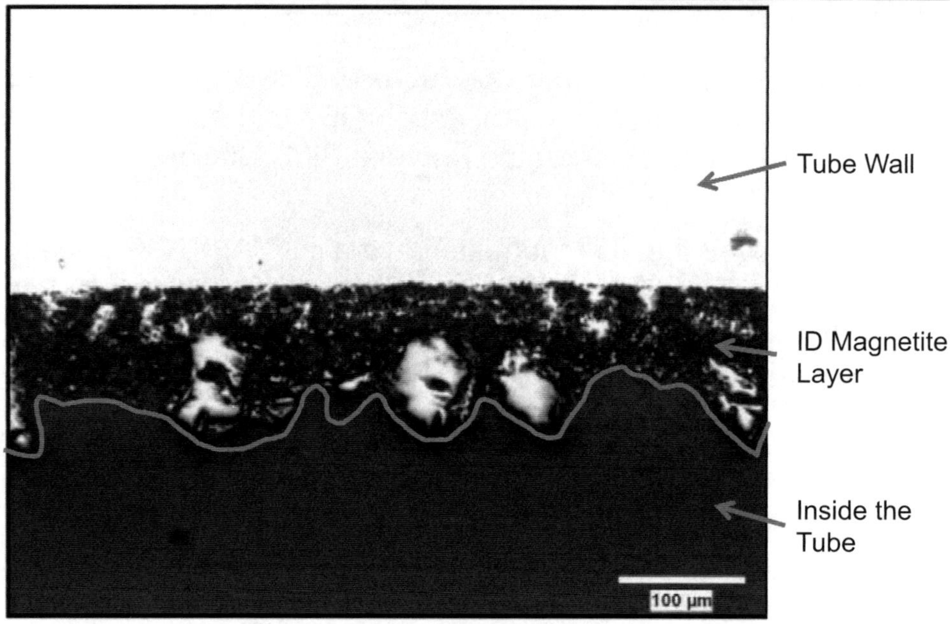

(180° orientation)

ET Signal Due to ID Magnetite Distribution

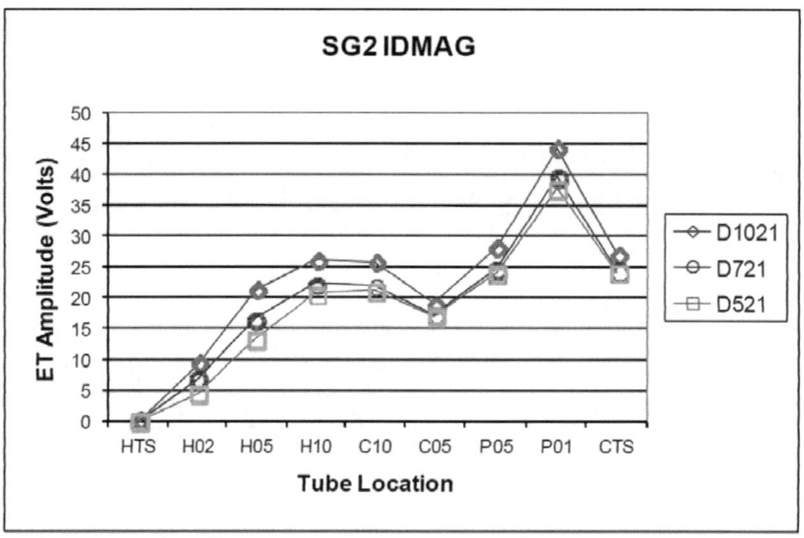

Bobbin Probe ET “P5 Channel” (480 kHz Absolute) Vertical Component Magnitude used for Estimating Quantity of Magnetite in Darlington Steam Generator Tubes

ID Magnetite Observations

- Maximum Thickness observed in pulled tube destructive examination was 162 microns. It is believed that a certain amount of this oxide was probably inadvertently removed during the tube removal process and destructive examination.

- ET estimate of ID Magnetite thickness based on the P5 voltage distribution in the section of the removed tube with the highest amount of magnetite is estimated to have been between 150 and 220 microns before the tube was removed from the steam generator.

Impacts of ID Magnetite on ET

- Shortened ET Probe Life due to wear on the abrasive ID Deposit

- Restrictions and Stuck Probes due to Loose Magnetite

- ET Signal Noise contributions from variations in ID Magnetite Layers distort and obscure ET signals from small flaws

- Drift from ID Magnetite Layer can saturate the signal in the amplifier or A/D converter. This is mitigated by strategic hardware nulling of the instrument.

- Magnetic Shielding from this ID deposit makes ET flaw signals weaker than they be would if no deposit was present.

Freespan Noise in Darlington SG Tube

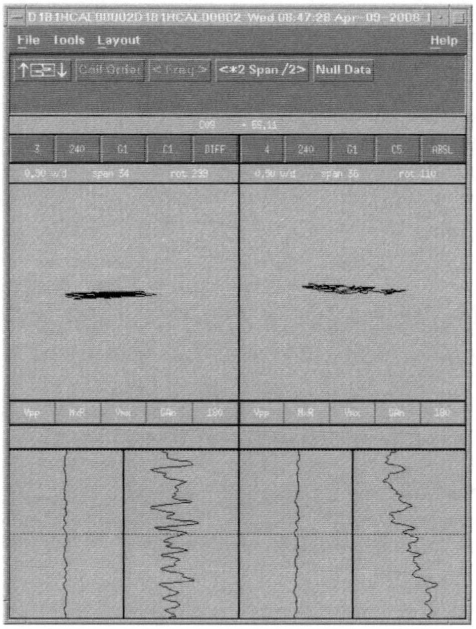

- A substantial amount of this background noise is considered to be from Variations in ID Magnetite Layer along the scan direction of the Bobbin Probe.
- This Noise is primarily horizontal, but it still has a small vertical component

Reference Flaw

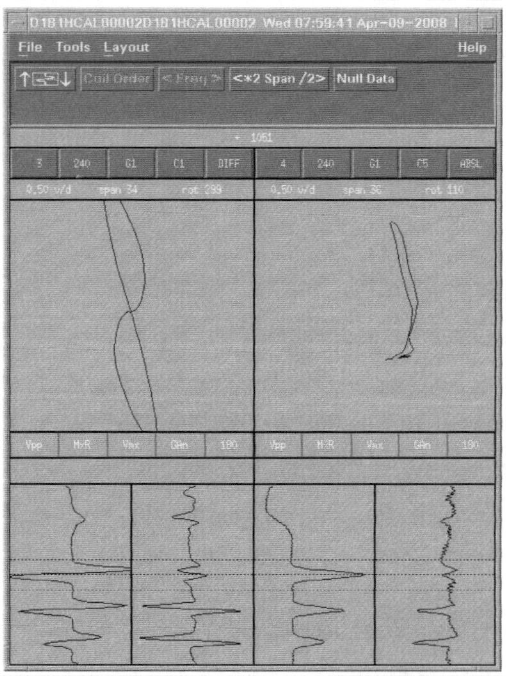

- 40% TW calibration flaw (flat bottom hole)
- Diameter = 4.8 mm

ET Noise from ID Magnetite

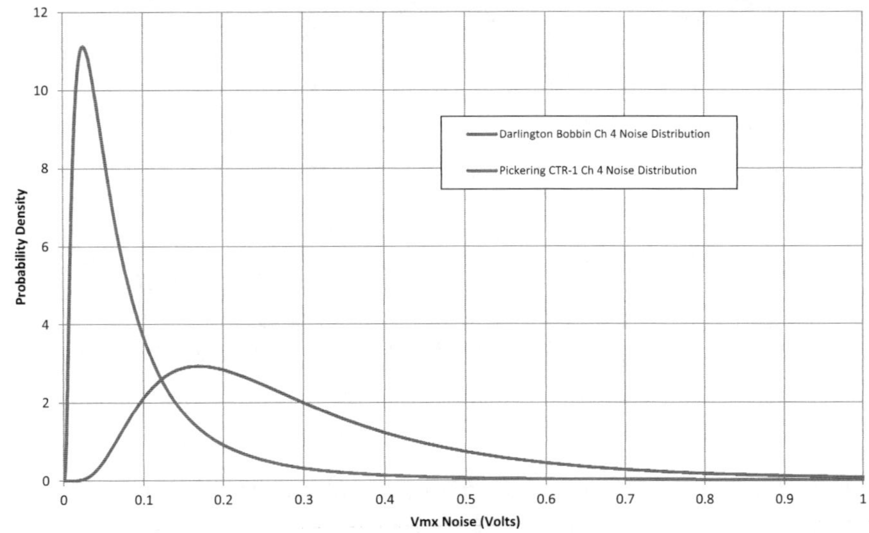

Comparison of Noise Distributions in Bobbin Probe Channel 4 Absolute Data in Darlington and Pickering Steam Generator Tubes

ET Noise from ID Magnetite

- Although there are many differences in the SG Bobbin Probe systems at Darlington and Pickering (e.g. Diameters, Test Frequencies, Calibration Tubes), it is believed that a **Substantial Contributor** to the increase in noise in Darlington Bobbin data is the **Probe's Sensitivity to ID Magnetite.**

- It is believed that Pickering Bobbin Probes (called "Carter-1" probes) are **Less Sensitive** to **ID Magnetite and Variations in ID Magnetite** because Magnetic Biasing of the ID Magnetite Layers due to the **Permanent Magnets in these Pickering Probes**.

ET Detection of ID Magnetite

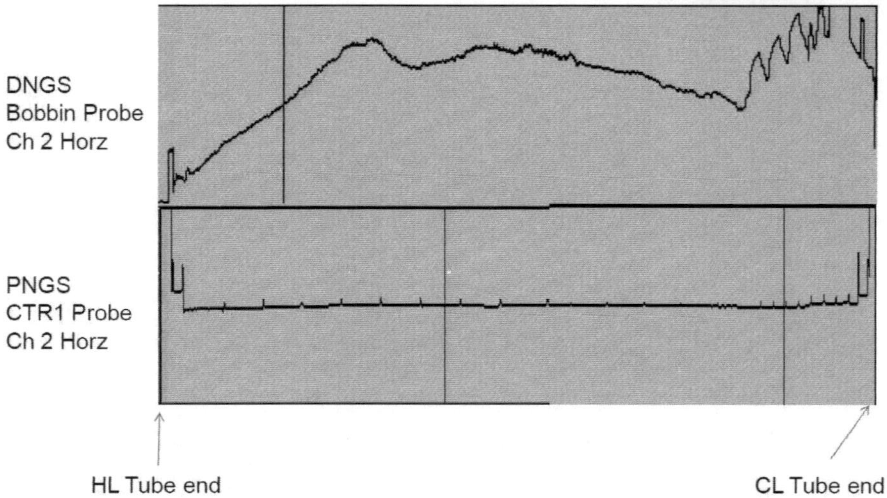

- Comparison of ET Bobbin Probe Sensitivity to ID Magnetite in SG Tubes in Darlington and Pickering.
- No drift from ID Magnetite in Pickering Tubes indicates magnets in CTR-1 probes have saturated the ID Magnetite.

ET Detection of ID Magnetite

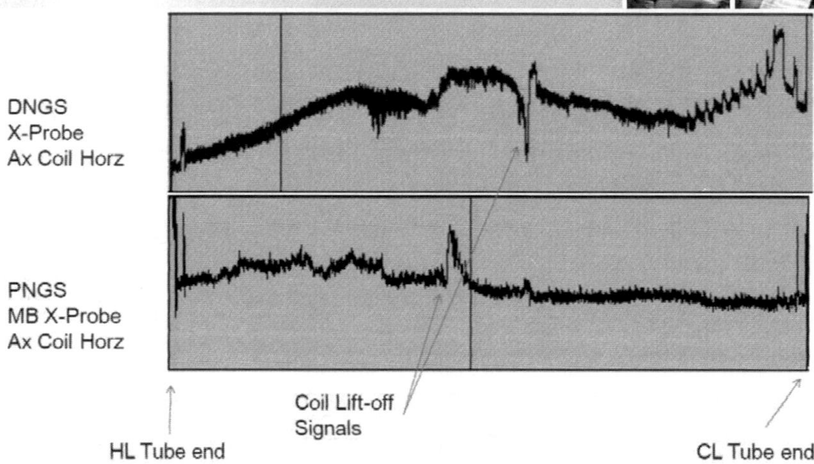

- Comparison of ET X-Probe Sensitivity to ID Magnetite in SG Tubes in Darlington and Pickering.
- No drift from ID Magnetite in Pickering Tubes indicates magnets in those X-Probes have saturated the ID Magnetite.

Shielding Effect of ID Magnetite

- Shielding Effect of ID Magnetite Layers Estimated using a Mathematical Modelling Approach.

- Signals and Signal Reductions due to ID Magnetite Layers Estimated using Multilayer Analytic Solutions to Maxwell's Electromagnetic Equations similar to the Solution developed by Dodd, Cheng and Deeds in Journal of Applied Physics, 1973.

- Magnetite Layers, Tubes and Support Plate Layers are assumed to be linear, homogeneous and isotropic.

- The Magnetic Permeability of a Magnetite deposit layer ($\mu_r \sim 4$) is much lower than the Permeability of individual Magnetite particles (typically $\mu_r \sim 50$ to 100)

Basic Elements in Analytic Model

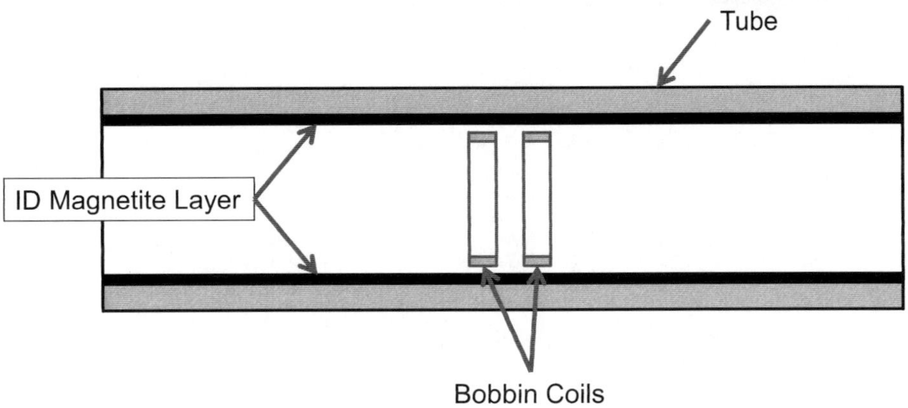

- Tube is assigned an Electrical Resistivity value equivalent to the nominal value of material used for Darlington SG Tubes (Alloy 800, $\rho = 98.9\ \mu\Omega$-cm).
- ID Magnetite layers are assumed to be nonconductive with relative magnetic permeability between 4 and 8.

Basic Elements in Analytic Model

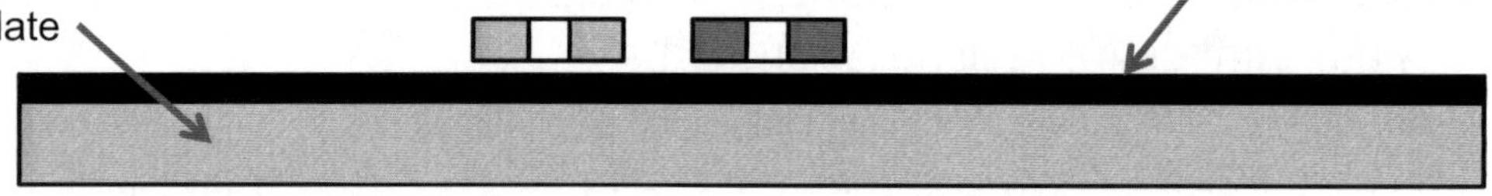

- Signals calculated using Analytic Solutions to Maxwell's electromagnetic equations.
- Centre-to-Centre separation of coils set to the separation of X-Probe coils as projected to the SG tube mid-wall.
- Flat Plate is assigned an Electrical Resistivity value equivalent to the nominal value of material used for Darlington SG Tubes (Alloy 800, ρ = 98.9 μΩ-cm).
- Plate Thickness is equal to nominal thickness of Darlington SG tubes.
- ID Magnetite layers are assumed to be nonconductive with relative magnetic permeability between 4 and 8.

Signal Responses to ID Magnetite

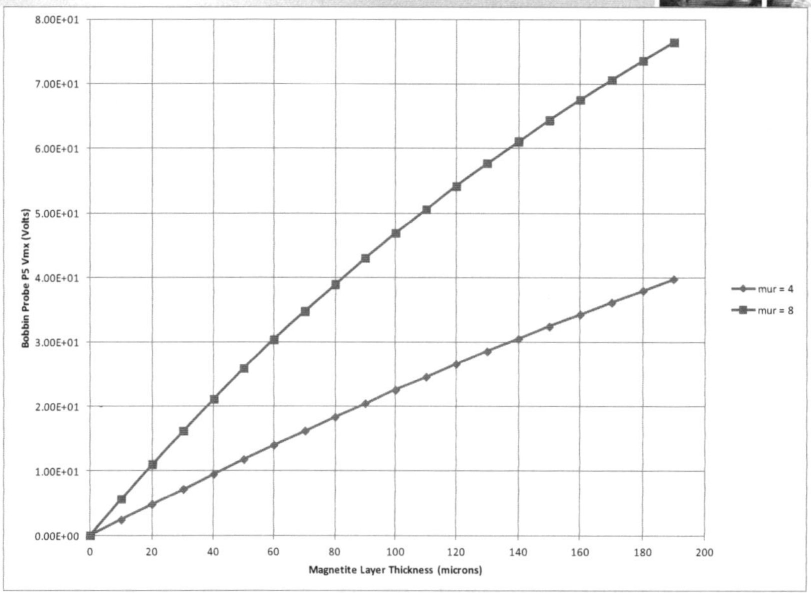

- Effect of ID Magnetite Permeability and Thickness on Bobbin Probe Signal.

Magnetic Shielding Effect on Bobbin Probe

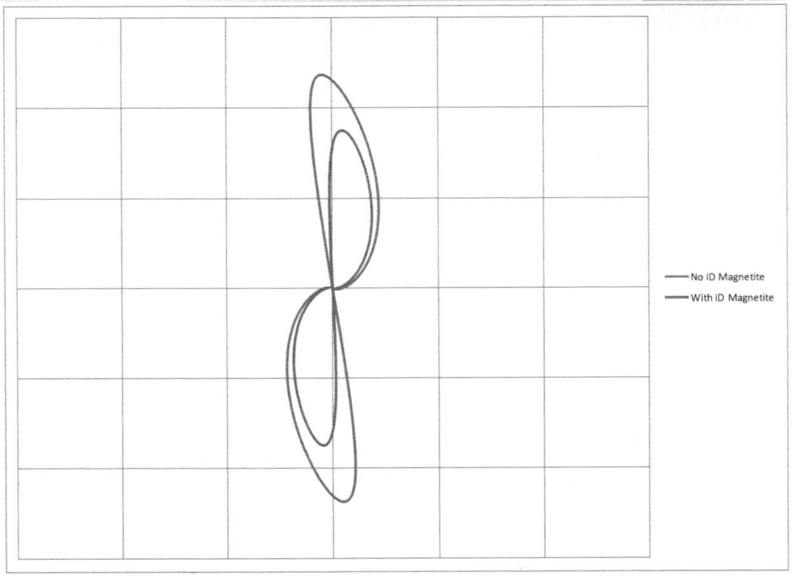

- 40% TW localized micropit-like flaw, Test Frequency = 240 kHz
- Typical Magnetite Layer close to P01 Location in Darlington SG Tube.

Magnetic Reduction – Volts peak-to-peak

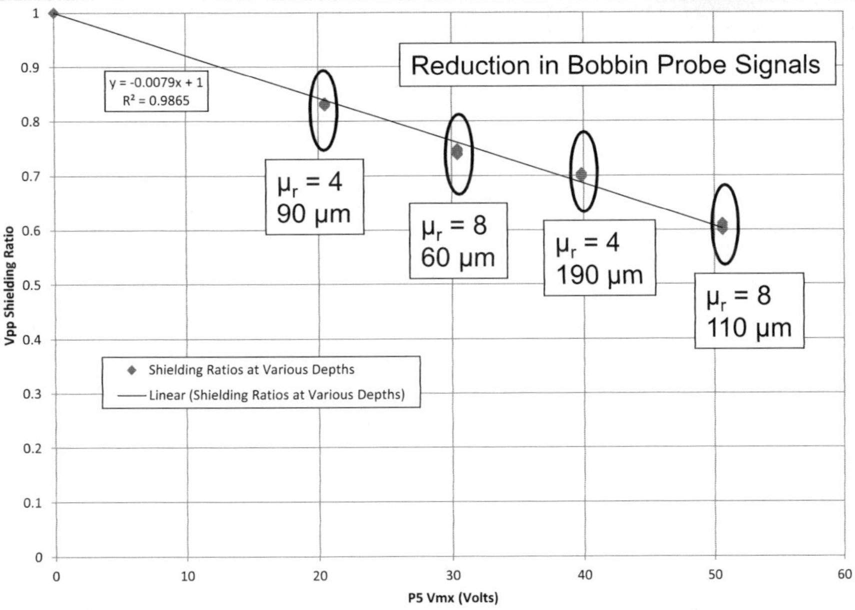

- Localized OD Pit-like Signals from Bobbin Probe at 240 kHz

Magnetic Reduction – Vertical Component Volts

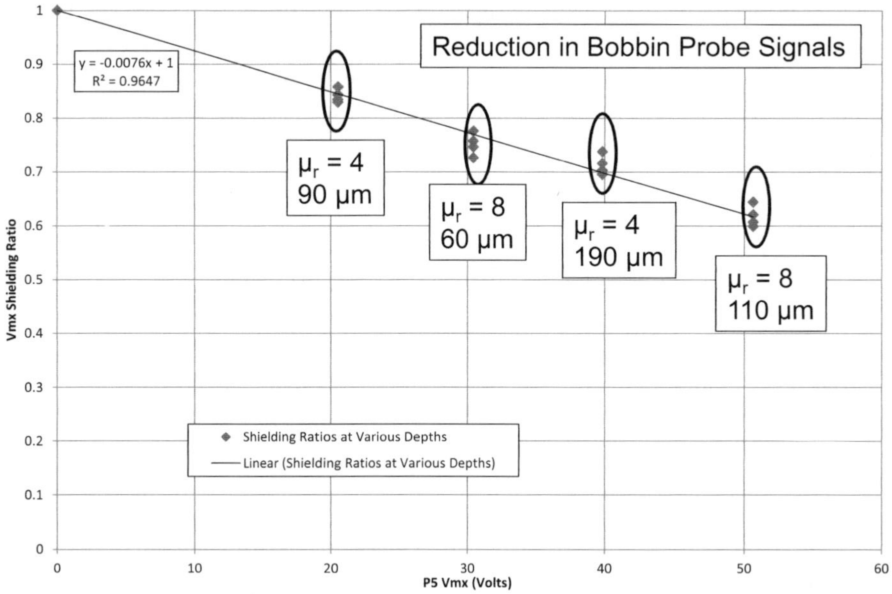

- Localized OD Pit-like Signals from Bobbin Probe at 240 kHz

Shielding of X-Probe Signals

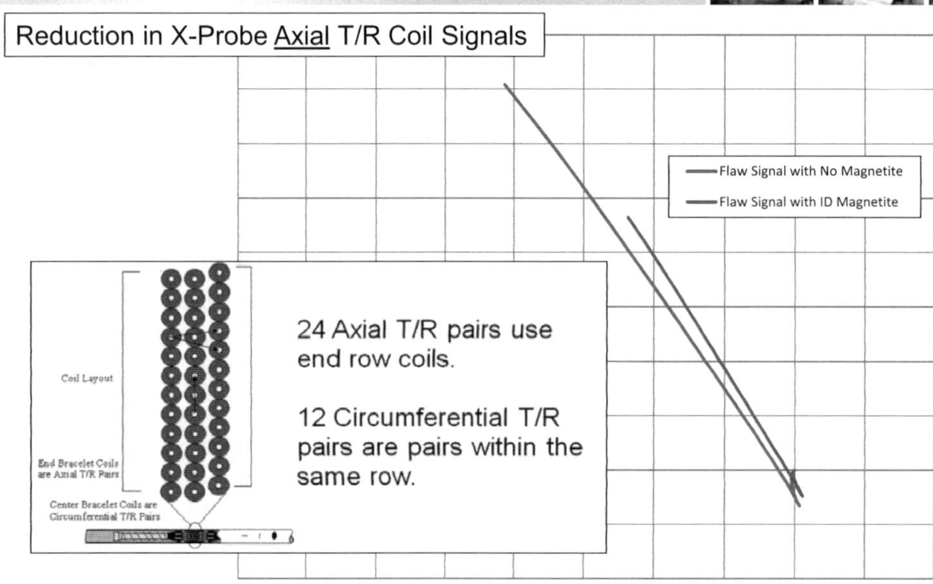

- 80% TW localized axial crack-like flaw, Test Frequency = 250 kHz
- Beyond worst case Magnetite Layer in Darlington SG Tube

Magnetic Reduction – Volts peak-to-peak

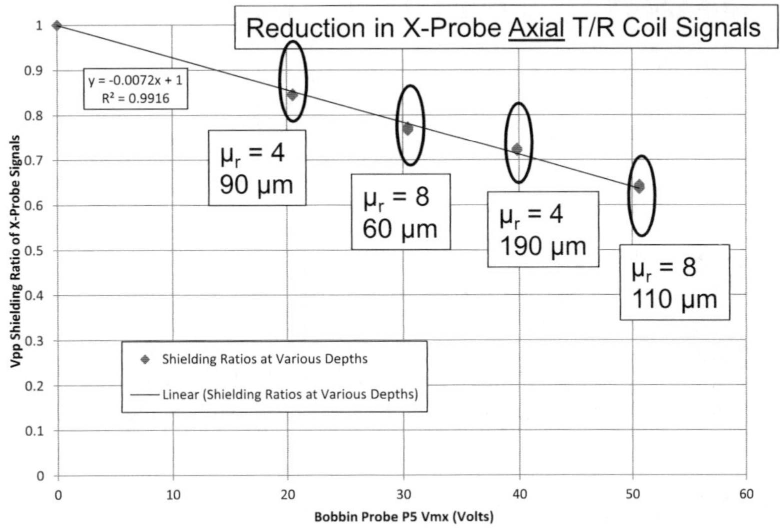

- Localized Axial OD Crack-Like Signals from X-Probe Axial Coils at 250 kHz

Magnetic Reduction – Vertical Component Volts

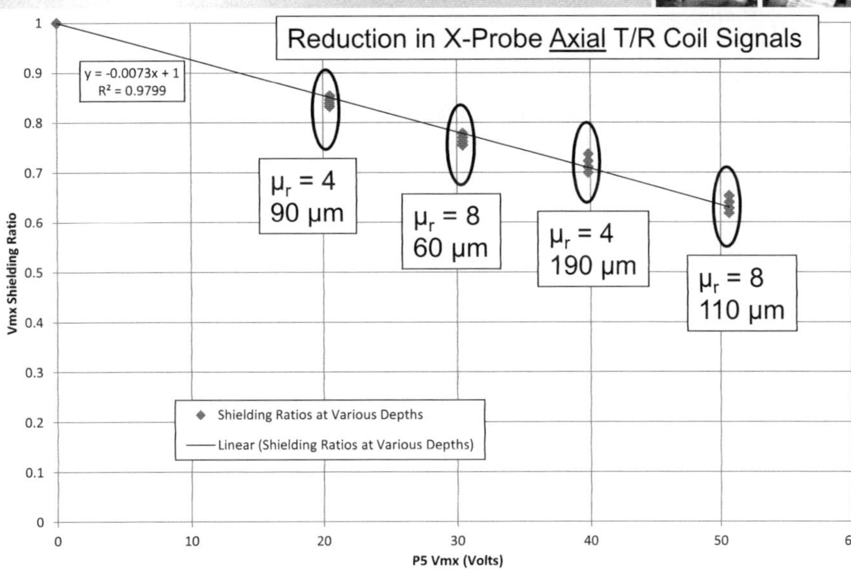

- Localized Axial OD Crack-Like Signals from X-Probe Axial Coils at 250 kHz

Shielding of X-Probe Signals

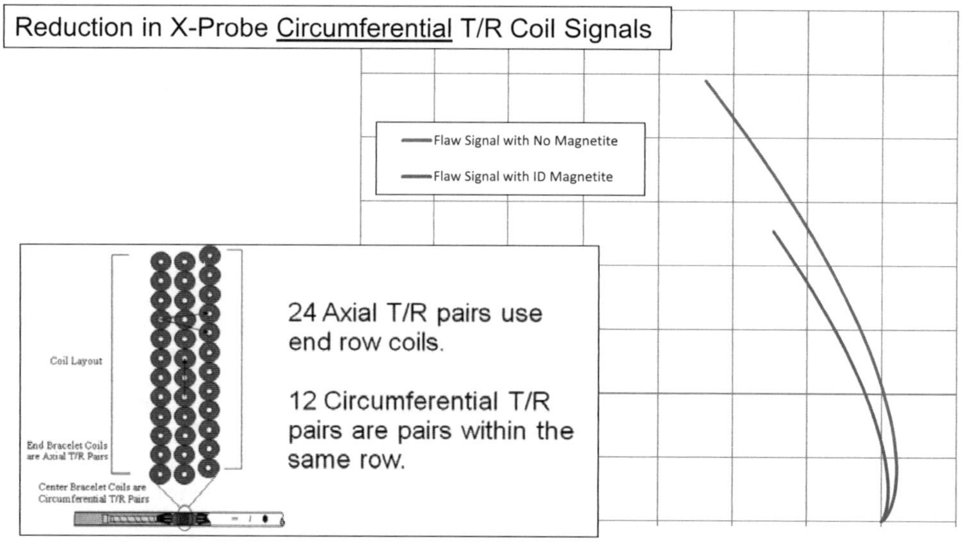

- 80% TW localized circumferential crack-like flaw, Test Frequency = 250 kHz
- Beyond worst case Magnetite Layer in Darlington SG Tube

Magnetic Reduction – Volts peak-to-peak

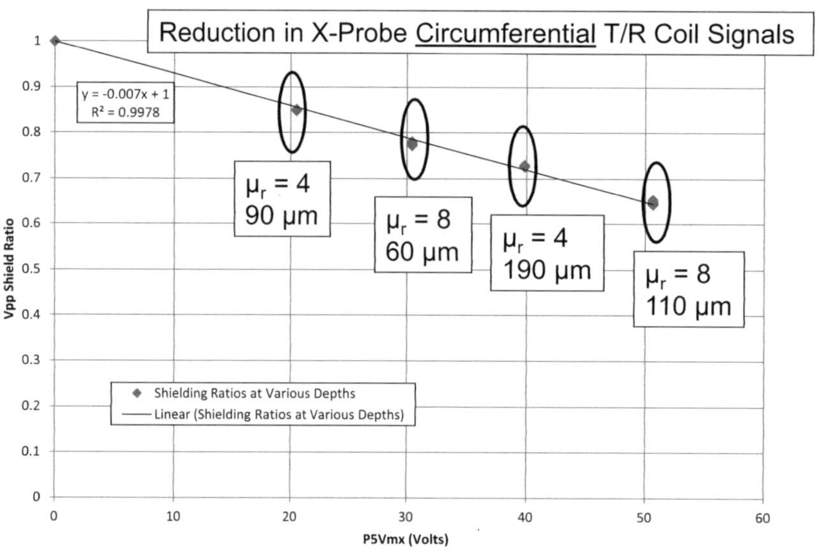

- Localized Circumferential OD Crack-Like Signals from X-Probe Circumferential Coils at 250 kHz

Magnetic Reduction – Vertical Component Volts

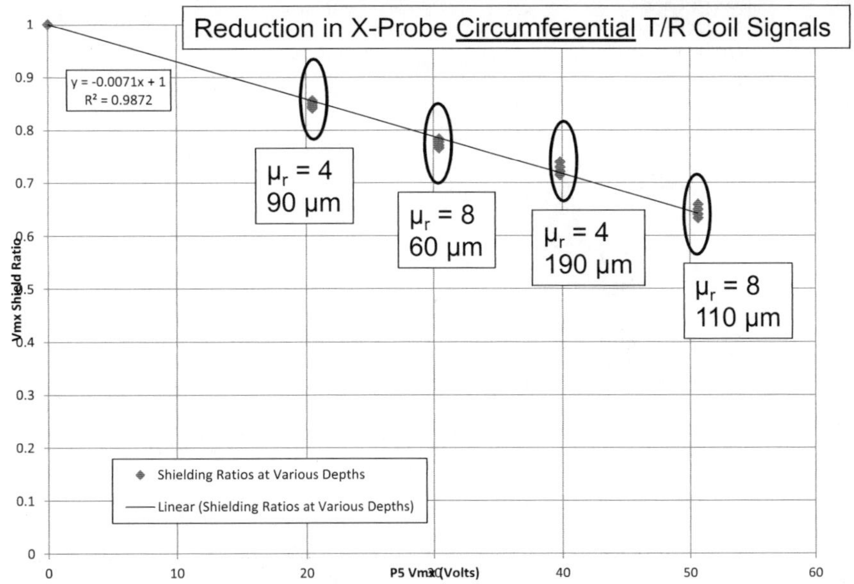

- Localized Circumferential OD Crack-Like Signals from X-Probe Circumferential Coils at 250 kHz

Summary/Conclusions

- Although, the presence of ID Magnetite in CANDU Steam Generator Tubes increases the difficulty of in-service ET inspections, reliable high quality inspections can still be achieved. IMS deploys UT to verify ET results.

- Mathematical modelling has been used to estimate the effects of magnetic shielding on Bobbin Probe and X-Probe signals due to the ID magnetite layers.

Questions?

?

Thank-you for Your Attention

Application of Bayesian Inversion Techniques on Multi Frequency Eddy Current Data for Sub Surface Crack Profiling In Steam Generator Tubing

Saad Ahmed Warraich, Tariq Mairaj Rasool Khan, Moez Ul Hassan [1]

[1]Department of Electrical and Power Engineering,
Pakistan Navy Engineering College, National University of Science and Technology,
Pakistan
E-mail: saadahmedwarraich@gmail.com, khan.tariq@pnec.nust.edu.pk, moezulhassan@gmail.com

Abstract: Steam generator (SG) tubing in nuclear plant offers a barrier between the radioactive material and steam. Effective NDT of the SG tubing is mandatory to ensure the safe operation of nuclear power plants. Multi-frequency eddy current (EC) NDT is being exercised in the nuclear industry. Detection and subsequent crack profiling from the acquired EC signals is a challenging task once dealt with subsurface and OD cracks. Data fusion based EC signal inversion techniques are proposed to address the gap. Measurement data was acquired from the simulation of sub surface cracks in steam generator tubing at three different scan frequencies. The proposed spatial correlation data fusion technique has been applied on the acquired multi frequency EC data. Flaw profile prediction from single frequency and data fused results has also been performed by using recursive Bayesian estimation procedure based on Sequential Monte Carlo method. To investigate the efficacy of the proposed data fusion method; quantitative analysis has been conducted. The robustness of the proposed technique was also assessed injecting artificial noise in the EC measurements.

Keywords: Non Destructive Inspection, Multi-Frequency Eddy Current Measurements, Data Fusion, Spatial Frequency Correlation, Particle Filter Inversion

1. Introduction:

Pipelines in various fossil-fuel power plants and nuclear facilities undergo effects of corrosion. Non-destructive testing allows for a significant reduction in inspection of such systems and so increases plant availability. Repair actions if necessary can be made at an appropriate time which also contributes to better maintenance planning [1]. Using NDT for evaluation of the material condition and forecasting its degradation provides a greater opportunity to reduce the amounts and costs of testing the materials. It also contributes to the possible early warning of excessive component degradation. NDT will also enable plant operators and owners to optimize the long term operation of their power plants. This optimization should certainly result in increased cost competitiveness. The situation in the future will be essentially based on diagnostic systems and degradation modeling [2]. Predictive or condition based maintenance requires analysis of collected data for deciding whether or not maintenance action is mandatory. This method not only results in the decrease of needless preventive maintenance actions, but also insures that actions are performed only when the condition of equipment justifies it. [3]

Eddy-current testing (ECT) is a popular NDT technique used for locating cracks in steam generator tubes of nuclear power plants. Therefore it is pivotal to validate the reliability and flaw detection accuracy of this method [4]. The probability of flaw detection and its sizing accuracy mainly depend on the way the crack is oriented and located on the material. The dependence is also based on tube support structures and corrosion deposits.

ECT information is used to determine tube integrity and for the selection of tubes which require repair or removal from service. Validation of the ECT method can be achieved by destructive examination of a defective tube pulled out from a steam generator during in-service inspection [5].

An eddy current probe is able to locate cracks based on the material resistivity and the frequency of the current flowing through the probe coil. Resistivity defines the conductivity of the material. Hence materials with higher resistivity values have a semi-conducting nature. The coil excitation frequency has an inverse relationship with the depth to which eddy currents can penetrate the material under test. This means that greater frequencies allow the current to have less penetration depth in sub-surface region of the test material [6]

With eddy current techniques there is no contact between the sample and the sensor, making it a desirable method for many practical situations. Conventionally, detection of flaws in conductive samples has been performed with room temperature coils, where the signal-to-noise ratio (SNR) is proportional to the frequency. However, normal coils have the required sensitivity only at high frequencies, limiting flaw detection to regions near the surfaces of the sample. Experimental results have shown that circumferential cracks can be better detected and sized using a plus point coil probe rather than a pancake or bobbin coil probe.

With the help of ECT, steam generator tubes having a non-magnetic nature can be inspected to ensure the safety of nuclear power plants. The inspection of tiny cracks formed due to stress corrosion cracking and fretting corrosion can become problematic. It is very hard to locate such flaws due to the lack of penetration and limited skin depth [7].

Subsurface damage to a structure cannot be visually observed, unless penetration of water and other agents results in surface rupture. Since early evaluation of tubes is desirable, passive methods of monitoring the inception of corrosion have been examined.

In ECT, a conventional technique involves a single frequency excitation source to detect cracks by changes in voltage or impedance. However, such techniques are sensitive to various characteristics inherent to the crack. Multi-frequency EC measurements have also been used to offer a more robust inspection of structural integrity. These measurements allow for a decrease in signal anomalies that can potentially mask the cracks [8]

Although crack sensing using ECT is a relatively mature technique, only crack existence has been a major concern in actual inspections of steam generator tubes. Reconstruction of crack profiles has been considered an extremely difficult challenge and thus has not been attempted to date. However, determination of a crack profile is indispensable for more reliable evaluation of defects and optimization of maintenance. The study of eddy current inversion, aiming to reconstruct crack profiles from eddy current signals, has advanced greatly in recent years [6].

An inverse analysis is much more difficult than a forward analysis because of ill-posedness of the problem. Chen proposed a new fast forward solver based on A−φ formulation to overcome this difficulty. The key idea of this new fast solver is that it is possible to compute eddy current signals taking only the flawed region into consideration using pre-computed unflawed potentials as a database [9].

Although the degree of acceleration using the solver strongly depends on the number of nodes inside the crack, it can compute eddy current signal within a few minutes in general once database is established. Chen applied the fast forward solver to ECT inversion problems. The scheme is based on a gradient method that needs forward analysis in every step in order to obtain the gradient of the total error function

Popa's approach to reconstructing crack profiles differed from Chen's. He tried to simulate the mapping that exists between crack profiles and eddy current signals using neural networks and succeeded in reconstructing complicated cracks from both simulated and experimental data [6].

In spite of recent advancements in inspection technology, there still exists a clear need to quantify and enhance the reliability of in-service inspection techniques. Keeping this background in mind, an attempt to reconstruct the subsurface crack profiles was carried out using an inverse scheme developed by the authors. The scheme and its results are reported in this paper.

2. Research Methodology

2.1 Modeling of Sub-surface flaw

A two-dimensional sub-surface flaw was modelled inside the steam generator tubing. This model is used to determine the EC measurements. These EC measurements are then used for the inversion. Different filtering techniques are used for the inversion process. However, in this manuscript, a Particle filter (PF) based inversion technique is used. There are two variables that govern the modelling of sub-surface crack i.e. width of the flaw and its depth. The depth of the flaw is further based on two factors: initiation depth of the crack and the depth of the sub-surface crack itself.Multiple rectangular sub-surface flaws of various depths and widths were modeled inside the tubing. The cross-sectional view of the tubing shows one of the modelled flaws in Fig.1.

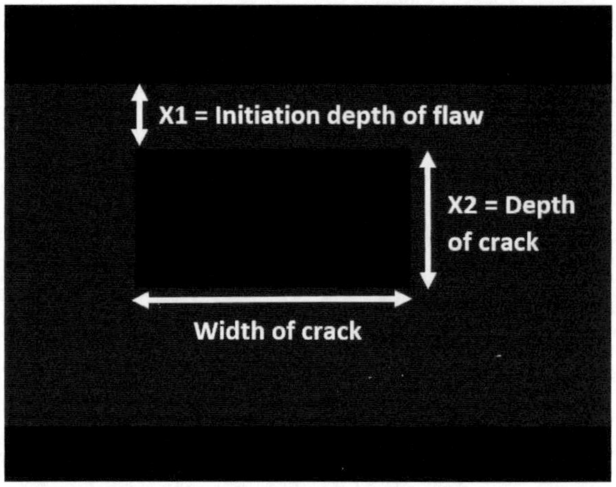

Figure 1 Cross-sectional view of the tubing (blue) showing the modeled sub-surface crack (black)

2.2 Data settings and description of FEM simulation

The modeled sub-surface cracks were simulated using a 2D finite element method (FEM) model [10]. Measurement data was acquired using eddy current (EC) signals at multiple frequencies from the sub-surface discontinuities on the steam generating tubing. Use of different excitation frequencies to acquire EC data provides complimentary information due to skin depth phenomena [11,12]. The coil used of EC measurements is of 100 turns and an inner and outer diameter of 16.4mm and 19.7 mm. The eddy current coil frequencies used for simulation of the cracks are 200 kHz, 400 kHz and 800 kHz. The thickness of the tube and its axial length was 2.5 mm and 3.75 mm, respectively.

2.3 Particle filter based inversion scheme

Particle filters are sequential Monte Carlo filtering methods. PF's are the generalization of Kalman filters which are not bounded by the selection of linear state transition models as well the Gaussian noise. Detailed discussion of PF is discussed in [11-14]. In this manuscript, we assume the measurements as yk from a single measurement mode and the probability density function (PDF) of state uk p(uk/ yk) may be obtained recursively in two stages i.e. prediction and update. In prediction stage, we predict the PDF from one measurement location to the next. Since the state is usually subject to random noise, the prediction step generally distorts the PDF [11-14] which is then corrected in the update stage using the available measurement.

In order to apply particle filtering algorithm, the desired posterior PDF is represented in terms of samples and their associated weights at each location. Let Ns be the total number of samples that are drawn from importance density, u_k^i are the set of support points with associated weights w_k^i where $i = 1 : N_s$. (Ns is equals to 1000 samples). According to the principle of Importance Sampling, the weights are normalized such that $\sum_{i=1}^{N_s} w_k^i = 1$ and the posterior density can be approximated as:

$$p(u_k \mid y_k) \approx \sum_{i=1}^{N_s} w_k^i \delta(u_k - u_k^i) \tag{1}$$

The samples can be drawn from the importance density $q(u_{1:k} \mid y_{1:k})$ and the weights are given by:

$$w_k^i \propto \frac{p(u_{1:k}^i \mid y_{1:k})}{q(u_{1:k}^i \mid y_{1:k})} \tag{2}$$

$p(u_{1:k} \mid y_{1:k})$ is approximated with a new set of samples when the observation y_k is known at the loading cycle k. Given the set of weights $w_{k-1}^{1:N_s}$ after loading cycle '$k-1$', the weights after loading cycle k can be calculated recursively using the weight update equation as:

$$w_k \propto w_{k-1} \frac{p(y_k|u_k)p(u_k|u_{k-1})}{q(u_k|u_{k-1}, y_k)} \tag{3}$$

In the Bootstrap version of PF, the prior density $p(u_k \mid u_{k-1})$ is used as importance density. Therefore (3) becomes:

$$w_k \propto w_{k-1}\; p(y_k|u_k) \tag{4}$$

(4) can also be written as:

$$w_k^i \propto p(y_k|u_k^i) \tag{5}$$

When multiple measurement modes are available, just as in our case, likelihood PDF corresponding to each measurement mode need to be considered in weight assignment to the sample.

Assume that $w_k^{i,q}$ is the weight of the sample i at the position index k assigned by the individual measurement mode q. The likelihood function corresponding to the qth measurement mode is given by:

$$w_k^{i,q} \propto p(y_k^q|u_k^i) \tag{6}$$

If the measurement process is assumed to be independent, then the joint likelihood due to measurement modes q=1,2,3,......,Q is the product of likelihood for the individual measurement mode, given by:

$$p(y_k|u_k^i) = p(y_k^1|u_k^i)p(y_k^2|u_k^i)........p(y_k^Q|u_k^i) \tag{7}$$

And the final weight assigned to the sample I at the position index k is as follows:

$$w_k^i \propto w_k^{i,1}, w_k^{i,2}, ..., w_k^{i,Q} \tag{8}$$

The acquired EC measurement data from the FEM simulations of the sub-surface cracks are related by a curve-fitting linear polynomial function. The measurement model is based on the linear function as given in (9)

$$y = Ax_1 + Bx_2 + c \tag{9}$$

2.4 Training database

The training of the measurement model was done by using multi-frequency EC measurement data acquired from different sub-surface crack profiles. Fig. 2 shows the simulated EC measurements of a sub-surface flaw profile with X1 and X2 as 0.2 mm and 0.46 mm, respectively:

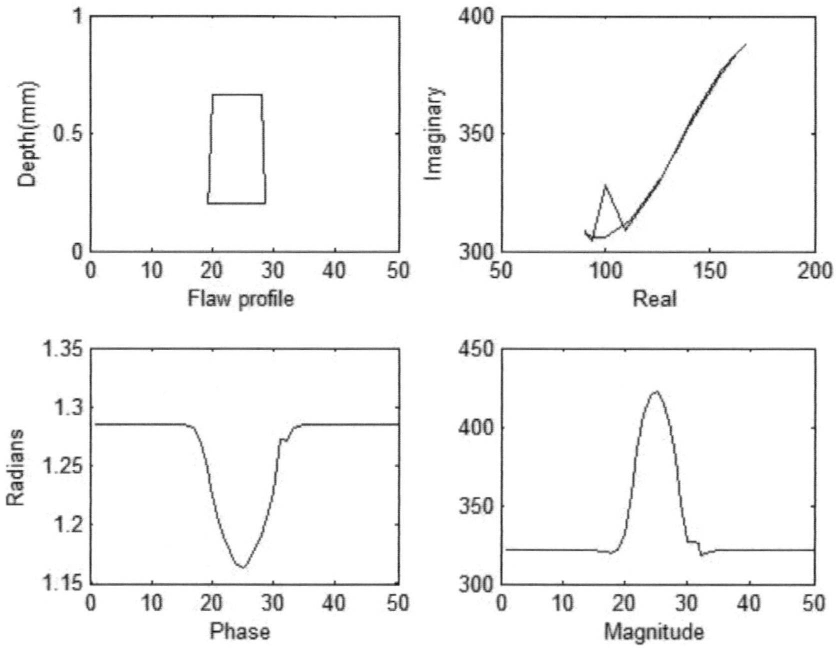

Figure 2 Eddy current measurements of a simulated sub-surface crack

3. Results

The multi-frequency EC measurement data has been acquired from simulation of three different crack profiles. The EC measurements were taken at 200 KHz, 400 KHz and 800 KHz scan frequencies. To perform the PF inversion technique, measurement data at the three scan frequencies were fused together. Data fusion has been done by multiplying together the weights corresponding to each single frequency measurement mode. PF based inversion results obtained after data fusion is shown in Fig.3.

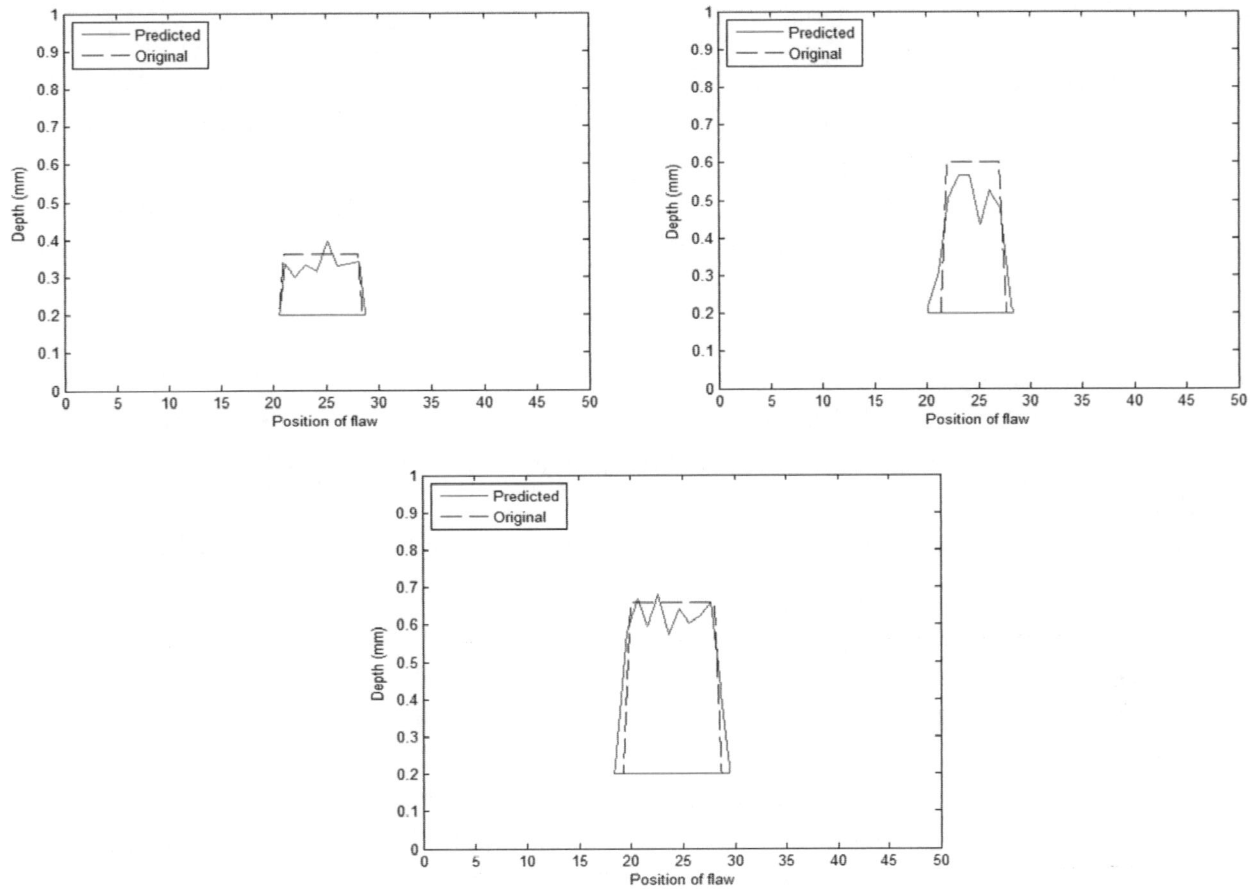

Figure 3 PF based inversion results

Table 1 shows the RMSE values of the inversion results at 200, 300,400 kHz and of the fused data. It has been observed that the RMSE values of the fused data are less as compared to the RMSE values of the individual frequency results. The results of RMSE show the efficacy of proposed data fusion technique. The proposed technique has been applied to other eddy current data for inversion of sub surface crack.

Table 1 RMSE Values of Single Frequency and Fused data of three different crack profiles

Frequency	**Crack 1**	**Crack 2**	**Crack 2**
200KHz	0.0743	0.0401	0.1524
300KHz	0.1350	0.1082	0.1247
400KHz	0.0746	0.0462	0.1453
Fused	0.0723	0.0353	0.1431

References

[1] E. N . Barrlett and R. E. Uhrig. Nuclear Power Plant Diagnostics using Artificial Neural Networks. Frontiers in Innovative Computing for the Nuclear Industry, Jackson Hole, Wyoming September 15-18,1991.

[2] J. H. Lee, Y. W. Park and M. H. Song. Evaluation of Plugging Criteria of Steam Generator Tubes and Coalescence model of Collinear Axial Through Wall Cracks. Journal of Korean Nuclear Society. Vol. 32, Number 5, pp. 465-476, October 2000.

[3] B.Mavko, L.Cizelj and G.Roussel. Steam Generator Tubes Rupture Probability Estimation - Study of the Axially Cracked Tube Case. Nuclear Energy Agency, 75 - Paris (France). Committee on the Safety of Nuclear Installations; 189 p; 1992; p. 129-139; Workshop on Reactor Coolant System Leakage and Failure Probabilities; Cologne (Germany); 9-11 Dec 1992

[4] D.R. Diercks, W.J. Shack and J. Muscara. Overview of steam generator tube degradation and integrity issues. Nuclear Engineering and Design. Volume 194, Issue 1, November 1999, Pages 19–30

[5] H. Fukutomi, T. Takagi, M. Nishikawa. Remote field eddy current technique applied to non-magnetic steam generator tubes. NDT & E International. Volume 34, Issue 1, January 2001, Pages 17–23.

[6] N. Yusa, Z. Chen and K. Miya. Quantitative profile evaluation of natural cracks in a steam generator tube from eddy current signals. International Journal of Applied Electromagnetics and Mechanics, vol. 12, no. 3,4, pp. 139-150, 2000

[7] H. Fukutomi, T. Takagi and M. Nishikawa. Remote field eddy current technique applied to non-magnetic steam generator tubes. NDT & E International. Volume 34, Issue 1, January 2001, Pages 17–23

[8] A. Sophian, G. Y. Tian, D. Taylor and J. Rudlin. A feature extraction technique based on principal component analysis for pulsed Eddy current NDT. Volume 36, Issue 1, January 2003, Pages 37–41

[9] N. Yusa, Z. Chen and K. Miya. Sizing of stress corrosion cracking on austenitic stainless piping in a nuclear power plant from eddy current NDT signals. Nondestructive Testing and Evaluation. Vol. 20, No. 2, June 2005, 103–114.

[10] Xuan, L. Zeng, Z. Balasubramaniam, S. and Udpa, L. (1999). Element-Free Galerkin Method for Static and Quasi-Static Electromagnetic Field Computation, IEEE Trans. Magn., 40,12-20

[11] T. Khan, P. Ramuhali, (2011) Particle filter based multi-sensor fusion for solving low frequency electromagnetic NDE inverse problems, IEEE Trans. of Measurement and Instrumentation,Vol: 60, No:6, pp. 2142-2153.

[12] T. Khan, P. Ramuhalli, (2010).Sequential Monte Carlo Based Data Fusion for Characterization of Corrosion on Aging Aircraft Structures, Electromagnetic Nondestructive Evaluation, 181-188.

[13] Khan, Tariq, and Pradeep Ramuhalli. "A recursive bayesian estimation method for solving electromagnetic nondestructive evaluation inverse problems." IEEE Transactions on Magnetics 44.7 (2008): 1845-1855.
[14] Khan, Tariq, Lalita Udpa, and Satish Udpa. "Particle filter based prognosis study for predicting remaining useful life of steam generator tubing." Prognostics and Health Management (PHM), 2011 IEEE Conference on. IEEE, 2011.

Nov 16, 2016

Finite Element Simulation of Laser-based Non-contact Nondestructive Evaluation

Seyed Mehdi Mirsadeghi

Supervisors: Dr. Simon Park, Dr. Ron Hugo
Pipeline Engineering Center (PEC)
Micro Engineering, Dynamics and Automation Lab (MEDAL)
Dept. of Mechanical & Manufacturing Engineering
Schulich School of Engineering at the University of Calgary

Outline

- Introduction
- Laser Ultrasonic Generation
- Objective
- Methodology
- Simulations
- Experiments
- Summary

2

Introduction

	Excitation Method	Target Structure	Detection Method
Contact		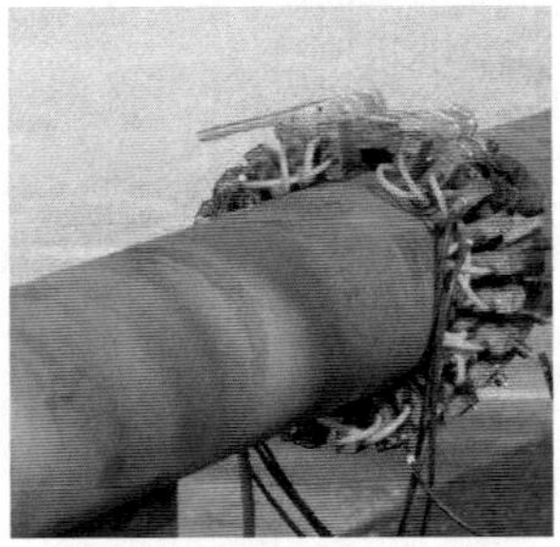	
Non-Contact	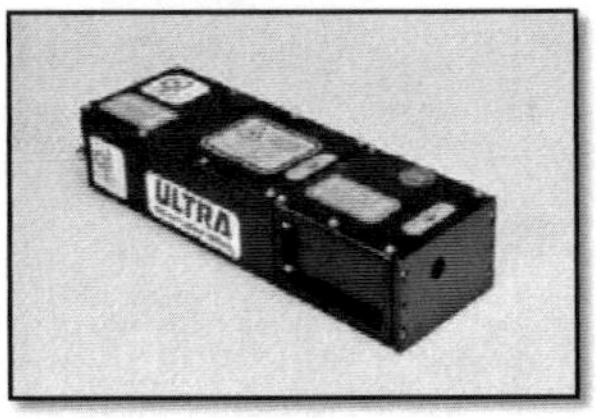		

Introduction

Why Non-Contact NDE?

- Zero Mass Loading
  - No additional damping
  - No shift of eigen frequencies
- High Bandwidth
- High Spatial Resolution
- Functional in Harsh Environment
  - High temperature, corrosive, etc.
- No transducer cabling and instrumentation
- Less Vulnerability

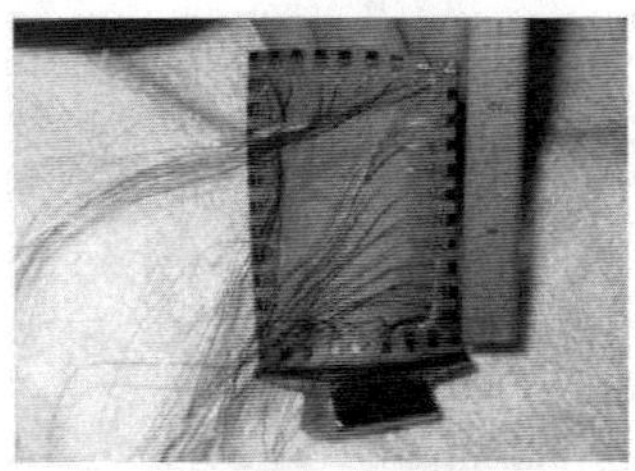

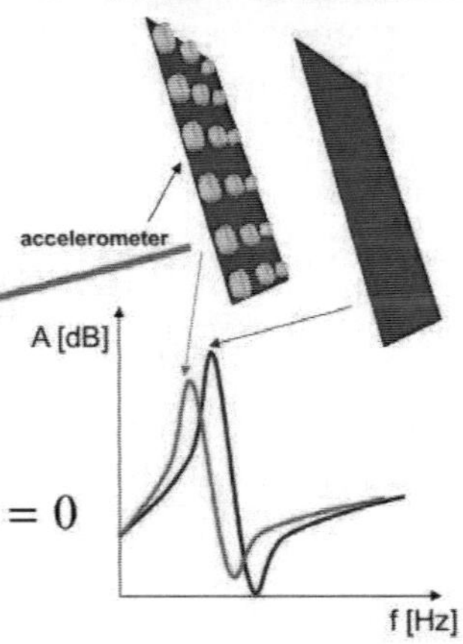

$$[M]\{\ddot{x}(t)\} + [D]\{\dot{x}(t)\} + [K]\{x(t)\} = 0$$

equation of motion

Laser Ultrasonic Generation

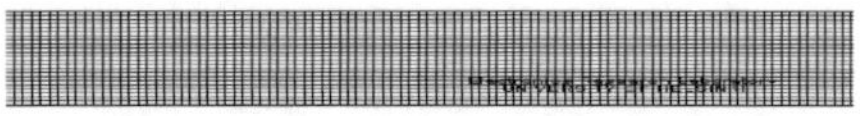

Laser Ultrasonic for Non-Contact NDE

http://electronics.physics.helsinki.fi/

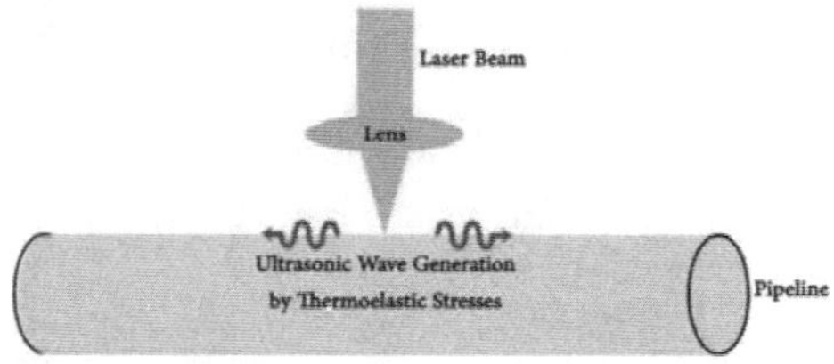

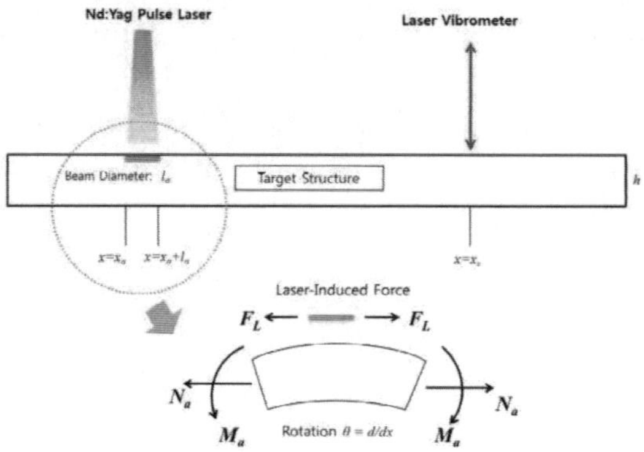

SCHULICH School of Engineering

SHM of pipelines in nuclear power plants Hyeonseonk 2014

5

Laser Ultrasonic Generation

Optical wavelength

- **The efficiency of absorption** (ultrasonic generation): inversely proportional to the square root of optical wavelength
- **Surface Condition:** rough/blackened surface increase the absorption/efficiency at all wavelength
- **Safety:** high power lasers are dangerous in the visible and near visible areas → CO2 the safest candidate.

Efficiency for AL

Carbon dioxide laser 10.6um

↓ x 2.7

Nd:YAG laser 1.06um

↓ x 3.5

Krypton fluoride laser 246nm

Beam Profile

- **Beam Profile:** Less important; can be changed by standard optical means
- **Irradiated area:** important; affects the power density
- **Gaussian or Flat profile for commercial lasers**

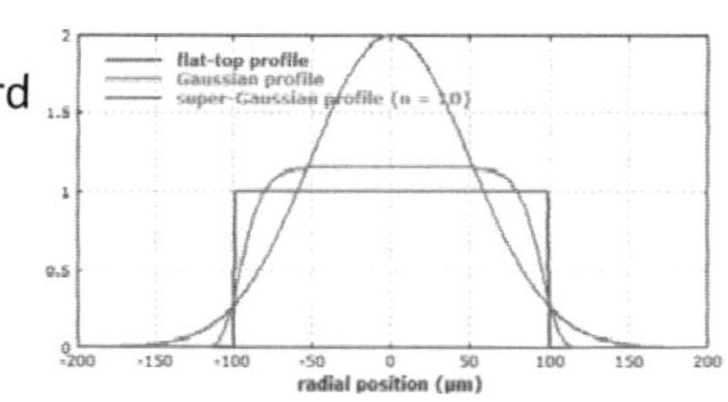

SCHULICH School of Engineering

6

Laser Ultrasonic Generation

Pulse Energy

- **High Pulse Energy:** Having enhanced ultrasonic amplitude
- **Low Pulse Energy:** Not damaging the surface (ablation, vaporization, melting, plastic deformation)
- **Threshold (in power density):** 1-10 $MWcm^{-2}$ for Al and Steel using the Q-switched pulsed laser

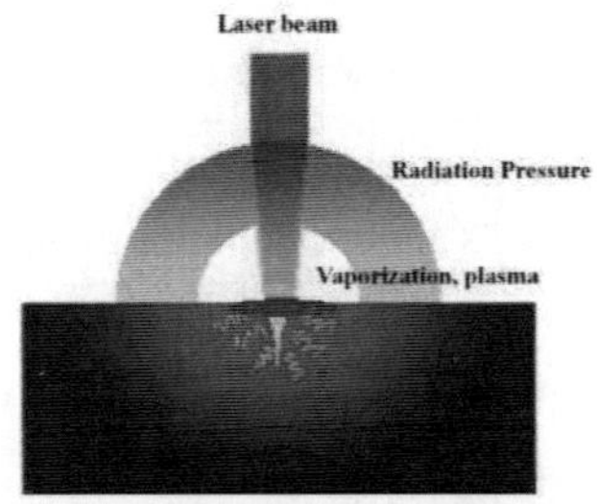

Pulse Duration

- **Important:** Affect ultrasonic waveform characteristics
- **Commercial lasers:** 20 ns or less, generating ultrasonic energy up to 20 MHz

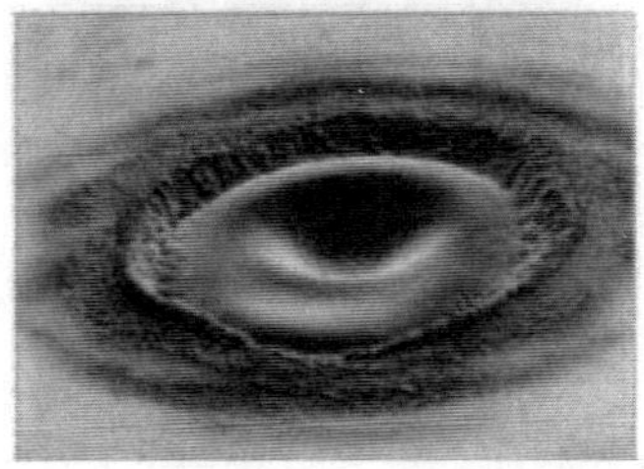

Repetition Rate

- **Flaw Detection:** high repetition rate → signal averaging → improved SNR
- **Scanning speed**

SCHULICH School of Engineering

7

Laser Ultrasonic Generation

Laser Type	$\lambda\ (\mu m)$	E (J)	$\tau\ (ns)$	Rep (Hz)	Comments
Nd:YAG (Q-switched)	1.06	0.01-10	10-50	<1000	Wide application, compact system, available for low repetition rate

Threshold (in power density): 1-10 $MWcm^{-2}$ for Al and Steel using the Q-switched pulsed laser

Beam Profile

- **Beam Profile:** Less important; can be changed by standard optical means
- **Irradiated area:** important; affects the power density
- **Gaussian or Flat profile for commercial lasers**

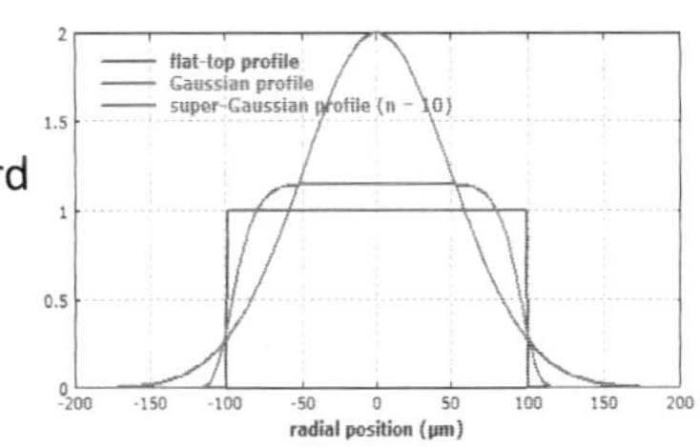

SCHULICH School of Engineering

8

Laser Ultrasonic Generation

Laser Type	$\lambda\ (\mu m)$	E (J)	$\tau\ (ns)$	Rep (Hz)	Comments
Carbon dioxide (Q-switched)	10.6	0.1-1	200	400	Low generation efficiency, bulky, less hazardous
Nd:YAG (Q-switched)	1.06	0.01-10	10-50	<1000	Wide application, compact system, available for low repetition rate
Nd:glass (Q-Switched)	1.06	1-10	10-50	<10	For highest power application
Ruby (Q-switched)	0.694	1-10	10-50	<50	High power application
Excimer (e.g. KrF)	0.249	0.1-1	10-20	10-200	High generation efficiency

Laser Ultrasonics – Scruby and Drain

SCHULICH School of Engineering

9

Laser Ultrasonic Generation

Physical Parameter	Laser	Piezoelectric Probe
Proximity to sample	Totally remote	Contact via solid/liquid medium
Couplant requirement	Couplant free	Couplant variability problems
Beam directivities	Valuable by optical means	Fixed
Bandwidth	Wider bandwidth	Narrow
Size and Weight	Large and Heavy	Small
Environment effect	Suitable for very hot, corrosive and radioactive environments	Limited applications
Other	More expensive Safety Considerations	

SCHULICH School of Engineering

10

Objectives

Development of Non-contact Nondestructive Evaluation System for detecting structural defects

- Enhancing the spatial resolution for detecting pipeline defects
- Creating reliable numerical model
- Running efficient and accurate algorithms for imaging defects
- Developing software for signal processing and damage identification

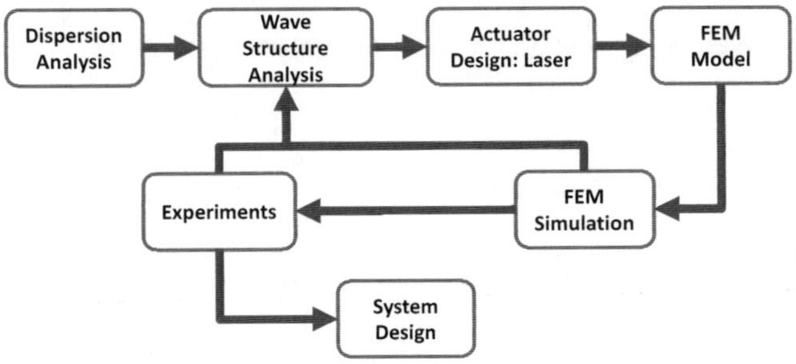

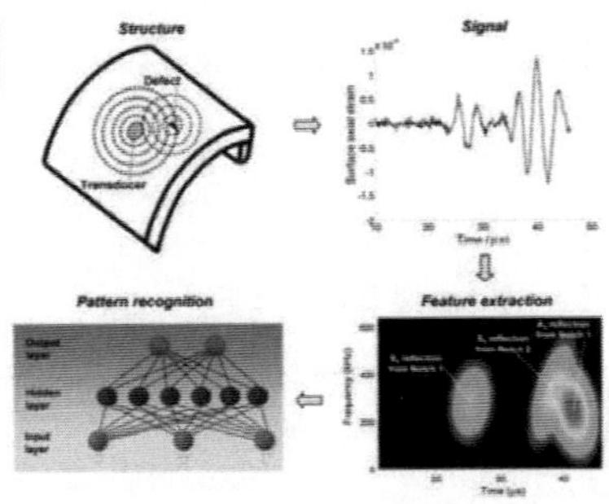

Guided wave structural health monitoring- Raghavan 2007

Methodology

Dispersion: Dispersive Wave Propagation

$$(\lambda + \mu)\nabla\nabla.\boldsymbol{u} + \mu\nabla^2\boldsymbol{u} = \rho\ddot{\boldsymbol{u}}$$
$$\boldsymbol{u} = \nabla\phi + \nabla \times \boldsymbol{H}$$
$$\frac{\tan\beta b}{\tan\alpha b} = \left[-\frac{4\alpha\beta k^2}{(k^2-\beta^2)^2}\right]^{\pm 1}$$

Navier Governing Equations
Helmhotz Decomposition
B.C. Zero Stress on Free Surfaces
Rayleigh–Lamb Frequency Equations

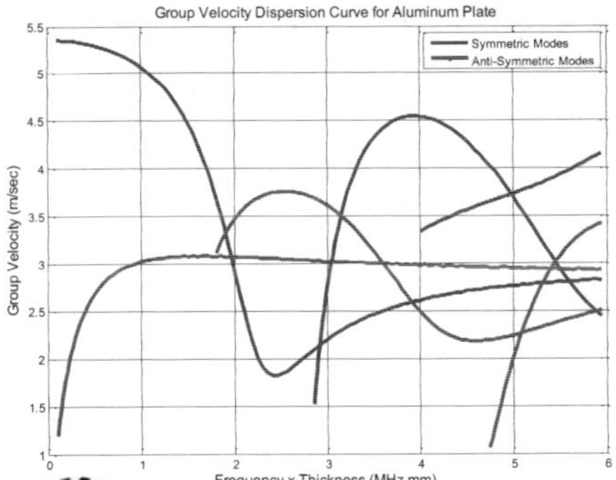

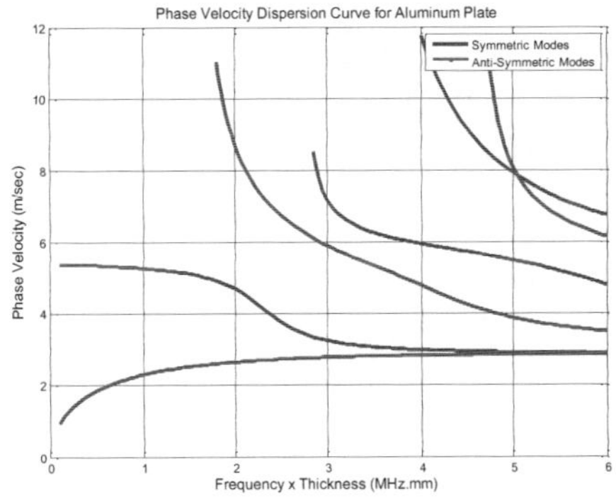

Methodology: Guided Waves

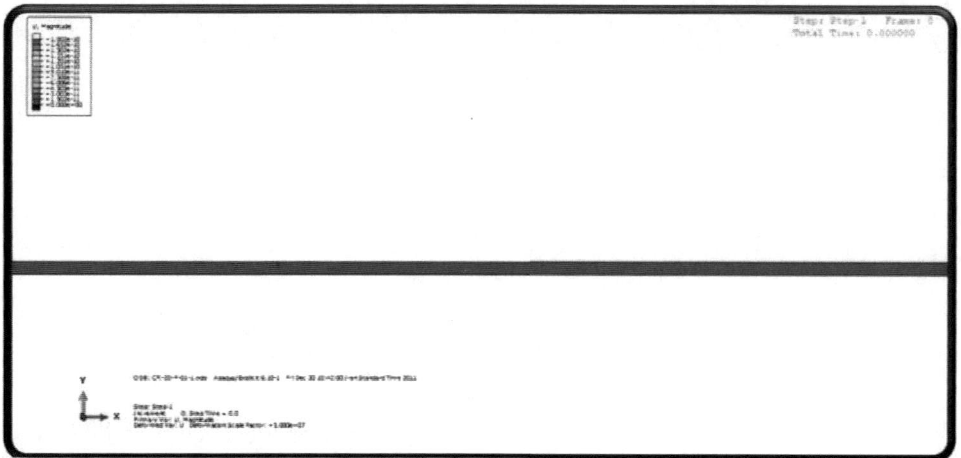

SCHULICH School of Engineering

Methodology: Guided Waves

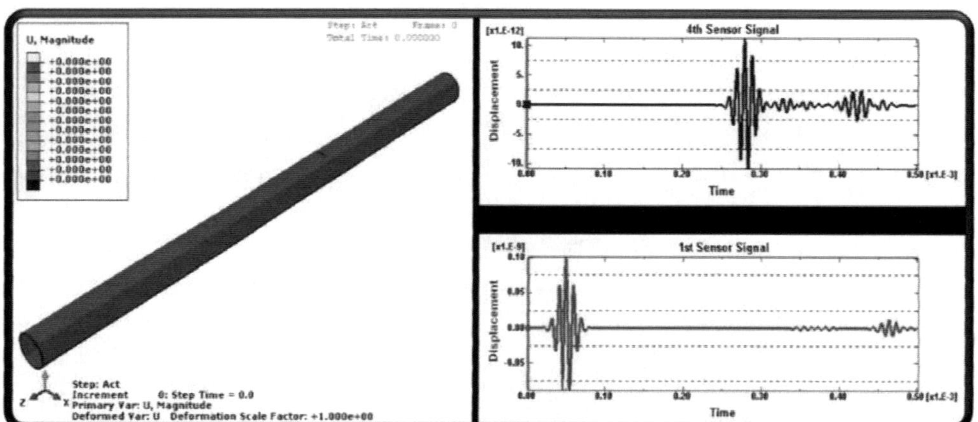

SCHULICH School of Engineering

Thermoelastic Wave Propagation

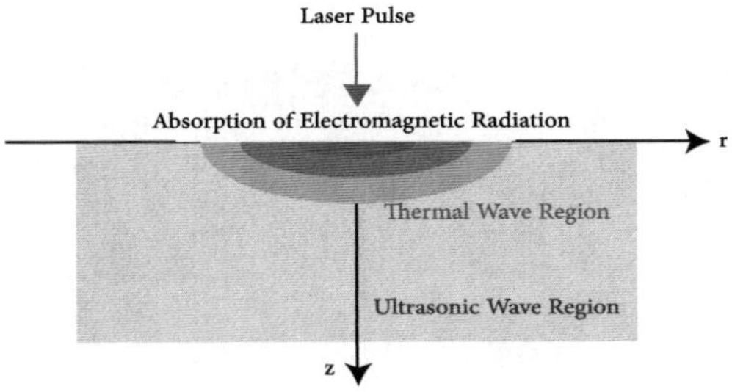

ρ: *Density* T: *Temperature* t: *Time* h: *Thickness*
c: *Specific heat capacity*
k: *Thermal conductivity*
r, θ, z : *Cylindrical coordinate*
λ *and* μ *are the Lamé constants*
$f(t)$: *Temporal distribution of laser pulse*
$g(t)$: *Spatial distribution of laser pulse*
I: *Power intensity*

Thermal Conductive Eq.

$$\rho c(T)\frac{\partial T(r,z,t)}{\partial t} = \frac{1}{r}\frac{\partial}{\partial r}\left(rk(T)\frac{\partial T(r,z,t)}{\partial t}\right) + \frac{\partial}{\partial z}\left(k(T)\frac{\partial T(r,z,t)}{\partial z}\right)$$

Boundary Conditions

$$-k(T)\frac{\partial T(r,z,t)}{\partial z}\bigg|_{z=0} = I_0 A(T) f(r) g(t) \qquad \frac{\partial T(r,z,t)}{\partial z}\bigg|_{z=h} = 0$$

Heat source at the front surface | Thermally insulated

Displacement field

$$(\lambda + 2\mu)\,\nabla\big(\nabla . U(r,z,t)\big) - \mu\nabla \times \nabla \times U(r,t,z) - \alpha_t(3\lambda + 2\mu)\nabla \mathrm{T}(\mathrm{r,t,z}) = \rho\frac{\partial^2 U(r,t,z)}{\partial t^2}$$

Initial Conditions

$$T(r,z,0) = 300\ \mathrm{K}$$

SCHULICH School of Engineering

15

[illegible]1K$^{6-1}$)

Simulation

Parameters	Value
Thermal expansion coefficient (K^{-1})	23.4×10^{-6}
Poisson's Ratio	0.33
Young Modulus (GPa)	68.9
Thermal Conductivity ($W.m^{-1}.K^{-1}$)	170
Specific Heat ($J.kg^{-1}.K^{-1}$)	900
Density ($kg.m^{-3}$)	2700
Reflection Coefficient	0.95

Fully-coupled thermal stress analysis – Abaqus/Explicit
Pulse energy: 10 mJ, 20 mJ, 40 mJ
Pulse duration: 6 ns
Simulation time: 20 us
Stable time increment: 0.27 ns – 0.45 ns

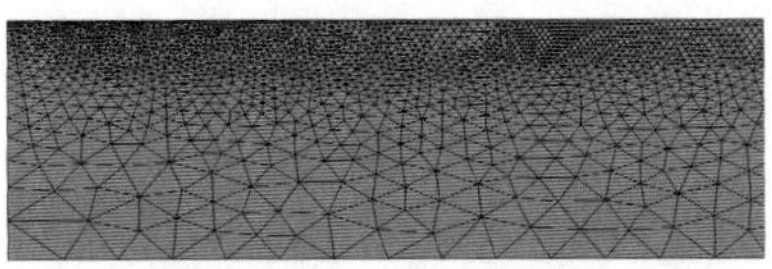

SCHULICH School of Engineering

Element Size
Max: 0.5mm Min: 5um

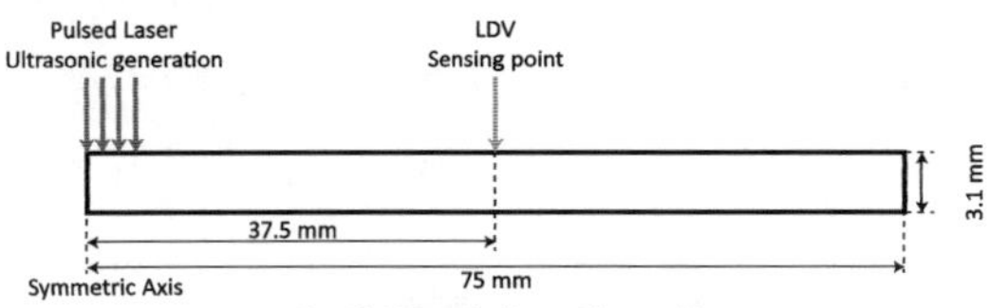

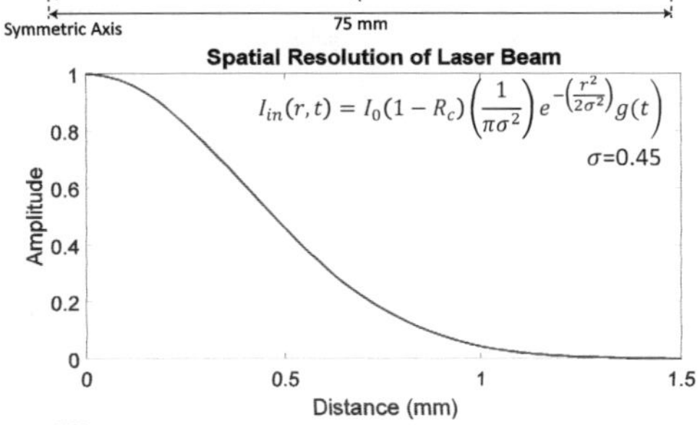

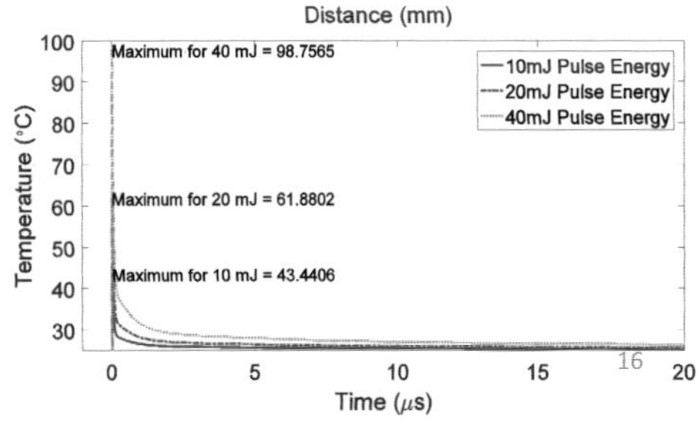

16

Simulations: Results

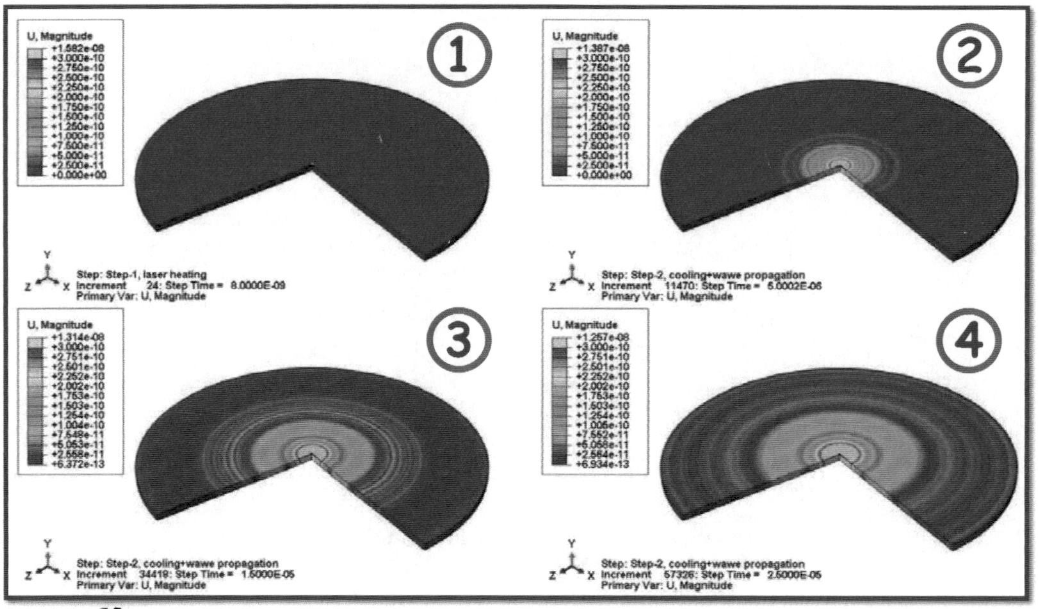

SCHULICH School of Engineering

Experiments

Parameters	Explanation
Structure	Aluminum Disc (R=75 mm, t=3.1mm)
Excitation	Nd:YAG Wavelength: 1064nm Pulse duration: 6ns Pulse Frequency: 1Hz Laser beam radius: 1.5mm Pulse energy density: 560 mJ/cm^2
Sensing	Laser Doppler Vibrometer Sensitivity: 10mm/s/V OR 20mm/s/V Frequency: 350 kHz OR 1 MHz

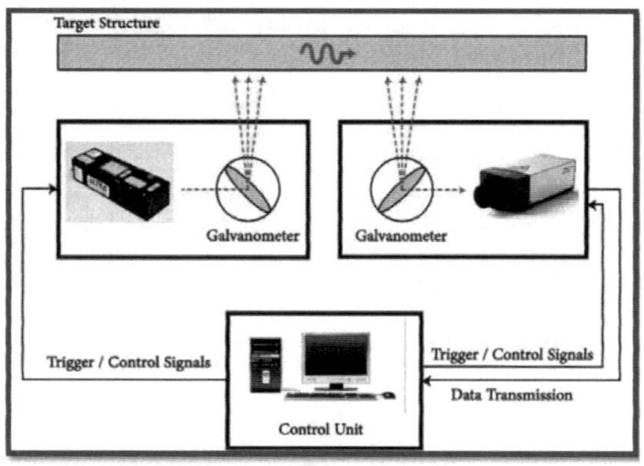

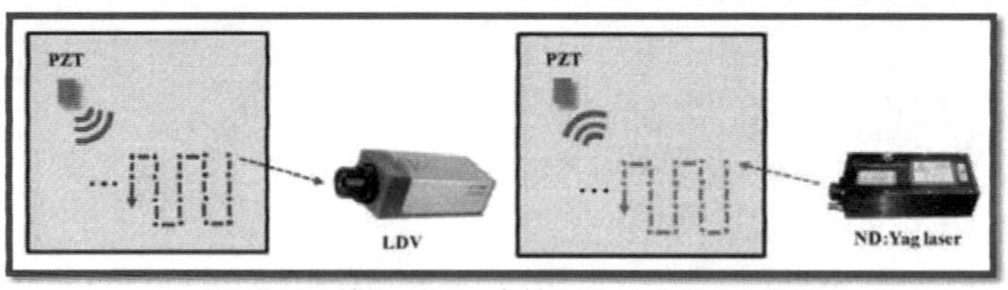

Non-Contact Ultrasonic Wavefield Imaging, Yun-Kyu An 2013

SCHULICH School of Engineering

Experiments

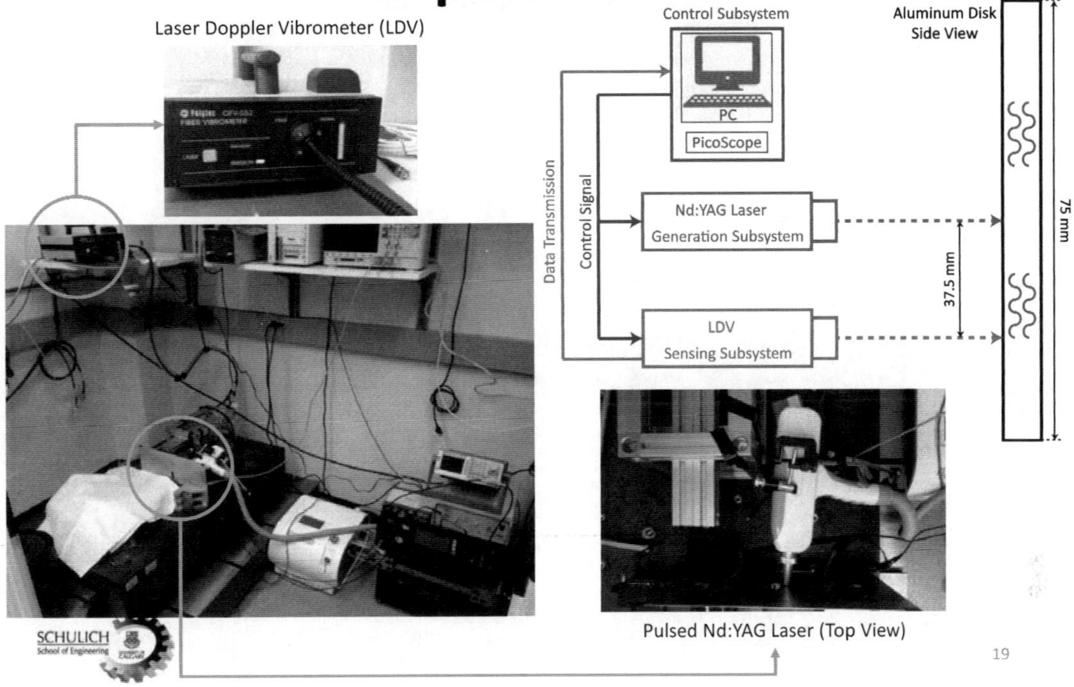

19

Experiments

Sensitivity: 10 mm/s/V
Maximum Frequency 350 kHz

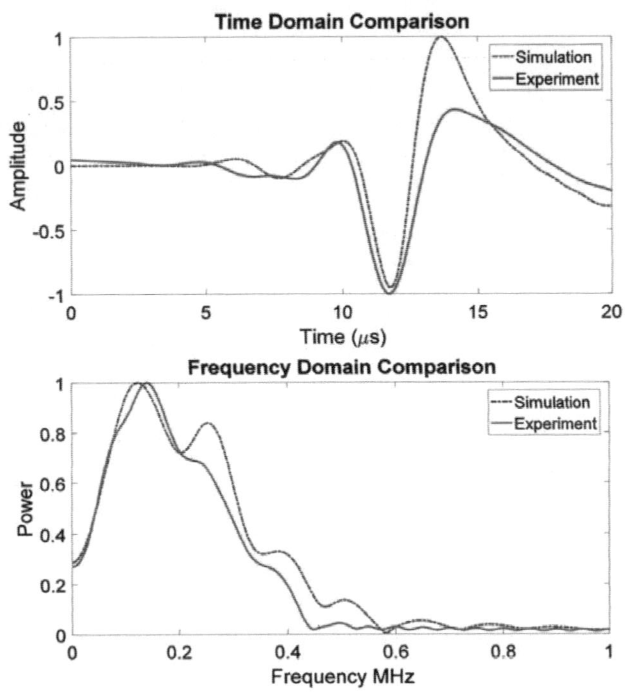

SCHULICH
School of Engineering

20

Experiments

Sensitivity: 20 mm/s/V
Max Frequency: 1MHz

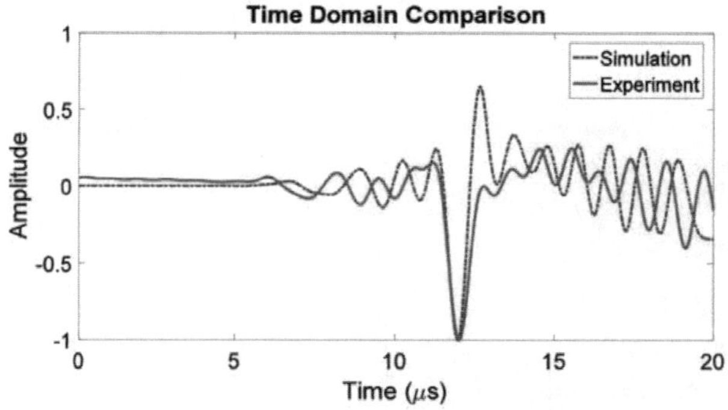

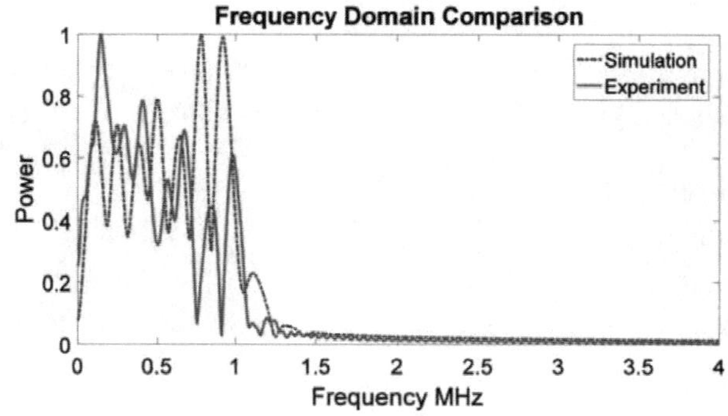

21

Defect Sensitivity Analysis

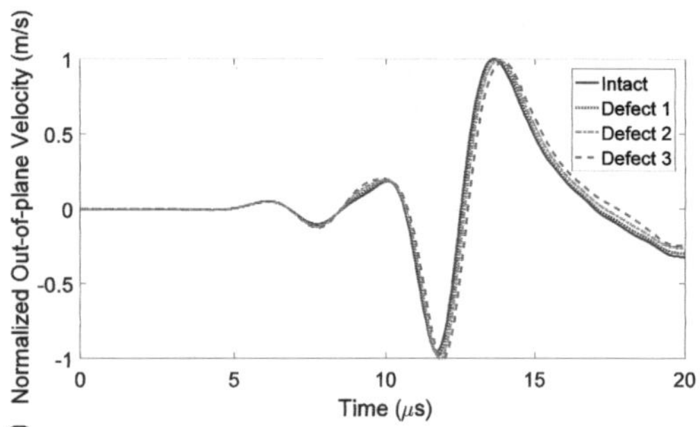

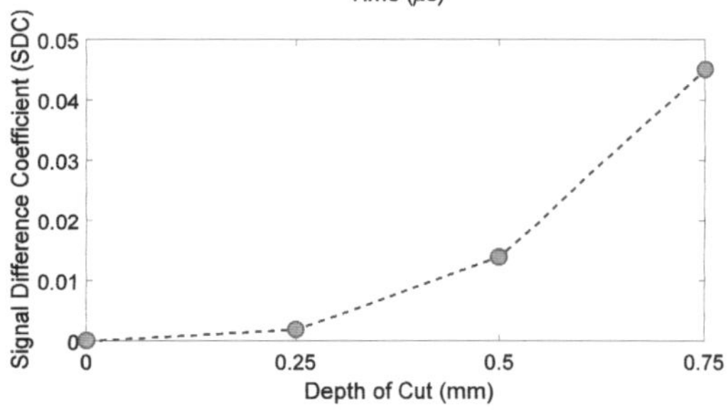

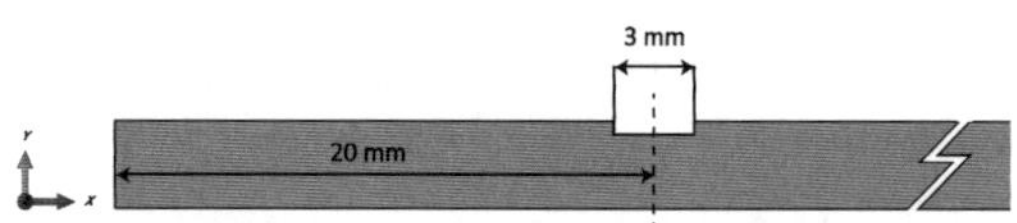

	Length	Depth
Defect 1	3 mm	0.25 mm
Defect 1	3 mm	0.5 mm
Defect 1	3 mm	0.75 mm

Signal Difference Coefficient

$$\mathbf{0} < SDC = 1 - \frac{C_{XY}}{\sigma_X \, \sigma_Y} < 1$$

X: Healthy
Y: Defected

22

Summary

Summary

- Dispersion characteristic and wave structure of the propagating waves in plates are studied
- Technical parameters requirements investigated for thermoelastic laser-based wave generation
- Ultrasonic guided wave propagation is numerically simulated in plates and pipeline
- Simulation results are validated using the experimental tests
- The sensitivity of the SDC to defect severity is studied

Future Work

- Baseline-free methods for crack identification
- Installing the scanning capability using the galvanometers
- Defect classification using pattern recognition methods

SCHULICH School of Engineering

23

Thank You for Your Attention

PEC Pipeline Engineering Center

SCHULICH School of Engineering

UNIVERSITY OF CALGARY

SCHULICH School of Engineering

24

Eddy Current Detection of Partially Conducting Flaws in Pickering NGS Steam Generator Tubes

6th International CANDU In-Service Inspection NDE Workshop and NDE in Canada 2016 Conference
November 15 to 17, 2016

Sean Sullivan – Inspection & Maintenance Services

BEHAVIOURS

- Say It, Do It
- Simplify It
- Think Top and Bottom Line
- Integrate and Collaborate
- Tell It As It Is

VALUES • SAFETY • INTEGRITY • EXCELLENCE • PEOPLE & CITIZENSHIP •

PEOPLE POWERING THE FUTURE OPG NUCLEAR

ONTARIO POWER GENERATION

Presentation Overview

- Introduction – Intergranular Attack Flaws (IGA)
- Interaction of Electrical Current with Volumetric Flaws and Partially Conducting Flaws
- Equations for Mathematically Modelling ET signals from flaws
- Comparison of Modelled Signals with Signals from confirmed IGA Flaws
- Estimation on influence of Flaw Conductivity on ET Detection Capability
- Summary/Conclusions

Intergranular Attack Flaws (IGA)

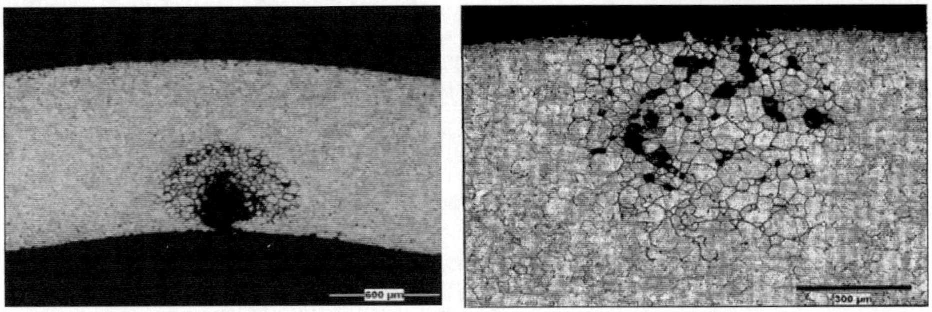

(a) **(b)**

Photographs of some typical IGA flaws:

(a) ID IGA flaw in a Pickering 1, SG tube.
(b) OD IGA flaw in a Pickering 1, SG tube.

Electrical Current Interaction with Flaws

- Hemispherical Flaws

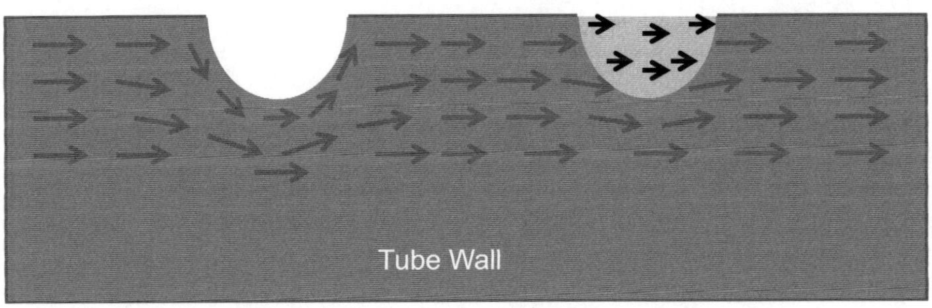

Theory

Equation for ET flaw signal (eddy current interaction) from a Transmit-Receive Coil Configuration:

$$\Delta V = \frac{1}{I_R}(\sigma - \sigma')\int_V \vec{E}^T \cdot \vec{E}_0^R dV$$

σ is the electrical conductivity of the material being inspected. σ' is the electrical conductivity of the flaw. σ' is equal to 0 for volumetric flaws. $\vec{E}^T$ is the electric field within the flaw region that was generated by the time harmonic magnetic field from the transmit coil of the eddy current probe. $\vec{E}_0^R$ is the electric field that would be generated in the flaw region (assuming that no flaw is present) by the receive coil if a time harmonic current in the receive coil (I_R) were flowing through it.

Theory (Continued)

For small spherically shaped or hemispherically shaped flaws, the interactions between the current and the flaws are modelled as small localized dipoles in opposition to the general flow of current.

$$\Delta V \approx \frac{(\sigma - \sigma')}{I_R}\left(\frac{\frac{3}{2} < \vec{E}_0^T \cdot \vec{E}_0^R >}{1 + \frac{\sigma'}{2\sigma}}\right) \times Flaw\ Volume$$

$<E_0>$ is the electric field calculated at an optimized location within the flaw region that is assumed to represent an average uniform field for the volume occupied by the flaw. This assumption has been validated by verifying that comparisons between predicted signals from this equation and measurements of calibration flaw signals are reasonable.

ET Probes Under Consideration

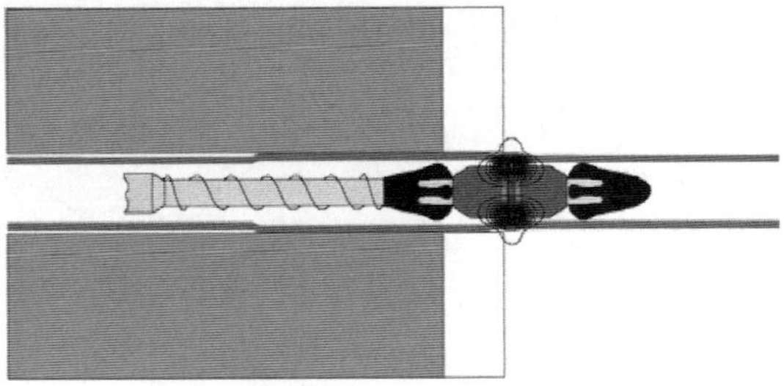

Magnetically Biasing Bobbin Probe
(CTR-1 Probe)

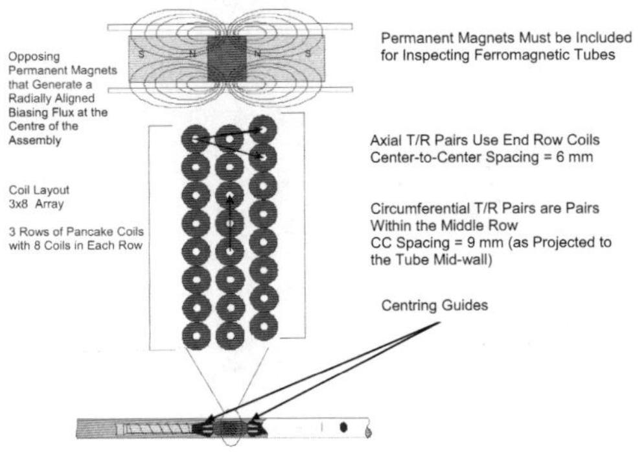

Magnetically Biasing X-Probe
(3x8 coil array)

Comparison of Model with ID IGA Signals

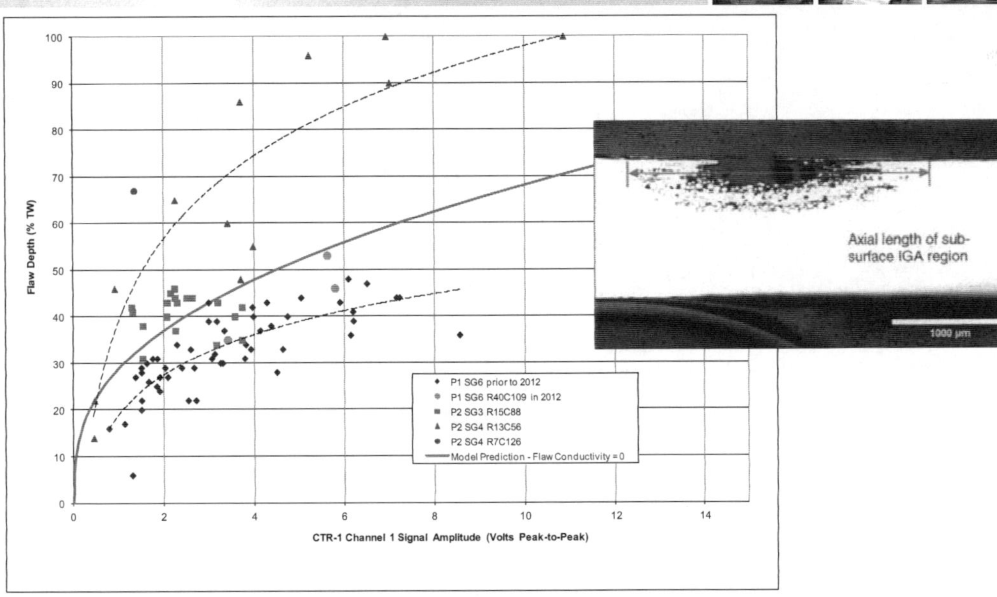

- Model is conservative compared to Pickering 1 ID IGA flaw signals. Real flaws had larger volumes than hemispheres.

Comparison of Model with OD IGA Signals

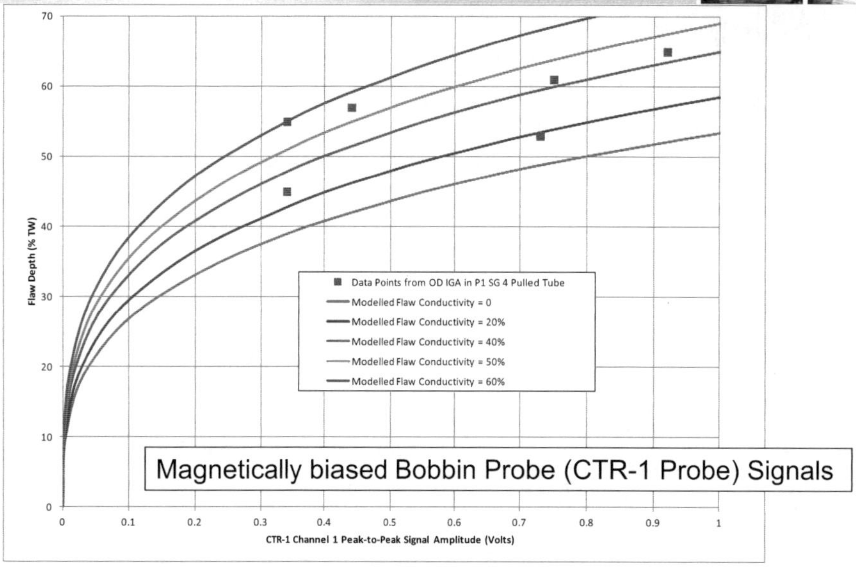

- OD IGA flaws were observed to be more consistent with hemispheres (length to depth ratios ~ 2).
- Model indicates observed OD flaws likely to be partially conducting.

Comparison of Model with OD IGA Signals

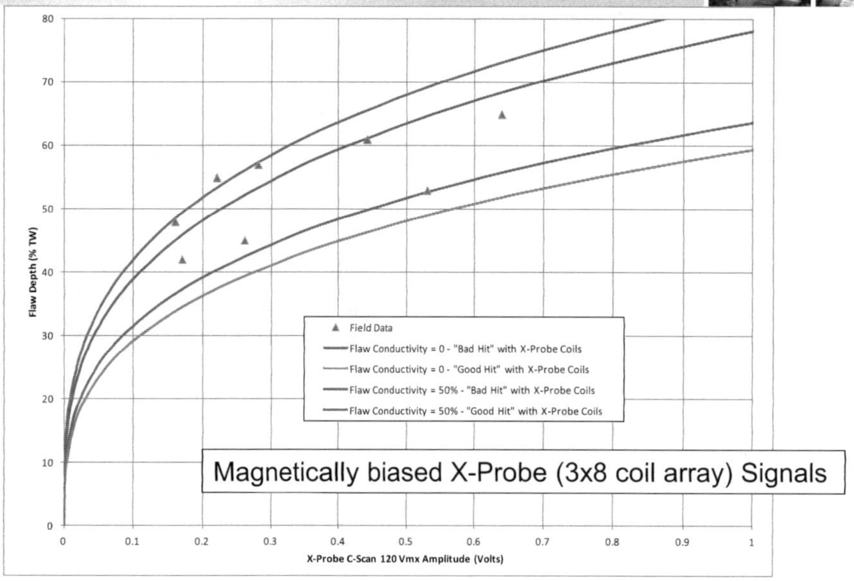

Estimated Detection Capability

- Probability of Detection (POD) estimates in free-span zones based on comparisons of modelled CTR-1 Probe signals with measured noise in inspection data.

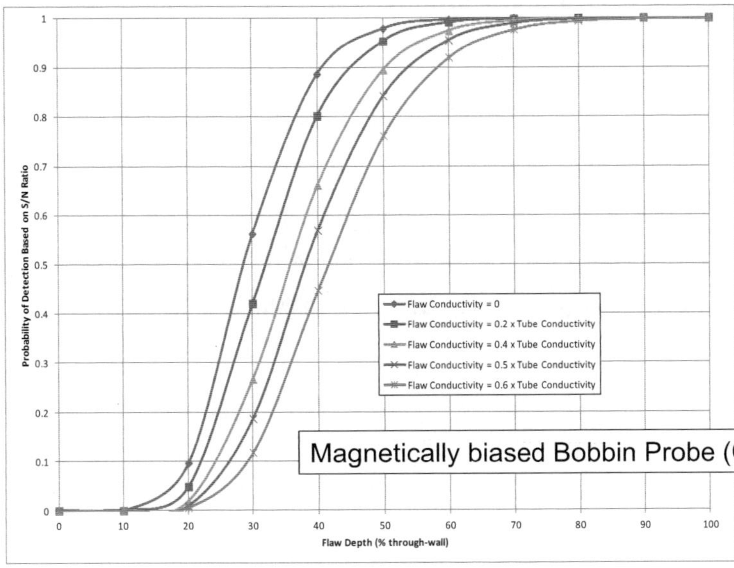

Estimated Detection Capability

- Probability of Detection (POD) estimates in free-span zones based on comparisons of modelled X-Probe signals with measured noise in inspection data.

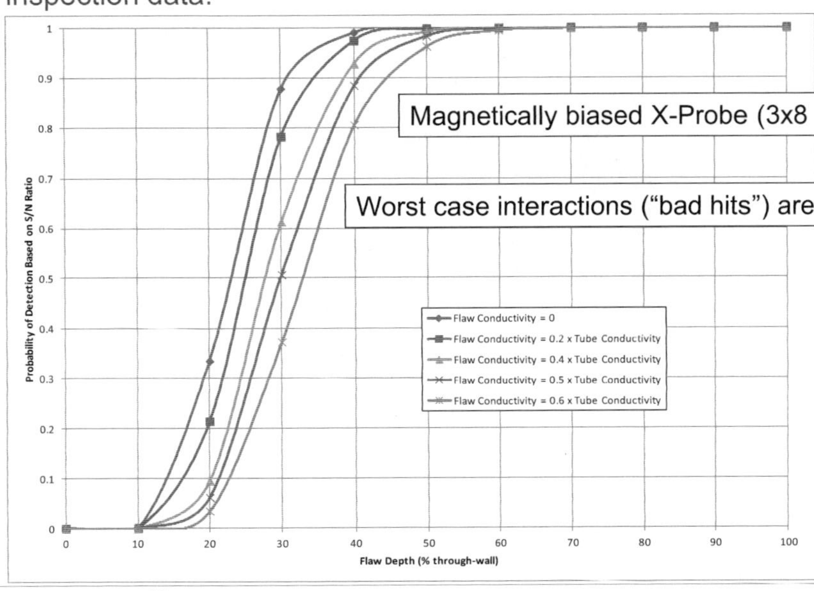

Summary / Conclusions

- A mathematical model has been developed to estimate the effect of electrical conductance within flaws on ET flaw signals.
- This model has been used to estimate ET detection capability for tight IGA flaws.
- Results suggest that ET probes may still have high sensitivity to IGA flaws even when the electrical conductivity within those flaws is up to 60% of the conductivity of the parent tubes.

Questions?

?

Thank-you for Your Attention

15 OPG INTERNAL USE ONLY **VALUES** • SAFETY • INTEGRITY • EXCELLENCE • PEOPLE & CITIZENSHIP •

ONTARIOPOWER GENERATION

Qualification of ultrasonic testing of Cat® ridged dump trucks ball studs from the inferior position of the stud

Neil K Young.
Asset Reliability Inspections Pty Ltd
Bunbury, WA, 6230, Australia
+61 8 97257550
neil.young@ari.com.au

Abstract

Background: Ball studs connect the steering linkages in Cat® Ridged Dump Trucks and are a critical component in these trucks. In the event of a cracked ball stud serious accidents can occur as well as very costly losses in service. Manufacturers require that ball studs be tested ultrasonically every 1000 hours of service. Guidelines prescribe that ball studs are tested from the superior position which requires two technicians and 1.5 hours per truck to test including access time to expose the top of ball stud. Some technicians test from the inferior end of the ball stud to reduce time (0.25 hours per truck) required for testing however accuracy of the assessment is questionable. This background has generated two project aims:

1) To determine the sensitivity and specificity of testing ball studs in dump trucks from the inferior versus superior position ultrasonically using usual methods and non-custom/conventional equipment.
2) Should the sensitivity and specificity of testing ball studs from the inferior position using usual methods not meet satisfactory standards, design a method of testing ball studs in dump trucks ultrasonically from the inferior position which meets sensitivity and specificity standards.

Method: A sample of 10 ball studs from Cat® Ridged Dump Trucks were tested ultrasonically inferiorly and superiorly using conventional ultrasonic equipment by two ISO 17025 Ultrasonic Level 2 certified technicians. They were blinded to the condition of ball studs. Inferior and superior testing were done a week apart so technicians could not compare results for each ball stud. The ball studs were then tested using magnetic particle testing to confirm if cracks present or not. As the results indicated unsatisfactory sensitivity and specificity for testing ball studs inferiorly using standard technique, a new procedure using a custom made angled wedge with a 13mm ultrasound probe was developed and tested, with technicians testing the 10 ball studs using this technique and comparing to magnetic particles testing results. Technicians were again blinded to the condition of ball studs.

Results: When testing from the manufacturer recommended superior position versus magnetic particle testing, sensitivity was 100% and specificity 100%. When testing inferiorly with standard probe sensitivity was 40% compared to magnetic particle testing. When testing inferiorly adding a custom made angled wedge with a 13mm probe the sensitivity is 80% and specificity 100%.

Discussion: Testing ball studs superiorly with a standard probe and inferiorly using custom angled wedge with a 13mm probe meet satisfactory sensitivity and specificity standards while testing inferiorly with standard probe does not meet satisfactory sensitivity values. Given that the time taken to test the ball studs in one truck superiorly versus inferiorly is 1.5 hours compared to 0.25 hours, the results for this new technique suggest using a custom angled wedge and 13mm probe is satisfactory and more feasible alternative.

1.0 Introduction

Ball studs connect the steering linkages together in Cat® Ridged Dump Trucks (Figure 1.) and are prone to cracking (Cat 2009). Manufacturing guidelines prescribe regular testing ultrasonically to assess the integrity of ball studs (Cat 2009). There are eight off ball studs per truck (Figure 2) and they come in different sizes depending on the truck size (Cat 2009).

Figure 1. Cat® Ridged Dump Truck

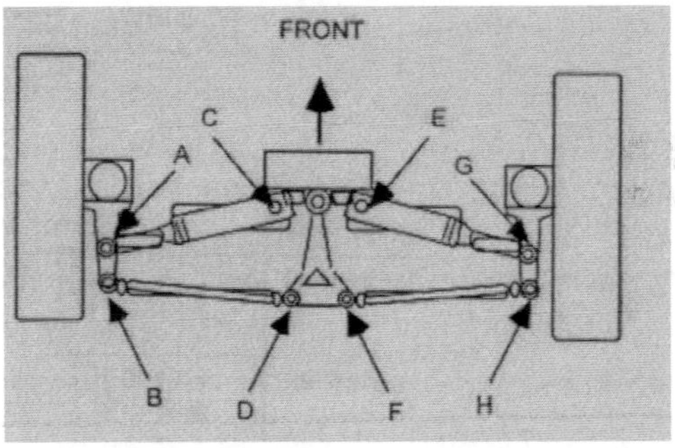

Figure 2. Cat® Ridged Dump Truck Ball Stud Locations (shown by arrows/letters)

The Cat® dump truck service manuals state that ball studs should be tested ultrasonically every 1000 hours (approximately testing them every 3 months if the trucks work day and night shift) for in-service cracking to prevent loss of steering in the event of one cracking through (Figure 3), (Cat 2006, p.1). Loss of steering is obviously a major safety concern for these giant vehicles (Department of Natural Resources and Mines 2001, Department of Industry and Resources 2004) and (haul) road side repairs are more difficult and labour intensive when compared to planned maintenance performed in the workshop. Government Safety Bulletins have been released in Australia regarding cracking of ball studs and detailing steering failures (Department of Natural Resources and Mines 2001, Department of Industry and Resources 2004).

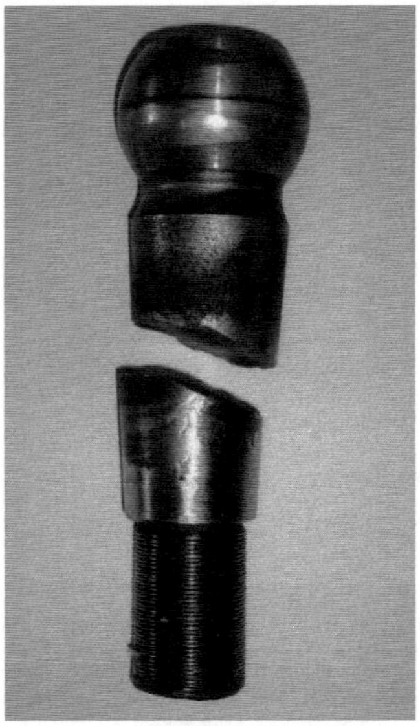

Figure 3 Failed Ball Stud

Figure 4 Access Point to the Top of the Ball Stud (position A & B from Figure 2.)

The Cat® testing procedure specifies removing the grease caps above the ball studs and testing the stud from the "top" (superior position) which is the non-threaded end (Cat 2006 p.6). To test from the superior position a technician is required to remove the grease caps, perform the test and replace the grease caps which on average takes two technicians 0.75-1.5 hours per truck (time specified is the average time taken to test trucks over the last 3 years) (ARI 2015). See Figure 4 to illustrate superior view of ball stud. The testing of the ball studs only takes ~0.25 hours per truck and therefore the majority of the time the truck is out of service is due to obtaining access to the test area.

The threaded end of the ball stud is naturally exposed and therefore does not require any labour or time to gain access to this end. Therefore if testing could be qualified from the inferior end of the stud this would reduce the testing time from ~1.5 man hours (including access time) to ~0.25 man hours.

The challenges are:

- the taper of the stud and attempting to get sound from a smaller diameter to the larger diameter where the studs crack and
- the loss of cross sectional area within the threaded section due to cotter pin holes in this area.

Cubberly and Bakerjihan (1989) explain that a disadvantage of ultrasonic testing is the difficulty testing irregular shapes which is a consistent description of the shape of a ball stud. Figure 5 following shows the area of interest (where the stud typically cracks) from the manufacturer inspection procedure (Cat 2006 p. 6) and the cotter pin locations (Figure 5 and 6).

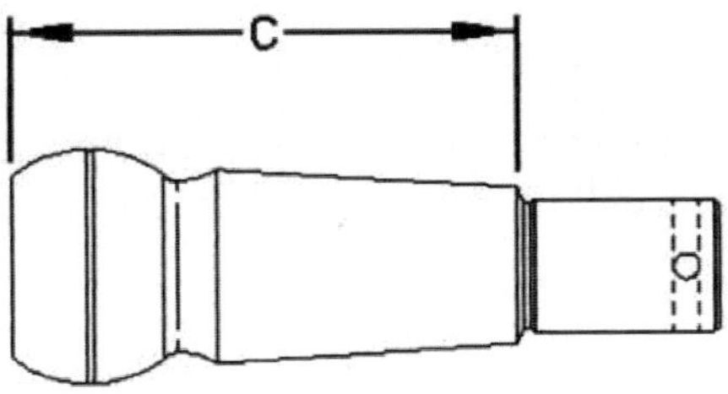

Figure 5. Area of Interest (Cracking) Shown as "C" as per Manufacturer Inspection Procedure

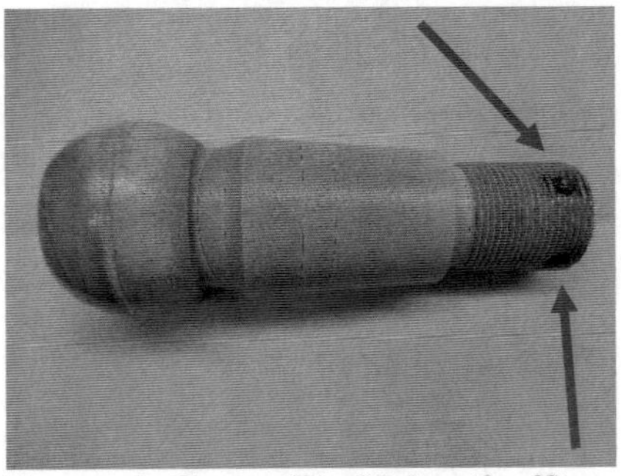

Figure 6. Cotter Pin Holes (2 off) on Cat 777 Ball Stud

Asset owners often enquire whether the studs can be tested from the inferior position and report that other companies agree to this method. ARI have numerous cracked reference studs at their workshop which technicians are unable to repeatedly find cracks from the inferior position using standard equipment and currently refuse requests to test from this position. This results in a loss of work with asset owners preferring to reduce costs by having to pay less labour costs for the quicker inferior testing. An example of a statement from another company's webpage confirms their use of procedures testing from the inferior position – in relation to testing ball studs:

> "Asset Management Engineers' procedure is a non-destructive straight beam examination using high frequency sound energy to conduct the examination and take measurements.
> A key advantage of the non-destructive UT inspection is that it does not require any fitters to remove covers or bolts." (Asset Management Engineers 2015)

Therefore the objectives of this project are to:

1) To determine the sensitivity and specificity of testing ball studs in dump trucks from the inferior versus superior position ultrasonically using usual methods and non-custom/conventional equipment.
2) Should the sensitivity and specificity of testing ball studs from the inferior position using usual methods not meet satisfactory standards, design a method of testing ball studs in dump trucks ultrasonically from the inferior position which meets sensitivity and specificity standards.

Within these objectives the term sensitivity refers to the proportion of abnormal samples (cracked studs) identified compared to the true number of abnormal samples (ie. Are all cracked studs correctly identified?) and specificity refers to the proportion of normal samples identified compared with the true number of normal samples (ie. Are all intact studs correctly identified?) (Wei et al 2015).

On a review of the literature, no previous studies have investigated this area of non-destructive testing (NDT). This study will aim to provide new knowledge to the NDT field in regards to testing ball studs and aim to confirm accuracy of current techniques and design new techniques if necessary to ensure time efficient and accurate test methods.

2.0 Experiment 1

Experiment 1 was designed to address the first objective of the study:
To determine the sensitivity and specificity of testing ball studs in dump trucks from the inferior versus superior position ultrasonically using usual methods and non-custom/conventional equipment.

2.1 Methods Experiment 1

2.1.1 Design

A singled blinded randomised sensitivity and specificity study comparing conventional ultrasound techniques testing ball studs for cracks from the superior and inferior position compared to "gold standard" magnetic particle testing.

2.1.2 Equipment

Ball Studs:
ARI have collected numerous studs over the years for training and reference. Not all studs collected are cracked as some have been removed by asset owners for wear during servicing and in some cases studs were not cracked but worn/damaged and when tested ultrasonically in situ were mistakenly called as cracked studs and removed. Ten studs were available for testing, n=10 (Figure 7).

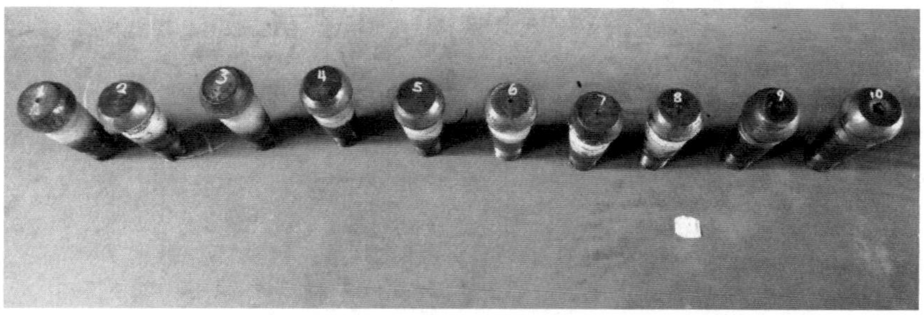

Figure 7. Ball Studs for Blind Testing

Ultrasonic Set:
The ultrasonic set was required to:

- have A-scan presentation
- have reserve of sensitivity of at least 20 dB at the maximum beam path used
- have a frequency range between 2 MHz to 10 MHz.

Probe:
5MHz 12mm single crystal probe was used.

Coupling Medium:
A satisfactory coupling medium was used to transfer the ultrasound from the probe to the surface of examination object. The coupling medium – polycell solution - had good wetting characteristics at the temperature of test.

Calibration Blocks:
An AS2083 No 1 (IIW V1) block was used for calibration. Blocks were constructed using material with similar nominal acoustic velocity to the material under test.

2.1.3 Procedure

Equipment was calibrated prior to testing being performed. Single blinded randomised ultrasonic testing (Salkind 2010, p.387) of 10 off ball studs (Figure 8) from the superior and inferior position of the stud was performed by two ISO 17025 certified Ultrasonic Testing (UT) Level 2 operators who had extensive experience of testing studs from the top. The studs included in the sample were worn studs that gave indications from wear but were not cracked. Superior and inferior position testing were conducted 1 week apart.

The test procedure used for testing from the superior position was the manufacturer's procedure (Cat 2006). The same procedure was then slightly modified with an additional 6db for use from the inferior position to compensate for the smaller test surface (diameter) and cotter pins with an increased test sensitivity to attempt to not bias the existing test methodology of testing from the superior position. The use of different frequency probes within the manufacturer's procedure given range (Cat 2006 p. 5) also did not affect the test results. The use of different size probes was not practicable due to the size of the studs being tested.

The cracking mode is from an in-service fatigue mode (ATTAR 2014) which means the studs crack from the outside surface in making them suitable for magnetic particle testing. The studs were all "gold standard" tested with magnetic particle testing to confirm the presence of cracking (Figure 8) using a surface test method following ultrasonic testing.

Fig 8. Failed Ball Stud Showing Fatigue Beach Marks (up to ~70% cross sectional area) Before Brittle Fracture

Magnetic particle testing was chosen as it is a simple but very effective test method that is very sensitive to surface breaking cracking (McKelvey 1980) such as the cracking in the ball studs being examined.

2.2 Results Experiment 1

5 off the 10 studs were confirmed to be cracked using magnetic particle testing. No false positives were detected ultrasonically (100% specificity). Typical cracking evident is shown below in Figure 9.

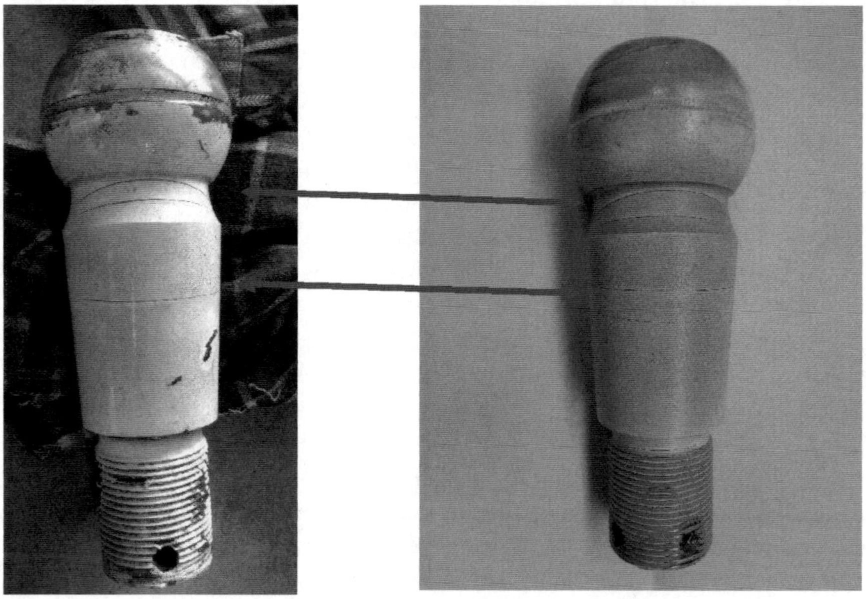

Figure 9. Typical Cracked Studs (black circumferential lines are cracking)

Summary of the test results that are shown in Table 1 are detailed below:

- 5 off the 10 studs were cracked (confirmed via magnetic particle testing)
- 2 off the 10 studs that were not cracked were worn/externally damaged
- All 5 cracked studs were detected ultrasonically testing from the superior position of the stud (100% sensitivity)
- Only 2 off the 5 cracked studs were detected ultrasonically from the inferior position of the stud (40% sensitivity)

Therefore there is only a 40% (2 cracks found via the bottom / 5 cracks evident) chance of detecting the cracks with standard (non-custom) test equipment when testing from the inferior end of the stud. Both ultrasonic test methods provided 100% specificity. Both technicians obtained the same results.

Table 1. Test Results – Whether Cracking Was Present and Found
Type of Test Vs Test Sample Stud Number

	1	2	3	4	5	6	7*	8	9^	10
Magnetic particle testing to confirm whether cracking was evident **(gold standard test)**	Yes	Yes	No	No (damaged stud)	No	Yes	Yes	Yes	No (worn stud)	No
Top of stud - Normal probe 12mm Ø / 5MHz	Yes	Yes	No	Reflector evident from damage	No	Yes	Yes	Yes	Reflector evident from wear^	No
Bottom of stud - Normal probe 12mm Ø / 5MHz	No	Yes	No	No	No	Yes	No	No	No	No

*Stud 7 crack was 10mm in length and directly in line with the cotter pin hole and was unable to be detected when tested from the bottom of the stud. This crack could be considered to be minor in length and an acceptable risk when compared to the production loss/expenses of highly time consuming testing from the top of the stud.

^Stud 9 wear reflectors were comparable to cracking reflectors and as a result wear to this level will be identified as cracked studs in the field. Ultrasonic reflectors are unable to differentiate from cracking and wear discontinuities. Note that the level of wear evident required the stud to be replaced to prevent damage to the steering arm from excessive movement.

The green highlight indicates cracking found from the bottom with standard equipment.

The red highlight indicates cracking not evident testing from the bottom with standard equipment.

2.3 Discussion Experiment 1

The results from Experiment 1indicate that there is a significant reduction from the 100% probability of detection of cracks in balls studs from dump trucks when testing from the superior position of the stud compared with testing from the inferior position using non customised equipment (40 % sensitivity). Malhorta (2016) reports that sensitivity below 70% or 0.7 is poor and unacceptable.

This initial testing indicates that the testing of ball studs from the inferior position with longitudinal conventional UT is not effective or ethical and supports the second stage of the study to design a more sensitive method for testing ball studs ultrasonically from the inferior position in dump trucks.

3.0 Experiment 2

Experiment 2 was designed to address the second objective of the study.

- Should the sensitivity and specificity of testing ball studs from the inferior position using usual methods not meet satisfactory standards, design a method of testing ball studs in dump trucks ultrasonically from the inferior position which meets sensitivity and specificity standards.

Justification for Experiment 2 is further strengthened by industry complaints regarding some companies testing from the inferior position and questionable accuracy of results. As a result of these complaints and technical discussions with an expert technician with was suggested using longitudinal waves with a small angled wedge. Lhemery et al (2002) reports that the use of angle probes or wedges enables a better detection of flaws in contoured and irregular shapes.

3.1 Methods Experiment 2

3.1.1 Design

A singled blinded randomised sensitivity and specificity study comparing an ultrasound technique testing ball studs for cracks from the inferior position with a custom angled wedge compared to "gold standard" magnetic particle testing. Using Snell's Law different angled wedges could be made to induce different longitudinal waves within the stud.
For example, as per Snell's law a 5 degree wedge will induce a 11 degree longitudinal wave in the test item, which calculates the angle of refraction into a medium based on the incident angle and velocity of sound waves in the two mediums.

Snell's Law: $$\frac{\sin \alpha}{\sin \beta} = \frac{V_A}{V_B}$$

Where α is the wedge angle; β is unknown (angle of sound in the stud); V_A is 2.68km/s; V_B is 5.9km/s

3.1.2 Equipment

Equipment is as for 2.1.2 except for differences in probe and addition of custom angle wedge to the probe (see Figure 10).

Figure 10. Custom Angled Wedge (made from a standard order 45 degree wedge)

3.1.3 Procedure

A single blinded randomised ultrasonic testing of 10 off ball studs (Figure 8) from the inferior position of the stud using the custom wedge and 0.5" singe crystal probe was performed by two ISO 17025 certified Ultrasonic Testing (UT) Level 2 operators. The same studs were used as for Experiment 1. Testing occurred 6 weeks after Experiment 1.

The test procedure used for testing from the inferior position was developed by ARI. The results of the ultrasonic testing in Experiment 2 were compared with magnetic particle testing results as for Experiment 1 and sensitivity and specificity were calculated. The significant details of the testing procedure with custom degree wedge are outlined below.

Method of Examination

- Manually with direct contact coupling, using the pulse-echo method.
- The stud was tested from the inferior position of the stud.
- Peak memory was used and reset for each stud tested.

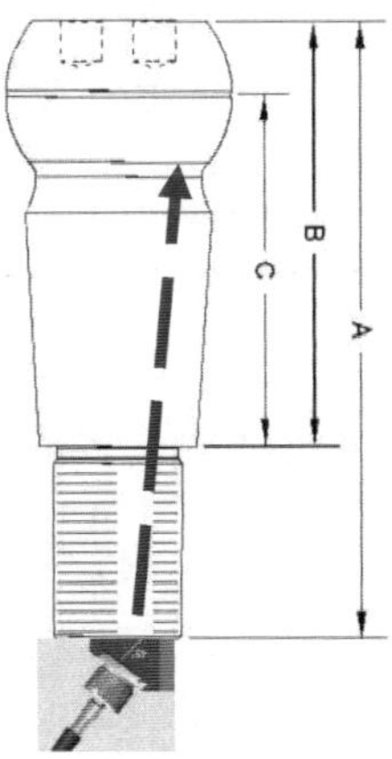

Figure 11.Positioning of probe for ultrasonically testing ball stud inferiorly with custom angle degree wedge

Interpretation of Signals

Areas of concern that are susceptible to cracking are highlighted below (Figure 12) in red and indications in these areas were assessed as acceptable or not as per the acceptance criteria detailed within the procedure (Appendix 1).

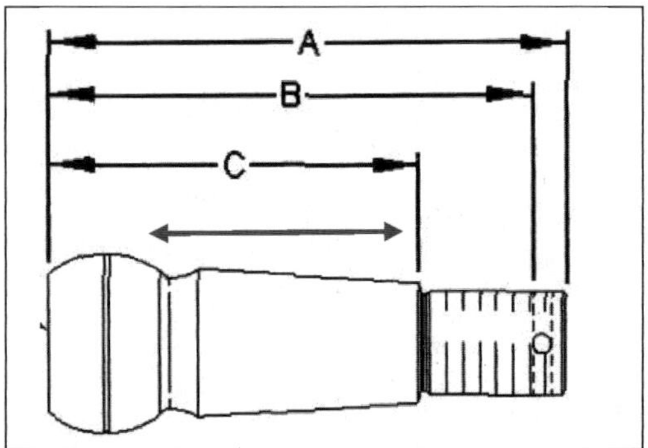

Model	A	B	C
768	161	149	112
769, 771	161	149	118
770, 772, 773, 775 B-F, 776, 777	174.8 or 168.15	158 or 155.4	121

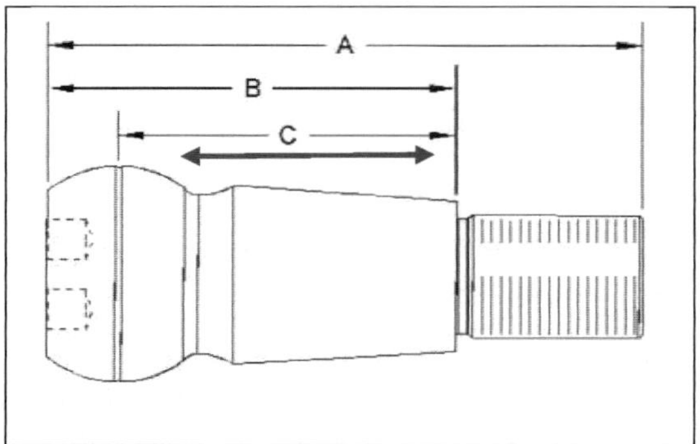

Model	A	B	C
784, 785	273	187.8	155.5
789, 793, 797B	295	210.6	178.3
795F AC, 795F, 797B S/N:797B259-UP	341.1	240.3	195.3

Figure 12. Area of ball stud susceptible to cracking highlighted in red

Acceptance Criteria

Results were recorded and assessed against the following criterion. Ultrasonic indications greater than 20% full screen height (FSH) when at evaluation sensitivity as detailed in ARI's test procedure, in the area identified as susceptible to cracking shall be rejected.

3.2 Results

The results in Table 2 on the following page include the magnetic particle testing of the 10 ball studs and the results indicating if cracking was evident using the 0.5" probe with custom angled wedge.

The sensitivity when ultrasonically testing ball studs using a 0.5" probe with custom angled wedge to detect flaws is 80% when compared with magnetic particle testing and specificity 100%.

Table 2 Test Results – Whether Cracking Was Found or Present
Type of Test Vs Test Sample Stud Number

	1	2	3	4	5	6	7*	8	9^	10
Magnetic particle testing to confirm whether cracking was evident	Yes	Yes	No	No (damaged stud)	No	Yes	Yes	Yes	No	No
Bottom of stud - Custom equipment 13mm Ø	Yes	Yes	No	No	No	Yes	Not evident*	Yes	Reflector evident from wear^	No
Crack reflector height at evaluation sensitivity % FSH	30%	35%	N/A	N/A	N/A	Shank 40% Groove 35%	Not evident*	Shank 25% Groove 10%	N/A	N/A

*Stud 7 crack was 10mm in length and directly in line with the cotter pin hole and was unable to be detected when tested from the bottom of the stud. This crack could be considered to be minor in length and an acceptable risk when compared to the production loss/expenses of highly time consuming testing from the top of the stud.

^Stud 9 wear reflectors were comparable to cracking reflectors and as a result wear to this level will be identified as cracked studs in the field. Ultrasonic reflectors are unable to differentiate from cracking and wear discontinuities.

Note that cracks found from testing from inferior position with standard equipment was half the amount of cracking found with custom made equipment shown in green.

3.3 Discussion Experiment 2

The use of a 0.5" probe with a custom angle wedge meets acceptable standards (Malhotra 2016) for sensitivity when testing ball studs ultrasonically for cracks from the inferior position. A sensitivity of 80% was achieved. Some critics may consider a higher sensitivity level should be required, however it should be noted that the crack in the ball stud in which the inferior wedge approach could not detect the crack is likely to be considered negligible risk.

The development of a proven testing technique from the inferior position of ball studs in dump trucks is beneficial to industry as it can save industry thousands of dollars in both the amount of truck down time and in testing costs. These feasibility issues need to be weighed up in relation to test accuracy to decide on the best approach for each client.

If the suggested solution had not achieved viable results then the use of phased array longitudinal probes that can be electronically steered of up to 5-10 degrees could have been considered as a potential solution. This would not be the preferred option as it would add additional problems such as higher level of training required for the testing operator and higher inspection equipment costs.

4.0 Conclusion

Testing ball studs superiorly with a standard probe and inferiorly using custom angle wedge with a 0.5" probe meet satisfactory sensitivity and specificity standards while testing inferiorly with standard probe does not meet satisfactory sensitivity values. Given that the time taken to test the ball studs in one truck superiorly versus inferiorly is 1.5 hours compared to 0.25 hours, the results for this new technique suggest using a custom wedge and 0.5" probe is satisfactory and more feasible alternative in relation to labour costs and truck down time.

Caution should be taken with results due to small number of ball studs tested (n=10) and further testing would strengthen evidence.

Acknowledgement

The author wishes to acknowledge the contribution of Mr Paul Grosser CEng, ISO9712 Level 3 UT, RT, MT, PT, ET, PAUT, ToFD for the idea of the angled wedge for Experiment 2.

References

ARI 2015: Audit 2013-2015 – Time taken to inspect ball studs in rear dump trucks, Asset Reliability Inspections, Western Australia

ARI 2015a: Standard working procedure for ultrasonic ball stud testing TP UT06, Asset Reliability Inspections, Western Australia

Asset Management Engineers 2015: Ultrasonic testing on truck ball studs, accessed 28/12/2015
http://asseteng.com.au/are-you-operating-machinery-with-cracked-ball-studs/

ATTAR 2014: Investigation into failed ball stud, Advanced Technology Testing and Research, Morley, Western Australia

Cat 2006: Inspection procedure for steering ball studs {4305, 4305}, Caterpillar Inc, Illinois, United States

Cat 2009: Location of ball studs – Cat ridged dump trucks, Caterpillar Inc, Illinois United States

Cubberly WH and Backerjihan R 1989: Tool and Manufacturing Engineers Handbook Desk Edition, Society of Manufacturing Engineers, Michigan, United States

Department of Natural Resources and Mines 2001: Cracks in steering ball joints in rear dump trucks -Mines safety alert no. 66, Queensland Government

Department of Industry and Resources 2004: Significant incident report no. 128 – Steering Failures on haul trucks, Government of Western Australia

Lhemery A, Calmon P, Chatillon S and Gengembre N 2002: Modeling of ultrasonic fields radiated by contact transducer in a component of irregular surface, Ultrasonics, vol.40, pp 231-236

Malhotra R 2016: Empirical Research in Software Engineering: Concepts, Analysis, and Applications, CRC Press, Florida, United States

McKelvey W 1980: Level III study guide: magnetic particle method, The American Society for Nondestructive Testing, Ohio, United States

Salkind 2010: Encyclopedia of Research Design Volume 1, SAGE, London, United Kingdom

Wei H, Chen H, Guo Y, Jing G and Tao J 2015: SOM-based intrusion detection for SCADA systems, Proceedings of the 2014 Asia-Pacific Conference on Electronics and Electrical Engineering, Shanghai, China, Dec 27-28 2014, pp 57-59

How to Implement a PoD into a Highly Effective Inspection Strategy

Michael WRIGHT [1]

[1] Eclipse Scientific; Waterloo, Ontario, Canada;
Phone: 1-(519)-886-6717, Fax: 1-(519)-886-1102; e-mail: mwright@eclipsescientific.com

Abstract

The probability of detection is a mathematical model that associates the likelihood, or frequency an NDT system, technique and/or inspector will detect a specific sized indication. For most engineering models it is very important to know what can be detected with a 90% probability. Also, what defect size yields a confidence level of 90-95%. The confidence level is very similar to the concept of repeatability.

Inspection strategies can be optimized several ways using PoD model values. Evaluating the PoD can be used as a way to quantify inspection capabilities of a technique, inspection system/equipment and the capabilities of the inspection personnel.

There are several types of PoD models. The most common are the â vs. a, binary or hit-miss, Model Assisted (MAPoD) and several advanced models.

Model Assisted PoD (MAPoD) is a newer model building technique that can be done using several methods. Data can be collected and analyzed statistically to build the final PoD model. Data that has been obtained using modelling software can be evaluated and compared with true inspection samples to create the final PoD model. Also, a mathematical simulation based evaluation can be performed to evaluate the final PoD.

There are also several advanced PoD models available for deeper mathematical evaluation or for rare inspection situations.

Keywords: Probability of Detection (PoD), Confidence Level (CL), Model Assisted PoD (MAPoD), optimization, qualification, technical justification.

1. Introduction

There are many levels to an inspection development and several experiments that are performed to gather supporting evidence. As the experiments are performed the analysis helps optimize the inspection itself. All experimentation begins by taking a single measurement. Experiments are designed to help quantify how repeatable measurements are. This helps reveal all of the sources of variation. These sources are called essential and influential parameters. The PoD model, Probability of False Call (PoFC) evaluation, and the noise study are directly linked to each other. Once all three studies are complete, an inspection can be optimized to help produce the highest PoD and the lowest PoFC.

The inspection development process is very similar to the steps in the scientific method.

2. Inspection Developments

Inspections start out as simple techniques. A technique is a method that produces a repeatable measurement or test. Techniques are incorporated into written instructions and are the building blocks of procedures. An instruction outlines the steps required to perform a technique and to produce the same values regardless of the inspection personnel performing the tests. Instructions are created to provide a solution to an inspection situation. As more evidence is gathered an instruction can be turned into a procedure. Procedures are much more detailed than an instruction since they address code and specification requirements. A good procedure has evidence that shows the instructions and techniques meet the requirements of the inspection. Figure 1 shows the standard process flow of a typical inspection development.

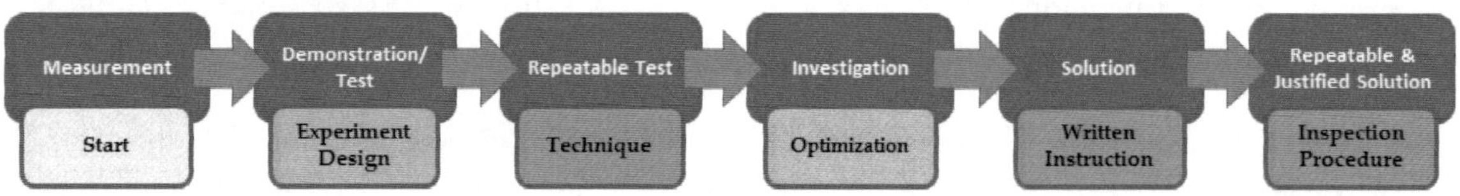

Figure 1 – The inspection development road map

With all the statements of a procedure meeting code and specification requirements, a procedure can be approved. Once a procedure is approved, it can be further reviewed and qualified by an independent qualification board.

2.1 The Qualification Roadmap

Qualification traditionally requires all of the supporting evidence to be gathered into a document called a technical justification (TJ). A TJ traditionally contains very specific NDT studies, evaluations and demonstrations. Figure 2 shows the inspection qualification road map.

The measurement of an NDT system's detection capabilities is evaluated using a mathematical model called the Probability of Detection (PoD). Due to the time and costs involved in evaluating a PoD model most inspections do not develop a full PoD model. When personnel, equipment, and/or procedure qualification is required a full PoD model is normally required to demonstrate an inspection or part of an inspection is meeting the inspection specification requirements. Qualification is demonstrated and/or captured in a technical justification document. This process has been standardized for nearly all inspection developments.

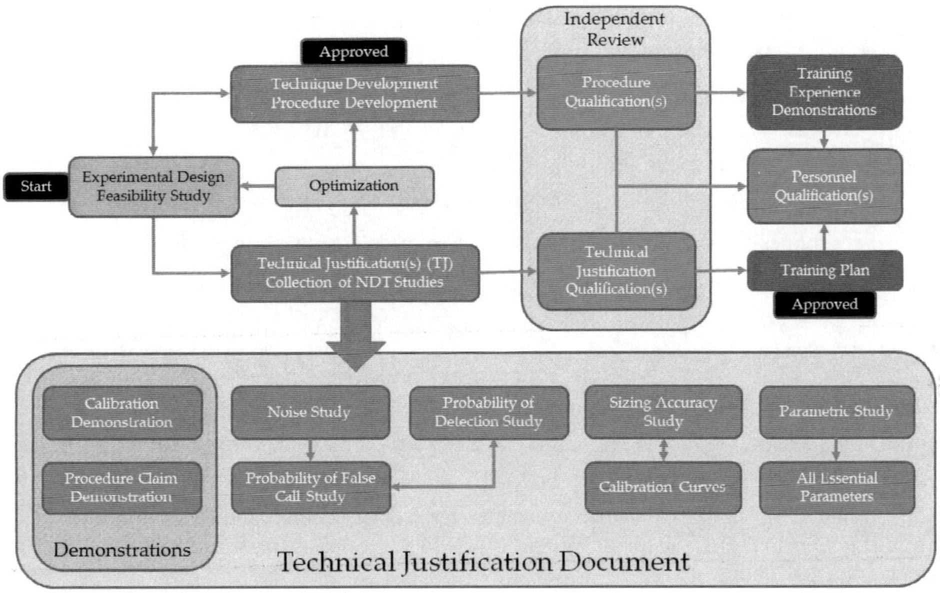

Figure 2 – The inspection qualification road map *[1]*

The TJ is a living document that acts as a canister to gather all the supporting evidence that a procedure meets code and/or specification requirements. Also, all the claims in the procedure are demonstrated and this evidence is included as part of the TJ document.

TJs normally contain several sizing accuracy studies. A sizing accuracy study is traditionally included for all reported variables that have tolerance requirements stated in a specification or by code. Sizing accuracy results are directly linked to calibration curve design.

Parametric studies are performed to evaluate how significant a parameter causes variation in the inspection results as well as to learn the parameter's boundary values for the inspection.

TJs normally contain modelling results as well as real inspection data. A PoD study is normally only composed of real inspection data. With the advancement of computers and the availability of modelling, MAPoD is becoming increasingly important.

The PoD study is considered to be an essential component of a TJ, but it is also the most complex of the studies to evaluate properly. A higher understanding of statistical and experimental techniques is normally required to perform a good PoD study. This knowledge is not commonplace in the NDT industry and it results in many PoD studies being weak or poor. The PoD mathematical model was derived from a binomial PoD model. PoD models have evolved and currently there are several mathematical PoD models available.

3. Evolution of the PoD Model

3.1 The Beginning of the NDT PoD Model

During the days of the space race, evaluating an inspection system's detection capabilities became very important.

Performing a full PoD was not feasible due to the availability of samples, computation requirements and costing. NASA created the first methodology for demonstrating an inspection system's capabilities. This was the NASA 29 of 29 test. The NASA 29 of 29 test evaluates a point of the traditional PoD curve using a binomial probability model and the exact Clopper-Pearson confidence interval.

The probability is measured using the following *[2]*: $$P = \frac{X}{N}$$

Where: $P = Experimental\ Proportion,\ N = Number\ of\ Trials\ \&\ X = Number\ of\ Successes$

The upper confidence limit is defined as: The upper limit P_u such that there is only a $(1 - \alpha)$ probability that the true proportion is $P' > P_u$.

The lower confidence limit is defined as: The lower limit P_l such that there is only a $(1 - \alpha)$ probability that the true proportion is $P' < P_l$.

Where, P' is the true proportion P_u and P_l are the upper and lower proportion confidence limits.

The upper and lower confidence limits are evaluated using the following steps:

- Calculate the degrees of freedom (v_1, v_2)

  Lower limit: $v_1 = 2(X + 1)\ \&\ v_2 = 2X$

  Upper limit: $v_1 = 2(N - X + 1)\ \&\ v_2 = 2(N - X)$
- State/specify the α limit required.
- Evaluate the appropriate value of the F-distribution $\left(F_\alpha(v_1, v_2)\right)$.
- The exact Clopper-Pearson confidence intervals are calculated using the following:

  For the lower limit:

  $$P_l = \frac{X}{X + (N - X + 1)F_\alpha(v_1 . v_2)}$$

  For the upper limit:

  $$P_u = \frac{(X + 1)F_\alpha(v_1, v_2)}{(N - X) + (X + 1)F_\alpha(v_1, v_2)}$$

3.2 Current Binary PoD Models [1]

A binomial PoD model uses 0 for a missed indication and a 1 for a hit or detection. This is a 0 or 1 based model called a binary PoD model.

PoD function is related to the binary data using Generalized Linear Models (GLM). Using a GLM, first a linear regression model is used to define the initial estimate $(f(a_i))$ of the binary response of different crack sizes. This estimate is in the form of Equation (1):

$$f(a_i) = \beta_0 + \beta_1 a_i \qquad \text{Equation (1)}$$

However, this initial estimation cannot be used to relate the indication sizes to the PoD model. The reason is that using the linear regression model $f(a_i)$ varies continuously with the indication size. This is not accurate since the data nature is binary and bounded. To overcome this problem, the GLM links the estimated values of the binary data to explanatory variables through the probability of detection, which varies continuously from 0 to 1.

Four common binary link functions, Logit, Probit, LogLog and cLogLog *[3]* are listed below:

Logit	$f(a_i) = \ln(p(a_i)/(1 - p(a_i))$	Equation (2)
Probit	$f(a_i) = \Phi^{-1}(p(a_i))$	Equation (3)
LogLog	$f(a_i) = \ln(-\ln(1 - p(a_i)))$	Equation (4)
cLogLog	$f(a_i) = -\ln(-\ln p(a_i))$	Equation (5)

Once any of the above options are used the values of β_0 and β_1 should be estimated properly to reach the maximum likelihood of the model results. The idea is to select model parameter estimates such that the likelihood is maximized based on the model, given the actual observed inspection data.

For hit/miss testing, the likelihood of P, based on a set of observations is:

$$L(\theta) = \prod_{i=1}^{n} P_i^{\,x_i} (1 - P_i)^{1 - x_i} \qquad \text{Equation (6)}$$

Where $\theta = \{a_1, a_2, a_3, \dots, a_n\}$ is a set of data, P_i, PoD(a), is the probability of detection of crack size a_i, and x_i is the inspection outcome, 0 for miss, 1 for hit.

3.3 Other PoD Models [1] [3]

There are other PoD models that use the inspection amplitude instead of a binary hit/miss detection. Amplitude response data has more information that can be applied to the PoD model development process. This was the initial reason for the development of an amplitude response PoD model, or as it is more commonly known, the $\hat{a}\ vs\ a$ PoD model *[1]*. Amplitude based PoDs are very powerful models due to the increase in information.

The relationship between â vs. a can be expressed:

$$y = \beta_0 + \beta_1 x + \varepsilon \qquad\qquad z = \frac{[y - (\beta_0 + \beta_1 x)]}{\delta}$$

Let the standard normal PDF (Probability Density Function) be represented as:

$$\phi(z) = \frac{1}{\sqrt{2\pi}} e^{-(\frac{z^2}{2})} \qquad \text{Equation (7)}$$

Let the standard normal survivor function be represented as:

$$Q(z) = \int_{z}^{\infty} \phi(\varphi) d\varphi \qquad \text{Equation (8)}$$

The PoD function simplifies to:

$$PoD(a) = 1 - Q\left[\frac{\log a - \mu}{\sigma}\right] \qquad \text{Equation (9)}$$

The likelihood function has the following form when accounting for censored data (misses and saturated data samples):

$$L(\theta) = \prod_{i=1}^{m} (1 - Q(z_{th})) \prod_{i=m+1}^{m+r} \frac{1}{\delta} \phi(z) \prod_{i=m+r+1}^{n} Q(z_{sat}) \qquad \text{Equation (10)}$$

Both the binary and $\hat{a}$ vs a models require a large set of sample points. With the availability of modelling software, there is a large desire to supplement the required number of samples needed to evaluate a PoD model.

3.4 Model Assisted Probability of Detection (MAPoD)

There are several methods available to join modelling with the PoD mathematical models. This enables building a PoD model using less real physical data. The two most common methods involve statistical and/or finite element methods. *[1]*

A statistical build MAPoD relies almost entirely on real samples that are scanned with repeated trials.

Building a statistically designed MAPoD begins with a statistical investigation phase. The first test to perform is the evaluation of each target type to determine if they are Independent and Identically Distributed (IID). The IID test is the most common step that is skipped or missed during a good PoD development. The IID test is not unique for MAPoD, but it should be applied for all PoDs. If different defect mechanisms are not identically distributed they will require separation during the evaluation of the model parameters. All targets that are not IID should have a separate qualification that includes a PoD model for each mechanism. If different target types have the same distribution response, that is if they are IID, then they can be grouped into the same PoD.

Most initial development has a small set of targets available. This is a set of 5 targets as a minimum, but 9+ targets are still recommended. This set of targets can be a simple feasibility set that shows the effect of a parameter, or enables building a calibration curve. This set of indications is scanned with repeated trials 30 to 35 times. This enables building response distribution models for each target in the small set. Once the distribution parameters for each target is available a function can be fit to the model parameters.

With the distribution parameters across the range of targets a random sample of points can be generated. Each set of points is generated using the parameters from the function that was generated using the statistical evaluation of the real target response data. All generated points should be weighted with regards to their distribution type, as well as the distribution parameters. This ensures a generated set of points will represent the possible inspection data.

Each set of points is inserted into the PoD parameter generating algorithm and the final model parameters are saved. The generation of points is repeated and the PoD parameters are solved. This point generating and PoD solving is repeated for many trials. The number of trials can be in excess of 10000+ iterations.

As the iterations are completed, each model parameter is plotted with regards to the other model parameters. Each pair of parameters has a likelihood from the generated set of points. The likelihoods are compared and the one model with the highest likelihood is used to represent all the possible points. This point is the generated set of data that represents all the possible data with the highest likelihood and this single set of data represents the best fit for the entire data range that is possible.

Figure 3 shows a MAPoD set of generated points and Figure 4 shows the $\beta_0\beta_1$ solution field and the generated set with the maximum likelihood.

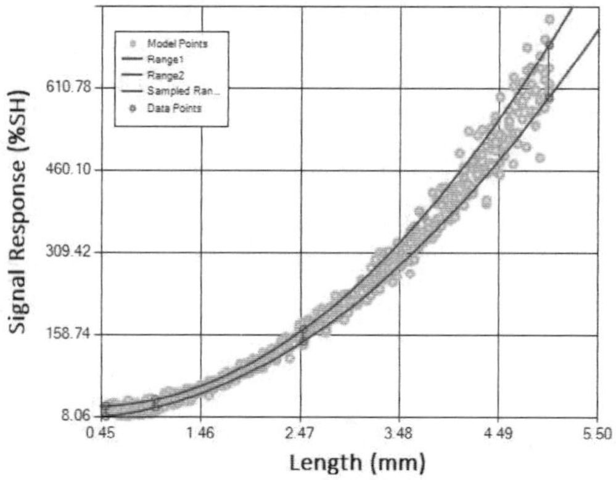

Figure 3 – Points used for MAPoD showing upper and lower bounds *[1]*

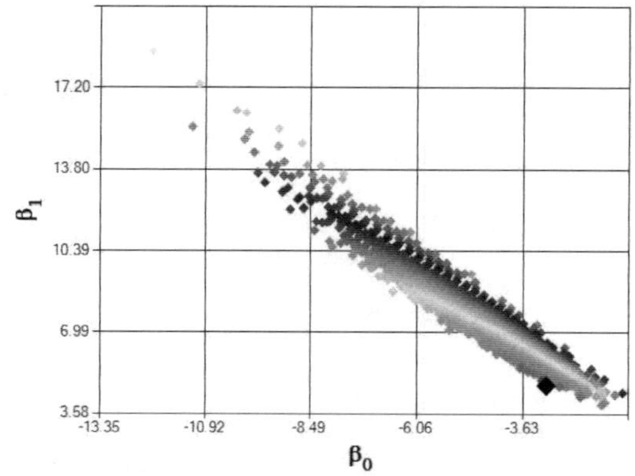

Figure 4 – Beta field showing model convergence and the point of maximum likelihood*[1]*

4. Building a PoD Model

4.1 PoD Model Fit Evaluation [1]

In statistics, many tests are available to measure how well a model fits the data it was constructed from. This is called measures of goodness of fit. The most common measures of model fit for PoD models are the AIC and/or the BIC.

AIC value is calculated using the following formula:

$$AIC = 2k - \ln(L) \qquad \text{Equation (11)}$$

Where k is the number of estimated parameters and L is the maximum likelihood of the model. AIC value indicates an estimation of the amount of information lost using a given statistical model. Therefore, lower AIC values indicate a more accurate model for a given data set.

BIC is another measure of the quality of a model to be used for a data set. BIC is very similar to AIC, with the difference of consideration of the number of parameters used for building the model into account, in addition to the data used for the model. Commonly, the likelihood of the model can be increased if more parameters are added to the model relative to the number of available data points, but this can have drawbacks such as over fitting. Therefore, increasing the relative number of parameters to the number of data points available can have penalties, in terms of measuring the quality of a model.

This penalty is considered in BIC as the following formula:

$$BIC = -2\ln(L) + k\ln(n) \qquad \text{Equation (12)}$$

4.2 PoD Model Selection

The goal of selecting the PoD model should be based on the model representing the inspection situation to the highest accuracy.

First, the model selection process is started by choosing the main model type. If the data is amplitude response based, the â vs. a model should be used. If the data is not amplitude based then a binary or hit/miss model should be used. Whenever there is any uncertainty if the data is amplitude based, the model should be selected based on the Akaike Information Criterion (AIC) and the Bayesian Information Criterion (BIC). The AIC and BIC values represent how well a set of data fits a model. When two models have the same AIC/BIC values, the logic of Occam's razor should be applied. Occam's razor states that the simplest explanation is likely the most correct explanation. Finally, if the data has a strong amplitude response, always default to the â vs. a model.

For an â vs. a model, if the target response is normally distributed, use the data type that produces the lowest AIC/BIC values excluding ln(a) & ln(â). If two data types have the same AIC/BIC values, apply the logic of Occam's razor. If the target response is not normally distributed, the best data type to use is ln(a) & ln(â). Whenever there is any uncertainty if the target response is normally distributed the data type should be selected using the lowest AIC/BIC values. When two models have the same AIC/BIC values, the logic of Occam's razor should be applied.

For a binary or hit/miss model, if the target response is normally distributed, the Logit or Probit model should be selected based on the model that produces the lowest AIC/BIC values. As stated above, when two models have the same AIC/BIC values the logic of Occam's razor should be applied. Whenever there is any uncertainty about the distribution type, the binary model should be selected using the AIC/BIC that produces the lowest value and also respecting Occam's razor for models with the same AIC/BIC values.

If the target set is not normally distributed, right skewed data (biased with too many large targets) should use the LogLog model. Left skewed data should use the cLogLog model. Whenever unsure how the data is skewed the LogLog and cLogLog models should be selected based on the model that produces the lowest AIC/BIC values. As with all model selection, when two models have the same AIC/BIC value, the model should be selected using the logic of Occam's razor (LogLog model).

4.3 PoD Model Confidence Level Evaluation [1]

There are several mathematical methods available to calculate a confidence interval. The most common are the method of Chen & Iles, Wald, Wilson-Score & Clopper-Pearson methods. Not every confidence interval calculation is appropriate for all situations. This is similar to the idea of model fit using the AIC's & BICs.

With model fitting contexts, the Wald interval about $PoD(a_i)$ can be used because an ideal $PoD(a_i)$ is modelled and the fit is computed from the ideal.

For MAPoD, the Wald test should be used, while for real PoD data, the Wilson interval will handle the variation of the data better. The Wilson-Score method has less impact from outliers as well. When plotting observed $PoD(a_i)$s, the Wilson interval should be applied.

For the binary PoD models, the Wald is the most common method due to its ease in calculation, but this is not always the method that produces the highest accuracy. The method that represents the data with the highest accuracy, with regards to the inspection requirements should be the method chosen.

With less than 100 samples, the Wald method can be either very liberal or very conservative in its approximation. For larger sample sets and probabilities between 10% and 90%, the Wald method will work with a high degree of accuracy.

For sample sets less than 100 samples, the Wilson-Score method can be used with a much higher degree of accuracy. This does not mean that the Wilson-Score method should be used for all cases.

For small sample sets (≤25 samples) and probabilities <10% and >90% the exact method (Clopper-Pearson) will produce the highest accuracy. For larger sample sets (>100 samples) and probabilities <1% and >99%, the exact method will produce <2% error. The exact method also works well with extremely small or large probabilities.

For situations where no error can be tolerated, the exact method should always be used. This requires the proper number of samples. The exact method is guaranteed to never fall below a nominal level. As the accuracy is increased, the required number of samples is increased as well.

5. Optimization of a PoD Model

A PoD model is directly linked to the noise study and the Probability of False Call (PoFC) study.

For amplitude and binary based PoD models, there is a threshold when a signal is classified as a defect or not. The threshold is the link value between the noise study, PoD and the PoFC.

By varying the threshold an inspection procedure can have its PoD maximized, while keeping the PoFC as low as possible. As the threshold is lowered, more signals are classified as defects. Since the noise response is fixed relative to the threshold, signals that are noise will be falsely reported as defect signals. This is one of the ideal ways a PoD can be used to help optimize an inspection technique.

Any indication response at or above 40% SH will be reported as detected or as hits (1 for a binary PoD).

An example PoD model is shown in Figure 5. This PoD model was constructed using a threshold that was set to 40% SH.

Figure 6 illustrates the effect on the PoD curve when varying the threshold from 15% to 80%. Each colour in the graph represents a new PoD model that was generated using a specific threshold value as listed in the graph's legend.

Figure 7 shows the size of indication that meets the required a90/95 qualification level. As the threshold is reduced, the size of the target that meets the a90/95 level also reduces.

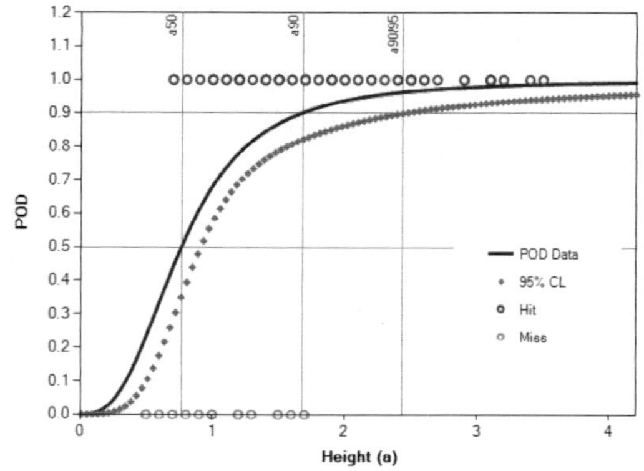

Figure 5 – Example PoD model with 40% SH threshold

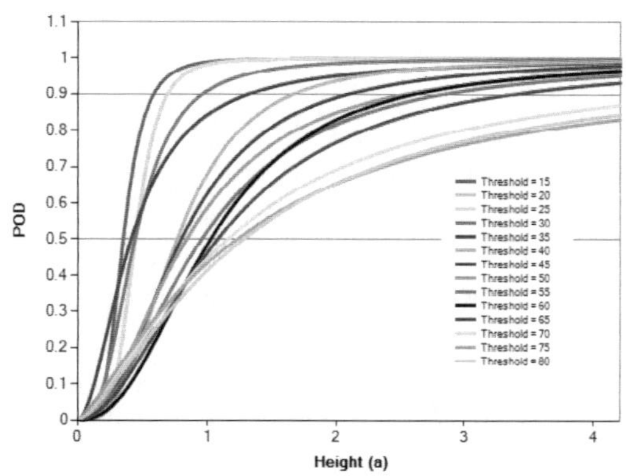

Figure 6 – All PoD models from various thresholds

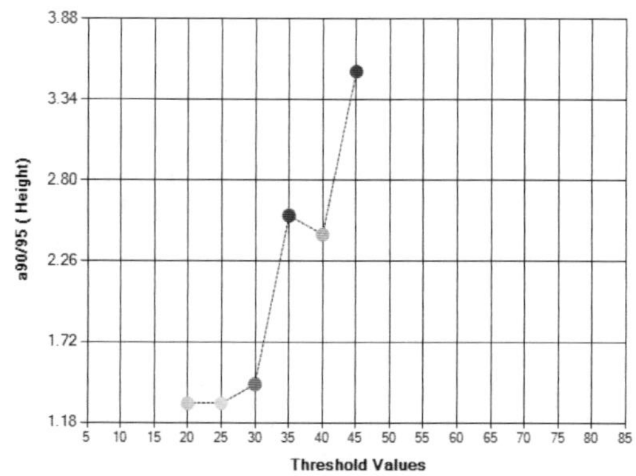

Figure 7 – Threshold vs. a90/95

As seen in Figure 7, this example shows that the lowest threshold of 25% SH has the smallest a90/95 value. 20% SH also has the same value for a90/95 and if choosing between 35% and 20%, using 25% would be the better choice, since it would produce a lower PoFC.

A value of 30% SH was selected for this procedure development, since the a90/95 value at 30% SH met the inspection requirements. Lowering the threshold would enable reporting smaller indications, but this would also increase the PoFC levels. Since the inspection requirement was met at the higher level, it was the best choice for this PoD model and procedure development.

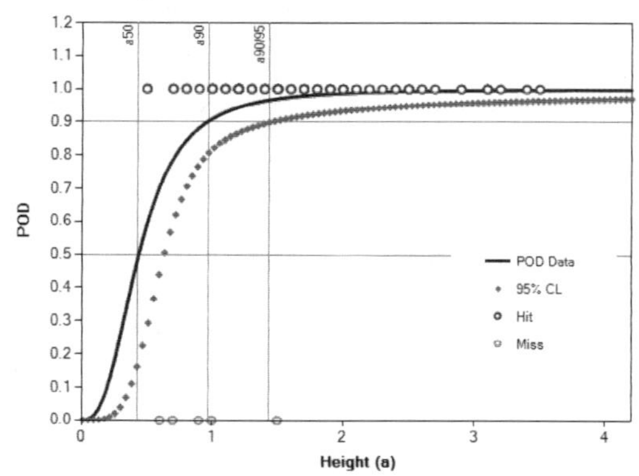

Figure 8 – Final optimized PoD model

Figure 8 shows the final PoD model built using 30% SH.

It is common for UT procedures to set a threshold at or around 40% Screen Height (SH). All signals at or above 40% SH will be reported as defect signals. Amplitude based PoD models can have greatly different threshold values and levels. For example, ET inspections can have voltage as the threshold measurement. 1.0 Volts peak-to-peak can be an example threshold level.

A threshold should be kept as high as possible and not reduced unless there are specific gains from the effect on the PoD levels. This always keeps the PoFC as low as possible, but it enables an inspection to possibly meet an inspection requirement a higher threshold might not be capable of reaching.

Most procedures do not have this optimization performed for various reasons. As the inspection industry is being driven to detect smaller indications, the advantages for optimizing the PoD and PoFC are becoming increasingly important. New inspection developments constantly demand the ability to detect smaller indications or to detect indictations with higher confidence and repeatability.

6. What makes a good PoD Study?

A good PoD study has the following characteristics:

- Uses proper inspection techniques that represent the insepection being evaluated as close as possible. If the technique being evaluated is not consistently used, the resulting PoD model will not represent the inspection it is being constructed for.
- Model selection is based on model fit paramters, such as AIC or BIC values.
- It is constructed using a fair statistical set of smaples that cover the entire variable range.
- It is constructed using IID samples only.
- The influence and/or significance of variables is known. If more than one factor directly influences the signal response, each factor should be covered using the appropriate range of samples.

A good PoD will be constructed fairly. Fair PoD evaluations can be compared with good accuracy. Comparing fair PoDs makes sense, while comparing two randomly generated PoDs does not make any sense at all.

Comparing the final values from the PoD model, such as the a50, a90 and/or the a90/95, can reveal important improvements during a high level inspection development. Comparing a PoD level is a great method to help quantify a change to a technique. This evaluation can enable a quantification for any inspection improvements.

7. References

1. Wright, Michael, 'Developing Qualified NDT Procedures and the Technical Justification Process', NDT Textbook, Eclipse Scientific, October 2016.

2. Rummel, Ward D. 'Recommended practice for demonstration of non-destructive evaluation reliability in aircraft parts.', Materials Evaluation, Vol. 40, 1982.

3. MIL-STD-1823A. s.l. : DoD, 2008.

CANDU Owners Group Inc.

"Excellence Through Collaboration"

Excellence through Collaboration

CANDU Inspection Qualification Bureau (CIQB)

The application of the European Qualification Method to Inspection of CANDU Nuclear components

Jeff Weed, Program Manager, CIQB, COG
KiSang Jang, Senior NDE Specialist, CIQB, COG

CINDE NDT in Canada Conference 2016 and
6th CANDU In Service Inspection Workshop

November 15 - 17, Holiday INN, Burlington, Ontario, Canada

What is Inspection Qualification?

- The standard governing periodic inspection of CANDU plants, CAN/CSA N285.4 requires "Performance Demonstration" of *Inspection Procedures, Inspection Equipment* and *Personnel*
- A pilot study performed by OPG considered two candidate inspection qualification schemes
  - ASME Section XI, Appendix VIII
  - European Network for Inspection and Qualification (ENIQ)
- In consultation with EPRI (Electric Power Research Institute) in the US, OPG selected the ENIQ methodology

Why chose ENIQ?

- The ASME scheme was solely focused on ultrasonic inspection, whereas the CANDU requirement included any inspection method /technique
- The ASME scheme was predicated on light water reactor designs, i.e. Pressurized Water Reactor (PWR) and Boiling Water Reactor (BWR) designs, and did not accommodate key CANDU components such as the horizontal fuel channels and feeder pipes.
- The ASME approach is focused on Piping, Reactor Vessel and Coupling only.

COG

Why chose ENIQ?

- Following the success of the OPG pilot, seven CANDU utilities elected to collaborate to produce a common approach to inspection qualification. This group followed the lead of OPG and selected the ENIQ methodology and established the Inspection Qualification Joint Project (IQJP)
- In parallel with the COG project, the Canadian Nuclear Safety Commission (CNSC) conducted an independent study that also concluded the ENIQ methodology was appropriate for the CANDU design in terms of meeting regulatory needs

COG

What is ENIQ?

- The European Network for Inspection and Qualification (ENIQ) is a utility driven network for harmonizing the approach to qualification of Inspection Processes in Nuclear Facilities.
- ENIQ was established under the auspices of the European Union's Joint Research Centre (JRC) in Petten, the Netherlands.
- All European nuclear jurisdictions, with the exception of the former USSR, are represented.
- Both COG and EPRI participate in ENIQ, as non-voting members.

COG Official Use Only 5

How do ENIQ and ASME XI, VIII differ?

- Both ENIQ and ASME methodologies have the same intent, to ensure that inspection processes applied to nuclear power plants return correct results.
  - ASME Section XI, Appendix VIII is very prescriptive. It defines the components, number of physical test pieces, flaw types, the numbers of flawed and non-flawed specimen areas, the distribution and size of flaws and the personnel pass/fail criteria. Sizing accuracy is also defined.
  - The ENIQ approach defines how to undertake inspection qualification. It is a *non-prescriptive* approach and, therefore, may be applied to any component and any inspection method/technique.

COG Official Use Only 6

How is ENIQ structured? - 1

- The Steering Committee of ENIQ formed two sub-committees:
  - Task Group Qualification (TGQ) focused on issues related to qualification of inspection procedures and personnel. This is the topic of this presentation, and
  - Task Group Risk (TGR) focused on issues related to risk informed in-service inspection
- ENIQ has now been incorporated into the European Union's NUGENIA program for the provision of an overall administrative structure, but maintains its distinct ENIQ identity.

COG Official Use Only 7

How is ENIQ structured? - 2

- From its inception, ENIQ has maintained an open system with documentation freely available, now via the NUGENIA web site. The documentation includes:
  - Evolution of the process
  - Recommended practice documents (RPs)
  - Guidelines
  - Reports, special studies

COG Official Use Only 8

How is ENIQ structured? - 3

- ENIQ has five major elements –
  - A definition as to what the inspection must achieve - the Inspection Specification (IS)
  - An Inspection Procedure (IP)
  - A Technical Justification (TJ)
  - Training and Testing of Personnel using the qualified procedure.
  - An independent body to conduct the qualification process – Inspection Qualification Body (IQB).

The Inspection Specification (IS)

- The Inspection Specification (IS) is the Licensee's responsibility
- It provides all the information needed by an Inspection Service Provider (ISP) to design and operate an inspection system that meets the requirements.
- It provides information on the component - material, dimensions, target flaw size, access, hazards, time limits, site environment, inspection volume, inspection reliability, etc. Perhaps most importantly it describes *experienced and postulated degradation mechanisms*, the required *minimum detection limit* and *sizing accuracy* required.
- It is an *engineering document*. Its requirements are set out in engineering units, meaning it does not dictate the inspection method to be used - however, in many cases, the choice of inspection method is self-evident.

The Inspection Procedure (IP)

- The Inspection Procedure (IP) is the responsibility of the Inspection Service Provider.
- It provides the details of the inspection system and the inspection process.
- It includes the required general certification of the personnel (per CGSB) and the specific qualifications of the personnel (per the training documents); the equipment and instrumentation; calibration procedures and detailed inspection process, recording and reporting requirements
- It may also contain an abbreviated description of the component.

COG

The Technical Justification (TJ)

- The Technical Justification (TJ) is the responsibility of the Inspection Service Provider.
- It provides the evidence that the inspection procedure meets the stated requirements of the specification. Typical 'evidence' may include, but is not limited to -
  - Technical argument
  - Reference to similar applications
  - Operating experience
  - Computer based modelling
  - Tests on physical representations of the component, often with implanted 'real' examples of degradation/flaws.
  - In-plant performance
  - Reliability

COG

Training and Testing of Personnel

- It is essential for the ISP to provide training of the inspection personnel specific to the IP.
- Testing of the personnel can take several forms:
  - independent examination by the IQB
  - independent examination by the ISP (with external oversight)
  - a combination where the IQB provides oversight of the ISP examination process.
- Testing must be consistent with the role of the personnel and may include blind testing on physical test pieces used solely for testing purposes
- Where the role of the personnel is data analysis, it is usual to use plant recorded data as a test format.

Inspection Qualification Body - 1

- The Inspection Qualification Body (IQB)
  - Is responsible for the design and conduct of the qualification process.
- ENIQ describes three types of IQBs: -
  - Type 1: A QB which is an independent third party organization
  - Type 2: A QB which is an independent part of the utility, set up on a permanent or long-term basis
  - Type 3: An ad hoc QB set up for a specific qualification.

Inspection Qualification Body - 2

- The basic requirements of the QB are:
  - Access to experts to review the documents provided by the ISP.
  - Maintenance of a Qualification Dossier to store all relevant documents, records and communications specific to a particular qualification.
  - A quality assurance program.
  - Testing of Personnel or the overview of testing

CANDU Inspection Qualification Bureau

- One of the principle outputs of the COG Inspection Qualification Joint Project was establishment of a body within COG to administer the inspection qualification process - the CANDU Inspection Qualification Bureau (CIQB)

- The CIQB is a hybrid of ENIQ Type 1 (permanent) and Type 3 (temporary) IQB, with a small core of managers within COG who contract-out to appropriate experts the task of examining the documents provided by the ISPs. Thus the experts selected vary according to the particular qualification at hand

Conclusions - 1

- The flexibility of the ENIQ inspection qualification methodology is both appropriate and beneficial for the CANDU plant design.
- The open availability of ENIQ reports, recommended practice documents and special studies has enabled the establishment of a successful inspection qualification process as administered by the CIQB.
- The Canadian nuclear regulator has accepted the CIQB process as meeting regulatory needs.
- The structure of the CIQB (Type 1 - Type 3 hybrid) has worked very well and has resulted in an efficient and cost effective inspection qualification process.

COG Official Use Only 17

Conclusions - 2

- The ENIQ process was developed to support the qualification of inspections in the Nuclear Industry in Europe and has proven itself to be adaptable to a variety of reactor configurations.
- The process provides considerable flexibility to adaptation to any component.
- While the focus to date has been on nuclear components, the process approach is equally applicable in petrochemical, gas, aerospace or any application when consequences of component failure are high and corresponding high confidence in inspection results are essential.

COG Official Use Only 18

CANDU Excellence through Collaboration

COG Official Use Only

EPRI | ELECTRIC POWER RESEARCH INSTITUTE

20 Years of Performance Demonstration Activities in the United States

"Where we have been and where we are going"

Carl Latiolais
Senior Program Manager NDE Reliability

6th International CANDU ISI Workshop/NDT in Canada 2016 Conference

November 2016

What is Performance Demonstration?

- Performance demonstration is the proof that NDE procedures, equipment, and personnel are capable of achieving a specified level of performance
  - Flaw detection reliability
  - Flaw sizing accuracy
  - Low number of "false calls"
- 'Performance demonstration' and 'qualification' are similar
  - Performance demonstration is the <u>practical</u> part of qualification

EPRI | ELECTRIC POWER RESEARCH INSTITUTE

Worldwide Demand for Performance Demonstration

- The demand for rigorous performance demonstration is increasing
  - New regulations and requirements
  - Long term operation
  - Expansion of the nuclear fleet
  - Increasing need for statistical measures of NDE reliability
- Impact of the demand
  - Utilities and NDE vendors are scheduling well in advance
  - Mockup manufacturers are operating at full capacity
- Expect this trend to continue

Generic Process for Developing New PD Programs

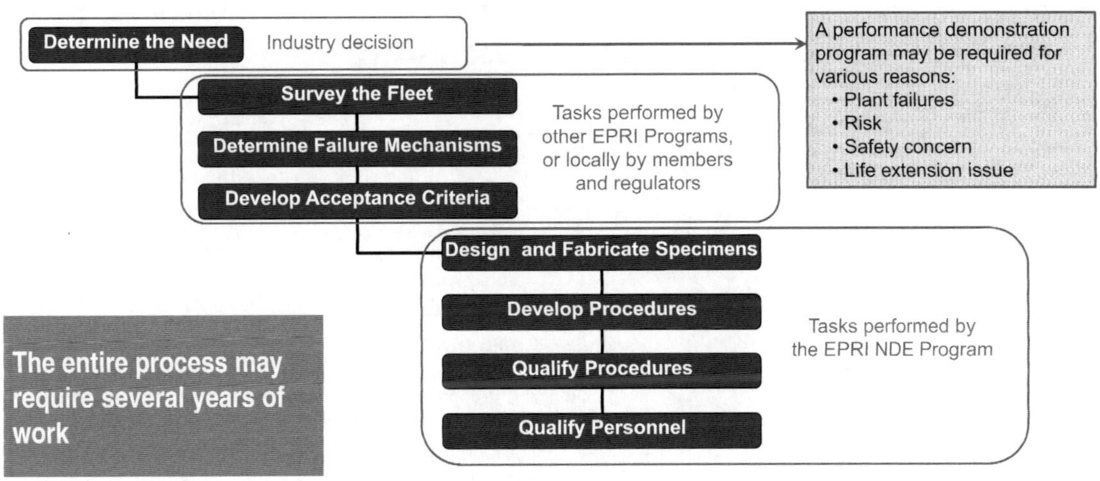

PD Performance demonstration

Generic Process for Developing New PD Programs

- Examples of time tables for some of the major activities
  - Code (minimum of 1 year for code case approval)
  - Survey of configurations
    - Could be up to 2 years depending on outage schedules
    - Depends on population and amount of data available from previous examinations
  - Procedure development (could be up to 2 years)
    - Funding needed
    - Depends on the complexity of the examination
    - Samples needed for development
  - Qualifications
    - Procedure qualifications take 2-6 months
    - Personnel qualifications take about one week per person, per procedure

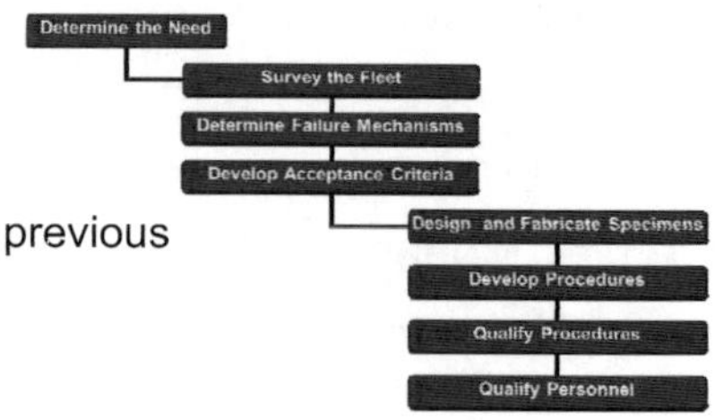

PD ***Performance demonstration***

EPRI | ELECTRIC POWER RESEARCH INSTITUTE

Types of Performance Demonstration

- Each country's requirements are different, but always it is some combination of two primary approaches:

Practical demonstration (physical trials) ⟷ **Technical justification** (theoretical argument)

USA: ASME Code

Europe: ENIQ

EPRI NDE Program is the original provider of performance demonstration services for compliance with ASME Code requirements for most of the primary pressure boundary

EPRI NDE Program has helped utilities in many nations, with both ASME and non-ASME requirements:

- Developing new performance demonstration programs, processes and organizations
- Providing justifications for applying EPRI's PD methods in the local regulatory environment

ASME *American Society of Mechanical Engineers*
ENIQ *European Network for Inspection Qualification*
PD *Performance demonstration*

EPRI | ELECTRIC POWER RESEARCH INSTITUTE

Scope of the EPRI NDE Performance Demonstration Team

- Qualifications of ultrasonic procedures, equipment and personnel
  - ASME Section XI, Appendix VIII
  - RPV upper head penetrations
  - International qualification support
  - Any future formal qualification
  - Mockup design and fabrication coordination for entire EPRI NDE group

Strategic Alliances

CRIEPI
(Japan Qualification Body)

KHNP Research
(Korea Qualification Body)

SQC
(Sweden Qualification Body)

QST
(Switzerland Qualification Body)

ASME ***American Society of Mechanical Engineers***
ENIQ ***European Network for Inspection Qualification***
PD ***Performance demonstration***

7

EPRI | ELECTRIC POWER RESEARCH INSTITUTE

Performance Demonstration Facility

8

EPRI | ELECTRIC POWER RESEARCH INSTITUTE

PD Program Review, Acceptance and Operation

- EPRI has been the PD leader since the 1970s
  - 1970s – Steam Generator Management Program NDE activities
  - 1980s – UT qualifications for BWR piping welds
    - Three-party agreement between NRC, BWR Owners Group and EPRI
  - 1990s – Administration of all qualifications in accordance with ASME Section XI, Appendix VIII, under the Performance Demonstration Initiative (PDI)
    - Pressure vessels, piping, bolting, nozzles
  - 2000s – Assistance to utilities worldwide in complying with local qualification regulations
  - 2010s – PWR RPV head penetration J-welds

Over 20,000 qualifications have been performed
Over 1,000 qualification candidates (personnel)
Over 100 procedures have been qualified

ASME	*American Society of Mechanical Engineers*
BWR	*Boiling Water Reactor*
NRC	*Nuclear Regulatory Commission*
PD	*Performance demonstration*
PDI	*Performance Demonstration Initiative*
PWR	*Pressurized water reactor*
RPV	*Reactor pressure vessel*
UT	*Ultrasonic examination*

9

EPRI ELECTRIC POWER RESEARCH INSTITUTE

ASME Code Section XI, Appendix VIII Defines Separate Demonstrations for:

- Piping Supplements
  - 2, Wrought austenitic piping welds
  - 3, Wrought ferritic piping welds
  - 9, Cast austenitic piping
  - 10, Bimetallic piping welds
  - 11, Overlaid wrought austenitic piping welds
- Reactor Vessel Supplements
  - 4, Clad/base metal interface of reactor vessel
  - 5, Nozzle examination from the outside surface
  - 6, Reactor vessel welds other than clad/base metal interface
  - 7, Nozzle examination from the inside surface
- Other
  - 8, Bolts and studs

And within each qualification there are:
- Procedure, personnel, and equipment qualifications
- Flaw detection, flaw length sizing, and flaw depth sizing qualifications

ASME American Society of Mechanical Engineers

10

EPRI ELECTRIC POWER RESEARCH INSTITUTE

Ultrasonic Examination Technique Categories

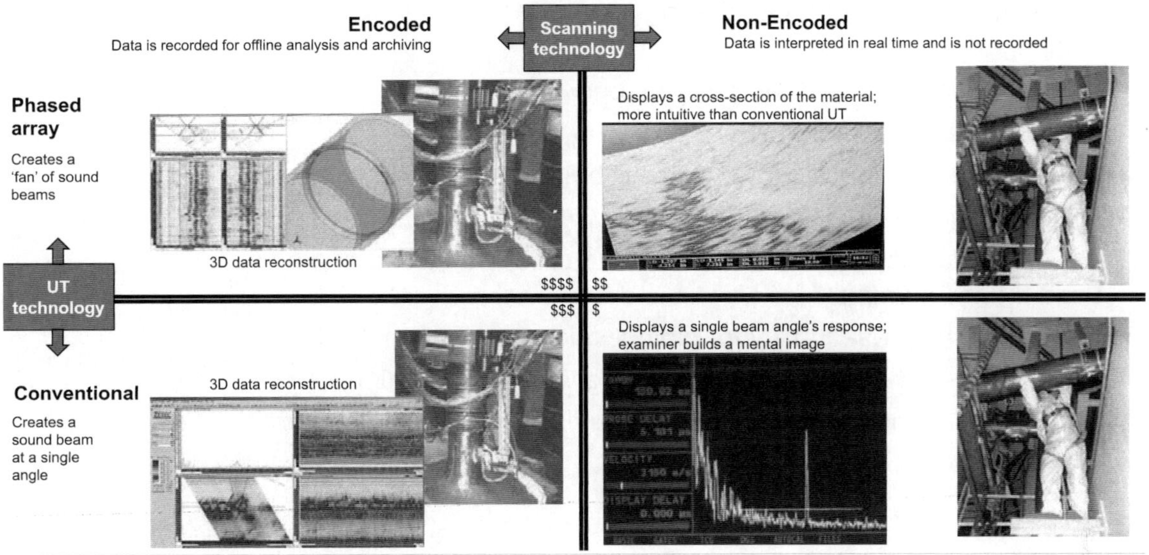

11

ASME Code Section XI, Appendix VIII

Qualified Non-encoded UT Procedures

- Generic procedures were developed and qualified for most manual examinations
  - Developed and agreed upon by entire industry
  - Cost effective
  - Qualifications are transferable from one organization to another
  - Almost all of the qualified manual examiners in the industry are qualified to these procedures

- PDI-UT-1: UT Ferritic Pipe Welds
- PDI-UT-2: UT Austenitic Pipe Welds
- PDI-UT-3: UT Through Wall Sizing in Pipe Welds
- PDI-UT-4: UT Bolts and Studs from the Bore
- PDI-UT-5: Straight Beam UT Bolts and Studs
- PDI-UT-6: UT Reactor Pressure Vessel Welds
- PDI-UT-7: Manual UT Through Wall and Length Sizing in Reactor Pressure Vessel Welds
- PDI-UT-8: UT Weld Overlaid Similar and Dissimilar Metal Welds
- PDI-UT-10: UT Dissimilar Metal Welds
- PDI-UT-11: UT Detection and Sizing Of Reactor Pressure Vessel Nozzle to Shell Welds and Nozzle Inner Radius

ASME ***American Society of Mechanical Engineers***
UT ***Ultrasonic examination***

12

ASME Code Section XI, Appendix VIII

Qualified Encoded UT Procedures

- Vessel, piping, vessel head penetrations
- Primarily vendor-owned
- Specialized and specific to equipment being used
- Expensive to qualify
- EPRI has developed and qualified some procedures that can be used by any member
  - Automated phased array for piping weld examination
  - Encoded manual phased array piping weld examination
  - Encoded manual phased array dissimilar metal weld examination

ASME ***American Society of Mechanical Engineers***
UT ***Ultrasonic examination***

13

EPRI | ELECTRIC POWER RESEARCH INSTITUTE

RPV Head Penetrations

- In USA UT of reactor vessel head penetration tubes must be performed with qualified procedures, personnel and equipment
  - Governed by NRC (10CFR50.55a), ASME Code (Code Case N-729-1), and the MRP Protocol
- EPRI Mockup Library
  - Control Rod Drive Mechanism (CRDM), Control Element Drive Mechanism (CEDM), and In-Core Instrumentation (ICI) Nozzles are represented
  - Can qualify rotating probes or blade probes
- Six vendors are fully qualified with both open-housing and blade UT probes
  - All USA configurations are qualified, with Technical Justifications for diameter tolerances
  - Exception: to date, no vendors have qualified for CE design ICI

ASME ***American Society of Mechanical Engineers***
CE ***Combustion Engineering***
ICI ***In-core instrumentation***
MRP ***Materials Reliability Program***
NRC ***Nuclear Regulatory Commission***
RPV ***Reactor pressure vessel***
UT ***Ultrasonic examination***

14

EPRI | ELECTRIC POWER RESEARCH INSTITUTE

New Development: Virtual Mockups

- Physical mockups are very expensive
- Virtual mockups will be less expensive
  - Add or delete flaws: "cut and paste"
  - Change flaw size or location
  - Insert simulated data
- This approach can generate an unlimited number of practice samples at a lower cost

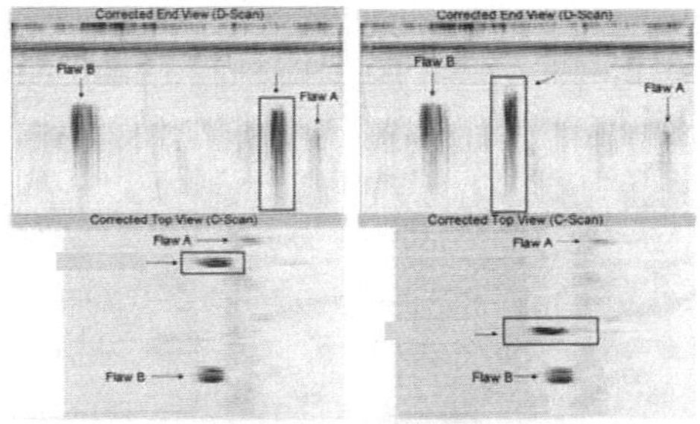

In the example above, one data file is real, the other is virtual.

Which one is real?

Virtual NDE Ultrasonic Data Players

- A windows-based software that allows users to simulate conditions of manual ultrasonic examinations by providing practice opportunities without requiring access to physical ultrasonic instruments or pipe specimens

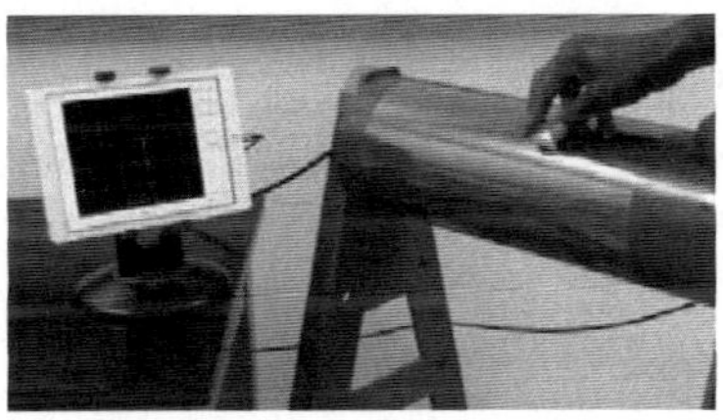

EPRI PD Program Summary

- Strong technical basis, developed over 20 years
- Proven and accepted in many nations
- The facility, inventory and staff expertise can be used to customize a demonstration to meet the requirements of any specific plant or regulatory requirement
- All fabrication and demonstration activities are performed in accordance with written instructions and procedures, which provides a clear QA audit path
- The program contributes significantly to the overall quality of ultrasonic examinations in the nuclear industry
  - Through extensive engagement with industry and regulators
  - Through outage support with urgent examination issues
  - Through research projects aimed at improving performance demonstration methods, advancing NDE technology, and improving NDE reliability

PD ***Performance demonstration***

17

EPRI ELECTRIC POWER RESEARCH INSTITUTE

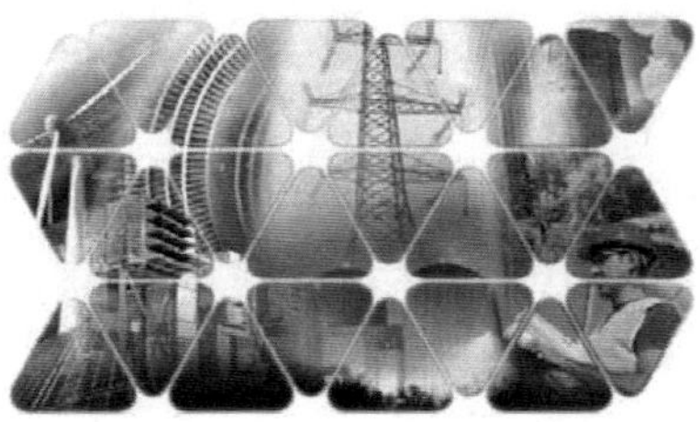

Together…Shaping the Future of Electricity

18

EPRI ELECTRIC POWER RESEARCH INSTITUTE

EPRI | ELECTRIC POWER RESEARCH INSTITUTE

NDE Reliability Update: A Guide for the Review and Implementation of Recent U.S. Nuclear Power Industry NDE Reliability Guidelines and Reports

3002009080

Carl Latiolais
Senior Program Manager
EPRI Performance Demonstration Program

6th International CANDU ISI Workshop
November 2016

Contact Information

- EPRI Project Manager: Ronnie Swain
  - Email: rswain@epri.com
  - Phone: +1-704-595-2514

EPRI | ELECTRIC POWER RESEARCH INSTITUTE

Problem Statement

- In recent years United States NDE operational experience (OE) has included cases of poor planning, execution, or data review practices that resulted in issues with the reliability or efficiency of NDE
- This has resulted in new guidelines being developed, governing the examination of dissimilar metal welds and, based on the strength of these offering, guidelines have been developed for planning and execution of all NDE in the plant
- So, how should nuclear power utilities, not subject to the same codes and regulations as the United States, consider implementation of recent EPRI NDE reliability guidelines and reports issued in response to U.S. industry NDE operating experience (OE) and lessons learned?

Background

- 2012 North Anna DM Weld Examination Event
  - PDI-qualified inspection was performed before pre-emptive mitigation
    - No indications reported in the pre-overlay exam
    - Subsequent machining revealed 5 deep, axial cracks
  - Plant Response
    - Performed Extent of Condition at the plant
    - Proceeded with weld overlay repair
    - Launched root cause evaluation
  - U.S. Industry Response
    - Incident was viewed by industry leadership as a "watershed" event
    - Industry executives called for the formation of an NDE Improvement Focus Group (NIFG) to review and improve upon all industry practices involved in the examination of DM welds

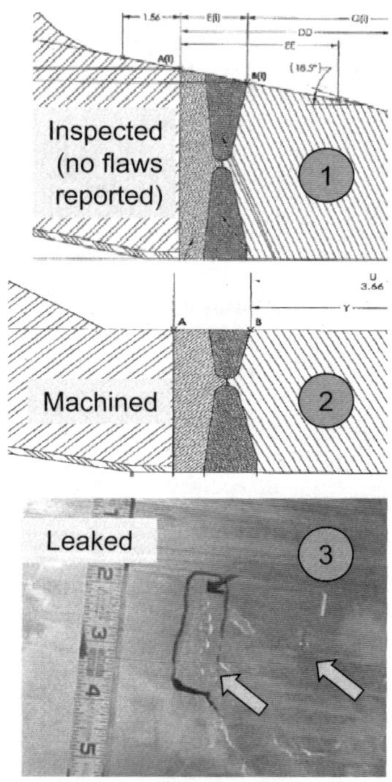

Total 5 cracks, all axial, all deep, **none detected**

1st Guideline Published by NIFG: Extent of Condition Process for DM welds

- *Nondestructive Evaluation: NDE Improvement Focus Group Extent of Condition Actions in Response to North Anna Dissimilar Metal Weld Operating Experience* (3002000041).
  - Published – Spring 2013
  - Provided a process for U.S. plants to categorize DM welds in terms of risk
    - The risk levels were determined by reviewing two key aspects
      - Whether the weld materials were considered susceptible to SCC (and had not been mitigated), and
      - Whether the weld had ever been examined by ASME Section XI, Appendix VIII qualified, encoded UT procedures, personnel, and equipment
  - For high risk welds, accelerated examination requirements were imposed
  - U.S. plants were required to use this process to categorize all their DM welds in the ISI program and report the results back to the NDE Action Plan Committee, along with examination plans for any high risk welds

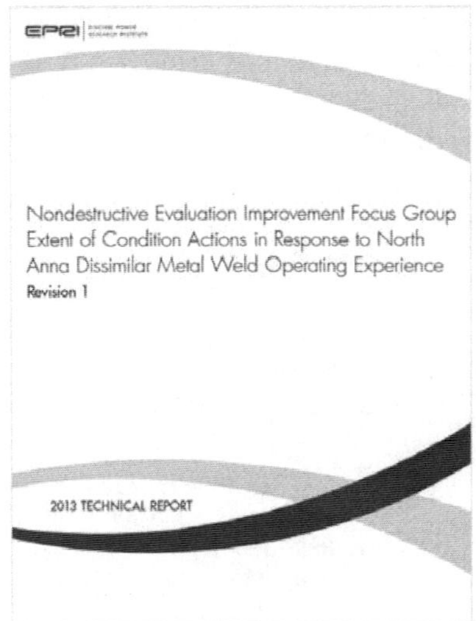

5

EPRI | ELECTRIC POWER RESEARCH INSTITUTE

2nd Guideline Published by NIFG: Revised Guideline for Conducting Examinations of DM Welds

- *Nondestructive Evaluation: Guideline for Conducting Examinations of Dissimilar Metal Welds, Revision 1* (3002000091)
  - Published in 2013
  - This report contains guidance on the planning, preparation, and execution of DM weld UT examinations
    - Although originally published in 2009, this report was thoroughly reviewed and revised by the NIFG to include the lessons learned from the North Anna OE
    - In addition, the NEI 03-08 implementation level of Revision 1 of this guideline was increased from *Good Practice* to *Needed*

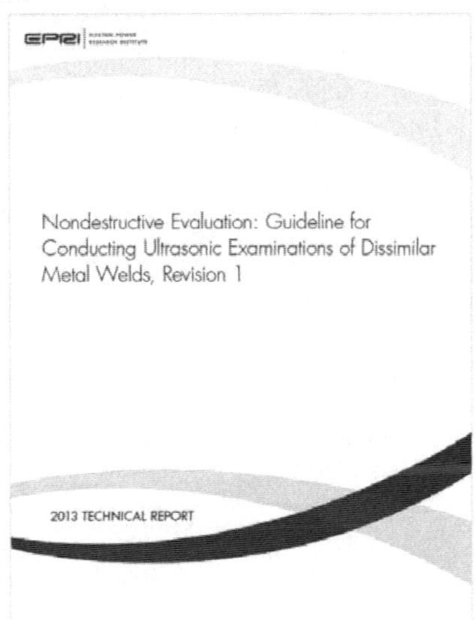

6

EPRI | ELECTRIC POWER RESEARCH INSTITUTE

3rd Guideline Published by NIFG: Revised Guideline for Conducting Examinations of DM Welds

- *Nondestructive Evaluation: Performance Demonstration Initiative (PDI) Guidance for Improved Reliability in Ultrasonic Examinations* (3002000204)
  - Published in 2013
  - This report contains revised versions of two pre-existing Performance Demonstration Initiative (PDI) documents
    - *Site-Specific Mockup Requirements for Dissimilar Metal Welds, Revision C,* and
    - *Guideline for Hands-On Practice (PDI-GL-001), Revision B*
      - Each of which was reviewed and revised by the NIFG, based on the North Anna OE and in an effort to improve the overall examination reliability associated with these products

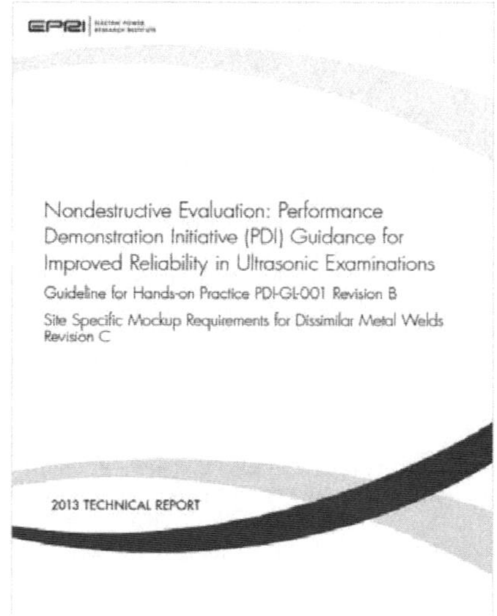

EPRI | ELECTRIC POWER RESEARCH INSTITUTE

4th Guideline Published by NIFG: Revised Guideline for Conducting Examinations of DM Welds

- *Nondestructive Evaluation: 2013 Team Scanning Assessment Conducted on Behalf of the NDE Integration Committee's NDE Improvement Focus Group* (3002002048)
  - Published in Fall of 2013
  - The NIFG developed a set of guidelines on the use of team scanning (Appendix B of *Guideline for Conducting Ultrasonic Examinations of Dissimilar Metal Welds, Revision 1*)
  - In an effort to test the reliability of team scanning, when employing these new guidelines, the NIFG conducted a blind testing experiment with three examiner teams using team scanning and following the new guidance
  - This report documents this experiment and the results and provides additional recommendations on the use of team scanning

EPRI | ELECTRIC POWER RESEARCH INSTITUTE

Additional Guideline Published by EPRI: Industry Best Practices for Performing Reliable NDE

- *Nondestructive Evaluation: Industry Best Practices for Performing Reliable NDE* (3002007329)
  - Published in 2016
  - Based on the reliability improvements for UT of DM welds, realized because of the NIFG products, industry executives requested that additional NDE guidance be developed to assist plant personnel with planning and executing <u>all</u> NDE with the same high standards of reliability
  - This report is formatted to allow quick reference to various aspects of the examination process, including:
    - Pre-Examination Preparation
    - Scheduling Examinations
    - NDE Staffing
    - NDE Staff Indoctrination
    - Examiner Preparation, Training, and Practice
    - Pre-Job Briefing
    - Use of Team Scanning
    - Oversight
    - Post-Job Debriefing
    - NDE Data Review
    - Examination / Outage Close-Out
  - In areas where more detail is warranted, the report provides references to other EPRI and industry documents containing additional information

9

EPRI | ELECTRIC POWER RESEARCH INSTITUTE

Additional Related Documents

- *NDE Reliability Update: A Guide for the Review and Implementation of Recent U.S. Nuclear Power Industry NDE Reliability Guidelines and Reports Product ID - 3002009080*
- Detailed descriptions of each reliability-related guideline
- A chapter acknowledging and discussing the fact that SCC susceptible materials may be identified differently, outside of the U.S.
- A discussion on the U.S. Materials Initiative *NEI 03-08*
  - Why it exists and how it affects U.S. utilities
  - Insights for international utilities considering how to utilize its guidance in their own plants
- Detailed instructions on how to access the EPRI Technical Reports and Guidelines discussed in this report

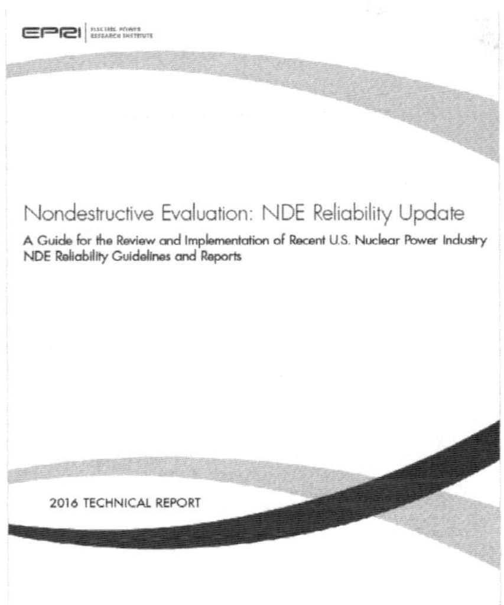

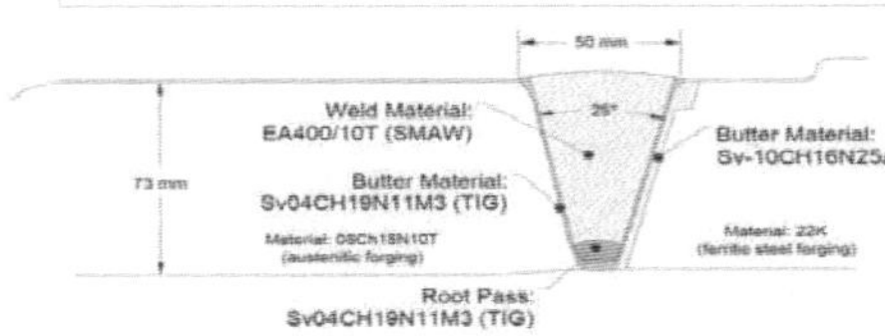

10

EPRI | ELECTRIC POWER RESEARCH INSTITUTE

Summary

- The U.S. nuclear industry has experienced some challenges with the consistency and reliability of NDE examinations, in recent years
- They have made a concerted effort to resolve these issues through development of guidelines and reports, all containing implementation requirements through their country-wide materials initiative *NEI 03-08*
- It has been recognized that international utilities are not confined by the same industry requirements and rules, but that plants with similar susceptibilities should consider how, and to what extent, to utilize these U.S. guidelines
- This report is intended to assist utilities, outside the U.S., to understand the guidelines that have been produced and in what ways they should consider applying them

EPRI | ELECTRIC POWER RESEARCH INSTITUTE

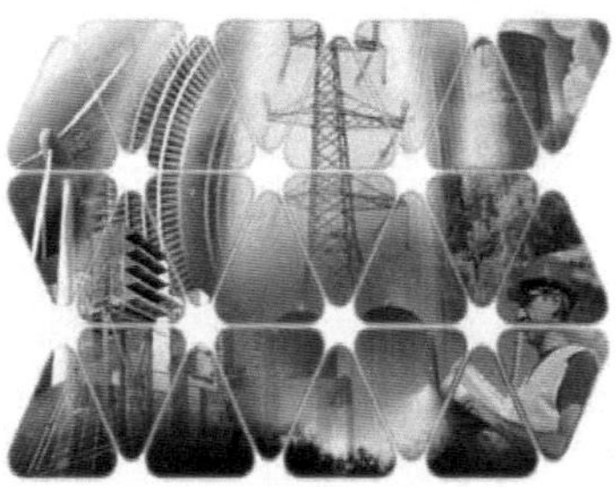

Together…Shaping the Future of Electricity

EPRI | ELECTRIC POWER RESEARCH INSTITUTE

Thick plate demonstration pieces with cracks on demand for PAUT scan to CSA W59-13 requirements

Viwek VAIDYA [1], Théo OUELLET [2], Jeremy CARIGNAN [3], Jocelyn MAGNY [3], Paul HOLLOWAY [4]

[1] Techno Vogue Inc., + 1 514 712 2801, e-mail: viwek@technovogue.com,
[2] CEGEP de Trois Rivières., + 1 819 376 1721 #2040, e-mail : theo.ouellet@cegeptr.qc.ca
[3] Centre de Métallurgie du Québec., +1 819 376-8707 #5026, e-mail :Jeremy.carignan@cegeptr.qc.ca,
[3] Centre de Métallurgie du Québec., + 1 819 380-6671, e-mail : Jocelyn.magny@cegeptr.qc.ca
[4] Holloway NDT & Engineering Inc., + 1 905 703 0201 , e-mail : paul@hollowayndt.com

Abstract

The CSA W59-13 standard permits the use of advanced ultrasonic testing methods such as PAUT on bridge structures, provided there is prior written agreement between the Engineer and the Contractor. Research and development was conducted at the Department of Metallurgy at the CEGEP de Trois Rivières, and Centre Métallurgie du Quebec, Canada to produce thick plate demonstration pieces with known discontinuities for PAUT assessment. The paper describes the findings of this experimental work.

Keywords: CSA W59-13, UT, PAUT, calibration blocks, DAC, TCG

1. Introduction

It is well known that technological advances take time to find their way into national and international standards, since most are developed through consensus between the members of the technical committees representing the stakeholders. At CSA, the technical committees have a balanced matrix between the User, Producer, Engineering Consultant & General Interest categories. It is important to note that members of these committees work benevolently to create these consensus standards sharing their expertise to protect public safety.

In 2013, the CSA W59-13 standard opened the door for the Canadian fabrication community to use advanced inspection techniques such as Phased Array Ultrasonic Testing (PAUT) and Computed Radiography (CR). In the last three years, some progress has been made to use CR inspection techniques for structural work replacing conventional film radiography. On the other hand, adoption of PAUT for mainstream structural work to CSA W59-13 requirements has not materialized due to several technical challenges.

This paper attempts to uncover these technical challenges within the current CSA W59-13 code for PAUT, and discusses experimental work and analyses required for general acceptance of PAUT for structural work in Canada.

2. Acknowledgments

The principal author wishes to thank all the sponsors for this unfunded R&D work. The experimental work completed in 2013 could not have continued without the generous help from Canadian Educational Institutions. Many thanks are due to the Department of Metallurgy and the professors at the CEGEP de Trois Riviers for providing the funding to purchase calibration blocks and contracted precision machining and inspection services. Thanks are also due to many students who participated in the preparation of the test plates for developing the "cracks on demand" methodology. Special thanks are due to Olympus - NDT division and their entire development and application team in Quebec City for providing valuable laboratory space and PAUT hardware and software for evaluating the welds. Leading up to the conference, the tireless and enthusiastic support of all co-authors and Level 3 experts from SGS Inc, Holloway NDT & Engineering Inc., and Torngats NDT was very much appreciated.

3. Background of PAUT

Ultrasonic examination uses the same principles as sonar used for the detection of submarines – a sound wave is emitted from a transmitter, bounces off any objects in its path and is reflected back to a receiver, somewhat similar to shining the beam of a torch at a mirror. Knowing the speed of sound in the material enables the distance of an object to be determined by measuring the time that elapses between the transmission of the sound pulse and detection of the echo. In welded components, the examination is generally performed by moving a small probe acting as both a transmitter and receiver over the reflector and displaying the echo on a screen. The oscillator sends pulses of electricity to a piezo-electric crystal, the pulse generator, embedded in the ultrasonic probe which causes it to vibrate at very high frequencies. Probes used for weld examination have frequencies generally between 2MHz and 5Mhz, the lower frequency probes being used for the examination of coarse grained and thicker materials, while the higher frequency probes are used for thinner materials and improved sensitivity to small defects.

The ultrasonic vibrations are transmitted into the material to be tested using a couplant such as water, oil, glycerine, and high-temperature paste. The better the surface finish the better the coupling – hence there is sometimes a requirement to grind the weld cap and adjacent areas of the welded joint smooth.

Once in the material, the vibrations travel in a predictable path until they encounter an obstruction or interface such as a line of slag, porosity or a crack. Depending on the angle at which the beam strikes the obstruction, some of the beam may be reflected back to the receiver in the probe. Here it vibrates a piezo-electric crystal and is converted back into electrical energy. This resultant signal is then amplified and displayed on a screen.

When the sound beam enters the object, it has a cross section approximately that of the transmitter. But like the beam of a torch, it will diverge. The beam divergence, in combination with beam scatter and absorption effects, effectively weakens the signal and is collectively referred to as attenuation. Attenuation effects need to be taken into account when evaluating signal amplitudes.

To obtain the strongest reflected signal, the beam should strike the feature ideally at 90° – flaws that lie parallel to the beam may be missed. This means that to examine a weld that may contain flaws in any number of orientations, a range of different angle probes and scanning patterns must be used. The CSA W59-13 code recommends shear wave probes with angles of 45°, 60° and 70°.

As with any measurement equipment, ultrasonic examination systems require calibration to confirm proper function and determine operating parameters. Calibration blocks must be made of the same material type as the test piece, and reference features such as side-drilled holes must be carefully controlled to ensure accurate and repeatable inspections. Ideally, a calibration block would be made from the actual test material to ensure the most accurate sound path and attenuation measurements possible. However, the practicality of this limited, so general material properties are assumed for most carbon and low-alloy steels, and thus only one calibration block is used for most cases.

CSA W59 assumes a fixed attenuation factor of 2dB per inch of sound path. This estimate is considered valid only while using a rectangular crystal over a specific range of sizes and at a nominal frequency of 2.25MHz. It is therefore understood that round crystals of different sizes and frequencies may differ from the assumed fixed attenuation factor and are generally not permitted.

The rapid progress of electronics and computing power has enabled complex methods of scanning and data processing to be developed. This has culminated in phased array ultrasonic testing (PAUT) which, as the name suggests, uses a transducer comprised of an array of individual elements. A single PAUT probe typically contains between 16 and 64 elements, although 256-element transducers are used for specialized applications. Each element in a phased array transducer can be pulsed independently in a set sequence or phase; the pulses of sound combining to produce a sound beam of a certain angle. By varying the time and pattern of the pulses, the angle and shape of the beam can be varied so that the beam can be steered electronically creating a sectorial scan or S-scan.

The benefits of this technology compared with conventional single transducer scanning are that the beam can be steered and focused with a single probe. Beam steering enables the beam to be swept through an object without moving the probe, the reflected data being processed to provide a visual image of a cone shaped slice through the object. Moving the probe enables a large number of slices to be assembled to provide a three dimensional image – a good example is the use in medical diagnostics to examine the functioning of the heart in real time.

For the non-destructive examination of welds, this ability to inspect a weld with multiple angle beams from a single probe means that the probability of detecting flaws is greatly increased. It is also possible to focus the beam electronically at multiple depths to improve the ability to accurately determine the size and position of weld flaws.

Methods developed now with PAUT enable flaw sizes to be determined with accuracy of better than ±1mm[1] used in determining fitness for purpose using fracture mechanics.

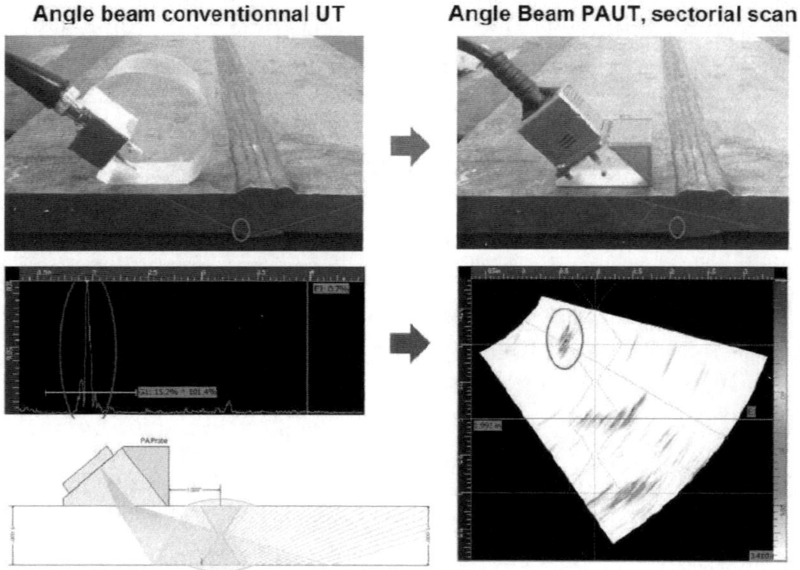

Figure 1

Figure 1 Standard 2.25 MHz W59-13 code probe versus a small 5 MHz PAUT probe scan plans for detecting a toe crack in 1 inch thick plate. PAUT sectorial scan displays the indication with the associated A Scan.

4. Experimental work to create " Cracks on Demand" methodology

In order to produce real cracks on demand, the author with the help of the technical staff at ESAB[2] designed a special FCAW wire such that, under restraint, the wire would produce cracking due to the higher levels of boron added to a base chemistry of a standard CSA E491T-9C or E71T-1C type wire. This was designed to produce a chemistry close to the C-Mn base metal of CSA 300W for the plates.

Earlier experimental work showed that it was difficult to produce cracks on demand in 0.75 inch thick plates using a U groove in the plate. This was directly related to the lack of restraint created on the weld bead due to the geometry of the preparation. Experimental work undertaken in this study was to create cracks on demand that were around 4 inches long and also create very small 3/16 inch long cracks in 0.50 inch thick plates.

a) Four inch long cracks

Under the guidance of professors at the CEGEP de Trois Rivieres, students succeeded to create these two categories of cracks while working over three work sessions in 2015 and 2016. Many geometries were tried, however, there was lack of consistency in producing the cracks on demand. Finally a 4 inch long machined groove approximately 0.50 inches deep in 1 inch thick

---

1 TWI reference

2 CSA W59 Advanced Inspection Methods, Vaidya et al .Cinde, NDT in Canada 2013, Calgary

plate with a 0.062” root opening produced the desired results. The preparation and presence of crack was confirmed with magnetic particle testing at the root and then by radiography of the finished weld.

b) Creation of 3/16 inch long cracks

There was a need to create very small cracks on demand, and this was achieved after number of trials on deep ground out grooves on 0.50 inch thick plates. The idea was to create a small tack weld that would crack on demand. After many trials with different groove dimensions and FCAW process, this approach was abandoned. Finally using the special Ti-B FCAW wire as filler with a GTAW process produced success. The results were confirmed with a PAUT scan. This technique now could produce cracks as small as 0.125” in test welds.

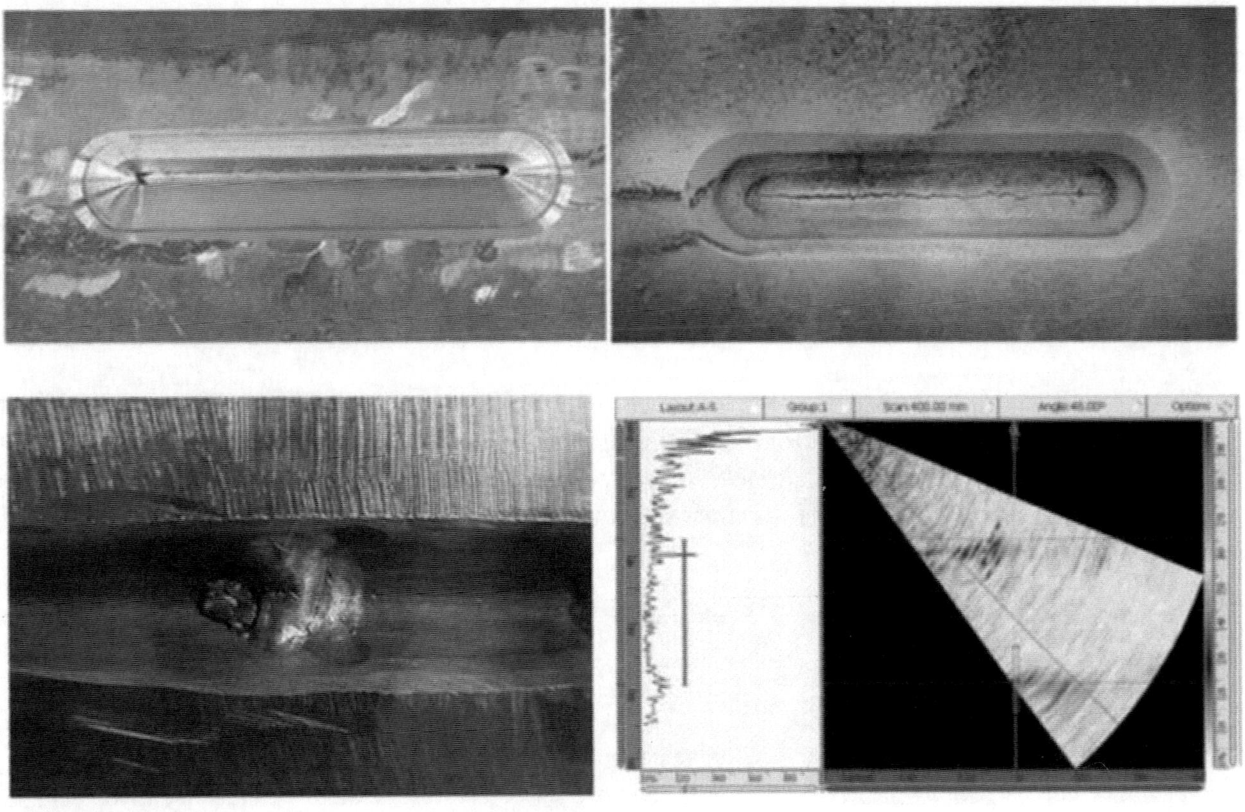

Figure 2 Methodology to create 4 inch long and 1/8 inch long cracks on demand

c) 3/4 inch thick plate welds with flaws for PAUT R&D

A large test plate 18” wide x 24” long x 0.75” thick was prepared in 300W material with a double V groove with 60 ° included angle, with 0.125” opening and a root face. Lack of Penetration was planned at 6.25” from side A, a round crater crack was created at 9.75 inches from side A and lack of penetration flaw approximately 4.5 inches was created towards the end of the side B. The plate was completely welded after flaw insertion by Submerged Arc welding at 500 amps, 34 volts at 10.2 inches of welding speed.

5. Results of the experimental work

The test plate was inspected with radiograpy, computed radiograpy and with PAUT encoded scan with two different laboratories. All techniques identified the areas and placement of the flaws in the experiment.

a) Radiographic testing

The test plate was inspected with radiography and detected three rejectable indications approximately at the location they were introduced. The report picked up a small 0.200" crack, at the location of the drilled hole to create a crater crack. No borderline indications could be produced acceptable to clause 11 but rejectable to clause 12. More work will be needed to create these in the future. The plate was rejected to both static and cyclic loading clauses.

b) CR testing by the CEGEP experimental

The results of CR testing in presented below. Once again, no code evaluation was performed on the CR radiographs represented below. This technique picked up 8 indications, some between 0.25" to 0.10" one at 5.6 inches length.

c) PAUT testing by lab 1 and 2

The plate was submitted to two PAUT inspectors for blind testing. The first lab produced a report indicating 5 areas of indications and the second lab produced 4 areas of indications. None of the PAUT results could be evaluated to CSA W59-13 requirements due to lack of suitable acceptable criteria for grading the indications, equivalent to, Table 11.3 and Table 12.5.

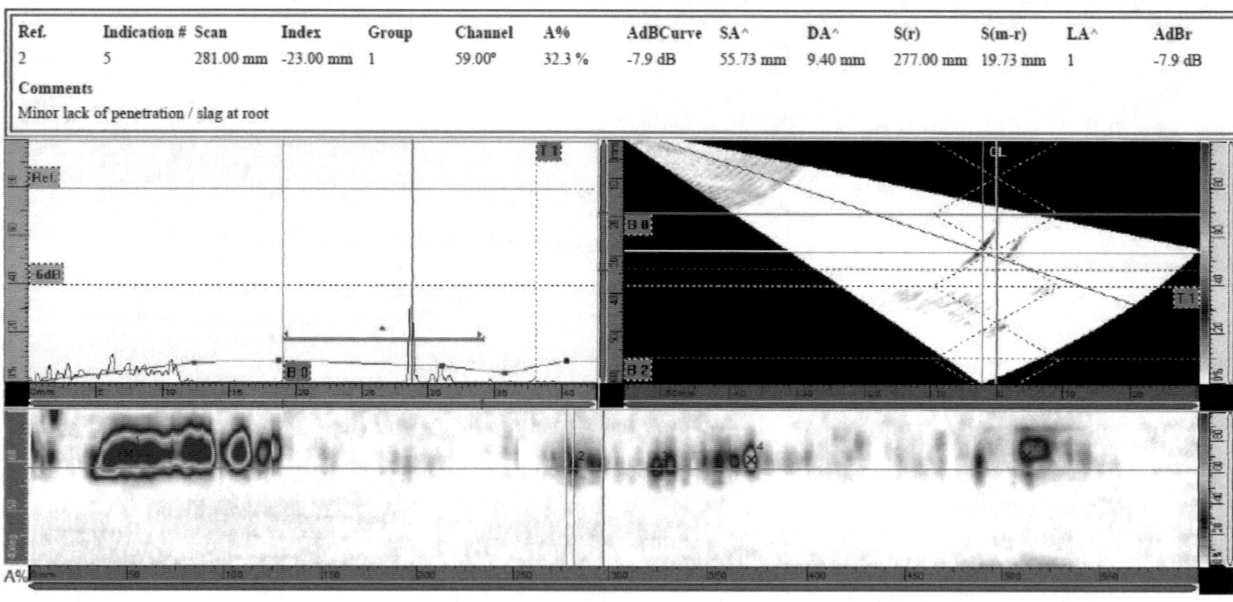

Figure 3 Encoded PAUT scan of the test plate identifying 5 indications

6. Discussion on CSA W59-13 – for application of PAUT

The authors reviewed the code[3] with the intent of applying manual, non-encoded PAUT for ultrasonic examination. The following paragraphs outline just some of the areas that would require revision to bring the code inline with modern practices; critical if PAUT phased array is to be accepted as a general inspection technique. These revisions would entail significant changes to the Clause 8.2 because the current procedures are based on outdated analog technology and methods.

a) Limitations based on analog equipment

The existing requirement for frequent 2-month or 40-hr periodic calibrations stems from the inherent electrical variances of older analog equipment. Digital equipment has long-since replaced analog equipment, yet the code still prescribes the same calibration and setup procedures in use since 1970[4]. The recommendation is to revise the code to accept only digital instruments, and to change the calibration procedures and intervals inline with modern standards.

b) Limitations of the 2.25 MHz large probe

The present practice involves using large, low frequency probes and a fixed attenuation estimate of 2 dB per inch. This process is an estimate and is not based on scientific study or physical laws of sound propagation. It was originally released as Appendix C in AWS D1.0 and D2.0 in 1969. The intent was to limit variability and attain consistent results by placing tight controls on equipment and processes, and to allow calibration using just the IIW-type block. The estimate is a poor representation of actual sound attenuation, requires the use of awkward large transducers, and requires adding high fixed scanning gains.

The following lists some of the drawbacks of the current 2.25 MHz transducer:

- Inconsistent coupling is common with large wedges
- Physical size limits access (especially in angled joints) and frequently prohibits evaluations on the 1st leg
- 2 dB per inch attenuation estimate is inaccurate given variations due to transducer bandwidths and sizes
- Fixed high scanning gains can produce excessive signal to noise ratios at long sound path distances
- Indication rating calculation is cumbersome to perform manually, or requires purchasing software options

The 2dB per inch assumption was based on a probe similar to a 20x22mm Krautkramer WB-series probe and is not typical of all probes within the allowable range.

[3] CSA W59-13 Welded Steel Construction (Metal Arc Welding)

[4] AWS D1.0 (1969) Appendix C – Ultrasonic Testing of Groove Welds

The mindset is "don't fix what ain't broke". Although W59 and D1.1 can be awkward to perform, the fact of the matter is that they are both extremely conservative and thus have a good track record for weld quality (consistent with most workmanship-based criteria). However, the system is indeed "broke" as it pertains to practicality and repeatability, resulting in numerous shortcuts taken in the field and a general lack of understanding of the limitations.

c) Other size probes and DAC/TCG calibration

It is time that the code be modified to allow use of more traditional, round probes of different sizes, and frequencies other than 2.25 MHz. When using these probes, a Distance Amplitude Correction (DAC) or Time Corrected Gain (TCG) calibration will be needed and the fixed 2dB/inch attenuation approach abandoned.

All of the following standards prescribe the use of a DAC for sensitivity calibration:

- API RP2X – Offshore Structural Fabrication
- ASME Sec. V
- ASTM E587 – Standard Practice for Ultrasonic Angle-Beam Contact Testing
- CSA Z662 – Pipelines (follows ASME guidelines)
- DNV Class Guideline 0051 – Non-destructive Testing
- EN 583-2 - Non-destructive testing, Ultrasonic examination, Part 2: Sensitivity and range setting
- ISO 17640 – UT of Welded Joints

Keep in mind that a CGSB Level 2 inspector performing inspections to W59 requires no code-specific training, and many inspectors shy away from structural work due in large part to the fixed attenuation estimates requiring the "d=a-b-c" calculation. And although the practical examination for CGSB Level 2 certification includes a W59 component, the transducer supplied for the examination is quite frequently the same ½" round transducer used for the ASME component, not the big red brick or snail wedge that the code requires. Thus, the misunderstanding and misapplication of the techniques is inherited "from birth".

d) Zero degree scan from Face C for fusion defects and lamellar tearing

Referring to Table 8.3, the code is directive in restricting the examination from Face A of all joints. Currently there is no mandatory instruction for inspection of the flat, unbeveled fusion faces opposite Face C on T-joints and corner welds.

This area is difficult to inspect with angle beams from Face A or Face B, especially for fusion and lamellar tearing type defects. It should be mandatory, if accessible, to scan T-joints with a zero-degree probe from Face C. There is an Annex Q regarding lamellar tearing, but currently the code prescribes no corresponding inspection technique.

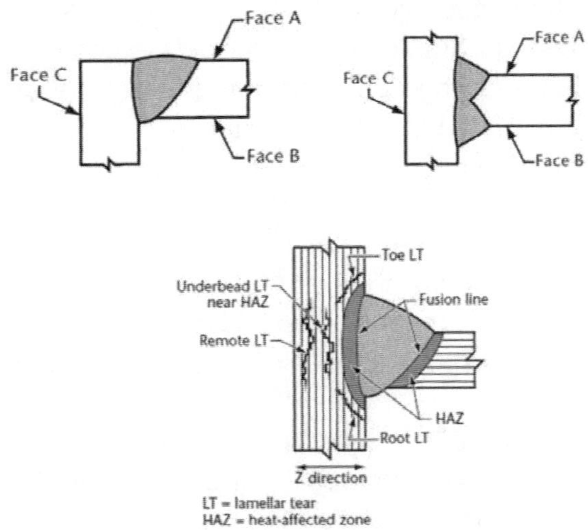

Figure 4 Tee and Corner weld inspection from face C

e) Limitations of angle specific procedure tables

W59-13 code describes specific procedures and probe angles in Table 8.3. However, there is no best angle for any general joint type and weld thickness. A joint beveled at 45° is best inspected with a 45° beam, if accessible, regardless of thickness. Yet the inspector is bound to use a single angle of 70° for most of the welds they are likely to see in the field. The probe angle should be selected based on the bevel angle and the sound path, amongst other considerations. This generally will be dictated by work at hand and actual joint specifics.

As the acceptance tables are also tied to specific angles, they would need to be addressed accordingly

f) Transfer correction for scanning surface

Calibration for sensitivity is performed on reference blocks which are typically shiny and clean, and usually nickel plated to resist corrosion. They do not accurately represent the condition of the test surface, which is usually ground, scraped, or somewhat irregular due to chipping and the typical surface roughness found in rolled plate or other product forms. The code should at least put the onus on the responsible Level 2 to evaluate the differences between the calibration and test surfaces to determine if a transfer correction should be carried out. This is common practice in ASME.

g) Permission to use manual PAUT in substitution of conventional UT

In 2013 when W59 was revised, the use of PAUT was included in the Clause 8.2.12, with an additional condition that phased array be considered an "alternative ultrasonic system" and may only be used if agreed to in writing by the Owner's Engineer and the Contractor. No differentiation was made in regards to manual versus encoded systems.

A clarification is needed for the word "Phased Array Systems". PAUT inspection can be performed manually or in an automated phased array system with an encoder. The encoded scan can then be read much like a radiographic film, in addition to indicating the length and depth of the flaws along the length of the weld. This clarification could open the door to use PAUT interchangeably with conventional UT.

Manual PAUT is in fact, exactly the same as traditional UT. Both techniques use electronic pulse generating equipment but different type of probes. PAUT is performed with many angles instead of just one. Portable PAUT equipment has been commercialized over the last two decades and is currently used on fabrications inspected to ASME, API, and other codes.

Additionally the following items need to be addressed in an Appendix to the current code, provided revised procedures and acceptance criteria, usable for alternate probes, frequencies, and manual PAUT, are demonstrated to be equivalent.

1. Define PAUT flaw detector requirements
2. Define calibration block requirements
3. Define transducer size limits
4. Define focal law requirements
5. Define jobsite calibration procedure (velocity, sensitivity, wedge delay, DAC/TCG)
6. Define scanning procedures for various joints
7. Define and demonstrate, acceptance criteria to be equivalent to static and cyclic requirements (Tables 11.3 and 12.5)

7. Conclusions

a) Experimental work undertaken during this work shows that small and long cracks on demand can be created using the specially designed Ti-B FCAW wire. Such demonstration pieces will be needed to establish equivalence for acceptance between PAUT and UT in the near future.

b) The paper identifies many areas of the current W59-13 code that need improvements in keeping with practicality and good science. Use of smaller transducers with higher frequencies will need the inclusion of DAC/TCG testing for sensitivity requiring significant revisions to the current Clause 8.2 in W59-13.

c) Cohabitation of PAUT in an appendix to the code may be one way to move forward.

8. References

1. TWI reference : http://www.twi-global.com/technical-knowledge/job-knowledge/ultrasonics-examination-part-3-129/

2. "CSA W59 Advanced Inspection Methods", Vaidya et al .Cinde, NDT in Canada 2013, Calgary Conference October 10th 2013

3. CSA W59-13 Welded Steel Construction (Metal Arc Welding)

4. AWS D1.0 (1969) Appendix C – Ultrasonic Testing of Groove Welds

A model-based study of the effect of semi-elliptical surface notch geometry on the signal of a Split-D eddy current probe

Ehsan MOHSENI [1], Martin VIENS [2], Demartonne R. FRANÇA [3]

[1] Département de génie mécanique, L'École de technologie supérieure (ETS), Montréal, Québec, Canada
Phone: +1 514 6605838; e-mail: mohseni.ehsan@gmail.com
[2] Département de génie mécanique, L'École de technologie supérieure (ETS), Montréal, Québec, Canada
Phone: +1 (514) 396-8681; e-mail: martin.viens@etsmtl.ca
[3] Engineering Technologies Department, John Abbott College, Montréal, Québec, Canada
e-mail: d.ramosfranca@johnabbott.qc.ca

Abstract

Eddy current testing (ECT) of tiny fatigue surface cracks are mostly carried out by means of differential probes. Among their various designs, Split-D reflection differential probes, named after their D-shaped internal cores and receiver coils, are particularly interesting because of their small footprint that makes them suitable for inspecting short surface cracks. In many of ECT theories and modelling trials that have been published on the interaction of fatigue cracks with the magnetic field of ECT probes, fatigue cracks are simplified and replaced by semi-elliptical notches. Therefore, in model-based ECT studies, electrical discharge machined (EDM) notches are frequently used since they are advantageous in terms of their low manufacturing cost as well as their usefulness in calibration and model verification procedures. Additionally, ECT signals obtained from EDM notches can roughly estimate those obtained from real fatigue cracks having the same size. Accordingly, in the present study a commercially available split-D surface probe is modeled based on its actual dimensions and material properties. The dimensions of the probe are extracted from a CT-scan reconstruction and inserted in the 3-D model. The probe scanning over 3 semi-elliptical notches having different sizes is simulated using the AC/DC module of COMSOL Multiphysics. A test frequency of 500 kHz is considered in the simulations, and the test block containing the notches is assigned with material properties of aluminum. The effect of simulation parameters, such as mesh size and distribution, is investigated, and hence the parameters are finely tuned to achieve consistent results. Afterwards, the reliability of the simulation outputs is assessed by comparing them to impedance measurements of semi-elliptical surface notches in an aluminum block. This validation study shows acceptable matching of the probe's impedance obtained from both simulations and measurements. After validation of the model, the sensitivity of eddy current signals to variations of the notch geometry (*e.g.*, notch opening, depth and length) is studied using the 3-D simulations. The importance of studying the notch opening originates from the fact that by decreasing this opening in the simulations, a better approximation of a fatigue crack shall be achieved. On the other hand, the ECT signals obtained from different notch lengths and depths establish a size dependent signal archive at the selected test frequency. This archive can be used as a basis for inversion purposes using artificial intelligence algorithms to be pursued in future studies.

Keywords: Eddy Current testing, reflection differential split-D surface probe, semi-elliptical surface EDM notch

1. Introduction

In the recent few decades, there has been an increasing demand from industries to employ quantitative non-destructive evaluations (QNDE) to optimize maintenance procedures. The quantitative evaluations developed in each domain of non-destructive testing (NDT) can efficiently provide a better understanding of the selected NDT technique, sensors and operational parameters used for the inspection as well as the imperfections that are subjected to test. Fortunately, the easy access to the diverse rapid computational tools in the recent years enabled users to employ them for the purpose of QNDE. As one of the widely known computational tools, finite element method (FEM) is becoming more popular for the analysis of the complex setup geometries in eddy current testing (ECT). Publications in the field of ECT has been pursuing different intentions in their investigations, some being devoted to either design or configuration improvement of ECT probes [1, 2], and many others being focused on modelling of the interaction of simple ECT probes with defective samples as well as developing the inversion algorithms based on a database of signals simulated by FEM [3-5]. Although the more complex configurations of ECT probe, such as non-axisymmetric split-D reflection differential probe, has been partially studied before using volume integral method [6-9], it would be advantageous to assess the capability of other computational methods for this task in order to investigate their potential strengths and shortcomings. Owing to the high importance of such probes in aerospace industries and considering the fatigue crack as a common defect occurring on both fuselage and wings [10], the modelling studies for these probes are majorly oriented toward explaining the interaction of ECT probes with fatigue cracks. Initiating such a study, however, always require introductory validation of the modelling approach [11]. Utilizing a narrow surface notch for approximating the ECT signal of a fatigue crack has been repeatedly used in the literature, since an EDM notch, despite its larger opening, can represent a fatigue crack to some extent if it is used with special care [12].
Accordingly, in the present contribution FEM is extensively used to simulate the interaction of a split-D differential probe with 3 semielliptical notches having different sizes; subsequently, a validation study is carried out to evaluate the reliability of the modelling approach. Moreover, the effect of changes of the notch opening on the probe's signal is investigated. The importance of studying the notch opening originates from the fact that by decreasing the opening in the simulations, a better approximation of the signal from a fatigue crack shall be obtained. Furthermore, dependency of the probe's signal to the notch geometry is discussed. These ECT signals collected from different notch lengths and depths establish a size dependent signal archive at the selected test frequency. This archive can be used as a basis for inversion purposes by training artificial intelligence algorithms to be pursued in future studies [13].

2. Impedance Measurements for The Split-D Probe

A commercial reflection differential split-D probe with a frequency range varying from 500 kHz to 3 MHz along with a Nortec 500S is used for impedance measurements. An aluminum 7075-T6 block containing three semi-elliptical notches, whose dimensions are tabulated in Table 1, is mounted on an encoded micrometric X-Y table. As depicted in Figure 1(a), split-D probe is installed within an alignment device and the probe's tip is leveled with the surface of the aluminum block. The orientation of the probe is also set in a manner that the flat surfaces of the D-cores become parallel to the notch side walls. An initial lift-off of 0.03 mm is used in all scans to avoid a tilt that might be caused by the contact between the probe and the block's surface during scans. As shown in Figure 1(b), each of the abovementioned notches is scanned perpendicularly to the central line passing through the notch length. All the scans are done across the middle of the length of the notch. While both the horizontal and vertical gains of the Nortec 500S are set equal, real and imaginary components of the probe impedance are acquired through the analog voltage outputs of the instrument (respectively proportional to the horizontal and vertical positions of the measured data on the instrument screen). The following steps are taken for scanning the notches:

1- In order to ensure the probe's perpendicularity to the surface, an aluminum block containing a through width notch is repeatedly scanned by the split-D probe. The probe is then finely aligned to get the best possible symmetry for the resulting 8-shaped impedance trajectory from the notch. Scanning of the calibration notch reveals that the 8-shaped impedance trajectories of this particular probe is slightly unbalanced. According to the X-ray computed tomography of the probe (to be shown in the following section), it is believed that the slight difference in the sizes of the internal D-cores is responsible for that unbalance.
2- After aligning the probe for perpendicularity, all three notches are scanned. For each test at the beginning of the scan, the probe's centerline is positioned 2 mm away from the notch's centerline, and the scan path continues for 4 mm. Hence, the probe has a total displacement of 4 mm with increments of 0.05 mm.

Table 1. Nominal dimensions of semi-elliptical notches in aluminum 7075-T6.

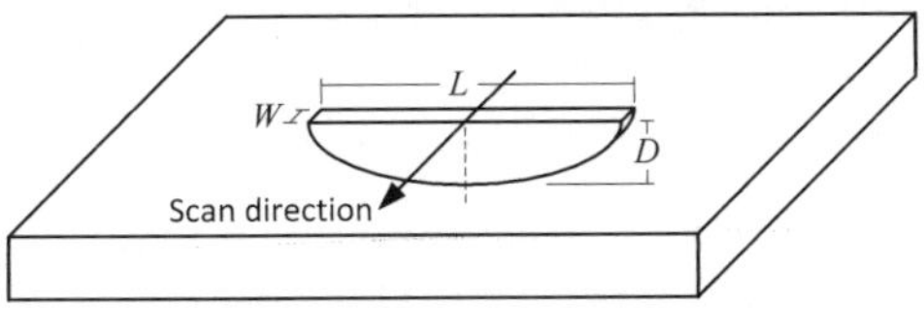

Notch	Length, L (mm)	Depth, D (mm)	Opening, W (mm)
A	2.84	1.11	0.1
B	1.62	0.63	0.1
C	0.81	0.31	0.1

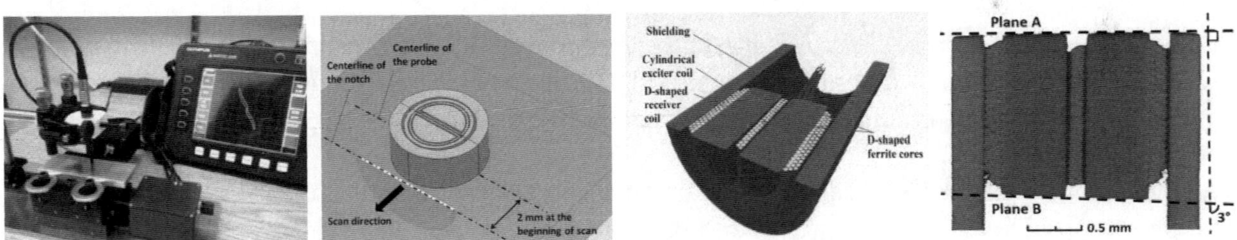

(a) (b) (c) (d)

Figure 1. (a) Setup for measuring the probe's impedance as it scans the surface notches. (b) Orientation of the probe and its scanning direction with respect to the notch length. (c) A 3-D model of the split-D eddy current probe reconstructed by X-ray micro computed-tomography. (d) Side view of the probe showing non-parallelism of the two planes A and B tangential to the top and bottom of the cores, respectively, as well as the dissimilarity between the two D-cores.

3. Finite Element Modeling/Analysis

Although the configuration of the split-D probe is known, the characteristics of the internal components have to be ascertained in order to model it properly for finite element analysis. For this purpose, the probe was scanned using a Nikon XTH 225 X-ray micro tomography. The 3 dimensional structure of the probe is shown in Figure 1(c). The dimensions of the internal components were measured from the reconstructed model and are presented in Table 2. According to the side view of the probe, which is illustrated in Figure 1(d), the slight difference between the D-cores is evident. If two planes are drawn, one located on the top and another tangential to the bottom of the ferrite cores, one can notice the non-parallelism of these two planes and the resulting difference in the dimensions of the cores. These imperfections happen during the manufacturing process and are inevitable.

Table 2. The dimensions of internal components of the probe extracted from the 3-D reconstructed model.

Wire diameter of the receiver D-coils	0.063 mm	**Gap between cores**	0.224 mm
Wire diameter of the driver's coil	0.055 mm	**Shielding inner diameter**	1.934 mm
Number of windings of D-coils	20	**Shielding outer diameter**	2.528 mm
Number of windings of driver's coil	37	**Shielding height**	3.000 mm
Height of the receiver and driver coils	1.260 mm	**Relative permeability of cores/shielding**	2500
Core diameter	1.254 mm	**Conductivity of cores and shielding**	0.2 S/m
Core height	2.000 mm	**Test frequency**	500 kHz

According to the dimensions of the notch geometry and the probe's components, a 3-D model for the assembly of the probe and block is built in COMSOL Multiphysics. Considering the plane symmetry of the assembly at every scan position, the model is cut in half to save time and computational resources. The most important material properties in the model, such as conductivity of 17.8 MS/m for the aluminum block and relative magnetic permeability of 2500 for the ferrite cores, are assigned to the related parts. A voltage of 6 V is applied to the driver's coil and the current flowing into the receivers is set to zero. AC/DC module and magnetic field physics formulation is chosen for this analysis. Equation (1), embedded in the magnetic field physics formulation, is based on Maxwell equations solving for electromagnetic vector potential:

$$(\nabla \times (\nabla \times \boldsymbol{A})) / \mu_0 \mu_r + (j\sigma\omega - \omega^2 \varepsilon_0 \varepsilon_r)\boldsymbol{A} = \boldsymbol{J}_e \ , \ j = \sqrt{-1} \tag{1}$$

In Equation (1), $\boldsymbol{A}$ is the electromagnetic vector potential, μ_0 is the magnetic permeability in vacuum, μ_r is the relative magnetic permeability, ω is the angular frequency, $\boldsymbol{J}_e$ is the externally applied current density, ε_0 and ε_r are the electrical permittivity in vacuum and the relative electrical permittivity, respectively.

The meshing process shall be carried out carefully, since the element size and size distribution are of high importance for the consistency and precision of the solutions. For this purpose, one should consider the standard penetration depth of eddy currents at the selected frequency in the block (*i.e.*, according to the conductivity of the sample, the standard penetration depth is 0.16 mm). Afterwards, there should be at least 2 elements across the penetration depth (*e.g.*, two elements per δ); however, it is more conservative to use 9 elements across the first 3δ [14]. It shall also be noted that the notch wall and tip are meshed with the same element size used in the first 3δ, since there might be current flowing from the sides and bottom of the notch depending on its geometry. After setting the element size and distribution for each domain, a free tetrahedral mesh composed of first order elements is generated for all domains. It is worth to mention that the first order elements are replaced by quadratic elements in all domains for notches as shallow as 0.3 mm deep. For all steps of the probe's displacement during a single scan, it is preferred to keep the mesh unchanged on the probe and block domains to reduce the noise caused by mesh regeneration.

A direct linear solver within COMSOL is used for electromagnetic field analysis of the probe scanning over each notch. This solver is changed to iterative for the cases where the elements are quadratic because of limitations in the server memory. The scan is simulated as 0.1 mm increments of the probe's displacement. The probe's center is initially positioned 1.5 mm away from the notch's centerline and it finishes scanning by traversing the sample 1.5 mm across, forming in this way the half of the original impedance trajectories. Subsequently, the resulting complex voltages of $\boldsymbol{V_{R1}}$ and $\boldsymbol{V_{R2}}$, which are induced across the primary and secondary differential receiver coils, respectively, are captured at each probe's displacement. In addition, the complex current flowing in the driver's coil, $\boldsymbol{I_D}$, is recorded for every probe position along the scan direction. The impedance ($\boldsymbol{\Delta Z}$) of the probe at each position of the scan is calculated through Equation 2. Accordingly, the probe's signal for each scan is plotted on the impedance plane.

$$\boldsymbol{\Delta Z} = (V_{R2} - V_{R1}) / \boldsymbol{I}_D \qquad (2)$$

The simulated scans are listed as follows:

1- Scans of the three notches, presented in Table 1, are simulated. The purpose of these simulations is to compare the measured and simulated signals from the notches in order to validate the simulations;

2- In order to gain a better insight into the effect of the notch opening on the signal from the probe, scans are simulated for the three aforementioned notches while their openings are varied from 0.01 to 0.09 mm.

3- A database comprised of the signals from the probe scanning semi-elliptical notches with a diversity of dimensions is created. The ratio of the notch depth to the standard penetration depth of eddy currents (D/δ) is varied from 1.88 to 6.88 by increments of 2.5. Besides, the ratio of the notch length to the driver coil's diameter (L/D) is varied from 1 to 3 with increments of 0.5 for each depth of the notch. The table instantly allows one to investigate the impact of the dimensional features and their proportionality on the signal and provides a set of data for training artificial intelligent based inversion algorithms.

4. Results and Discussion

The simulated scan results are organized here according to the numbered sequence presented in the previous section. Owing to the signals' symmetry, only half of the complete signal is given in the remaining sections of this study. For comparison purposes, the horizontal and vertical gains are compensated for the signals recorded from Nortec 500S. Figures 2(a), 2(b) and 2(c) compare the simulated and measured signals from the probe for half of the scan line defined over the notches A, B and C, respectively. In order to facilitate the comparison, the scale of these planes are identical. It should be noted that measured impedance trajectories of all the three notches are rotated by 23 degrees on the impedance plane to be in-phase with simulated signals. This initial discrepancy of the signal phase observed between the measurements and the simulations could be connected to the biased phase angle of the Nortec 500S circuitries. The signal amplitude is overestimated by the simulation for each notch; however, these differences at their highest, happening for notch C, do not exceed 6 percent. This could be related to deviations of the material properties used in simulations from the actual values, such as the core's permeability. It also could be caused by the notch geometries utilized in the modelling, since the nominal dimensions are inserted into the model. In terms of the signal shape, the agreement between the simulations and the measurements is satisfactory for the notches A and B. Regarding the signal from notch C, slight disagreement between simulation and measurement can be noted for the probe's positions at the beginning and the end of the scan line. Since the depth of notch C is comparable to the penetration depth, the sensitivity of the simulation outputs to the modelling parameters, such as mesh size and distribution, increases. In general, the signal shape related issues can be resolved by both mesh refinement around the notch geometry and increasing the scanning probe's positions along the scan line. Both of the aforementioned strategies have been optimized in these scans to favour the shorter simulation runs. Based on comparisons, the predictions of Comsol simulations are reasonably accurate for all three notches considering them having different sizes. Based on the results of the validation study, all simulations are assumed to be fairly reliable and will be further used to study the effect of notch opening, depth and length on the differential probe's signal.

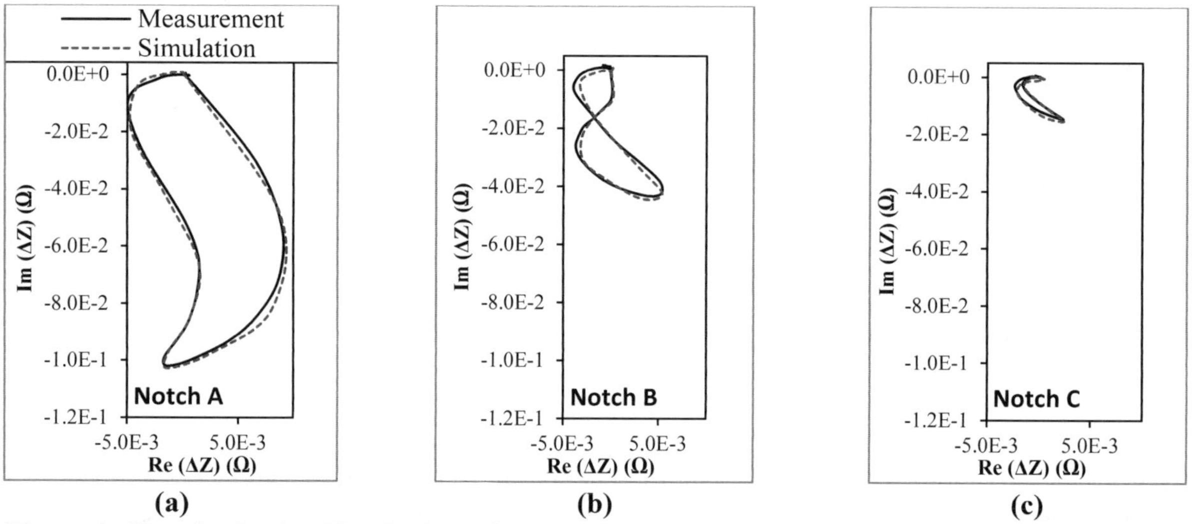

Figure 2. Signals obtained by the impedance measurements and Comsol simulations are plotted together for a) Notch A b) Notch B and c) Notch C.

The impedance trajectories of the notches A, B and C as the notch opening varies are presented in Figures 3(a), 3(b) and 3(c), respectively. As demonstrated by the corresponding figures, the shape of the signals remains unchanged as the opening of the notches varies. The magnitude of the probe's impedance is calculated through Equation 3 for each of the scan positions and subsequently, their maximum is taken as the signal amplitude. The angle Θ which corresponds to the maximum impedance is also calculated as presented in Equation 3, and used as the signal phase.

$$|\Delta Z| = \sqrt{(\mathrm{Im}(\Delta Z))^2 + (\mathrm{Re}(\Delta Z))^2}, \theta = Arctan(\mathrm{Im}(\Delta Z_{max}) / \mathrm{Re}(\Delta Z_{max})) \quad (3)$$

The variations of the signal amplitude and phase are plotted in Figure 4 versus the changes of notch opening for all the three notches under investigation. According to these plots, it is evident that both the signal amplitude and phase decrease with a quasi-linear trend as the notch gets tighter. In order to explain the relative slope of these curves, it is necessary to consider the probe position with respect to the notch when the probe impedance reaches its maximum value. These positions are given in Figure 5. As it can be seen, for notch A, probe impedance reaches its maximum value when the notch is located underneath the D core while, for notches B and C, probe impedance reaches its maximum value when the tip of the notch is positioned beneath the driver coil. Such a difference is attributed to the fact that notch A has a length longer than the probe diameter while notches B and C are similar or even shorter than the probe diameter. This difference has an impact on the way the notch interrupts or deviates the eddy current flow. As shown in Figures 5 (d) and (e), in the first situation (notch A), eddy current bypass the notch mainly by flowing under it while, in the second situation (notch C), eddy current bypass the notch by flowing beside it, along its length. In the case of notch B, a combination of both of the aforementioned situations occurs.

As it can be seen in Figure 4(a), notch A has the highest changes in the signal phase compared to the others. The phase of the probe impedance is directly related to a weighted average of the phase lags of eddy currents flowing at different depths below the part surface. When a notch disturbs the eddy current flow, the aforementioned weighting function is modified accordingly. For notch A, any widening of the notch opening results in deeper eddy current flow and, since eddy current phase lag grows linearly with depth, the phase of the probe impedance increases. For notches B and C, even if notch opening is widened, most of the current flow stays on the surface resulting in a smaller variation of eddy current phase.

In Figure 4(c), the signal amplitudes of each notch having different openings is normalized by the amplitude of the same notch with an opening of 0.02 mm. This figure shows that the sensitivity of the signal amplitude to notch opening is higher for notch C. This trend can be explained by the fact that, for notch C, a widening of notch opening broadens the gap between the two streams of current flowing parallel to the notch sides, thus heavily increasing resistance to such a flow. On the other hand, for notches A and B, an increase of the notch opening mildly lengthens the path of eddy currents flowing below the notch. They have, therefore, a slight impact on the signal amplitude.

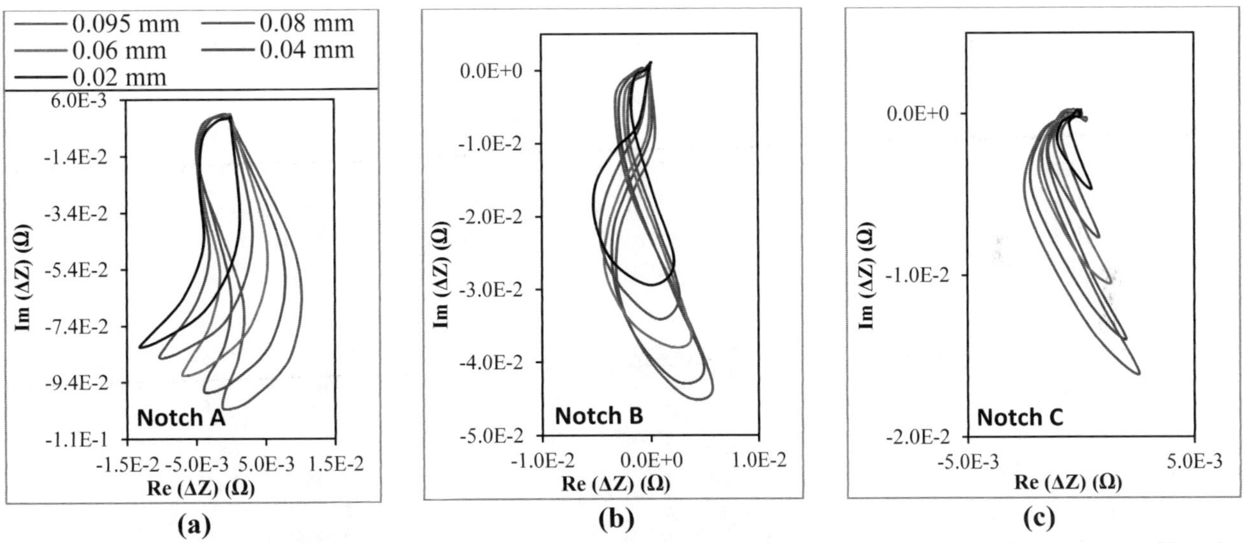

Figure 3. Simulated impedance trajectories as the notch opening changes for a) Notch A, b) Notch B, and c) Notch C.

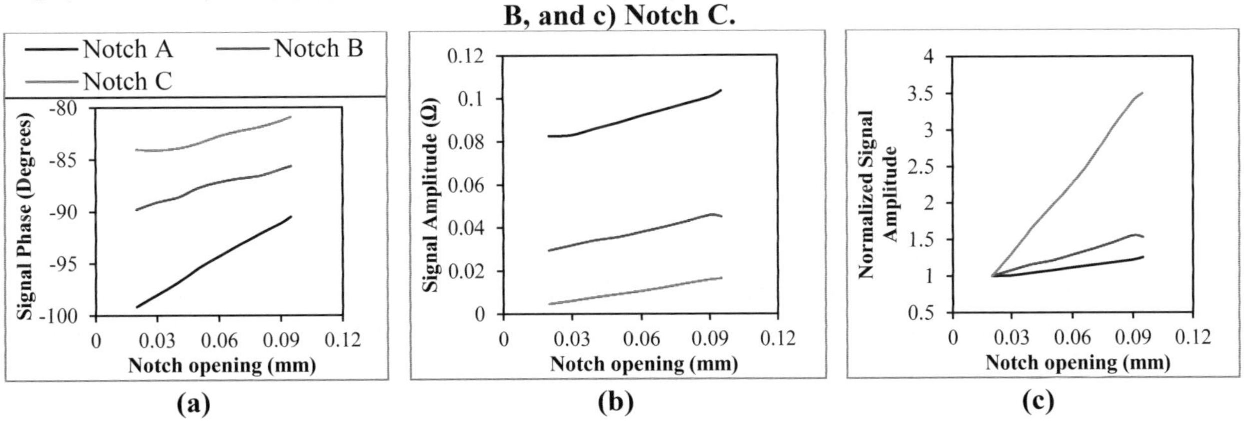

Figure 4. Model based results for a) Variations of the signal phase with changes in notch opening, b) Signal amplitude versus the notch opening and c) Normalized signal amplitude versus the notch opening.

The signature of the signal is highly dependent on the ratio of the driver's coil diameter to the notch length. According to the dimensions provided in Table (1) and (2), the length of notch A exceeds the diameter of the driver's coil, and the half signal shown in Figure 2(a) has one loop. Since the length of notch B is comparable to the coil's diameter, the resulting signal has 2 loops and again the signal reduces to one loop as the length of the notch gets smaller than the driver's coil diameter for notch C.

Figure 6 demonstrates the changes in trajectories of the probe's impedance when the length of the notch varies with a fixed D/δ ratio. In Figure 6(a), D/δ ratio is 0.3 mm and the corresponding L/D ratio is varied from 1 to 3. The notch length at its largest L/D ratio is 0.9 mm, which is still smaller than the driver's coil diameter. Therefore, the shape of the signal does not change whereas the signal amplitude increases with notch length. It is interesting to point out that the maximum amplitude of the probe's impedance occurs at a fixed probe's position of 0.8 mm away from the notch center for all L/D ratios (Figure 5(c)). Figure 6(b) presents the impedance trajectories of the notches having the D/δ ratio of 4.38 and, similar to the former case, the L/D ratio is varied from 1 to 3. In this figure, the shape of the signal transforms from one to double loops as the L/D ratio reaches 2.5, where the length of the notch is equivalent to the inner diameter of the receiver's coil (*i.e.*, 1.7 mm). By further increase of the L/D ratio to 3, the notch length exceeds the driver's coil diameter and the resulting signal possess one loop again with a different phase. Furthermore, in Figure 6(c) the D/δ ratio is taken as 6.88 mm and the ratio is swept in the same manner. According to this plot, the shape transformation starts at a lower L/D ratio of 1.5, where the 1.65 mm length of the notch once more reaches the driver's coil diameter. Following the observed behaviour of the impedance trajectories depicted in Figure 6, it can be concluded that the signal shape is mainly dictated by the notch length regardless the notch depth. Furthermore, the effect of the growth of the D/δ ratio on the increase of the signal amplitude is evident.

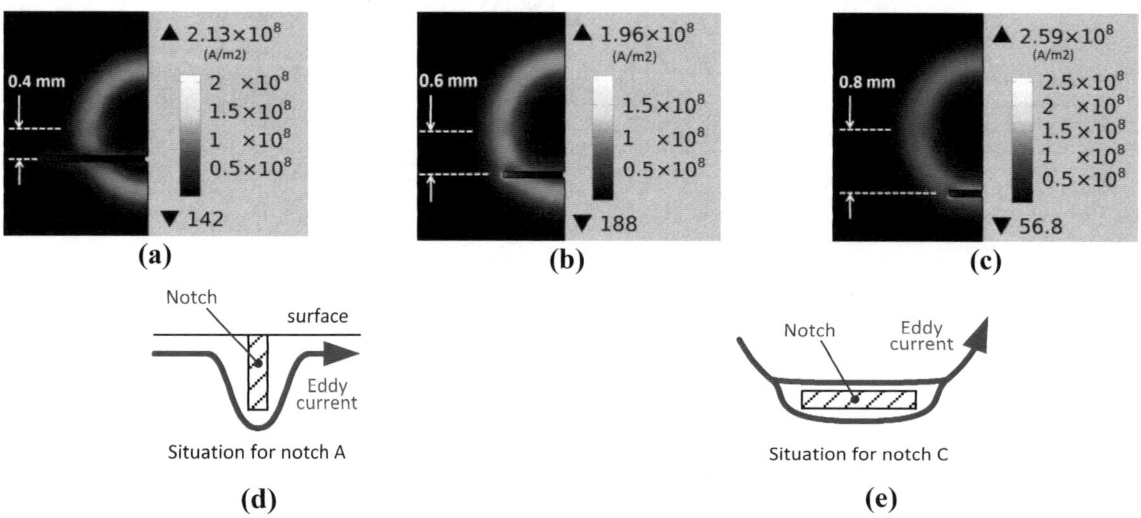

Figure 5. Current density on the surface of the block for a) Notch A, b) Notch B, and c) Notch C as the relative impedance of the probe ΔZ is maximized through optimal position with respect to the notch. How Eddy currents flow to bypass d) Notch A, and e) Notch C.

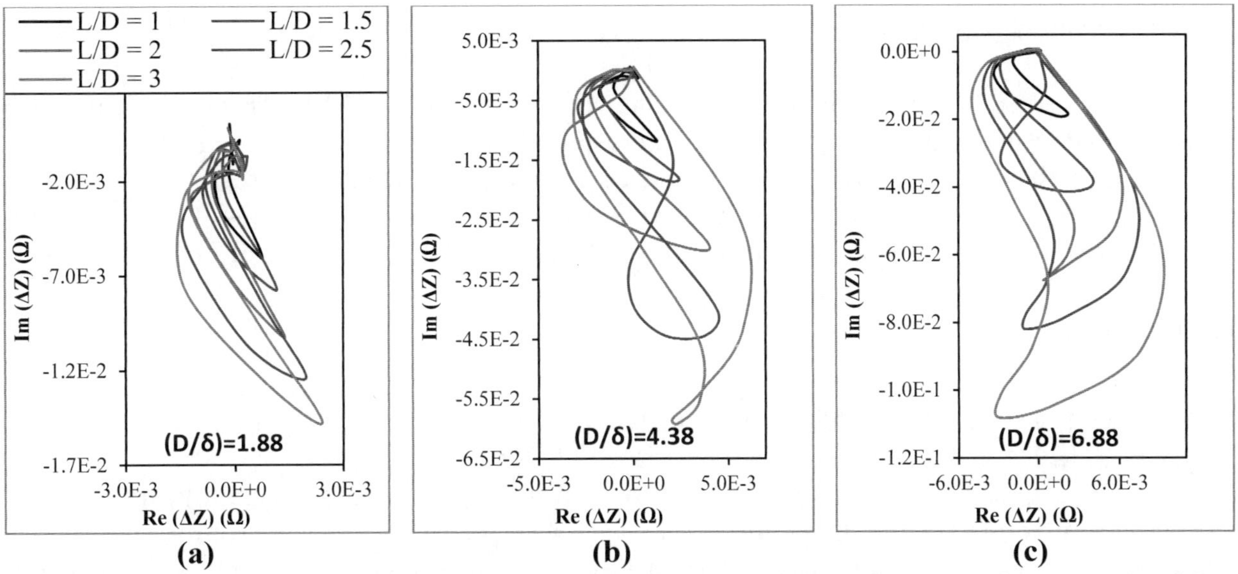

Figure 6. Changes in the impedance trajectories as the L/D ratio varies for a notch with D/δ ratio of a) 1.88, b) 4.38, and c) 6.88.

5. Conclusions

The impedance trajectories generated by the numerical modelling of the split-D probe within Comsol Multiphysics closely match the eddy current impedance measurements of the same semi-elliptic surface notch using the commercial Nortec 500S. It has been observed that modelling of notches having a length smaller than the probe's diameter shall be treated more carefully in terms of element size and distribution in order to achieve satisfactory results.

The signal amplitude increases as the notch opening becomes wider. Variations of both phase and amplitude of the impedance trajectories versus changes of the notch opening is linear for all the notches investigated here, suggesting that signals from notches having an opening within this interval can be estimated by interpolations. However, extrapolation of signals for notch openings falling outside the simulated interval shall be done cautiously, since this linear behaviour might not hold true beyond the range considered in the present investigation.

For notches shorter than the probe's diameter, the signal amplitude is highly sensitive to variations of the notch opening. Due to this high sensitivity, numerical simulations might not be the most suitable approach to reproduce signals from very tiny fatigue cracks, as the use of extrapolation of model based signals from EDM notches might result in large errors. On the other hand, larger notches have lower sensitivities to opening changes, and hence could be efficiently used for estimating signals from fatigue cracks having a similar geometry.

There are three possibilities related to the shapes of signals from the split-D probe scanning semi-elliptical surface notches, and these shape variations are only influenced by the ratio of the notch length to the driver's coil diameter. The shape of a signal is thus a characteristic that complements the analysis of a measurement in order to assess the geometry of a fatigue crack. Moreover, numerical signals obtained for different sizes of notch enable one to establish a training set for inversion algorithms. This procedure is advantageous as it saves time and cost related to the manufacturing of calibration blocks.

6. Acknowledgement

The present research is performed in the framework of CRIAQ-MANU418 project defined by Consortium de Recherche et d'Innovation en Aérospatiale au Quebec (CRIAQ), and it is supported by Natural Sciences and Engineering Research Council of Canada (NSERC) and industrial partners: CNRC Aerospace, Pratt & Whitney Canada, and L-3 Communications. The authors wish to acknowledge the CMC Microsystems for providing access to computational FEM software.

References

1. Huang, H., et al., "Design of an eddy-current array probe for crack sizing in steam generator tubes". NDT & E International. 36(7): p. 515-522, 2003.
2. Chen, Z. and K. Miya, "A new approach for optimal design of eddy current testing probes". Journal of Nondestructive Evaluation. 17(3): p. 105-116, 1998.
3. Song, S.-J. and Y.-K. Shin, "Eddy current flaw characterization in tubes by neural networks and finite element modeling". NDT & E International. 33(4): p. 233-243, 2000.
4. Babbar, V., et al., "Finite element modeling of second layer crack detection in aircraft bolt holes with ferrous fasteners present". NDT & E International. 65: p. 64-71, 2014.
5. Rosell, A. and G. Persson, "Finite element modelling of closed cracks in eddy current testing". International Journal of fatigue. 41: p. 30-38, 2012.
6. Mooers, R.D. and J.C. Aldrin., "Effects of angular variation on split D differential eddy current probe response". in 42nd Annual Review of Progress in Quantitative Nondestructive Evaluation: Incorporating the 6th European-American Workshop on Reliability of NDE. AIP Publishing, 2016.
7. Mooers, R., J. Knopp, and M. Blodgett., "Model based studies of the split D differential eddy current probe". in Review of Progress in Quantitative Nondestructive Evaluation. Volume 31, AIP Publishing, 2012.
8. Mooers, R.D., J.C. Aldrin, and J.S. Knopp., "Model the effects of core/coil size and defect length on eddy current response". in 41st Annual Review of Progress in Quantitative Nondestructive Evaluation. Volume 34, AIP Publishing, 2015.
9. Mooers, R.D., et al., "Split D differential probe model validation using an impedance analyzer". in 40th Annual Review of Progress in Quantitative Nondestructive Evaluation: Incorporating the 10th International Conference on Barkhausen Noise and Micromagnetic Testing. AIP Publishing, 2014.
10. Boller, C., "Ways and options for aircraft structural health management". Smart materials and structures. 10(3): p. 432, 2001.
11. Nakagawa, N., T. Khan, and J. Gray., "Eddy current probe characterization for model input and validation". in Review of Progress in Quantitative Nondestructive Evaluation. Volume 19, AIP Publishing, 2000.
12. Yusa, N., et al., "Numerical modeling of general cracks from the viewpoint of eddy current simulations". NDT & E International. 40(8): p. 577-583, 2007.
13. Xu, B., et al., "Intelligent Eddy Current Crack Detection System Design Based on Neuro-Fuzzy Logic". in International Workshop on Smart Material and Structures/ NDT in Canada conf./ NDT for the Energy Industry. Calgary, Alberta, October 7-10, 2013.
14. Santandréa, L. and Y. Le Bihan., "Using COMSOL-Multiphysics in an Eddy Current Non-Destructive Testing Context". in Proceedings of the Comsol Conference. Paris, 2010.

ELECTRIC POWER RESEARCH INSTITUTE

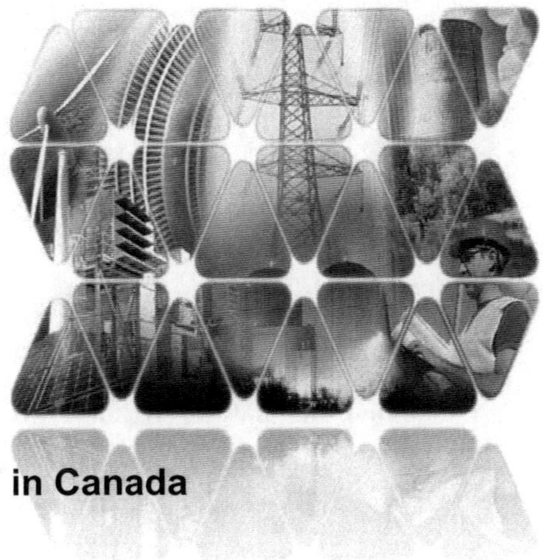

Alternative Examination Coverage

ASME Code Case N711

6th International CANDU ISI Workshop/NDT in Canada 2016 Conference

Patrick O'Regan
poregan@epri.com

Background

- Existing ASME Section XI Code examination figures in IWB and IWC developed over 30 years ago
- Minimal to no change to these examination figures since that time
- Parameters associated with defining effective inservice inspection requirements have matured due to:
  - Accumulated operating experience (root cause analyses)
  - Advances in fracture mechanic evaluations
  - Increased material susceptibility awareness
  - Emergence of risk-informed methodologies

Background (Cont'd)

- These factors lead to a more complete understanding of piping system integrity
- The Code is now in a position to identify more informed examination requirements
- Examinations can be improved by focusing on locations and associated volumes/areas where potential degradation is most likely to occur
- Implementation of this "inspection for cause" approach ensures component reliability and improves plant safety

Current Requirements

- In the US, most plants currently implement Code Case N-460 if > 90% examination coverage is achieved thereby negating submittal of a relief request
- In the US, in those cases where ≤ 90% examination coverage is achieved, a relief request (alternative) is submitted per 10CFR50.55a
- Preparation and submittal of relief requests requires resources of both the power plant and the regulator
- Seldom are additional examination requirements imposed beyond the coverage achieved and documented in the relief request
  - <u>That is, the best effort was already expended</u>

Alternative Requirements

- Case N-711 provides an alternative approach to determine the acceptability of the examination even if it is less than, or different than, the "deterministic" SXI guidance
- Case N-711 provides a prescriptive, step by step process, that enables the user to determine the Volume of Primary Interest (VPI)
- The Volume of Primary Interest (VPI) is dependent upon the component configuration and its susceptibility to degradation, as defined in the Case

TABLE 1
DEGRADATION MECHANISMS CRITERIA

Degradation Mechanisms		Criteria	Susceptible Regions
TF	TASCS	— piping > NPS 1 (DN 25); and — pipe segment has a slope < 45 deg from horizontal (includes elbow or tee into a vertical pipe); and — potential exists for a low flow in a pipe section connected to a component allowing mixing of hot and cold fluids, or potential exists for leakage flow past a valve (i.e., in-leakage, out-leakage, cross-leakage) allowing mixing of hot and cold fluids, or potential exists for convection heating in dead-ended pipe sections connected to a source of hot fluid, or potential exists for two phase (steam/water) flow, or potential exists for turbulent penetration into a relatively colder branch pipe connected to header piping containing hot fluid with high turbulent flow; and — calculated or measured $\Delta T > 50°F$ (10°C); and — Richardson number > 4.0	nozzles, branch pipe connections, safe ends, welds, heat affected zones (HAZ), base metal; and regions of stress concentration

EPRI | ELECTRIC POWER RESEARCH INSTITUTE

Alternative Requirements (Cont'd)

- The Case enables the user to determine if the limited examination coverage achieved on that particular component, captured the Volume of Primary Interest (VPI)
- Successful examination coverage of the Volume of Primary (VPI) will ensure detection of the postulated damage mechanism, if the degradation mechanism is operative at the component

Degradation Mechanisms	Process Decision Point [Note (1)]	If Decision Point is "Yes"	If Decision Point is "No"
TASCS TT	(c) Is the inspection location on a horizontal run to a steam generator or BWR vessel, including feedwater nozzle?	— see decision point (d)	— see decision point (e)
	(d) Was the weld, pipe side heat affected zone and pipe side counterbore transition region captured?	— document examination limitation and coverage achieved and verify examination performed to the examination requirements of B-F, B-J, C-F-1, C-F-2 or R-A, as applicable	— volume of primary interest not sufficiently examined — coverage requirement not met

EPRI | ELECTRIC POWER RESEARCH INSTITUTE

Alternative Requirements (Cont'd)

- If the applicable Volume of Primary Interest (VPI) is fully examined, a request to the regulatory body is not required
- Regulatory notification (i.e. not approval) is required via submittal of Form NIS-1 or Form OAR-1 for all piping weld locations that meet the alternative examination coverage requirements of the Case
- In those cases where the applicable Volume of Primary Interest (VPI) is NOT fully examined, a submittal would be required (e.g. 10CFR50.55a)

FORM N-711-A
ABSTRACT OF WELDS SATISFYING ALTERNATE EXAMINATION COVERAGE REQUIREMENTS OF CASE N-711

Examination Category	Weld Number	Weld Description	Percent Coverage	Description of Limitation

Current Status

- Case N-711 was approved by the ASME Section XI Subcommittee on Inservice Inspection in 2006
- Case N711 has not been endorsed by USNRC in Regulatory Guide 1.147
- Case N711-1 developed to address USNRC and other stakeholder input
- Case N711-1 currently out for ballot at Section XI Standards Committee
- No substantial technical change
  - ➢Clarifying text,
  - ➢Additional technical bases / references

Volume of Primary Interest

- The Volumes of Primary Interest (VPI) defined in the Case are depicted in the figures that follow for each damage mechanism and component configuration combination
- Where depicted in the applicable figures, the Volume of Primary Interest (VPI) includes the entire counterbore transition region if any portion of it is located within ½ in. of the weld fusion line
- TASCS and TT – Figures A1, A2, B1, B2, C1 and C2
- PWSCC – Requirements governed by Owner controlled Program (e.g. 10CFR50.55a)
- IGSCC (PWR) – Figures G1, G2, H1, H2,

EPRI ELECTRIC POWER RESEARCH INSTITUTE

Volume of Primary Interest (Cont'd)

Degradation Mechanisms: TASCS, TT

Component Configuration:
Horizontal Run to SG or BWR Vessel

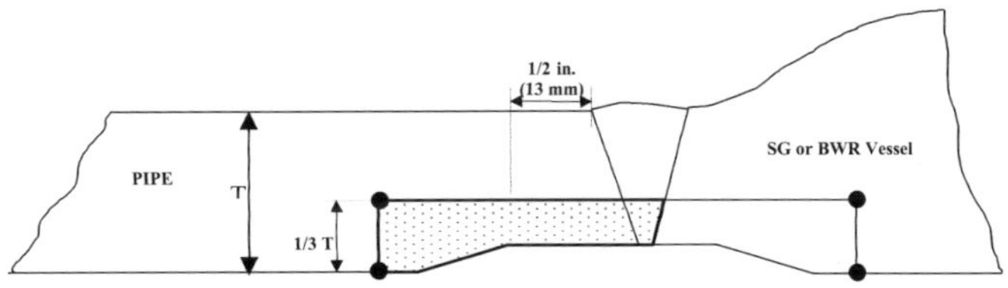

Figure A1

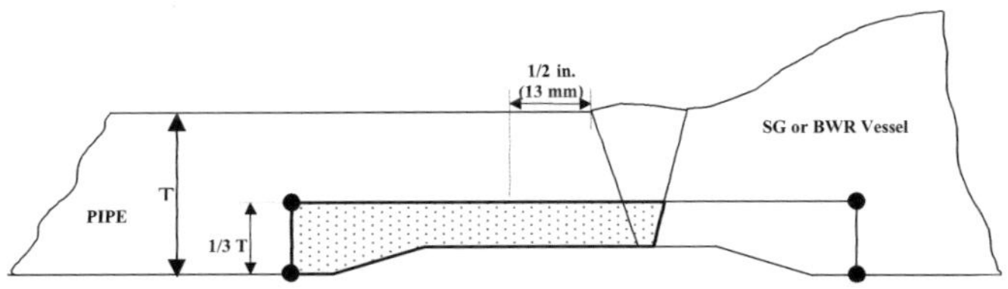

Figure A2

EPRI ELECTRIC POWER RESEARCH INSTITUTE

Volume of Primary Interest (Cont'd)

Degradation Mechanisms: TASCS, TT

Component Configuration:
Pipe to Component

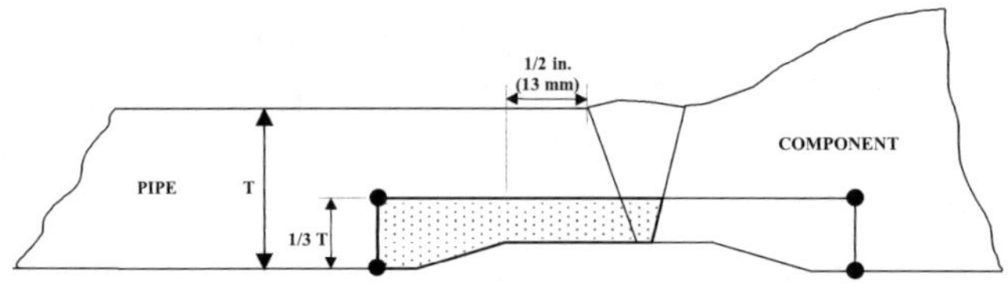

Figure B1

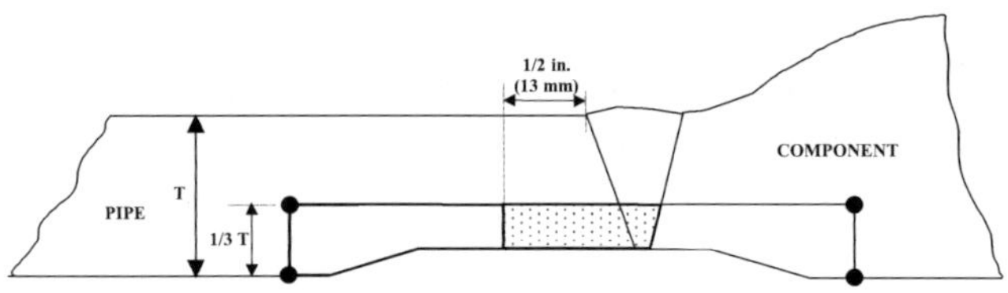

Figure B2

Volume of Primary Interest (Cont'd)

Degradation Mechanisms: TASCS, TT

Component Configuration:
Pipe to Pipe

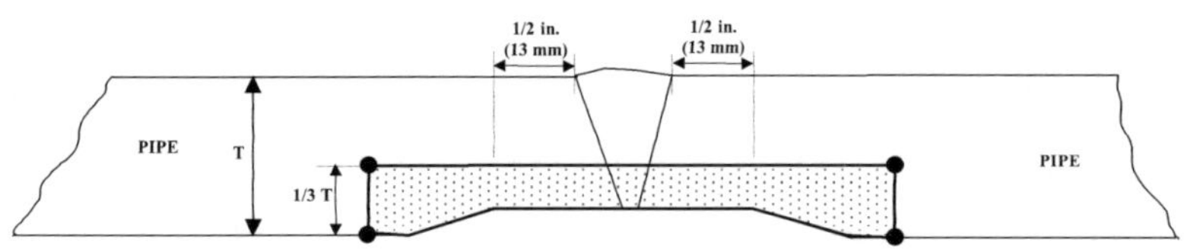

Figure C1

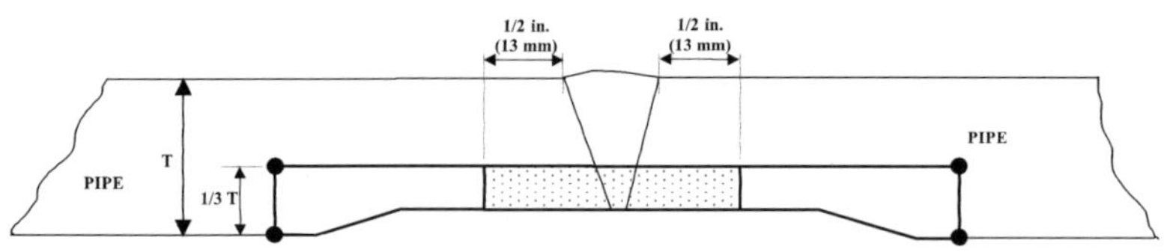

Figure C2

Volume of Primary Interest (Cont'd)

Degradation Mechanisms: IGSCC

Component Configuration:
Pipe to Cast Component

Ferrite/Carbon Content
within Resistant Region of Figure 1

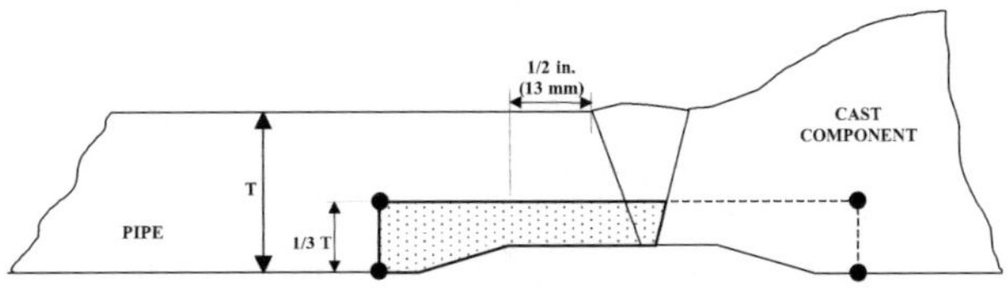

Figure G1

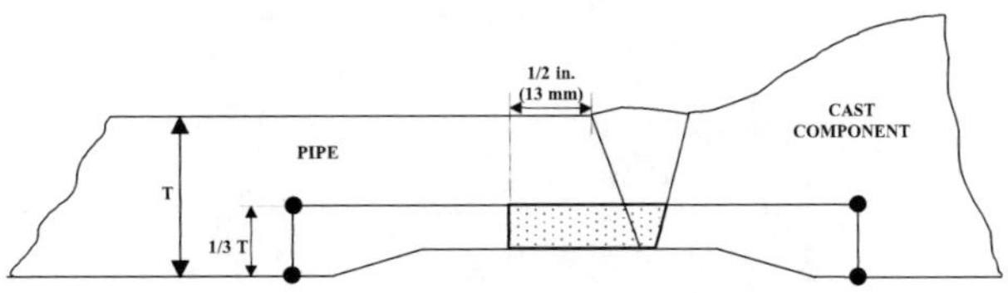

Figure G2

Volume of Primary Interest (Cont'd)

Degradation Mechanisms: IGSCC

Component Configuration:
Pipe to Cast Component

Ferrite/Carbon Content
NOT within Resistant Region of Figure 1

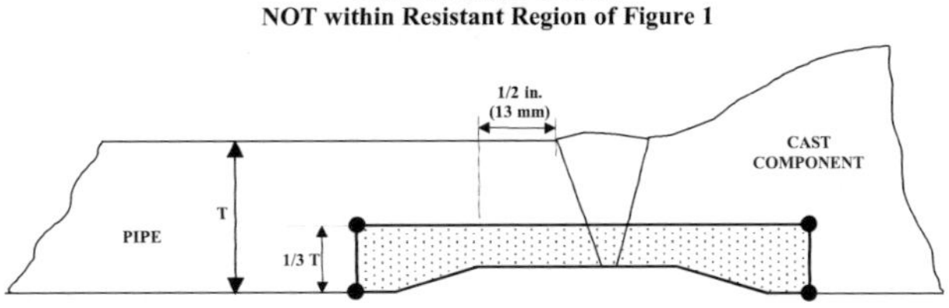

Figure H1

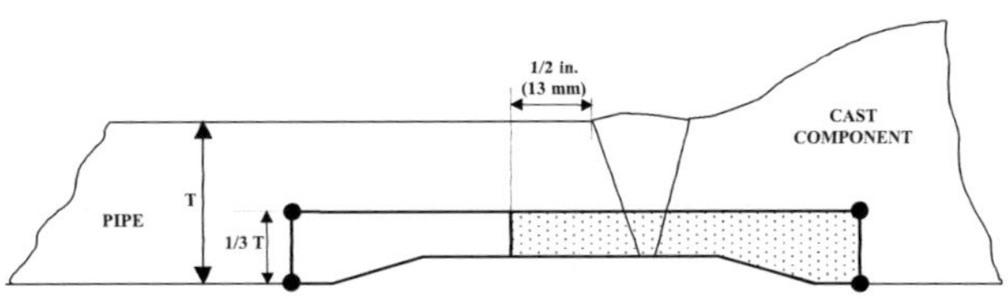

Figure H2

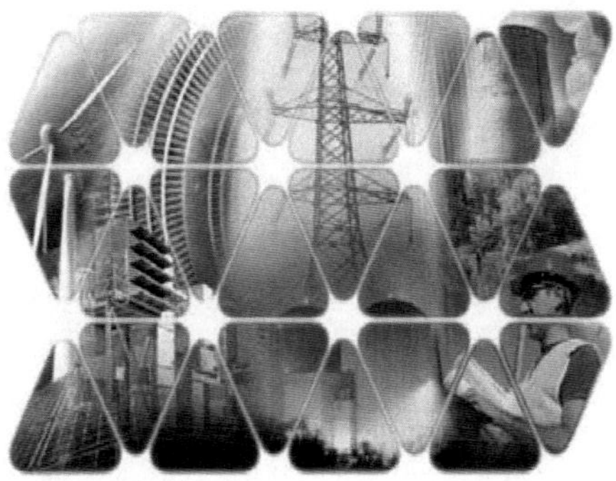

Together…Shaping the Future of Electricity

15

EPRI ELECTRIC POWER RESEARCH INSTITUTE

Development of Target Flaw Sizes for CANDU Pressure Vessel Welds for use in Inspection Qualification

Songyan Yang[1], Richard Yee[2], KiSang Jang[3], John Baron[4], Jeff Weed[5]

[1] Chair, Piping and Vessel Technical Committee, COG CIQB/NDE Joint Projects. Also Senior Design Engineer, Engineering Mechanics Department, Ontario Power Generation – Nuclear
[2] Senior Technical Expert, Inspection and Maintenance Engineering Station Support Programs Directorate, Amec Foster Wheeler
[3] Senior NDE Specialist and Project Manager, COG CIQB Joint Project
[4] Program Manager, COG CIQB Joint Project
[5] Program Manager, COG CIQB Joint Project

ONTARIO POWER GENERATION

Layout of Presentation

- Target Flaw Size (TFS) and Acceptance Flaw Size (AFS) in performance demonstration of NDE system and disposition of detected flaws
- Current TFS revisit. Rationales, scope and purpose of developing new TFS and AFS
- Development of new TFS and AFS

ONTARIO POWER GENERATION

Page 3

Background

- Inspection Specification (IS) defines inspection requirements
  - CSA N285.4
  - IS provides sufficient info for ISP to design, operate and qualify inspection system (technology, equipment, procedure and personnel)
- Target Flaw Sizes (TFS)
  - The size of a flaw (character, shape and dimension) used to qualify inspection system. Qualification requires demonstration of high accuracy in sizing and high probability of detecting of flaws ≥ TFS. Missed indications smaller than TAFS would not challenge the structural integrity of the component.
- Acceptance (Allowable) Flaw Size (AFS)
  - A screening criterion to separate inspected flaws to either Acceptable (fitness for service) or Un-acceptable (requires further analysis)

ONTARIOPOWER GENERATION

Page 4

Comply with ENIQ Recommendations

- European Network for Inspection Qualification (ENIQ)
  - European Methodology for Qualification of Non-Destructive Testing (3rd issue), EUR 22906 EN, ISSN 1018-5593
- Inspection Qualification Input Information (ENIQ Sec. 4.2)
  - 3. types, dimensions, orientations, locations and morphologies of defects to be detected and/or sized, depending on the defect situation considered
  - Defects are determined from applicable codes and standards, or by metallurgy and fracture mechanics calculations
- TFS and AFS are defects obtained from upfront, preemptive fracture mechanics analysis
  - Follow Elastic-Plastic Fracture Mechanics method
  - Under bounding loading conditions and material properties
  - Current looking, unconditionally stable until next inspection
  - TFS are used for qualification, AFS are for screening

ONTARIOPOWER GENERATION

Target Flaw Size (TFS)

- To demonstrate performance of NDE process
- Smallest of all AFS, therefore unconditionally stable until next inspection
- Requires high confidence in detection and sizing by NDE

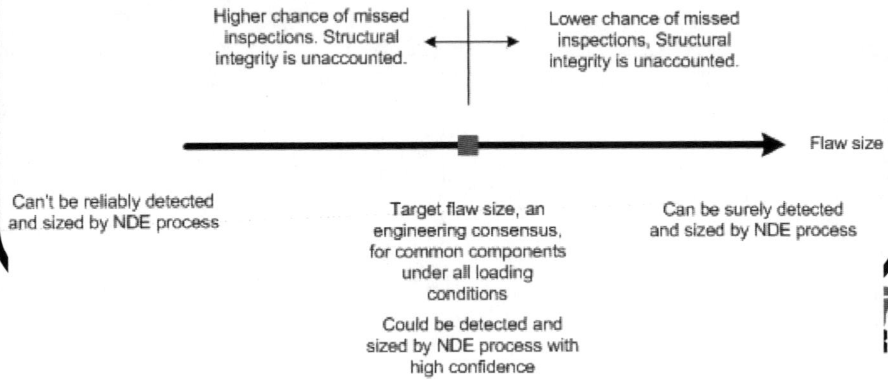

ASME Code Section XI Table IWB-3510-1 or Table IWB-35120-1 Could be Potential TFS

- IWB-3510-1 (Shell-to-Shell Welds)

Table IWB-3510-1
Allowable Planar Flaws
Material: Ferritic steels that meet the requirements of NB-2331 and G-2110(b) of Section III

Aspect Ratio, [Note (1)] a/ℓ	Volumetric Examination Method, Nominal Wall Thickness, [Note (1)], [Note (2)] t, in. (mm): $2^1/_2$ (65) and less — Surface Flaw, [Note (5)] a/t, %	$2^1/_2$ (65) and less — Subsurface Flaw, [Note (3)], [Note (4)] a/t, %	4 (100) to 12 (300) — Surface Flaw [Note (5)], a/t, %	4 (100) to 12 (300) — Subsurface Flaw, [Note (3)], [Note (4)] a/t, %	16 (400) and greater — Surface Flaw, [Note (5)] a/t, %	16 (400) and greater — Subsurface Flaw [Note (3)], [Note (4)] a/t, %
0.0	3.1	$3.4Y^{1.00}$	1.9	$2.0Y^{1.00}$	1.4	$1.5Y^{1.00}$
0.05	3.3	$3.8Y^{0.96}$	2.0	$2.2Y^{0.90}$	1.5	$1.7Y^{0.91}$
0.10	3.6	$4.3Y^{0.72}$	2.2	$2.5Y^{0.69}$	1.7	$1.9Y^{0.69}$
0.15	4.1	$4.9Y^{0.48}$	2.5	$2.9Y^{0.47}$	1.9	$2.1Y^{0.43}$
0.20	4.7	$5.7Y^{0.50}$	2.8	$3.3Y^{0.47}$	2.1	$2.5Y^{0.45}$
0.25	5.5	$6.6Y^{0.65}$	3.3	$3.8Y^{0.61}$	2.5	$2.8Y^{0.57}$
0.30	6.4	$7.8Y^{0.84}$	3.8	$4.4Y^{0.77}$	2.9	$3.3Y^{0.75}$
0.35	7.4	$9.0Y^{0.99}$	4.4	$5.1Y^{0.93}$	3.3	$3.8Y^{0.90}$
0.40	8.3	$10.5Y^{1.00}$	5.0	$5.8Y^{1.00}$	3.8	$4.3Y^{1.00}$
0.45	8.5	$12.3Y^{1.00}$	5.1	$6.7Y^{1.00}$	3.9	$4.9Y^{1.00}$
0.50	8.7	$14.3Y^{1.00}$	5.2	$7.6Y^{1.00}$	4.0	$5.6Y^{1.00}$

ASME Code Section XI Table IWB-3510-1 or Table IWB-35120-1 Could be Potential TFS

- **IWB-3512-1, Nozzle-to-Shell Welds**

Table IWB-3512-1
Allowable Planar Flaws
Material: Ferritic steels that meet the requirements of NB-2331 and G-2110(b) of Section III

	Volumetric Examination Method, Nominal Wall Thickness, [Note (1)] t, in. (mm)			
	$2\frac{1}{2}$ (65) and less		4 (100) through 12 (300)	
Aspect Ratio, [Note (1)] a/ℓ	Surface Flaw, [Note (2)] a/t, %	Subsurface Flaw, [Note (2)] – [Note (4)] a/t, %	Surface Flaw, [Note (2)] a/t, %	Subsurface Flaw, [Note (2)] – [Note (4)] a/t, %
0.00	3.1	$3.4Y^{1.00}$	1.9	$2.0Y^{1.00}$
0.05	3.3	$3.8Y^{0.96}$	2.0	$2.2Y^{0.90}$
0.10	3.6	$4.3Y^{0.72}$	2.2	$2.5Y^{0.69}$
0.15	4.1	$4.9Y^{0.48}$	2.5	$2.9Y^{0.47}$
0.20	4.7	$5.7Y^{0.50}$	2.8	$3.3Y^{0.47}$
0.25	5.5	$6.6Y^{0.65}$	3.3	$3.8Y^{0.61}$
0.30	6.4	$7.8Y^{0.84}$	3.8	$4.4Y^{0.77}$
0.35	7.4	$9.0Y^{0.99}$	4.4	$5.1Y^{0.93}$
0.40	8.3	$10.5Y^{1.00}$	5.0	$5.8Y^{1.00}$
0.45	8.5	$12.3Y^{1.00}$	5.1	$6.7Y^{1.00}$
0.50	8.7	$14.3Y^{1.00}$	5.2	$7.6Y^{1.00}$
Inside corner region	2.5	Not applicable	2.5	Not applicable

GENERATION

Acceptance Flaw Sizes (AFS)

- **To safeguard the structural integrity when flaws are left without repair until next inspection**
- **Obtained from Critical Flaw Sizes (CFS) by considering subcritical crack growth between two consecutive inspections**

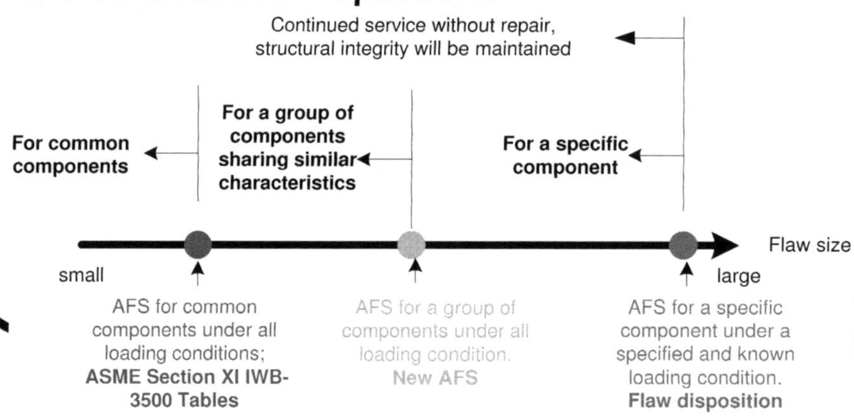

VER
ATION

Critical Flaw Sizes (CFS)

- From fracture mechanics calculations
  - Demarcation between stable (safe) and unstable (failure)
  - Forward looking, focus on next inspection
  - Margins factorized in calculations
- CFS and AFS
  - AFS are obtained from CFS by considering subcritical crack growth between two consecutive inspections
  - CFS is forward looking, AFS is current looking
  - AFS are dependent on inspection interval

ONTARIOPOWER GENERATION

TFS, AFS, CFS and Inspection Limit Size (ILS)

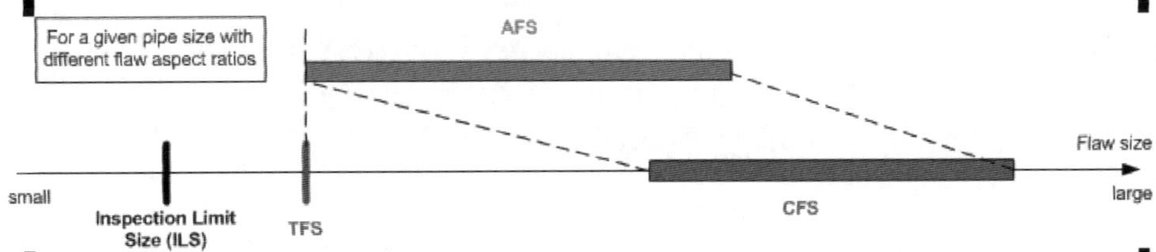

ONTARIOPOWER GENERATION

Page 11

CANDU Pressure Vessel Welds

- **Shell-to-Shell and Nozzle-to-Shell welds of CANDU Pressure Vessels from owner survey**
  - **Nuclear Class, Full penetration, ferritic steel welds subject to PIP as per CAS285.0**
- **Vessel nominal wall thickness from 12.7 mm to 108 mm**

Component [1]	Bruce A	Bruce B	Darlington	Pickering A	Pickering B	Pt Lepreau (CANDU 6)
Pressurizer	✓	✓	✓	✓	✓	✓
Bleed Condenser	✓	✓	✓	✓	✓	N/A
Degasser	N/A	N/A	N/A	N/A	N/A	✓
Maintenance	✓	✓	N/A	N/A	N/A	N/A
Steam Generator	✓	✓	✓	✓	✓	✓
Preheater	✓	✓	N/A	N/A	N/A	N/A
Shutdown	✓	✓	✓	✓	✓	✓

[1] Currently installed components (original, refurbished, or replaced)

WER
...ERATION

Page 12

Representative Basic Weld Configurations

- **Defects are idealized as Surface Planar Flaws (ASME B&PV Code Section XI IWA-3310)**
  - **Having sharp edges, most dangerous flaws**
- **Flaw Aspect Ratio 1:6 (depth to length) as per ASME B&PV Code Section XI Appendix L**

Case #	Weld Configuration		Flaw Orientation
	Geometry	Loading	
1	longitudinal seam welds in cylindrical shell	internal pressure	axial
2	cylindrical shell-to-spherical head girth welds	internal pressure	circumferential
3	flange-to-cylindrical shell welds of reinforced openings (e.g., manway)	internal pressure	circumferential
4	nozzle-to-cylindrical shell welds	internal pressure nozzle loads	circumferential
5	nozzle-to-spherical head welds	internal pressure nozzle loads	circumferential

R
)N

Page 13

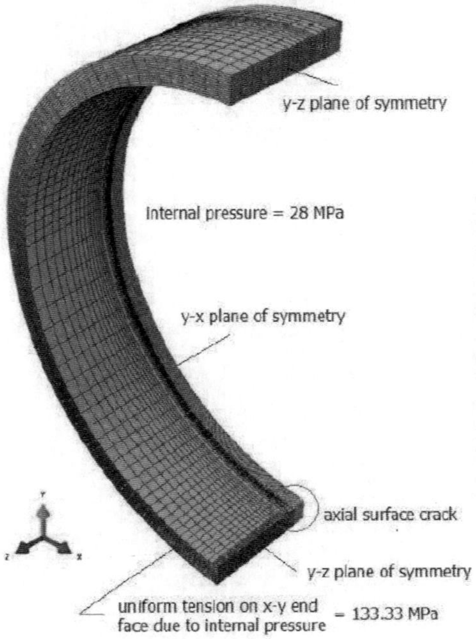

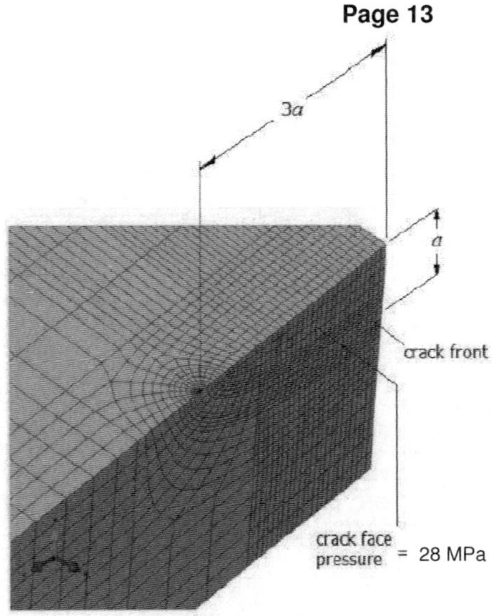

Longitudinal Seam Welds in Cylindrical Shell

Page 14

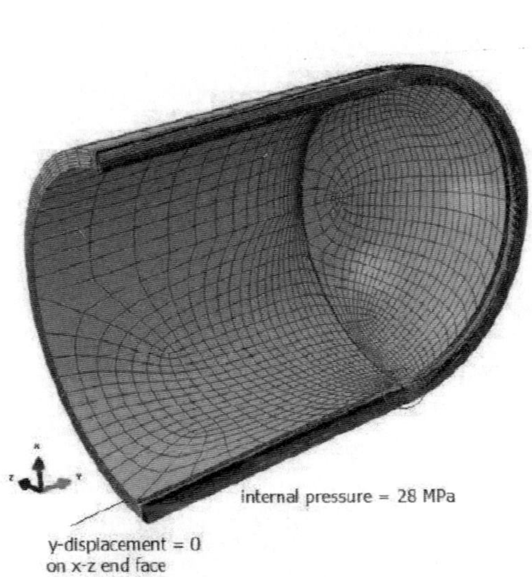

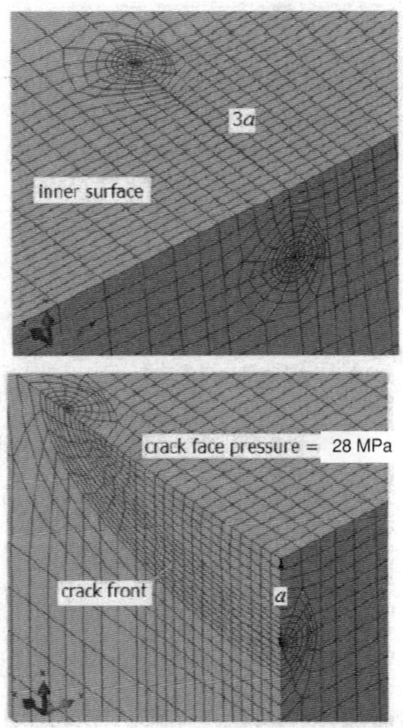

Cylindrical Shell-to-Spherical Head Girth Butt Welds

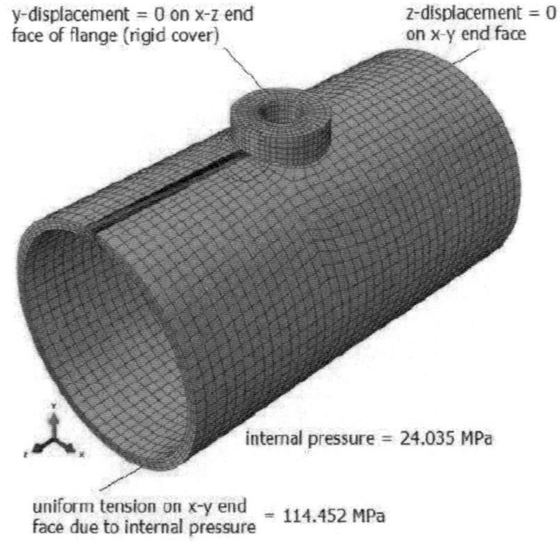

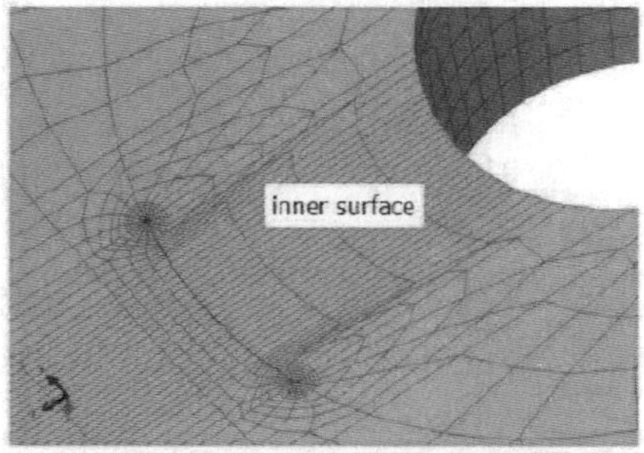

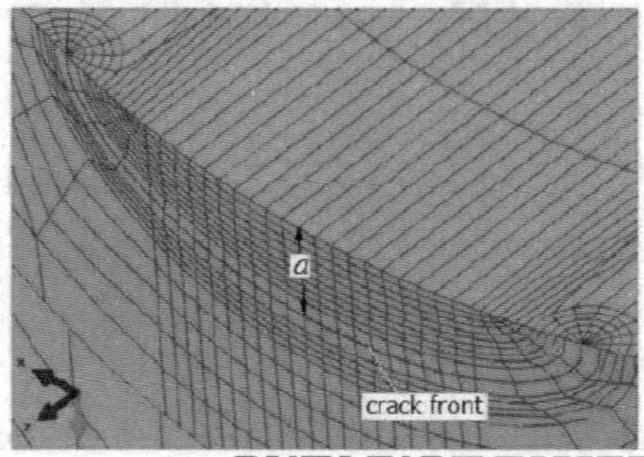

Flange-to-Cylindrical Shell Welds of Reinforced Openings (e.g. Manway)

ONTARIOPOWER GENERATION

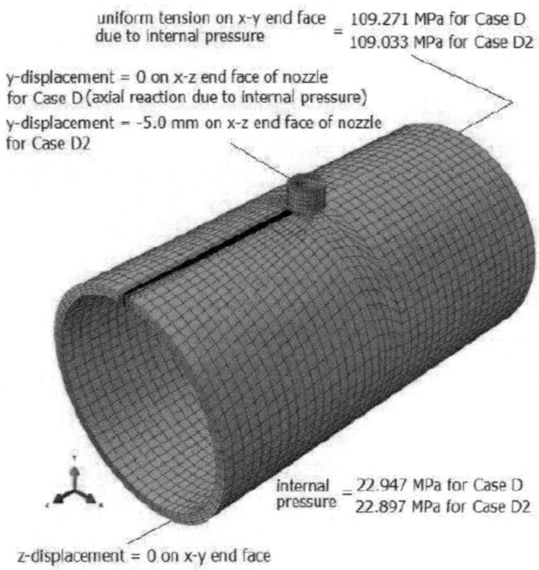

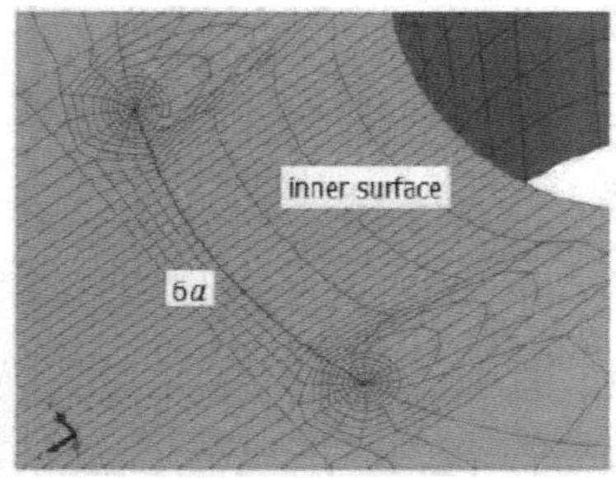

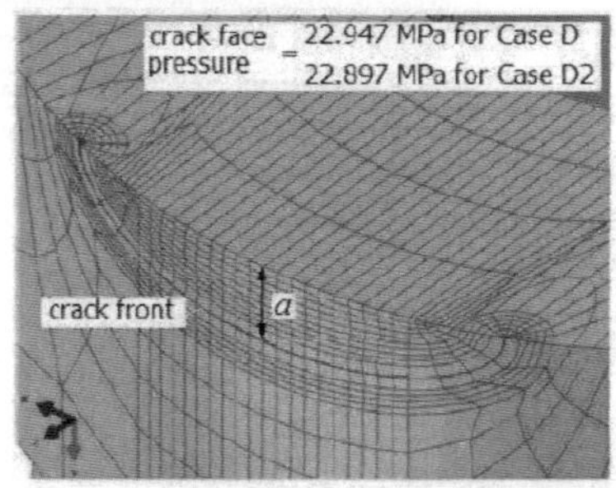

Nozzle-to-Cylindrical Shell Welds

ONTARIOPOWER GENERATION

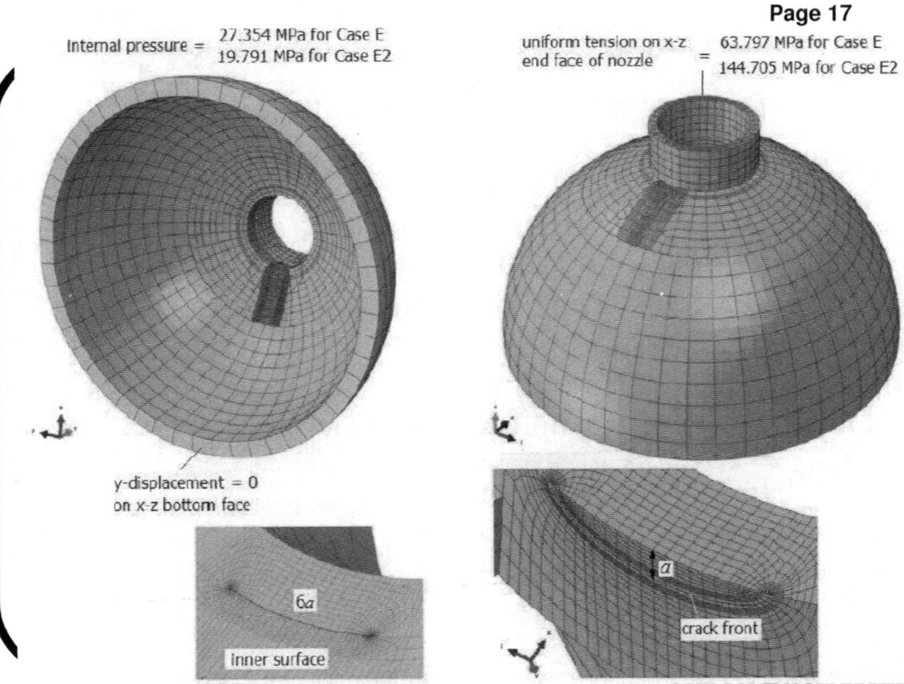

Nozzle-to-Spherical Head Welds

ONTARIOPOWER GENERATION

Highlights of Fracture Mechanics Analysis

- **Fracture Tolerance Analysis to determine CFS**
  - **Elastic Plastic Fracture Mechanics method of J-integral ductile tearing (J-T)**
  - **Under largest design limit loads (ASME Section III Level D)**
  - **Finite element for J-integral, lower bound resistant curve**
- **Fatigue Crack Growth Analysis to determine AFS**
  - **Paris formulation (ASME Code Section XI Appendix A)**
  - **12 year Inspection interval: given CFS, inverse calculate AFS, i.e. FCGA from initial size= AFS to final size= CFS**
  - **Under 200 full pressure cycles ($\Delta\sigma=2\sigma_y$) and 2200 significant thermal transients ($\Delta\sigma= \sigma_y/3$)**
- **TFS is smallest AFS**

ONTARIOPOWER GENERATION

Results of t=108 mm, Ri=1080 mm

- J-R Curve A is for Charpy V-Notch (CVN) energy of 50 ft-lbs, Curve B is for CVN of 90 ft-lbs
- Lower bound tensile is for SA-350 Gr. LF1 at 350°C, for Pickering A bleed condenser nozzles
- Tensile properties of SA-516 Gr. 70 represents most vessel and nozzle materials
- Aspect ratio 1:6 (depth:length)
- Case A (axial flaw in longitudinal seam welds in cylindrical shell) result is bounding

Case	Tensile Properties	J-R curve	Crack depth at failure (mm)	CFS (mm)	AFS (mm)	AFS/wall thickness
A	Lower Bound	A	4.570	3.341	2.961	0.027
A	SA-516 Gr. 70	B	9.720	6.758	4.740	0.044
B	Lower Bound	A	32.95	28.14	22.207	0.206
C	Lower Bound	A	50.56	44.89	36.142	0.335

ONTARIOPOWER GENERATION

Bounding Case A: Finite element J-integral Calculations Agrees With handbook

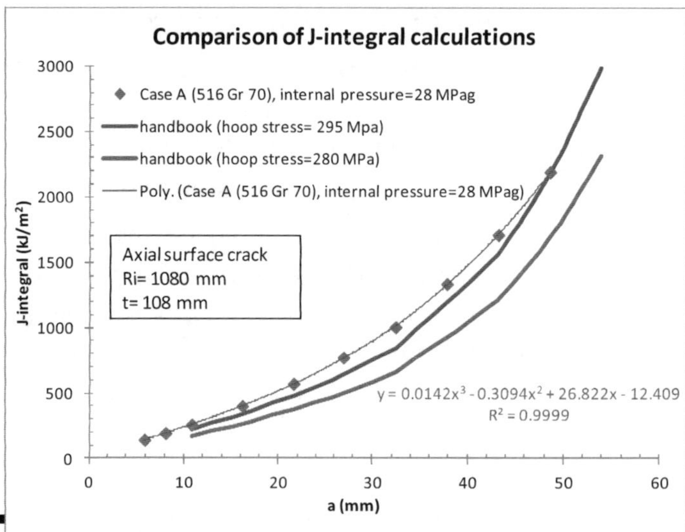

ONTARIOPOWER GENERATION

Bounding Case A: Nonlinear Dependence of J-Integral on Internal Pressure

- Small decrease in internal pressure (i.e. -4.97%) results in large decrease (about -22%) in J-integrals

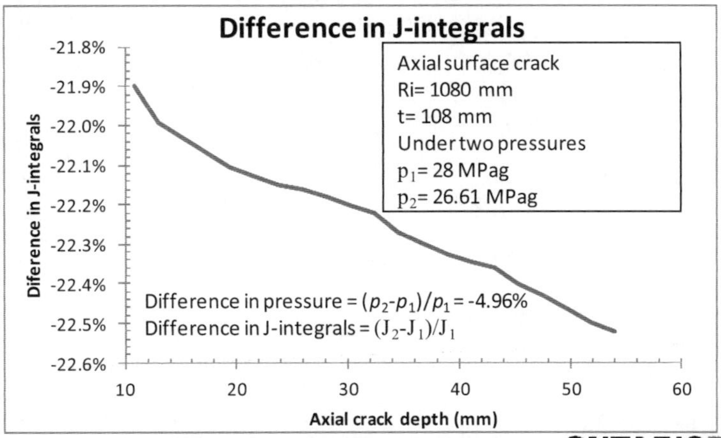

ONTARIOPOWER GENERATION

Bounding Case A: Nonlinear Dependence of CFS on Internal Pressure

- Small decrease in internal pressure (i.e. -4.97%) results in large increase (5%-45%) in CFS
- Nonlinearity is more significant for large wall thickness

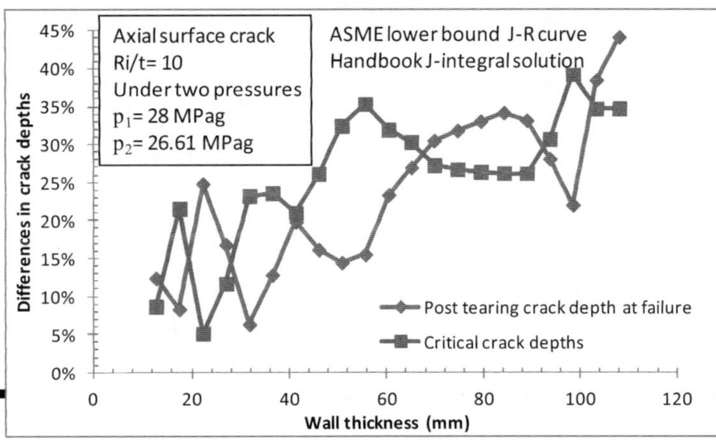

POWER GENERATION

Comparison of J-R Curves

- The JR curve corresponds to CVN=90 ft-lbs is about 6% (at JR at about 300 kJ/m²) lower than ASME lower bound JR curve (ASME B&PV Code Section XI Appendix C-8321).
- A slightly increase of CNV value from 90 ft-lbs to 95 ft-lbs would equate the two JR values at or beyond 300 kJ/m².

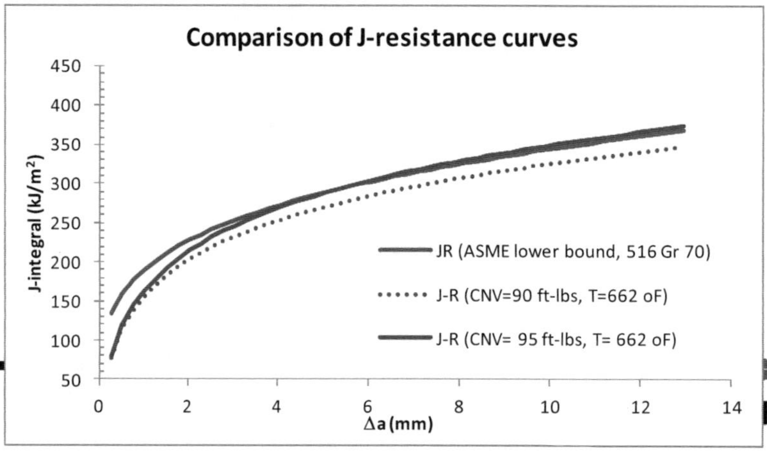

ONTARIO POWER GENERATION

Bounding Case A: Nonlinear Dependence of CFS on J-R Curves

- A small increase of JR curve would result in larger CFS, i.e. a 6% increase of JR curve would increase CFS by 15-20%.

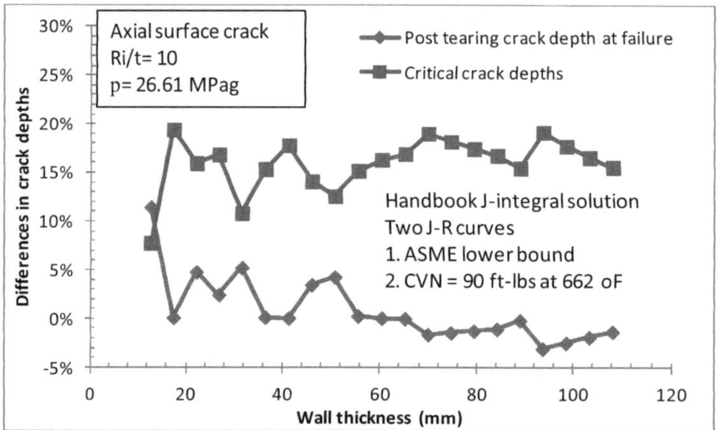

ONTARIO POWER GENERATION

Bounding Case A: J-Integral at failure Barely Change with J-R Curves

- The post-tearing crack depth and J-integral at failure barely changes with the increase of JR curve.

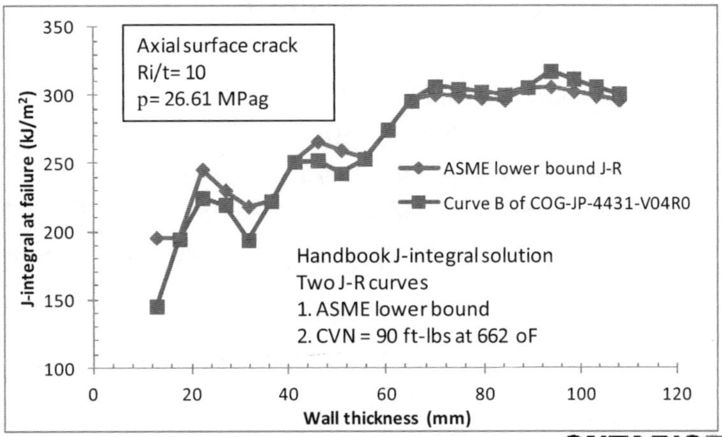

ONTARIOPOWER GENERATION

Calculation of CFS as a Function of Wall Thickness

- Calculations are for Case A (axial flaw in longitudinal seam welds in cylindrical shell)
  - Results applicable to other cases of vessel/nozzle welds
- OD range [279.4, 2376.0] mm, thickness [12.7, 108] mm, ID axial flaw, flaw aspect ratio 1:6 (depth:length)
- EPFM method of J-T to calculate CFS
  - J-integral is from Ductile Fracture Handbook (EPRI NP-6301)
  - J-Resistance curve is ASME lower bound (ASME B&PV Code Section XI Appendix C-8321).
  - Internal pressure=26.61 MPa(g), hoop stress= 306.6 Mpa, 0.7*Su=280 MPa

ONTARIOPOWER GENERATION

Calculation of AFS and TFS as a Function of Wall Thickness

- **AFS as calculated from CFS**
- **Fatigue Crack Growth Analysis**
  - Paris formulation
  - Initial size= AFS, final size= CFS
  - Under cyclic Stress (hoop stress)
    1. 200 cycles, from -150 MPa to 150 MPa ($-\sigma_y$ to σ_y)
    2. 2200 cycles, from -25 MPa to 25 MPa ($-\sigma_y/6$ to $\sigma_y/6$)
- **TFS are equal to AFS (for a given aspect ratio of 6)**

ONTARIOPOWER GENERATION

TFS as a Function of Vessel Wall Thickness (in mm)

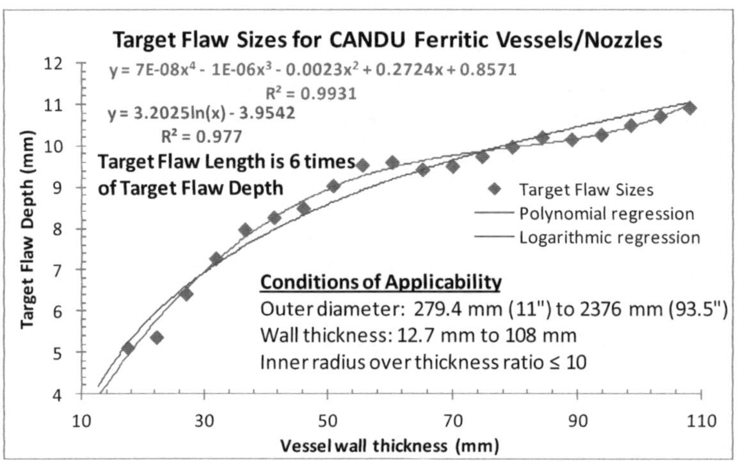

ONTARIOPOWER GENERATION

TFS as a Function of Vessel Wall Thickness (in % of thickness)

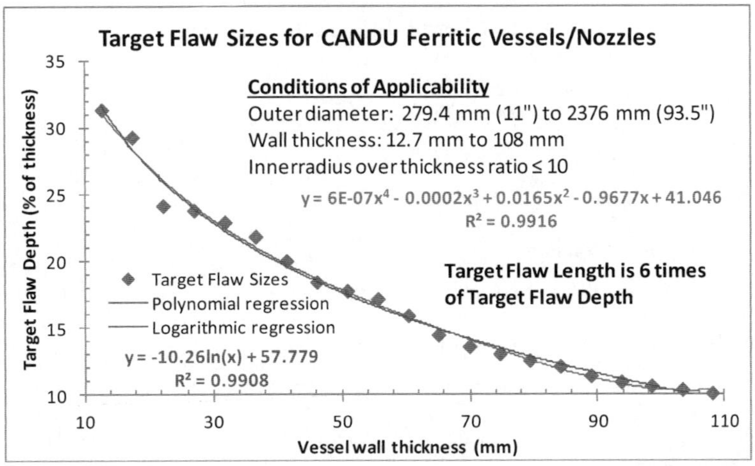

ONTARIOPOWER GENERATION

Comparison of new TFS with Section XI Table IWB-3510-1 of ASME B&PV Code

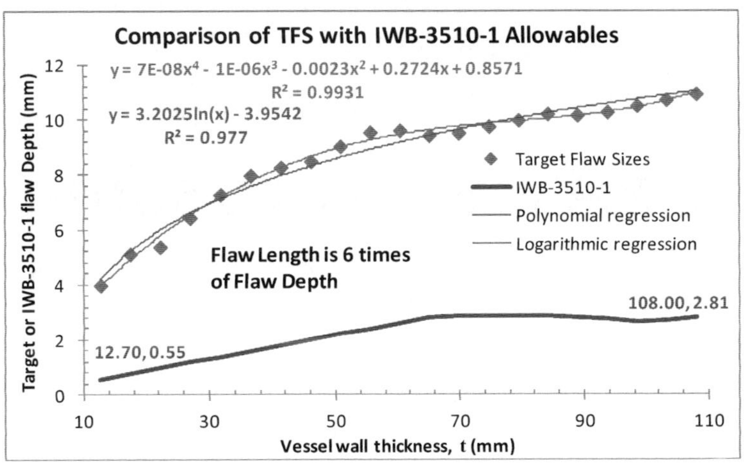

ONTARIOPOWER GENERATION

Comparison of new TFS with Section XI Table IWB-3510-1 of ASME B&PV Code (cont'd)

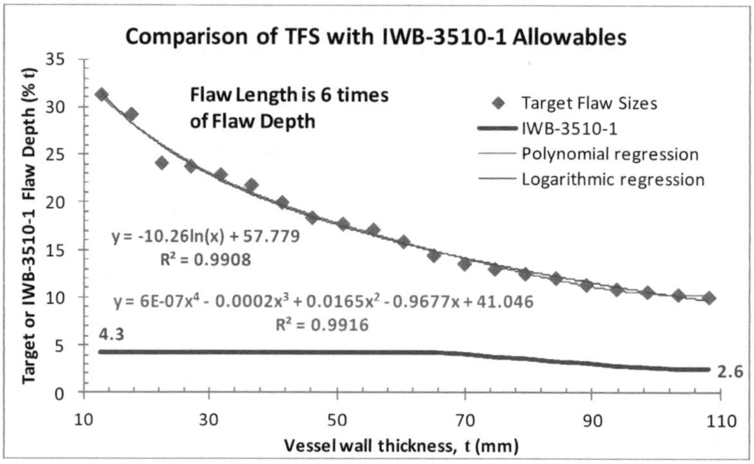

ONTARIO POWER GENERATION

Q&A and Discussion

ONTARIO POWER GENERATION

ANDE-1 Comparison With Current NDE PQ&C Processes

6th International CANDU ISI Workshop
Burlington, ON
November 15-17, 2016

Michael L. Turnbow, Chairman
ANDE Industry Sector Committee

2

Why ANDE?

Industry events, 30 years of round robin studies and recent operating experience (OE) indicates industry action needed to improve the performance and reliability of non-destructive examination and inspection.

3

NDE Issues Overview

- Decline in qualified workforce due to attrition
- Increasing demand due to aging plant issues and competing industries
- Variations in employer based qualification and certification
- Human performance issues
- Existing PQ&C processes do not align with INPO guidelines and best practices used in nuclear, the military and other industries.

4

NDE Performance Issues History

- 2008 EPRI Report 1016969 summarizes 30 years of round-robin studies and performance demonstration results; identifies common performance weakness root cause:
  - "The pass rate obtained by experienced (NDE) certified personnel is rarely 50% on first attempt at a demonstration test that simulates the exam conducted in the field"
  - ".... exposure to training and job experience (in accordance with the current ASNT employer based self-certification process) does not necessarily lead to job proficiency."
  - "Results such as these are not puzzling if one recognizes that unstructured training and job experience may or may not include the key factors essential for learning."

ASME SETTING THE STANDARD

5

Regulatory Perspective

January 12, 2010 NRC Letter to ASME:

"The issue regards questionable qualifications of Nondestructive Examination (NDE) personnel which relate to the ASME Boiler and Pressure Vessel Code (Code)....The Code addresses qualifications of NDE personnel; thus, these requirements become regulatory requirements with implications across the entire nuclear fleet."

ASME SETTING THE STANDARD

6

NDE Certification Program Comparison

	(A) ASNT-TC-1A/CP-189 (B)	ISO 9712	ANDE -1
Vision	(A) J-2 / (B) J-1	J-1	J-1
High School	N/A	N/A	High School or GED
Training	Time based general training-hours CP-105 No Job Task Analysis No Industry Evaluation Varies from employer to employer	Time based general training-hours ISO/TR 25108 No Job Task Analysis No Industry Evaluation Varies from trainer to trainer	Performance based Systematic Approach to Training (SAT), Specific Industry Sector (SIS) Job Task Analysis (JTA) identifying skills and knowledge, No time reqts.
Experience	Time Based (no guidance) No JTA, No Qual/Card, Varies from employer to employer	Time Based (no guidance) No JTA, No Qual/Card, Varies from candidate to candidate	Qualification Card based on JTA identifying skills and knowledge, No Time reqts.
Written Exams	Employer controlled based on general Body of Knowledge, No standard level of difficulty, Varies from employer to employer, No provision to evaluate effectiveness	Central Certification Body (CB) controlled based on general Body of Knowledge, Varies from CB to CB, No provision to evaluate effectiveness	Exams based on JTA and Psychometrics establishing skills/ knowledge/difficulty. Controlled by ASME CB with SIS oversight to assure performance

	(A) ASNT-TC-1A/CP-189 (B)	ISO 9712	ANDE-1
Practical Exam	Employer controlled, No JTA for skills criteria, no standard level of difficulty, Varies from employer to employer. Costly-limited sample set, may not include all conditions	Central Certification Body controlled, No JTA for skills criteria, Varies from CB to CB and may not include all applicable conditions	Developed by Specific Industry Sector-Based on JTA established skills. Includes both in-service and fabrication flaws, controlled by ASME CB
Certification	Employer Non-Transportable	Central CB Transportable	ASME CB Transportable
Annual Continuity	(A) Periodically (B) 12 Months	12 Months	12 Months
Recertification	(A) 5 Years Satisfactory Performance (B) 5 Years Satisfactory Performance w/10 Re-Exam	5 years Satisfactory Performance w/10 year Re-exam	Option 1-annual performance evaluation and Qual/Card, Option 2-Reexamination at 5 years
Independent evaluation of program effectiveness	No provision to share effectiveness between separate employer based programs. No provision to share operating experience or address individual performance issues.	No provision to require industry and CBs to share performance experience or individual performances issues.	Step 5 of the SAT process requires continuous feedback for training, experience and examination improvement. ASME CB monitors prog. perf. with SIS oversight

8

SETTING THE STANDARD

Systematic Approach to Training

- **SAT Process – Five Phases**
  1. **A**nalysis
  2. **D**esign
  3. **D**evelopment
  4. **I**mplementation
  5. **E**valuation

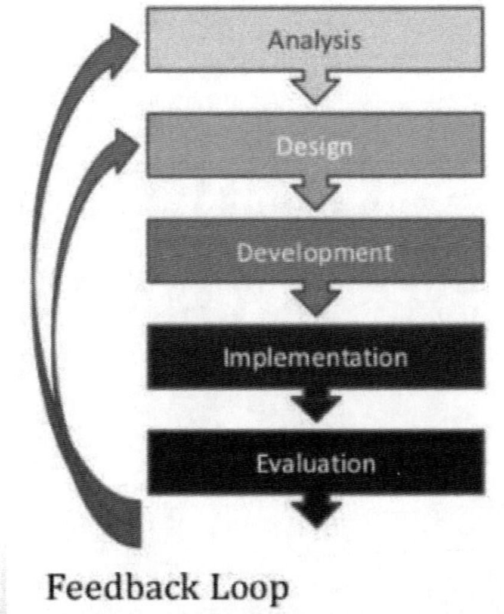

9

ANDE Program

Sample of ANDE Ultrasonic JTA

JOB/TASK ANALYSIS
ULTRASONIC

Duty Code: 100	Job Duty Area: Ultrasonic (Basic)							Date: 08/15/2013
Task ID Number: 1.0	Task Description: Review Historical Documentation							
Element No.:	**Performance Steps (Elements)**	***D**	***I**	***F**	**S**	**K**	**Skill (S) Knowledge (K)**	**Level I II or III**
1.01	Inservice - Obtain data If documentation does not exist-obtain weld profile of the applicable component FAB/Construction – Receive instruction on examination requirements from Traveler/Process Sheet	2	2	2	S	K	Ability to understand required data How to obtain weld profiles	I, II
1.02	Verify data is for applicable component Refer to supervisor/job coordinator if discrepancies exist	2	4	2	S	K	Comparison of data against assigned component	I, II
1.03	Review Technique	3	4	2		K	Understand ultrasound theory and techniques	I, II
1.03A	Compare previous technique to current requirements	2	5	4		K	Ability to distinguish new requirements	I, II
1.03B	Determine impingement angles/Review procedure	4	5	5	S	K	Understand the purpose/requirement of impingement angles Ability to perform calculations (geometry and trigonometry)	I, II
1.04	Review Coverage Limitations	3	5	5		K K	Comprehension of UT theory (geometry, trigonometry, algebra) Understanding the requirements for coverage calculations	I, II
1.04A	Determine component configuration	2	5	5		K	Understanding process pipe and component installations	I, II
1.04B	Ascertain if configuration/assembly of adjacent components interfere with examination	2	5	5	S		Capability to determine potential component restrictions based on technique and incident angle to be utilized	I, II

*DIF Determination (see attached definition and numerical rating)

Page 2 of 31

10

Qualification Cards (Experience)

11

ASME SETTING THE STANDARD

Background

- Current prescriptive on-the-job (OJT) training/experience does not provide performance measurement for meaningful learning opportunities
- Time based only and does not recognize training effectiveness and learning abilities
- Effective feedback when detecting and evaluating actual flaws is seldom provided since they are rarely encountered.
- Varying knowledge and skill of mentors

12

ASME SETTING THE STANDARD

Qualification Cards

- Qualification cards are used to specifically identify the essential elements of each task
- Provides a process to demonstrate skills proficiency at a specified performance level, time is not a factor
- Field OJT is structured and, where applicable, complemented by structured laboratory experiences.
- Timely and effective feedback is provided and assessed by an experienced and knowledgeable mentor.

ASME
SETTING THE STANDARD

13

ANDE UT Qualification Card

- UT Carbon Steel Pipe-to-Pipe, Page 2 of 23

QUALIFICATION CARD

CANDIDATE: ______ ID: ______

METHOD: Ultrasonic (Basic) JOB DUTY/ ENDORSEMENT AREA: Carbon Steel 6" or 12" Pipe to Pipe Job Duty Area: Pre – examination

ACTIVITY TITLE: Inspection Request/Component Information

Applicable Procedures, Drawings and/or Reference Documents:
Need to add reference documents

TASK ELEMENT NO*	ACTIVITY	Action Code	Candidate Initials/ Date	Level II & III Initials /Date	COMMENTS / REMARKS
	TASK: Obtain Component Information • Pipe to Pipe (6" or 12" diameter),				Qual card for each component configuration for a total of nine if candidate completes all activities.
1.01, 1.02	Assignment Sheet: Review: request, work order document(s), component specifics (configuration, material type and thickness, limitations, etc.), Code requirements, applicable drawings,	P			Will require a typical assignment sheet/ work request document. Component configuration drawing
2.05	Select/Review ultrasonic procedure: Ensure: applicable code requirements, scope vs. component to be examined etc.	P			Need procedures for each component configuration

* See JTA "Ultrasonic Basic" for identified elements and the associated skills and knowledge.

I have received adequate experience for the task as listed above and can perform this task in a proficient manner.
Candidate: ______ Date: ______
I have evaluated the subject Candidate in the above task and the performance results are acceptable.
Final Evaluator: ______ Evaluation Action Code(s) ______ Date: ______

Action Code (P) Perform (S) Simulate (D) Discuss Page 2 of 13

ASME
SETTING THE STANDARD

14

ANDE Program

The ANDE Program provides a new and unique approach to NDE and QC personnel qualification as compared to currently used national and international schemes

It implements the most effective currently know elements, including:

- Systematic Approach to Training (SAT)
- Detailed Job Task Analysis (JTA)
- Defining Experience Requirements (Qualification Cards)
- Independent, psychometrically validated third-party written and practical qualification examinations

ASME
SETTING THE STANDARD

15

Funding

- Funding Support:
  - Utilities $600K
  - Department of Labor $1.5M
  - NRC $200K ($ and Materials)
- Current Total Funding $2.3M
- Initial Project Estimate for Funding $2.5M

ASME
SETTING THE STANDARD

16

ANDE Program Status

- November 18, 2015 ANDE-1 approved by ANSI and published by ASME as an American National Standard. A unique performance based approach unlike any other NDE standard or guideline.
- Practical demonstration test pieces designed by a special industry task group and the Pacific Northwest National Laboratory to allow the measurement of Probability of Detection (POD).
- UT exam data base developed in accordance with JTA/Psychometrics beta tested.
- The UT Qualification Card approved.
- Other NDE/QC JTA's, examinations and qual cards are in different stages of development, review and approval.
- UT on-line training and web based simulation to be beta tested during CSCC Associates Degree Program class.
- Implementation scheduled for 1th quarter 2017.

17

ASME

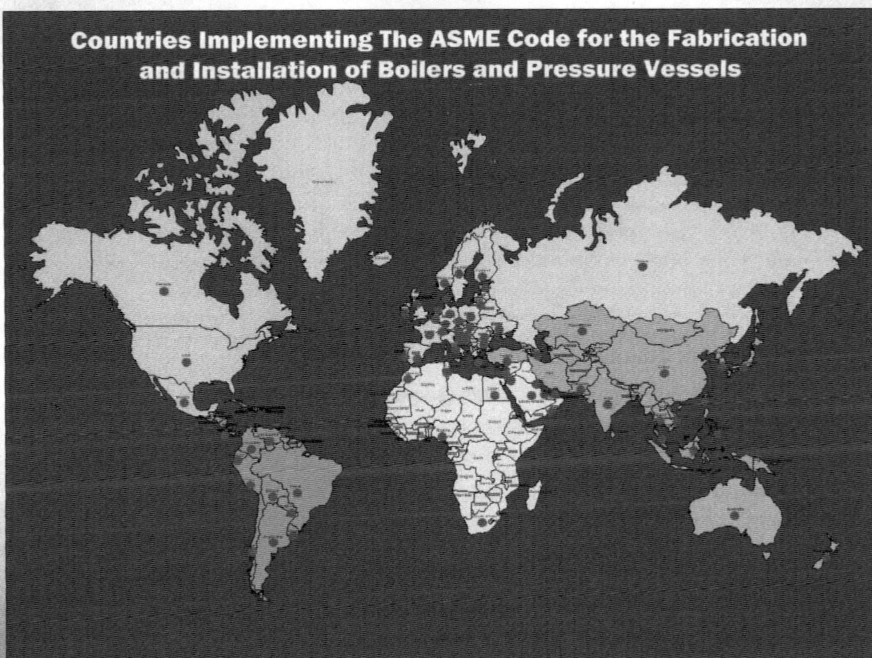

18

SAT/QUAL CARDS

(Structured Performance Based Training/Experience)

"To Me It Seems That All Sciences Are Vain And Full of Errors That Are Not Born of Experience, Mother of all Certainty"

Leonardo Da Vinci

(1452-1519)

ASME SETTING THE STANDARD

19

ASME

SETTING THE STANDARD

True Advancements for Longitudinal Weld Pipe Inspection in PA

Philippe RIOUX [1], Jonathan TURCOTTE [1], Philippe CYR [2]
[1] Sonatest AP, Quebec City, Canada
[2] Eclipse Scientific Inc., Quebec City, Canada

Contact e-mail: riouxp@sonatest.com; turcottej@sonatest.com; pcyr@eclipsescientific.com

Abstract. Phased Array inspection of curved parts brings some real technical challenges. Evaluation of seam welds on small diameter and heavy wall pipes is difficult to accomplish using actual tools that, most of the time, are based on a flat surface model. This often leads to erroneous and imprecise flaw location and sizes. Furthermore, interpretation of the S-Scans becomes very difficult in these circumstances. Trying to take into consideration the effect of the convex surface reflection is something the current instrument imagery is not taking into account and is very difficult for the operators to correct themselves. The typical tools like presenting an overlay of the weld just can't be used in a practical way.

This paper presents a totally new approach of inspecting seam welds and more generally curved parts. The novel method rethinks almost all aspects of the traditional implementation. Starting from an original beam focusing that allows the technician to define a specific focal points in a zone of interest. As well, this paper describes how the live data is represented in a live 3D scan plan in order to facilitate interpretation and identification of flaws. Corrected A-Scan views are also produced, allowing to perform accurate measurements. This paper finally presents a solution to calibrate accurately the instrument prior to performing these inspections on a curved surface.

Introduction

Longitudinal welded pipes are manufactured according to different specifications and with various wall thicknesses. These products are utilized in a number of diverse industries and are quite frequently used, would it be for power plant construction, water supply, nature gas and petroleum transport.

As for other welded parts, these pipes need be inspected for defects, both at the time of manufacturing and for maintenance purposes. Using phased array (PAUT) for flaw detection poses numerous problems, as the contact surface is curved. Moreover, not only the outside diameter (OD) causes some difficulties for the interface, the inside diameter (ID) will reflect each beam of the S-Scan at changing angles, as the reflection comes from a convex surface.

From our experimentation, we were able to conclude that the usual PAUT imaging was a less appropriate model to use in the case of curved surfaces because of beam divergence. This beam spread is obviously due to the curved OD interface but mostly to the convex ID reflection. In the end, this namely influences:

- The spatial coverage or acoustic density in the zone of interest

- The precision of the beam focusing
- The interpretation of data and imaging in S-Scans and A-Scans
- The measurements accuracy
- The calibration methods

1. Theory

1.1 Geometry Challenge

Nowadays, displaying a PAUT S-Scan with an overlay representing the weld is common, broadly used and widely accepted. This is a great aid in order to locate echoes produced by the part geometry itself or by potential flaws. This approach is simple and easy to understand because the reflections are coming from a flat surface, hence producing a perfect reflection as the one produced by a flat mirror.

In the case of flat surfaces, the skips of the sound path are dealt with by representing the weld overlay as a series of folded images. In the case of a curved surface, this method is more challenging to use, as the reflected image formed by the convex surface is distorted. An alternate representation would instead use the profile of the part, and draw folded beams, taking into account the specific reflection angle of each beam. Figure 1 visually shows these expected behaviours.

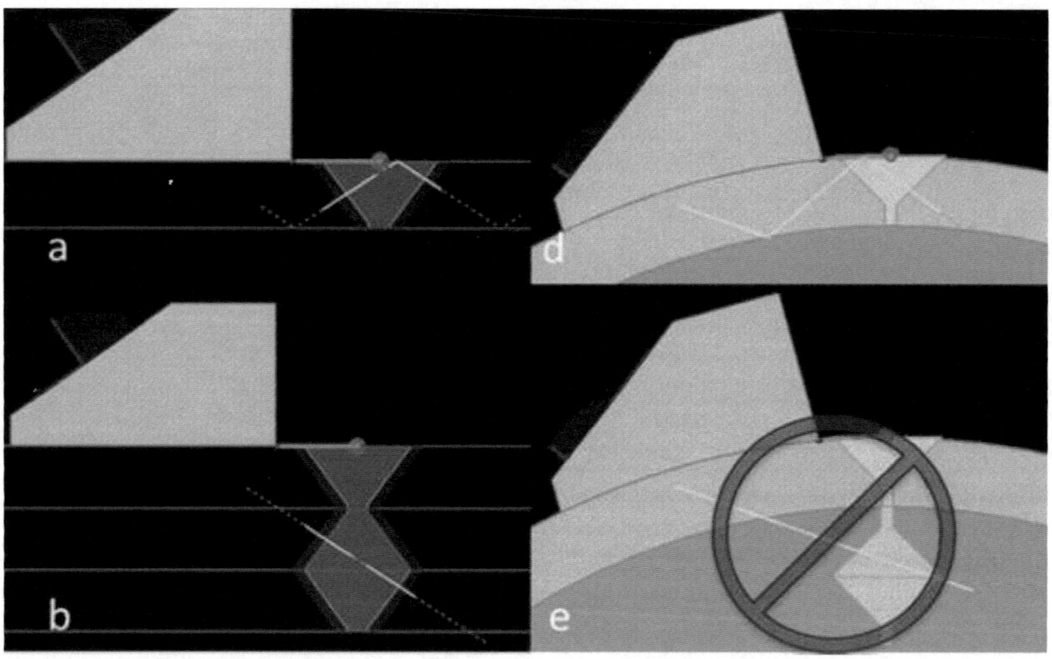

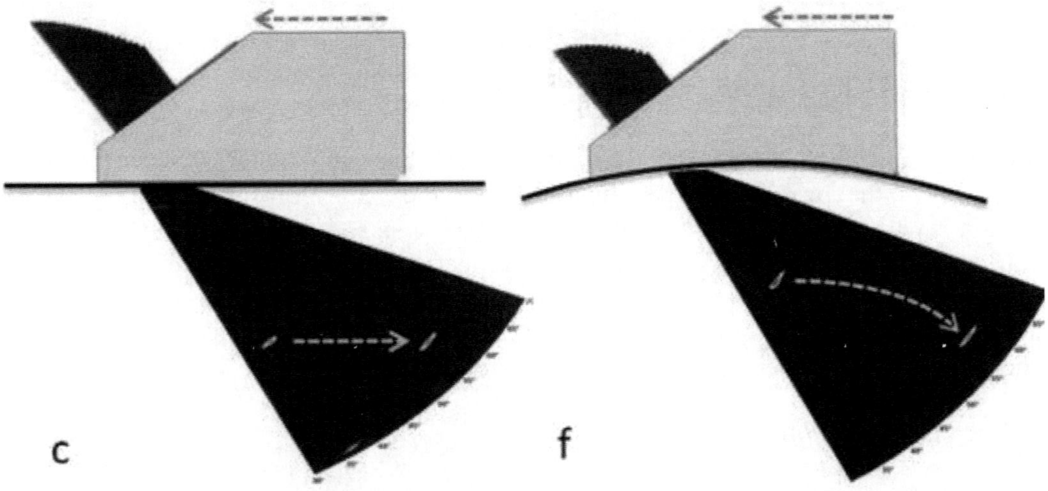

Fig. 1. Showing important differences between curved and flat surfaces imaging

Legend:

1a) Real sound path in the flat part
1b) Unfolded representation of the sound path (same as image 1a – as usually represented on instruments)
1c) The way an indication will "travel" in the S-Scan when the probe is moved backwards on a flat part
1d) Real sound path in the curved part
1e) Invalid representation of the sound path when using an unfolded overlay on a curved part
1f) The way an indication will "travel" in the S-Scan when the probe is moved backwards on a curved part

1.2 Beam Resolution versus Convex Surface

It shall also be understood that the beam spread, and hence the angular resolution, will greatly vary depending of the diameter of the part. For a reduced diameter pipe, the angular resolution in the second leg is significantly and unevenly reduced, lowering the spatial coverage, which may result in poor probability of detection. The smaller the diameter of the pipe, the larger the inter-beam distance in the second leg. Moreover, after the first reflection, the beams' energy density is also decreased due to this convex lens effect. For the same reasons, when producing focal laws using the traditional constant path focus method, the focusing points are affected, decreasing the image sharpness.

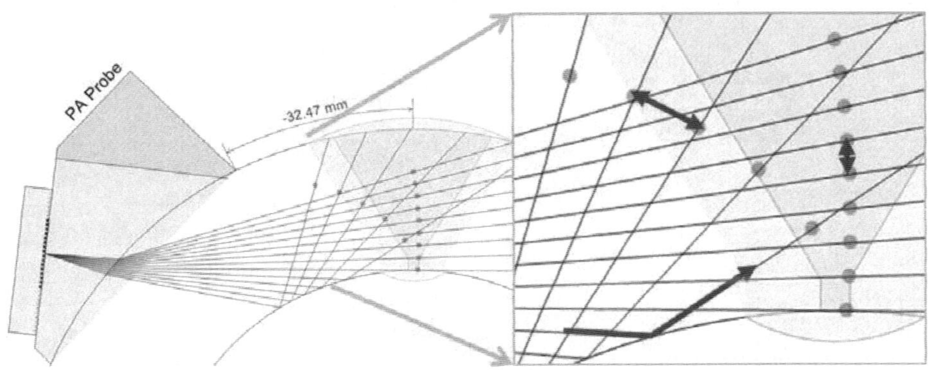

Fig. 2. Effect of ID reflections

2. Proposed Approach

2.1 Using the Part's True Representation

The new approach uses a true representation of the curved part. By using the main characteristics of the pipe, a precise illustration of the part's profile is drawn. Then, employing a ray tracer, a correct illustration of the beams is achieved, both in the first and consecutive skips, taking into account a different reflection angle for each beam. The above tools allow the technician to assess full coverage of the zone of interest is achieved. By offering these tools directly on the instrument, the inspection's configuration and required adjustments can be performed live, as needed.

2.2 Focal Law Calculations

As mentioned above, the convex nature of the part causes the S-Scan beams to diverge, reducing the beams' spatial density in the second leg. Of course, this can be compensated by increasing the angular resolution of the S-Scan. Yet, this does not guarantee a known spatial resolution in the zone of interest, most often the weld itself.

A better methodology would be to define the required spatial resolution of the beams in the region of interest, and use this parameter as a key input to the focal law calculator. It consequently ensures a proper coverage of the beams, with a known spatial resolution, totally independent of the curvature of the ID.

Interestingly enough, this method would be applicable in all circumstances. For example, for a flat specimen, in the case of longer sound path, the exact beam density might be difficult to figure out or calculate. By defining focus points with a specific spatial resolution, the focal law engine automatically provides effective coverage and beam density, as shown in figure 3. Here again, all tools required to define the spatial resolution, specify the focus points and calculate all resulting focal laws are embedded in the instrument for maximum flexibility.

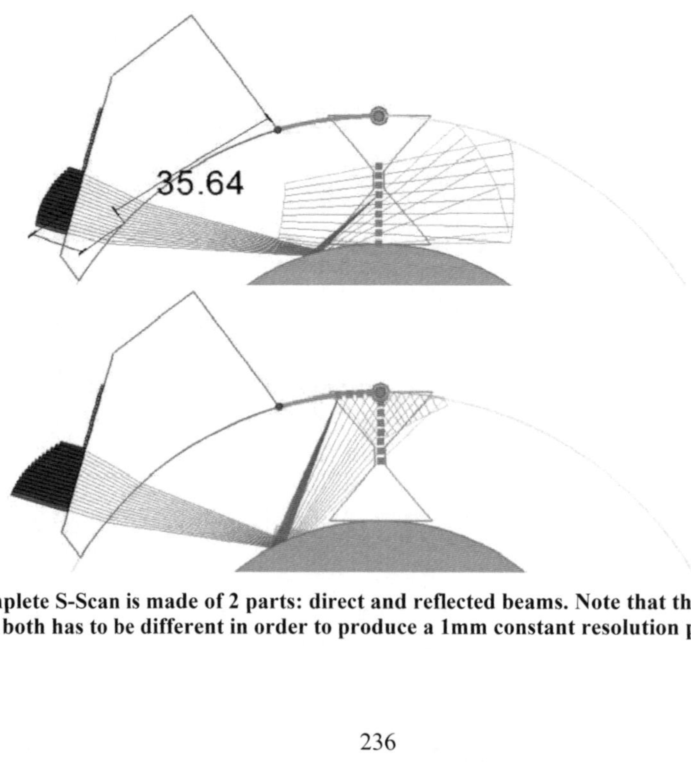

Fig. 3. The complete S-Scan is made of 2 parts: direct and reflected beams. Note that the angular resolution of both has to be different in order to produce a 1mm constant resolution pattern.

3. Data Interpretation

3.1 S-Scan

The basic sectorial scan will remain the same in the way it will be displayed. On the other hand, it shall be noted that the depth scale (usually the vertical scale) cannot be shown since, above the fact it is not linear, it is different for each beam of the scan. Again, this is a result of the curvature of the part. As an example, figure 4 below shows that for beam number 2, the depth to path distance equation is non-linear.

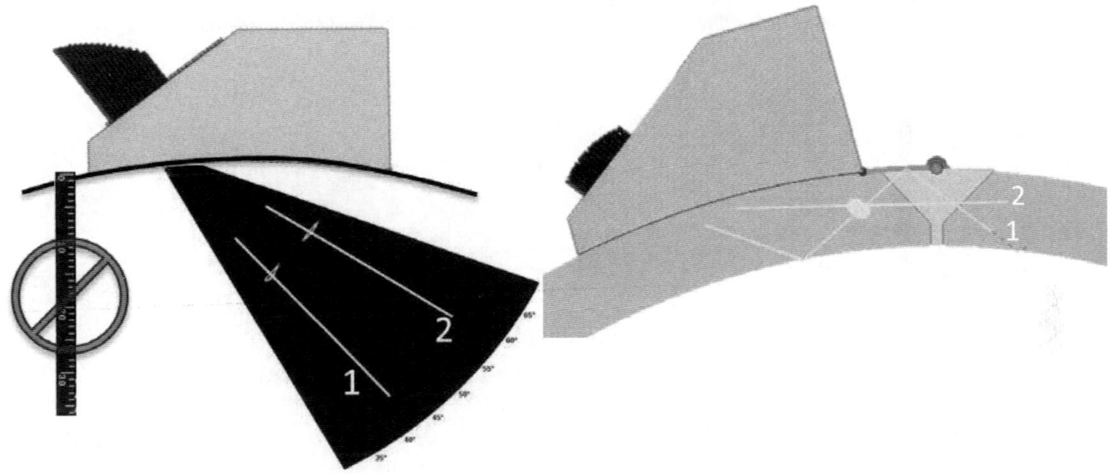

Fig. 4. Sectorial Scan view showing depth ruler not applicable

When using the standard sectorial scan on curved surfaces, it is also more challenging to interpret the position of the echoes that are revealed. The analogy of the distorted image produce by a convex (or concave, as a matter of fact) mirror applies. For one to be able to correctly interpret positions in the S-Scan taking onto account the effect of the convex bottom of the part would require remarkable cognitive abilities.

3.2 A-Scan

In the case of the A-Scan, the rulers need be adjusted. In depth mode, the ruler's scale is specific to each A-Scan. Of course, the sound path ruler remains standard. Additionally, the surface distance needs be calculated properly, taking into account the circumference. This allows to physically spotting the indication properly. Refer to figure 5, below. Note that a vertical dashed line separates the different legs in the A-scan view.

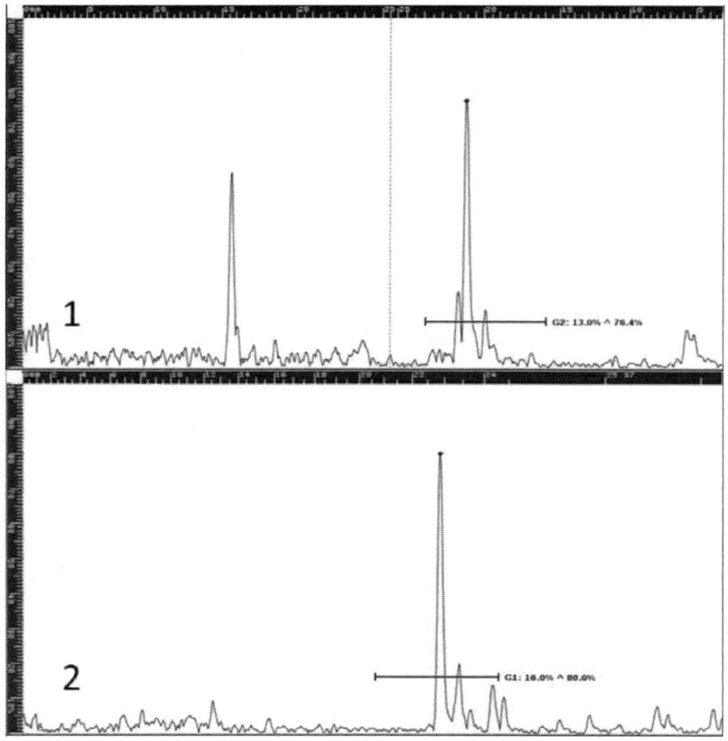

Figure 5 A-scan #1 refers to beam #1 in figure 4 and A-Scan #2 to beam #2

3.3 3D View with Ray Tracing

As a complement to the A-scan, the 3D view renders the specimen. It also shows the beam extractor(s) from the S-Scan. That is, the selected beams are represented in the part using the embedded ray tracer. The ray tracing also highlights the gates defined in the A-Scan, displaying these at the proper distance(s) in the sound path. The triggering point's exact position is also rendered and appears at the correct location in the part's model. Hence, by establishing a direct link between the S-scan and the 3D view, the instrument provides invaluable help to the user. Echoes, and namely flaws, could be located much more easily and with confidence. An example of the 3D view is presented in figure 6.

4. Results Presentation

4.1 Effect of Focusing Method

The experimentations conducted showed that the traditional focusing method (constant path) was perfectly usable and not greatly affected by the curved ID. This is mainly related to the fact that the limited number of active elements provide a much greater depth of field. On the other hand, additional results confirmed that a focusing degradation was noticeable using more than 22 elements.

As well, scans performed with 32 elements, producing a shallow depth of field, we providing poor image quality. In that case, the advantage of the new focusing method presented earlier is obvious. Figure 6 presents the results obtained.

On the other hand, it is sometimes desirable not to focus, namely in the case of huge volume inspection. This "steer only" scan guarantees a large depth of field and a constant energy distribution along the beam. This focusing method has also been implemented for completeness, as it addresses different needs.

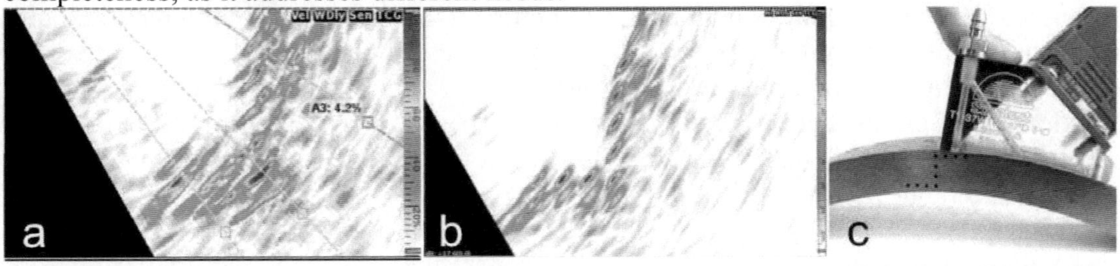

Fig. 6. Comparison of conventional method of focusing and new method proposed. Using 32 active elements. 6a) Constant path focusing pattern; 6b) Constant resolution focusing; 6c) Demo block used

4.2 Complete Layout of Data

The acquisition layout implemented, as presented in figure 7, should ease the interpretation of the seam weld inspection data. The different views complement each other. The S-scan offers a general view of the data that is being recorded. The A-scan displays the amplitude at one specific angle and offers a depth ruler. Quick measurements such as percentage full screen height (%FSH), true depth, surface path, sound path are available. The 3D view gives the ray tracing corresponding to the extractors shown in the S-Scan. The S-Scan view is not folded into the pipe in figure 7. The legs are drawn where it bounds. Hence, multiples reflections are not conflicted each other if it would have been folded.

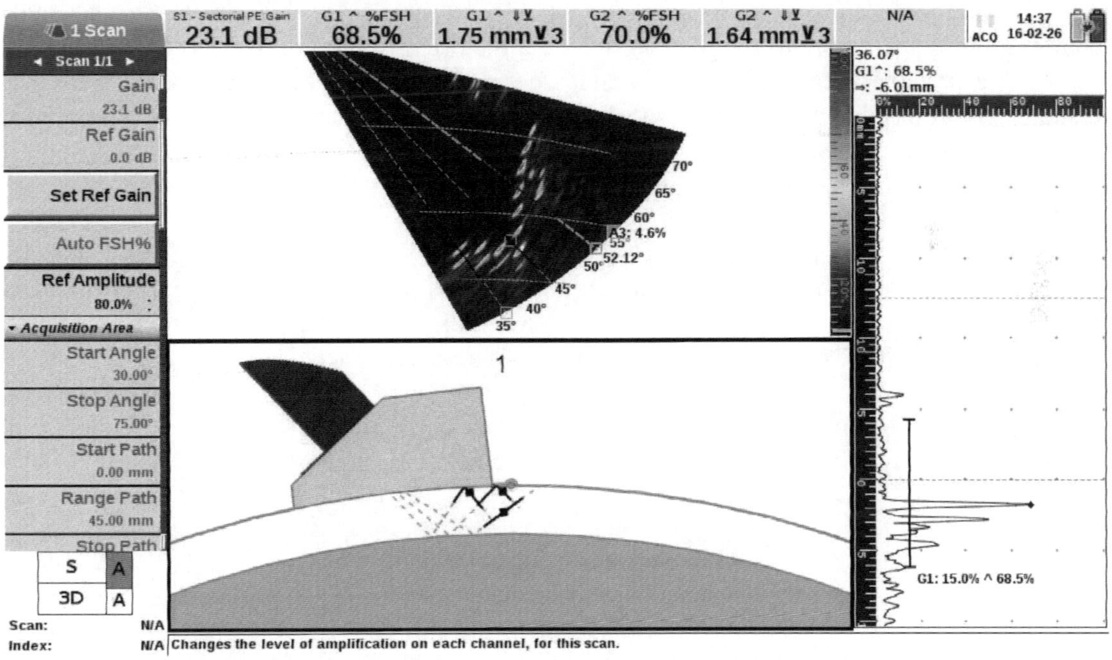

Figure 7. Recommended Acquisition Layout; ID indication on the A-scan and Confirmed in the 3D View

5. Calibration Method

The usual techniques to calibrate velocities and delays remain the same. Calibration using a radius gives very accurate results. Notches and SDH calibration are nonetheless possible, yet not as good as the previous method.

It is actually a different story in the case of the DAC/TCG calibrations. The quality of the calibrated or corrected curves is improved when a curved version of Navships test block is used. The number of points for each half skip makes the DAC more precise as this helps to better consider the curvature. The 3-point calibration using bottom and top notch in a pipe is the minimum acceptable in the industry. The closest sound path datum is direct, the second comes from the inner rebound, and the third is produced by both the inner and outer reflexions. This means that to cover a full skip, up to 6 DAC/TCG points shall be obtained and recorded. Nine-point calibration shall be quite common, considering many of the pipes will present a limited thickness.

Please note that at the time of writing this paper, the exact and optimal calibration technique was not yet determined.

6. Conclusion

In this paper, a method to improve the PAUT inspection of seam welds was presented. This approach is in fact applicable more generally to any curved surface component. The proposed practise namely offers a proper coverage of the region of interest, by guarantying accurate focus at predefined points. Moreover, the distance between these defined positions define the beams' spread, insuring a sufficient beam density, hence providing a known spatial resolution.

As the interpretation of the S-Scan in the second and third leg proves to be quite challenging by the nature of the part producing a distorted reflection, it was also shown that by using a true representation not only helps define the coverage, but also can be used to visualize more easily the location of defects. By representing specific A-Scans associated to the S-Scan extractors, and by displaying the gates and gate trigger exact position, the 3D view provides an additional and valuable interpretation tool.

Experimental results have shown that the method provides accurate and precise measurements. These tests have been conducted on a number of parts with different diameters and wall thicknesses. At the time of writing, real life testing, in the field, is just to begin and the final calibration strategy shall also be determined.

Finally, as the concept shall also include the proper calibration methods, it is believed to potentially comply with industry standards, like ASME V as an example.

References

[1] GINZEL, Ed, To Focus or Not To Focus, ndt.net forum discussion topic, August 2015

Flaw Detection and Characterization Using Ultrasound

Alex Karpelson

Kinectrics Inc., Toronto, Canada

Abstract
Results of the ultrasonic inspection of tubes sometimes are not satisfactory due to inability to characterize a flaw (i.e. determine its type, shape, orientation and dimensions). It is therefore worthwhile to develop some special simple approaches, which could significantly improve inspection capabilities. Such approaches include new techniques and novel ultrasonic transducers (e.g. a large normal beam probe, a few transducers working simultaneously in parallel and "looking" at the flaw from a variety of angles and directions, non-rotating probes, "tangential beam" technique, collimating probes, guided wave technique, and non-contact laser-UT technique) and also modification of the existing software for data analysis. As a result, it is possible to improve various inspection capabilities, identify flaw type, "reproduce" its shape and orientation, characterize flaw, determine its main parameters, and size it.

Keywords: ultrasonic inspection, novel techniques and probes, improved flaw characterization and sizing.

1. INTRODUCTION

Ultrasonic (UT) methods, currently used for tube inspection, are not always adequate for flaw characterization. Utilities occasionally experience problems trying to characterize a flaw (particularly, crack or an off-axis flaw) and define its shape, orientation, and size. Typically, the Normal Beam (NB) longitudinal waves and angle shear waves employing the Pulse-Echo (PE) and Pitch-Catch (PC) techniques are used for tube testing to detect, characterize, and size flaws located within the tube wall, on the Inside Diameter (ID) or on the Outside Diameter (OD) [1]. (Terms ID and OD are also used in context of the inside and outside surfaces, respectively). Modified or newly-developed inspection techniques, UT probes, and software should be employed to improve flaw identification and characterization, increase sensitivity and accuracy of measurements, decrease analysis time, improve efficiency of analysis, and decrease number of required replicas [2-9].

2. EQUIPMENT AND SPECIMENS

A computerized scanning rig with a calibrated UT system was employed for experiments. The UT system includes the Winspect™ data acquisition software, a SONIX STR-8100 digitizer card, and a UTEX UT-340 pulser-receiver. Standard techniques and probes [1] and novel techniques and probes [2-9] with different center frequencies, various aperture diameters and different focal lengths were used for testing. The rationale was to determine techniques and probes, which could provide the best and fast flaw characterization and high sizing accuracy. The scanning rig was configured to mount the tube specimen filled with water on the turntable. Transducers were put in the appropriate positions inside the tube. Then 3D B-scans, covering flaws and gating the respective NB, PE and/or PC responses, were performed.

A large number of specimens of different tubes (Pressure Tube (PT), Steam Generator (SG) tube, Feeder Pipe (FP), and Heat Exchange (HX) tube) with various types of flaws were used for testing. These specimens contained pits, cracks, hydrided areas, rectangular and V-notches (symmetric and asymmetric) with various widths, depths, orientation angles (axial, circumferential and off-axis notches), root angles, inclination angles, and root radii. Rationale was to determine the ability and sensitivity of different techniques and probes to characterize various flaws (particularly cracks and off-axis flaws), to determine their orientations and shapes, to estimate their root radii and inclination angles, and to size these flaws (particularly, the small ones). Special software, presenting 3D (isometric) images of flaws, based on the time-of-flight NB, angle PE, and angle PC 3D B-scans, was developed. The rationale was to produce flaw 3D-images, based on different UT techniques and probes.

3. "QUASI-TOMOGRAPHY" - COMBINED TECHNIQUE

The ability to combine information, obtained by using different UT techniques and probes, and then "reconstruct" the flaw, is the main advantage of the classic tomographic method [2]. During the last thirty years there has been considerable interest in UT computed tomography, because of its significant potential abilities. The X-ray computed tomography (CT), photon emission tomography, positron emission tomography, and magnetic resonance imaging (MRI) tomography are well known techniques, which produce very accurate images of an object's internal features, as well as information about the object's geometry and material characterization. At the same time, the most popular X-ray CT and MRI tomography provide only one type of images in the transmitted waves depicting, respectively, either the X-ray absorption index distribution, or proton relaxation time distribution in the object cross-section.

UT tomography is based on the insonification of each area of the test object by acoustic waves, originating from transducers, positioned around the object and able to transmit and receive signals from different directions and at various angles. Every cross-sectional 2D image can be obtained using e.g. the procedure, at which the first transducer transmits a short UT pulse and all other transducers (including the transmitter) receive the reflected and transmitted signals. Then the second transducer begins to transmit the UT pulses, and the procedure described above is repeated. Then the entire process is applied to the third transducer as a transmitter and so on, until each of the transducers had acted as a transmitter. The main advantage of the UT computed tomography is its ability to generate cross-sectional images (tomograms) of internal structure of the test object accurately depicting distributions of four different material parameters in the transmitted and reflected waves: acoustic impedance, UT velocity, material density, and attenuation index, thus providing high resolution 2D and 3D images. This method can be named "multi-parametric UT tomography". It is obvious that technique, which gives images depicting simultaneously distributions of four different physical parameters across a cross-section, has significant advantages in detectability, sensitivity, accuracy, resolution, and reliability, in comparison with method using only one physical parameter, thus significantly increasing the probability of detection and greatly improving characterization and sizing of "abnormal" areas. On the other hand, one should not forget that resolution of the UT tomographic system is always lower than resolution of the X-ray tomograph, because the "ideal theoretical resolution" is about half of the wavelength, and the UT wavelength is always much greater than electromagnetic wavelength in the X-ray range. However, UT tomography is a very complicated and expensive technique, because such approach is based on the sophisticated hardware (mechanical and electronic) and very special software. Therefore, UT tomography is still on the development stage only, and commercial UT tomographs are not available. So, it is worthwhile to try to develop some special simplified "approximate" tomographic method, i.e. "quasi-tomographic" approach for the tube testing, which will be, on the one hand, rather accurate, and on the other hand, comparatively simple. This method will significantly improve the inspection capability by employing only a few probes, "looking" at the flaw at various angles and from different directions, a few rather simple tomographic techniques, and software combining information obtained from these probes.

Techniques and probes, currently used in the inspection systems for tube testing, allow already "seeing" flaw at various angles and from different directions. Longitudinal and shear waves propagating in the axial and circumferential directions, PE and PC techniques, NB and angle probes are used to detect, characterize and size flaws in tubes. In order to develop a "quasi-tomographic" method, the information, obtained using different techniques and probes, should be combined together. To realize a "quasi-tomographic" technique, one can use a simple method, which just combines information from different transducers [2-3], e.g. by using special software superimposing different responses. The other idea is to connect simultaneously e.g. two angle probes (say, two circumferentially positioned transducers (Clock-Wise (CW) and Counter-Clock-Wise (CCW) shear wave probes) to pulser-receiver working in the PE mode. As a result, both transducers will simultaneously transmit UT signals and both will receive the responses. Each transducer will receive its own signals, reflected from the tube ID and OD, and also signals, transmitted by other probe and reflected from tube surfaces. Subsequently, three techniques will be realized simultaneously: angle CW PE, angle CCW PE, and angle PC. The obtained "combined" image will contain responses typical for these three techniques. In other words, the obtained image will look like three interposed images: CW PE, CCW PE, and PC. This "combined technique" can be realized as 3D or 2D B-scanning at different incident angles and at various probe positions [2-5].

Responses from different probes can be easily distinguished (see Figs. 1-9 below). While the PC response shows depth and width of the notch tip, the lengths and durations of the angle PE responses (CW and CCW) can characterize two inclination angles of the notch and even its root radius, which can be assessed rather accurately, because this technique is based on high axial measurement resolution in the direction of the UT beam trajectory (due to importance, this issue is analyzed separately in Section 13). Typical 2D combined circumferential B-scans of different ID notches with cracks in the PT are presented in Figs. 1-4, which clearly demonstrate that combined technique allows detecting and characterizing even tight and shallow cracks (e.g. delay hydride cracks). Fig. 5 shows combined circumferential B-scan of V-notch in the PT ID. In this combined scan three probes are connected in parallel [2-4] to pulser-receiver working in the PE mode (NB + CW PE + CCW PE + PC). Circumferential 2D combined (CW PE + CCW PE + PC) B-scan of 2.5" diameter FP with axial 1mm deep segment OD notch is presented in Fig. 6. Circumferential 2D combined (Forward (FW) PE + Backward (BW) PE + PC + NB) B-scan of SG calibration tube, containing seven OD circumferential notches 0.1mm wide with different depths, is presented in Fig. 7. Another combined technique (BW PE + "quasi-PC"), used for a small diameter SG tube testing, is presented in Fig. 8 (schematic of this technique) and Fig. 9 (axial combined BW PE + "quasi-PC" 2D B-scan). In this so-called "quasi-PC" technique, one can use one small standard probe working in the PE mode, and attach it to the special two-step acoustic mirror. This mirror allows exciting shear UT wave within the SG tube wall and receiving the UT wave, coming out of the tube wall, by the same probe.

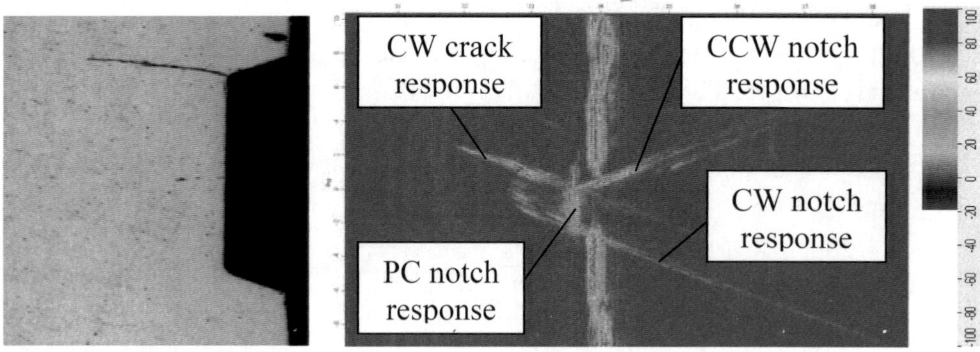

Figure 1. Picture of radial-circumferential PT cross-section and circumferential 2D combined (CW + CCW + PC) B-scan of ID rectangular axial notch 2.54mm wide 0.5mm deep in the PT with fatigue crack 0.8mm deep. Probes: focal length FL=40mm, center frequency f=20MHz, aperture diameter D=9.5mm, water-path WP=19mm, and incident angle $\alpha=27^0$.

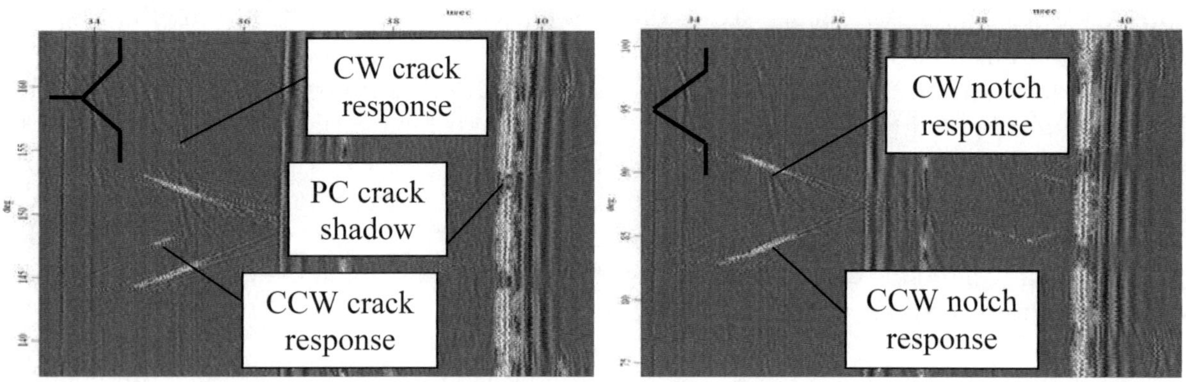

Figure 2. Circumferential 2D combined (CW + CCW + PC) B-scans of the ID V-notch (1mm deep and 45° tip angle) in the PT with delay hydride crack 0.5mm deep (left) and without crack (right). Probes: FL=40mm, f=20MHz, D=9.5mm, WP=18mm $\alpha=27^0$. Color scale is in Fig. 1.

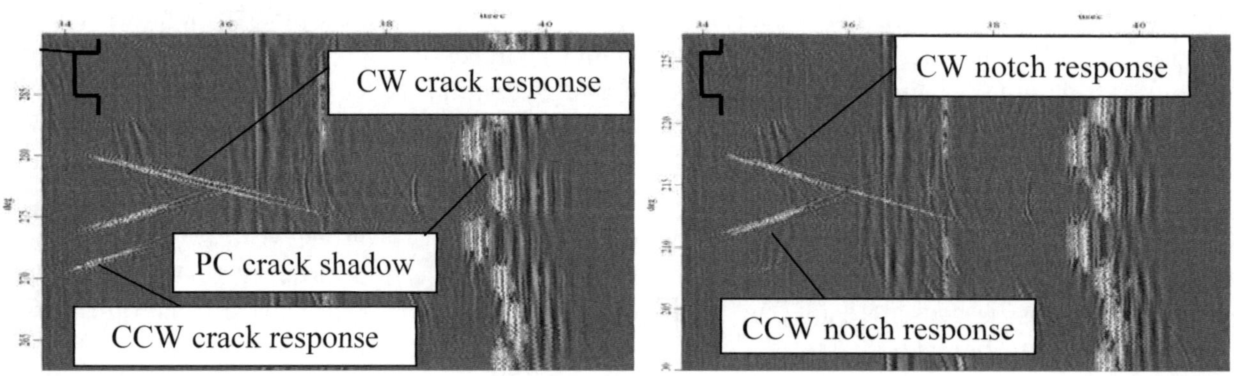

Figure 3. Circumferential 2D combined (CW + CCW + PC) B-scans of rectangular axial notch (1mm wide and 0.7mm deep) with delay hydride crack 1mm deep (left) and without crack (right). Probes: FL=40mm, f=20MHz, D=9.5mm, WP=18mm, $\alpha=27^0$. Color scale is in Fig. 1.

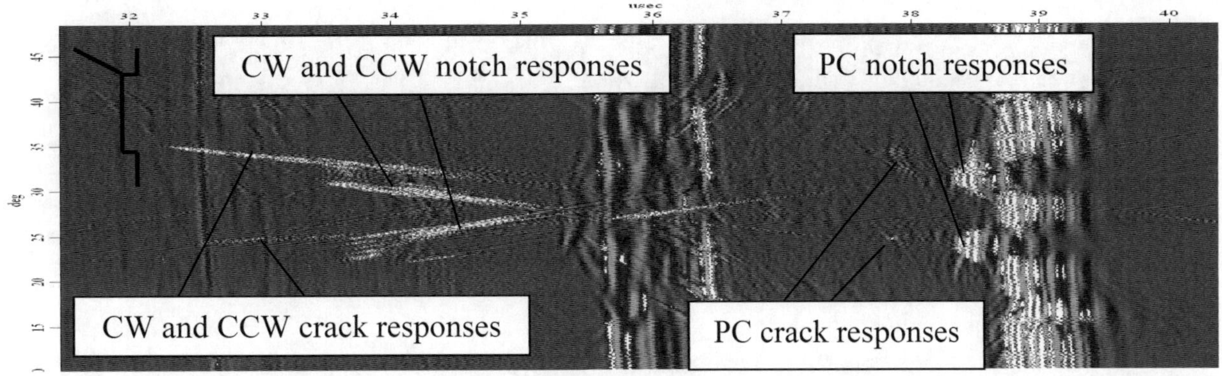

Figure 4. Circumferential 2D combined (CW PE + CCW PE + PC) B-scan of axial rectangular ID notch 0.5mm deep and 2.5mm wide in the PT with fatigue crack ~1mm deep. Probes: FL=40mm, f=15MHz, D=0.5", WP=18mm, $\alpha=27^0$. Color scale is shown in Fig. 1.

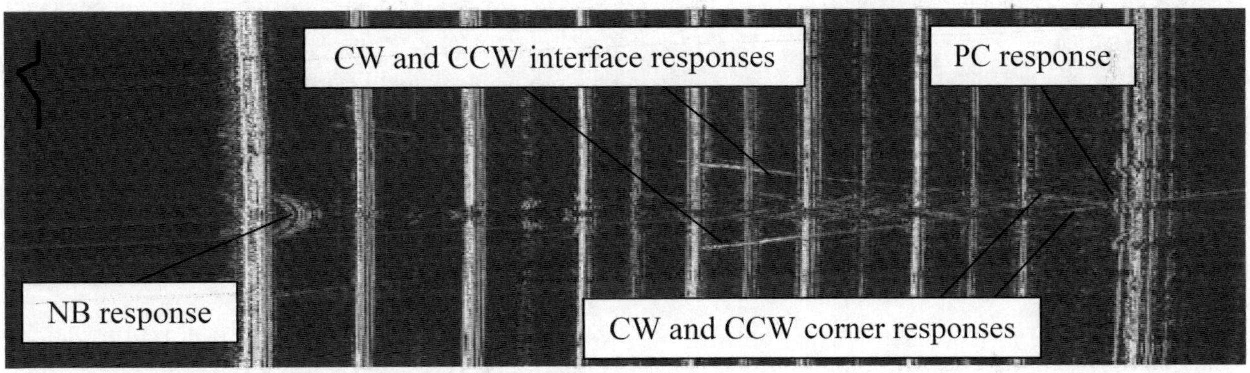

Figure 5. Circumferential 2D combined (NB + CW + CCW + PC) B-scan of $90^0/60^0$ ID axial V-notch 0.5mm deep in the PT. CW and CCW probes: FL=40mm, f=15MHz, D=9.5mm, WP=18mm, $\alpha=30^0$. NB probe: FL=100mm, f=15MHz, D=0.5", FL=4", WP=20mm. Color scale is shown in Fig. 1.

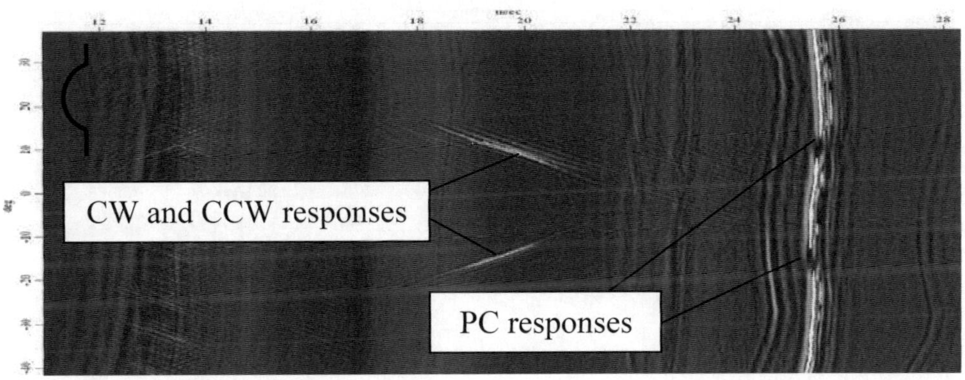

Figure 6. Circumferential 2D combined (CW + CCW + PC) B-scan of 2.5" diameter FP with calibration axial segment OD notch (1mm deep and 1mm diameter). Probes: f=10MHz, FL=38mm, D=6.35mm, WP=15mm, $\alpha=16^0$. Color scale is shown in Fig. 1.

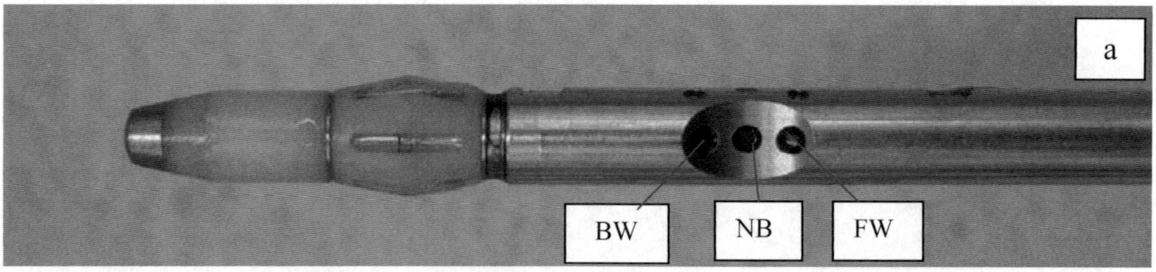

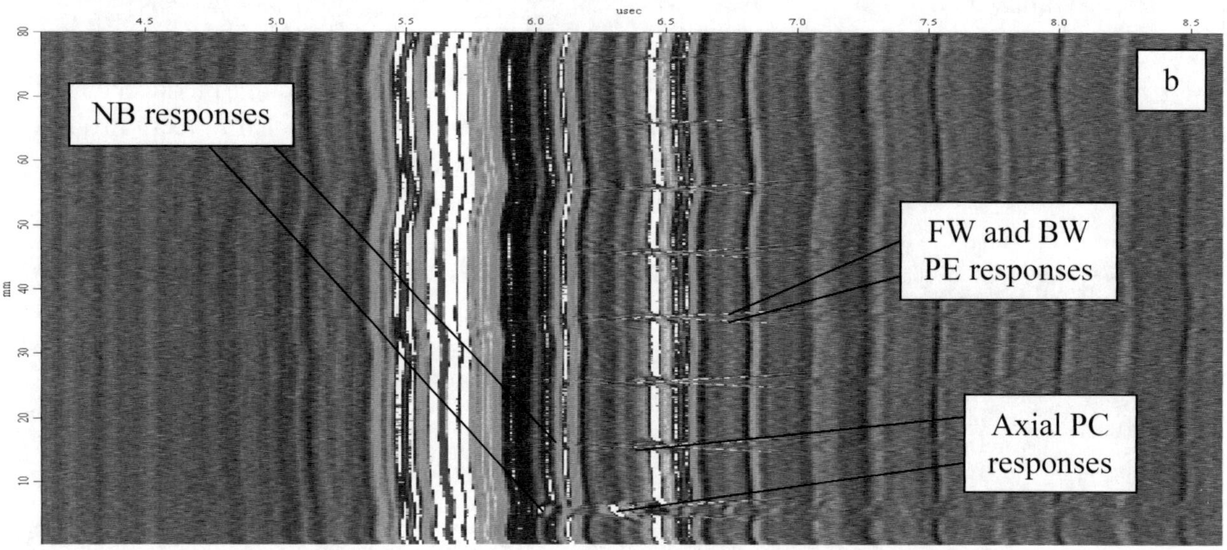

Figure 7. a - small "hybrid" probe with three piezo-elements. b - circumferential 2D combined (FW + BW + PC + NB) B-scan of 7 OD circumferential notches 0.1mm wide with different depths in SG calibration tube (ID=11.4mm, wall thickness 1.24mm). Probes: f=15MHz, FL=6mm, D=2mm, WP=4mm, $\alpha=21^0$ for FW and BW angle probes. Color scale is in Fig. 1.

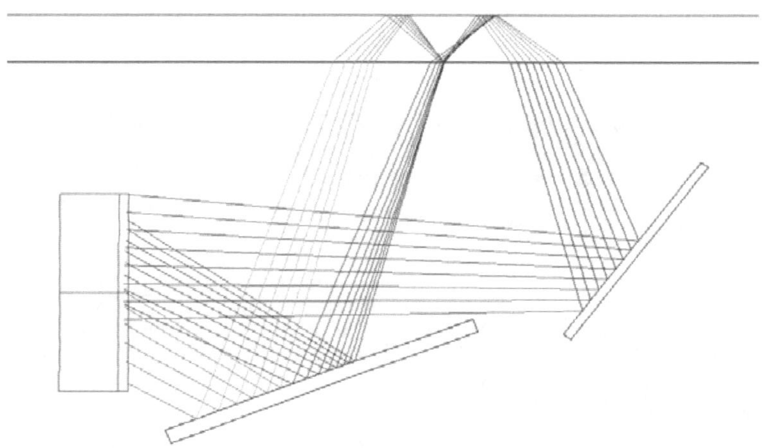

Figure 8. "Imagine-3D" beam tracing simulation of combined technique (BW PE + "quasi-PC"), showing acoustic beams transmitted from probe, reflected from the 1st mirror, refracted and reflected within the SG tube wall (ID=11.4mm, wall thickness 1.24mm), reflected from the 2nd mirror, and received by probe.

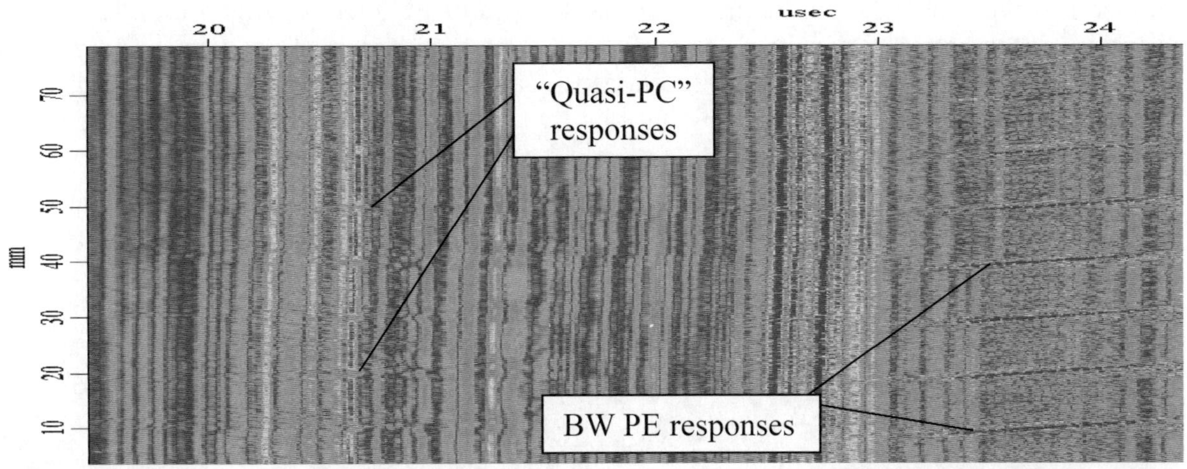

Figure 9. Axial combined (BW PE + "quasi-PC") 2D B-scan of seven circumferential OD rectangular notches 0.15mm wide with depths 0.05, 0.1, 0.2, 0.3, 0.4, 0.5, and 0.6mm in SG tube (ID=11.4mm, wall thickness 1.24mm). Probe: f=15MHz, FL=10mm, D=3mm, WP=9mm, $\alpha=22^0$. Color scale is shown in Fig. 1

4. NORMAL BEAM TECHNIQUE USING LARGE DIAMETER PROBE AND MULTIPLE REFLECTIONS

The large diameter NB probe works not only as a NB transducer, but also, due to its large diameter, this probe transmits and receives the UT waves at various angles, particularly after multiple reflections from the ID and OD. When UT beam passes many times through the area containing a shallow flaw, then even the very weak changes lead (due to the "accumulation effect" after multiple reflections within the tube wall) to some also weak but already distinguishable features of the received UT response. This effect can be also called the "self-amplification" of a weak feature: it is similar to the constructive interference.

Emphasize that during detection and characterization of shallow flaws, the efficiency of "feature accumulation" effect achieves maximum usually at the 3rd reflection, because of the distortion, shift and inclination of the flaw responses after each reflection. Therefore, the flaw image (i.e. flaw shape, orientation and dimensions), obtained as a result of "feature accumulation" effect after multiple reflections, becomes vague, fuzzy and unclear with composite shape and contour. At the same time, during material characterization, the "feature accumulation" effect achieves maximum efficiency approximately at the 7th reflection, because the UT beam, passing through the area of material with specific physical parameters, "feels" their influence many times. Thus, it works like a constructive interference. For example, if the UT beam passes through the area of material with stress concentration (created e.g. due to fatigue), where speed C of propagation of the UT wave is slightly greater than speed C in a "normal" area, then after each reflection from the tube wall the UT response arrives earlier and earlier in comparison with the response related to the "normal" area. Of course, the concept of multiple reflections and "feature accumulation" is valid not only for a NB probe, but also for any angle probe.

The large diameter transducer can be represented as a number of small probes stuck together. The central part of a large transducer works as a small NB probe, while the peripheral parts of this transducer work as small angle probes in the PE and/or PC modes of operation [2-5]. That is why transducer with a large diameter can "see" flaw simultaneously at various angles and from different directions [5]. 2D circumferential B-scans, performed, by using a large diameter probe, on a few PT specimens, containing various ID flaws, are presented in Figs. 10-11.

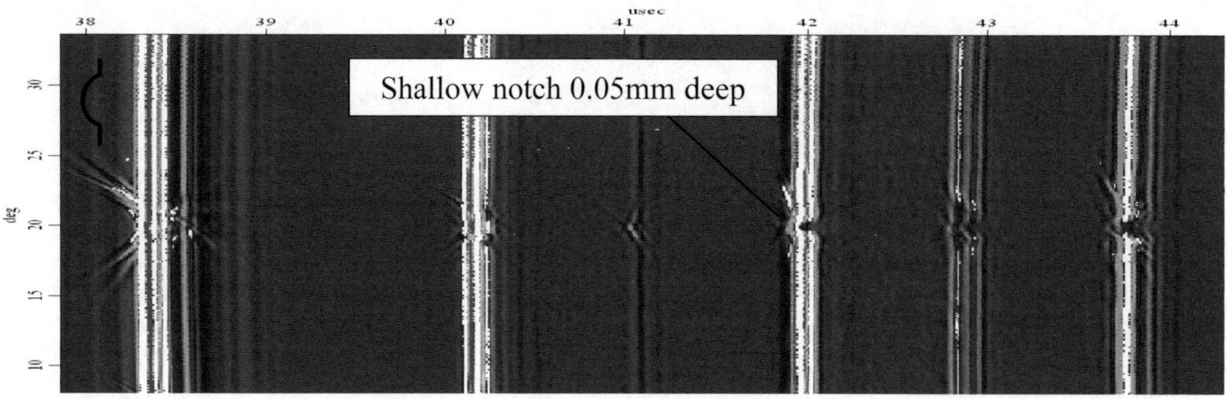

Figure 10. Circumferential 2D NB B-image of axial shallow ID segment notch 0.05mm deep in the PT. Probe: NB, D=9.5mm, FL=50mm, f=15MHz, WP=29mm. Color scale is in Fig. 1.

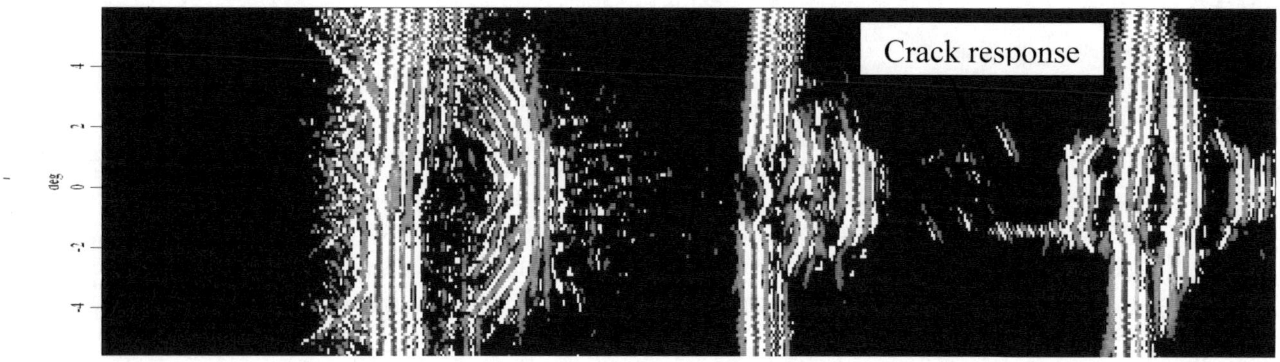

Figure 11. Circumferential 2D NB B-image of ID axial rectangular notch 0.5mm deep and 2.54mm wide with fatigue crack 0.9mm deep in the PT. Probe: NB, FL=55mm, f=15MHz, D=0.5", WP=15mm. Color scale is shown in Fig. 1.

2D circumferential B-image of multiple OD reflections, obtained using PT specimen containing area of material with stress concentration (created due to fatigue), is presented in Fig. 12. The stress (fatigue) area (around 0°) can be distinguished, because of the "feature accumulation" effect, after multiple reflections (starting approximately at the 9th OD reflection), as amplitude drops and distortions of the vertical lines slightly convex to the left due to a greater speed of propagation of the UT wave in the stress area (circled in red). Based on the known formulae for elastic parameters and measuring the amplitude decrease from one reflection to the other and time shift between neighboring responses, a few material parameters (acoustic impedance, UT wave longitudinal and shear speeds of propagation, density, various moduli of elasticity, Poisson's ratio, and Lame constants) can be determined for the normal and stressed areas. The obtained values can be used for stress (fatigue) level characterization and in general for material characterization: stiffness, compliance, compressibility, strength, flexibility, and so on.

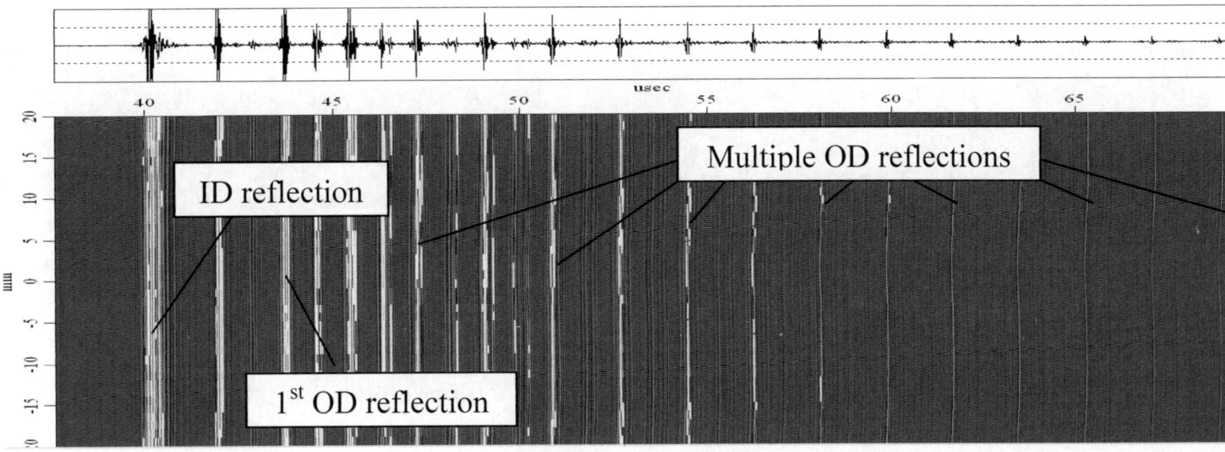

Figure 12. Circumferential NB PE B-scan of the PT with fatigue (stress) area around 0° (circled in red). Probe: NB, D=9.5mm, FL=50mm, f=15MHz, WP=30mm. Color scale is in Fig. 1.

2D circumferential B-scan, performed on a FP specimen with OD axial 1mm deep segment notch by using a large diameter probe, is shown in Fig. 13.

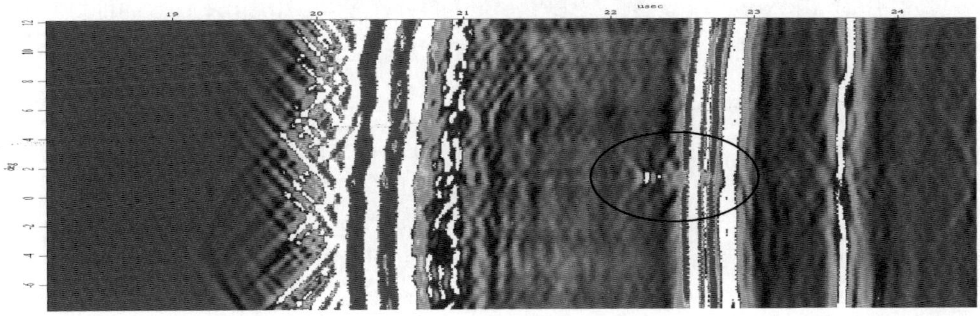

Figure 13. Circumferential 2D NB B-scan of 2.5" diameter FP with OD axial 1mm deep segment (1mm diameter) notch. Probe: NB, f=10MHz, FL=38mm, D=6.35mm, WP=15mm. Color scale is shown in Fig. 1.

The effect of the "feature accumulation" due to the multiple reflections can be also used to detect a "dry contact" between two metal surfaces, as it was described in [6] for detection of areas with dry contact between PT and calandria tube (CT) (see Fig. 14). Another example is detection of "dry contact" area between PT and spacer (see Fig. 15). After multiple reflections within the PT wall, the area with a dry PT/CT contact in Fig. 14 was detected, because at each reflection from the PT outside surface, a very small portion of UT energy from the impinging longitudinal acoustic wave leaks into the CT due to the dry contact between roughness peaks on the PT and CT surfaces. That is why amplitude of the UT wave, reflected from the area with dry contact, should be slightly lower than amplitudes of the waves, reflected from the "clean" areas. Multiple reflections "amplify" this effect, and that is why the dry PT/CT contact can be detected approximately after twelve reflections, see a small amplitude drop (a few percent of the absolute amplitude) in the central area about 0^0 on the circumferential 2D B-scan in Fig. 14.

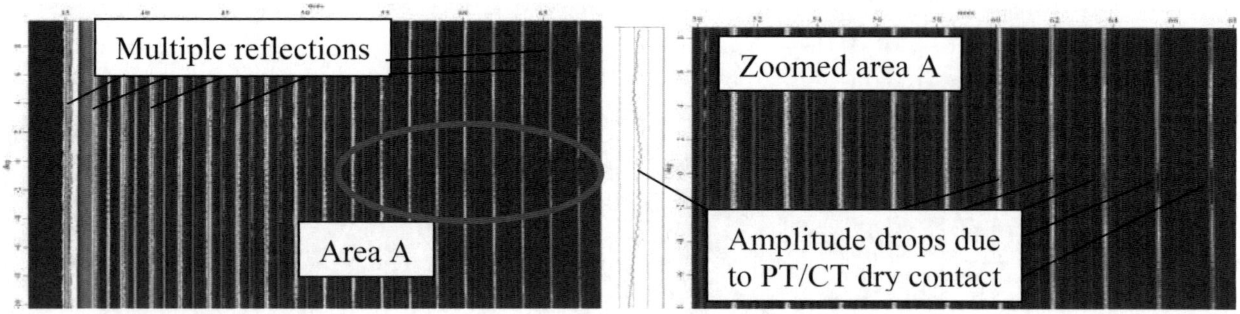

Figure 14. Circumferential NB PE 2D B-scan of the PT/CT area with dry contact. Probe: NB, f=10MHz, FL=40mm, D=9.5mm, WP=27mm. Color scale is shown in Fig. 1.

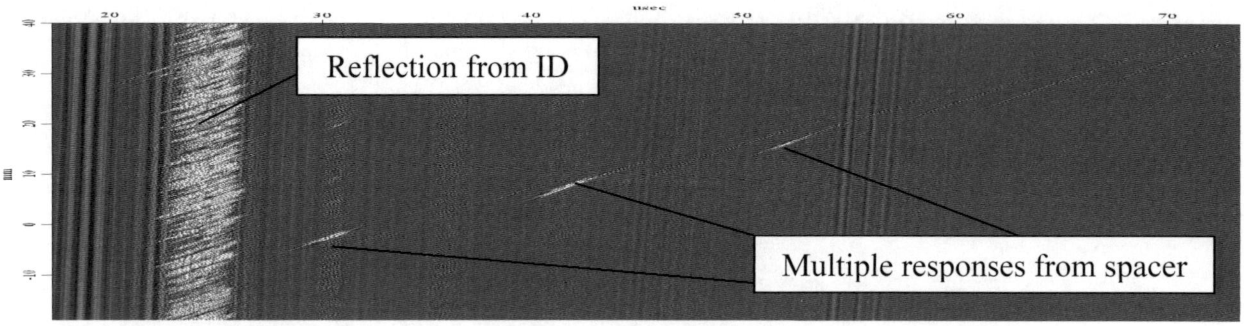

Figure 15. Axial PE 2D B-scan of the PT with spacer touching the outside surface. Angle cylindrically focused probe (f=5MHz, D=9.5mm, FL=40mm, WP=18mm), $\alpha=21^0$. Color scale is shown in Fig. 1.

Fig. 15 clearly demonstrate that garter spring, just touching the PT OD (i.e. without any load) can be detected as an amplitude peak by using angle beam PE UT response from cylindrically focused probe. Large size angle probe provides multiple reflections from area of "dry contact" between PT OD and spacer. Only transducers, "stretched" in the circumferential direction, can detect spacer without load, because these probes receive reflections from contacts between many spacer coils and the PT OD. Standard spherically focused probes cannot detect the contact between only one coil and the PT OD, because such contact provides a very weak UT response.

Also note that indirect UT assessment of spacer wear can be performed based on the direct detection and sizing of the "fretting marks" on the PT OD. It is logical to assume that spacer wear marks are strongly correlated to the PT OD marks, because wear should be mutual on the both friction surfaces: spacer coils and PT OD. Currently used UT techniques and probes (angle probes in the PE mode and/or NB probe 1st, 2nd or multiple OD reflections) can provide a good detectability and measurement accuracy of the PT OD flaws.

The effect of the "feature accumulation" can be also used to detect the PT area with increased hydride content and assess its concentration (i.e. material characterization). Such technique is based on changes in the amplitude and spectrum of the UT pulse, which occur due to scattering on the hydride platelets, when UT pulse passes through the hydrided area, see Figs. 16-18.

It is obvious that response changes, related to the hydride concentration, are weak (see Figs. 16-18). At the same time, signal fluctuations due to various changes in surface roughness and material microstructure (e.g. grain size, material morphology, imperfections, inhomogeneities, etc.) might be significant. This type of noise is inherent to the PT, and can never be completely suppressed. So the challenge is to distinguish a weak signal variation, related to hydride concentration, on the background of fluctuations due to the structure noise. Note that more sensitive and reliable results can be obtained using changes, caused by the hydride platelets scattering in signal spectrum, particularly when resonant scattering occurs. When UT frequency equals to the resonant frequency of hydride platelets (i.e. when the average platelet length equals to the half-wavelength of UT signal), then maximum scattering on the platelets occurs. It leads to the maximum difference between acoustic pulses (amplitude and spectrum) passed through the hydrided and non-hydrided areas. When the UT waves with resonant frequencies of the platelets are scattered on these platelets, then contribution of these frequencies in the spectrum of the received signal is weaker for the hydrided area. Respectively, the obtained spectrum is shifted to the low-frequency area, see example presented in Fig. 19 below.

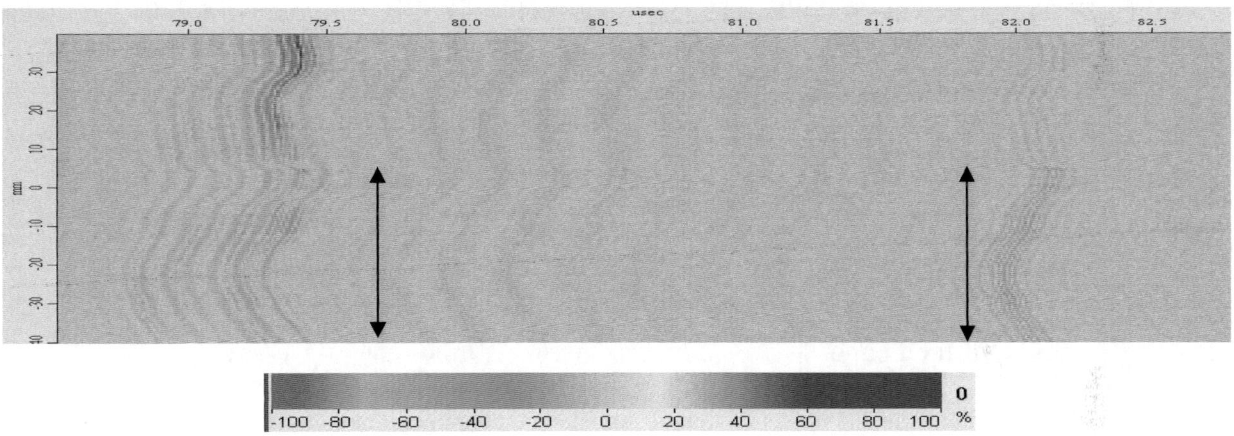

Figure 16. PC shear wave axial B-scan of the PT sample with 80ppm hydride concentration band (see arrows) located from -40mm to +5mm.

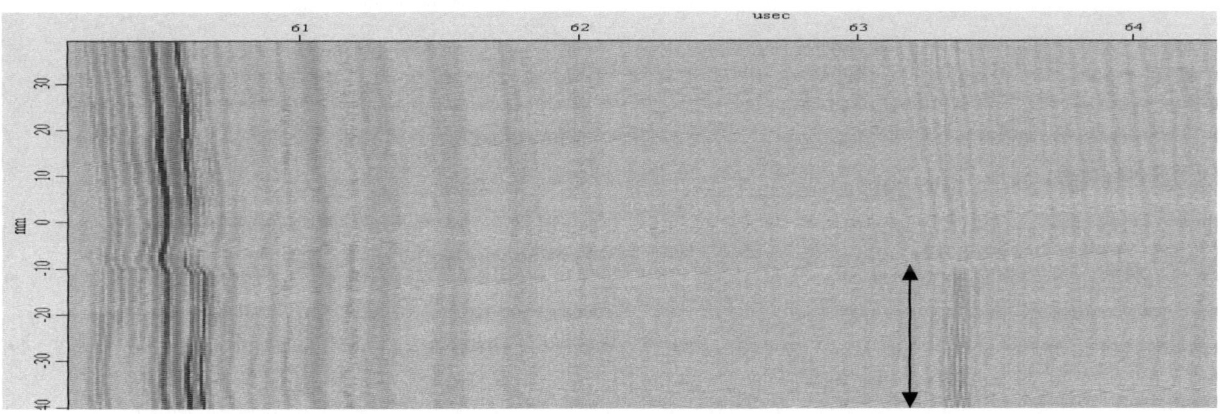

Figure 17. PC shear wave axial B-scan of PT sample with 40ppm hydride concentration band (see arrow) located from -40mm to -10mm. Color scale is shown in Fig. 16.

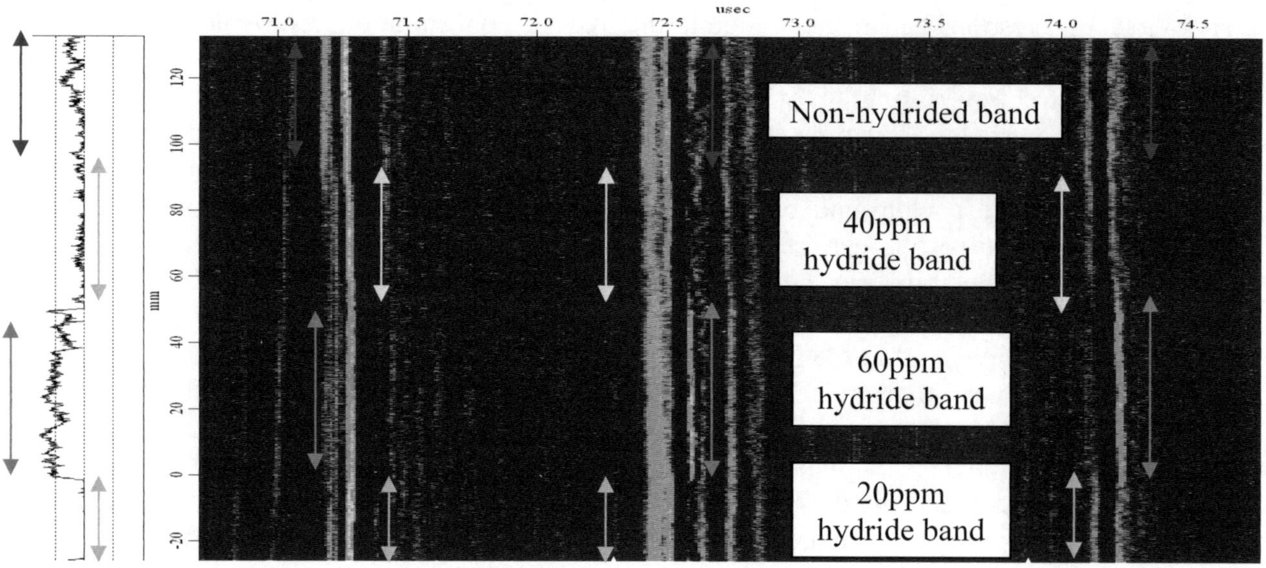

Figure 18. PC shear wave axial B-scan (signal is rectified) of PT sample with three hydrided bands: 20ppm hydride concentration band located at z<0mm, 60ppm band at 0<z<50mm, 40ppm hydride concentration band at 50<z<100mm, and non-hydrided band at z>100mm. Color scale is shown in Fig. 1.

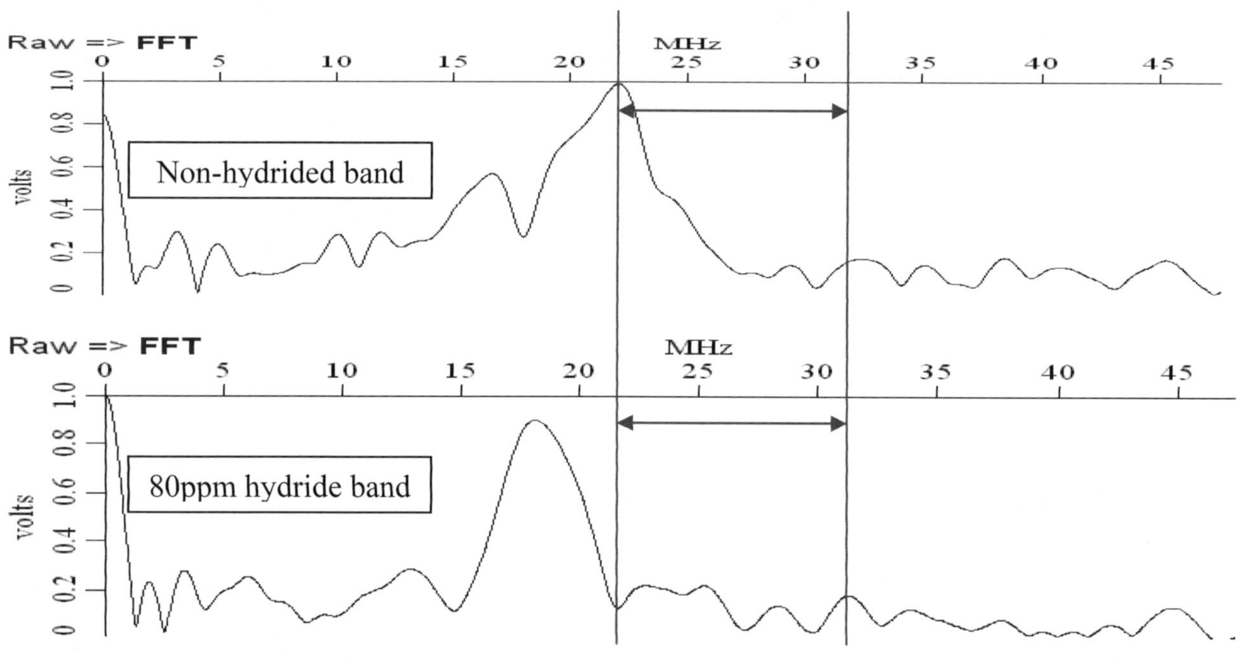

Figure 19. Spectra of two A-scans derived from NB axial B-scan of PT with 80ppm hydride concentration band. Probe: NB, f=20MHz, D=8mm, flat, WP=20mm. For hydrided area the high-frequency contribution of the 12th shear wave OD reflection within the range 22-32MHz is approximately 3 times smaller than for the non-hydrided area.

Analyzing Figs. 10-15, one can clearly see that it is quite possible to characterize flaw and evaluate flaw dimensions and/or perform material characterization.

5. CHARACTERIZATION AND SIZING OF ANGLE FLAWS

A standard shear wave probe in the PE mode can hardly detect the off-axis (angle) flaws (particularly flaws oriented approximately at $\pm 45^0$ to the tube axial axis), because the UT beam after reflections from the bottom and lateral faces of such flaw, propagates at angle and therefore does not return to the probe. Such a flaw can be detected in the NB and PC modes of operation, but only as a shadow in the main NB or PC response, because the UT wave after reflections from the bottom and the lateral faces of the flaw propagates at angle and therefore does not return to the probe. Recall also, that obtained shadow does not allow measuring depth of the flaw. A few novel methods [5], such as technique employing cylindrically focused probe and others allow distinguishing angle orientations of the off-axis flaws and sizing their depths.

As an example, a technique employing cylindrically focused probe, allows distinguishing angle orientations of the off-axis flaws and sizing their depths on a single B-scan, see Fig. 20. Indications in Fig. 20 demonstrate that, unlike the standard NB and PC amplitude shadowing C-scans, which can also represent flaw orientation, this method allows determining simultaneously the flaw angle orientation and also flaw depth on a single 2D circumferential B-scan. This method is very reliable, because it has a few indications correlated to the orientation of a flaw.

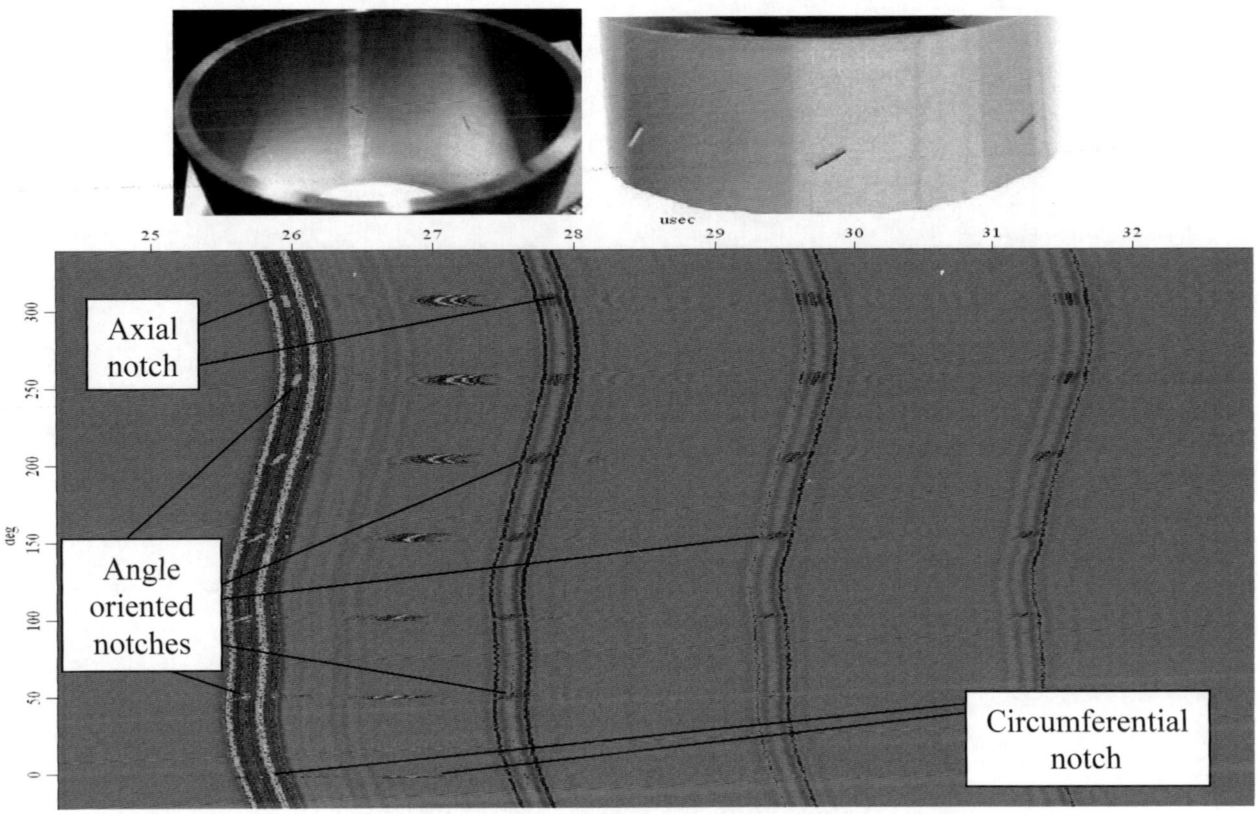

Figure 20. Pictures of PT and its replica, and circumferential angle 2D PE B-scan of seven ID off-axis V-notches (7mm long and 0.75mm deep with 45^0 root angle) oriented at the following angles to the tube axial axis: 90^0 (circumferential notch), 75^0, 60^0, 45^0, 30^0, 15^0, and 0^0 (axial notch) in the PT ID. Probe: cylindrically focused FL=25mm, f=15MHz, D=9.5mm, WP=27mm, $\alpha=6^0$. Color scale is shown in Fig. 1.

6. TUBE TESTING WITHOUT PROBE ROTATION

Ability to perform UT inspection of tubes without probe rotation significantly simplifies the delivery system and decreases the inspection time by increasing axial speed of the inspection system. Moreover, rotation always leads to mechanical vibrations, electromagnetic noise from the rotating motor, radial shifts, possible jams, and other problems. Different approaches were analyzed in [7]. As an example, schematics of two techniques, employing the single element (not a phased array!) circular normal beam tube-probe and circular angle cone-probe, both covering simultaneously 360^0 in the circumferential direction are presented in Figs. 21-22. Picture of the respective probes are shown in Fig. 23.

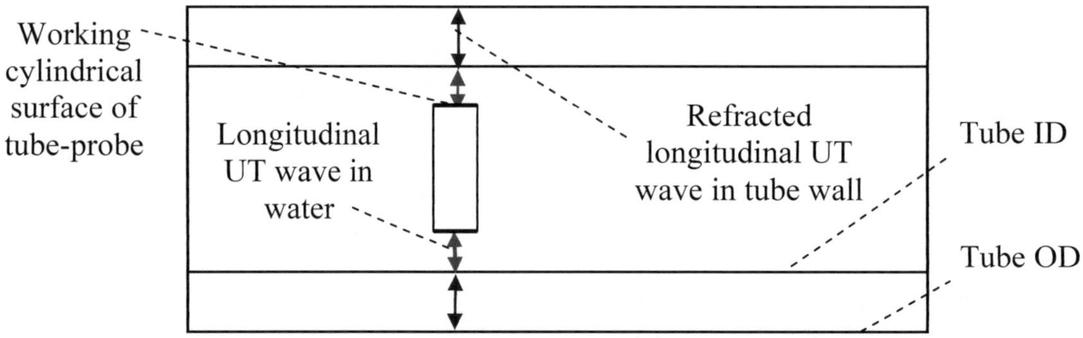

Figure 21. Schematics of NB circular tube-probe.

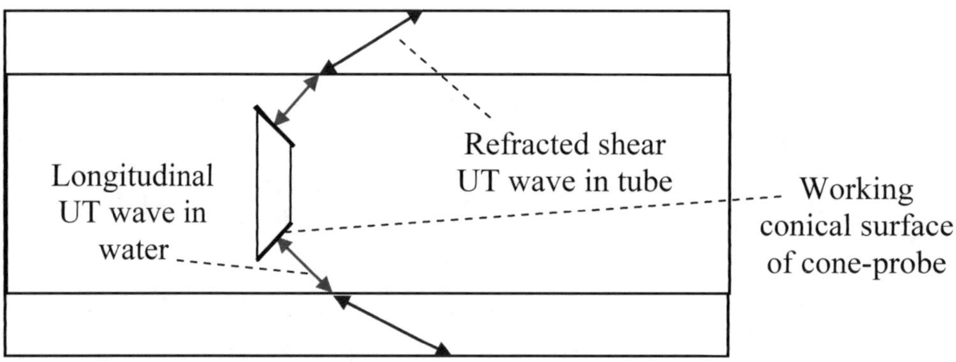

Figure 22. Schematics of angle shear wave circular cone-probe.

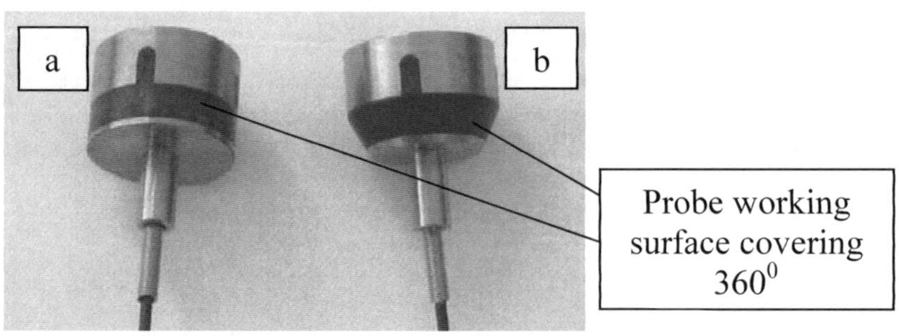

Figure 23. Circular probes covering 360^0: NB tube-probe (a) and angle cone-probe (b).

Solutions, similar to ones presented in Figs. 21-22, can be also obtained using standard focused axially-positioned probe with attached conical mirrors, see Figs. 24-26.

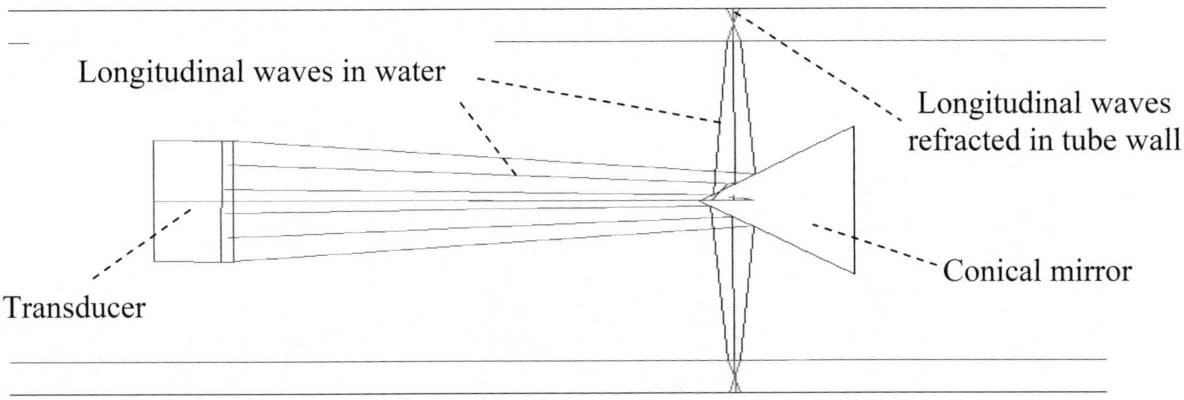

Figure 24. Schematic of NB probe, containing axially positioned standard focused transducer with attached 45^0 conical mirror.

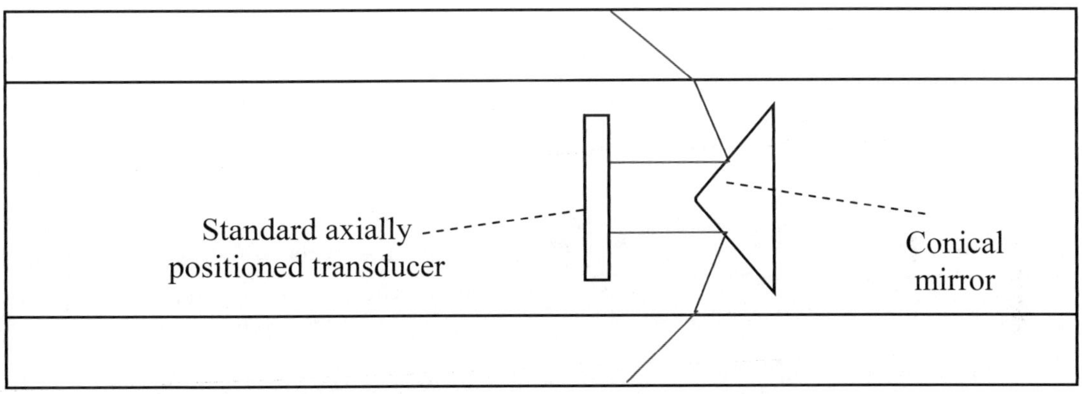

Figure 25. Schematic of angle shear wave probe, containing axially positioned standard transducer with attached 35^0 conical mirror.

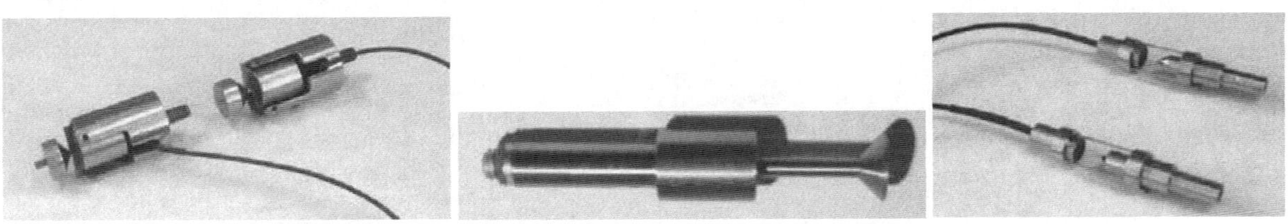

Figure 26. Circular probes covering 360^0, containing axially positioned standard focused transducer with attached 45^0 and 35^0 conical mirrors to realize NB and angle probe techniques.

2D axial PE NB and PE BW B-scans performed on a HX tube specimen, containing various ID and OD flaws, by using circular tube-probe and cone-probe, are presented in Figs. 27-28.

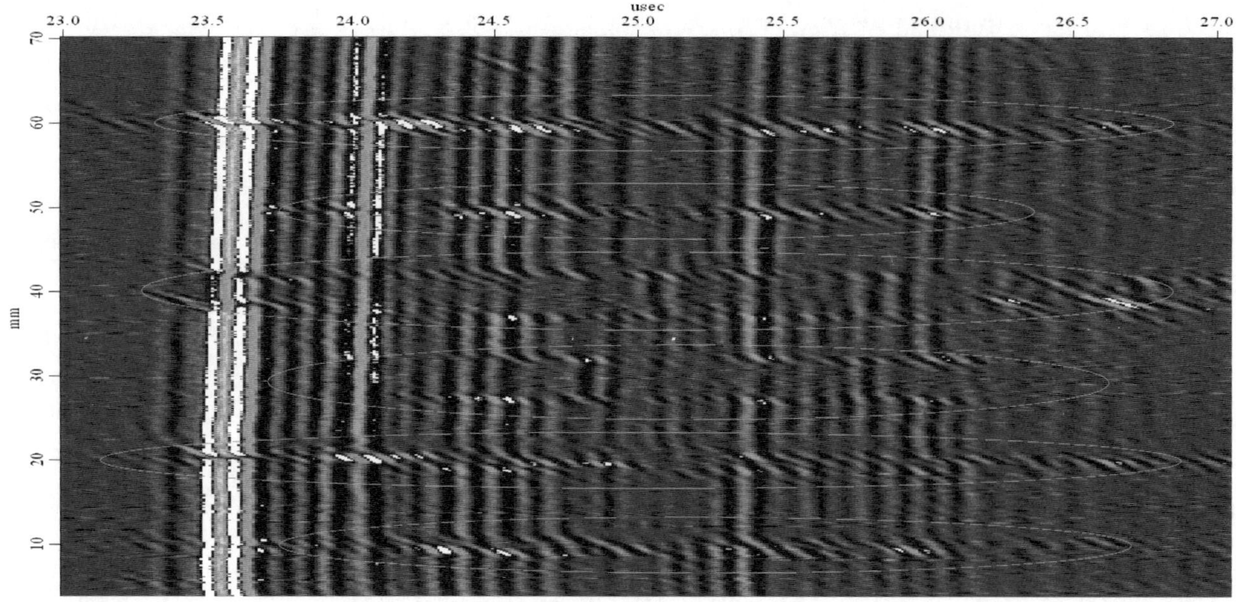

Figure 27. 2D PE NB axial B-scan of steel HX tube (ID=13.5mm, WT=1.2mm) with six flaws: OD pit (D=1.57mm and d=0.8mm) at axial coordinate z=10mm, similar ID pit at z=20mm, OD axial notch (l=5mm, w=0.36mm and d=0.8mm) at z=30mm, similar ID axial notch at z=40mm, similar OD circumferential notch at z=50mm, and similar OD circumferential notch at z=60mm. Probe: NB circular tube-probe, flat, f=10MHz, WP=18mm. Color scale is shown in Fig. 1.

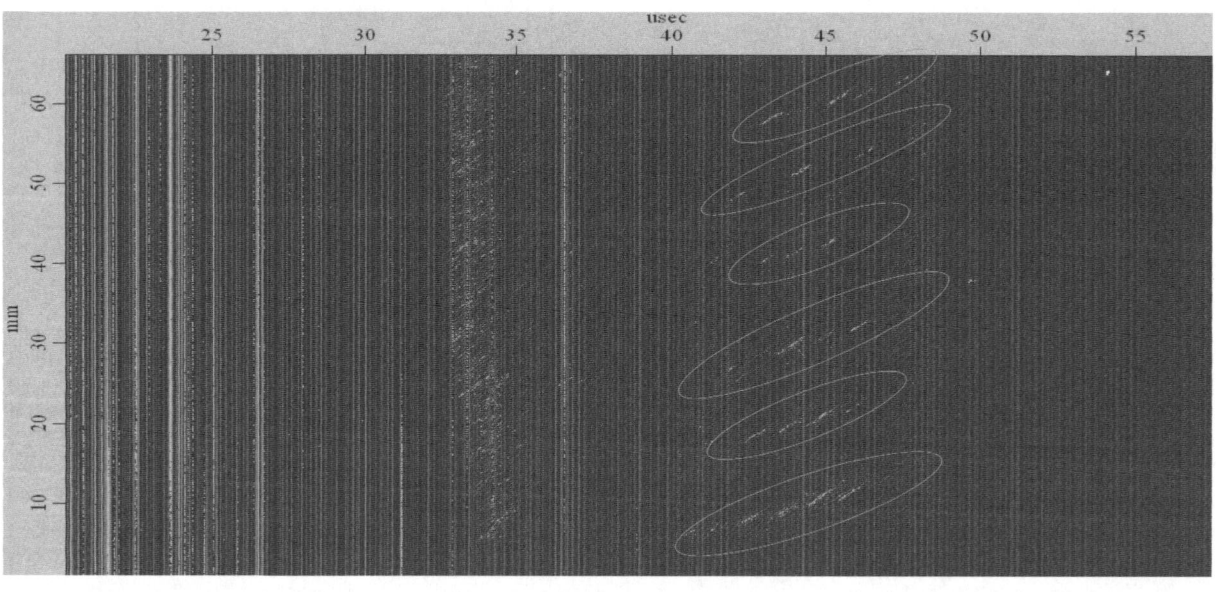

Figure 28. 2D PE angle BW axial B-scan of steel HX tube (ID=13.5mm, WT=1.2mm) with six flaws: OD pit (D=1.57mm and d=0.8mm) at z=10mm, similar ID pit at z=20mm, OD axial notch (l=5mm, w=0.36mm and d=0.8mm) at z=30mm, similar ID axial notch at z=40mm, similar OD circumferential notch at z=50mm, and similar OD circumferential notch at z=60mm. Probe: angle circular cone-probe, flat, f=10MHz, WP=18mm, $\alpha=19^0$. Color scale is in Fig. 1.

2D axial PE NB B-scans and PE BW B-scans performed by using circular tube-probe and cone-probe, on the PT specimen with row of OD pits, are presented in Fig. 29.

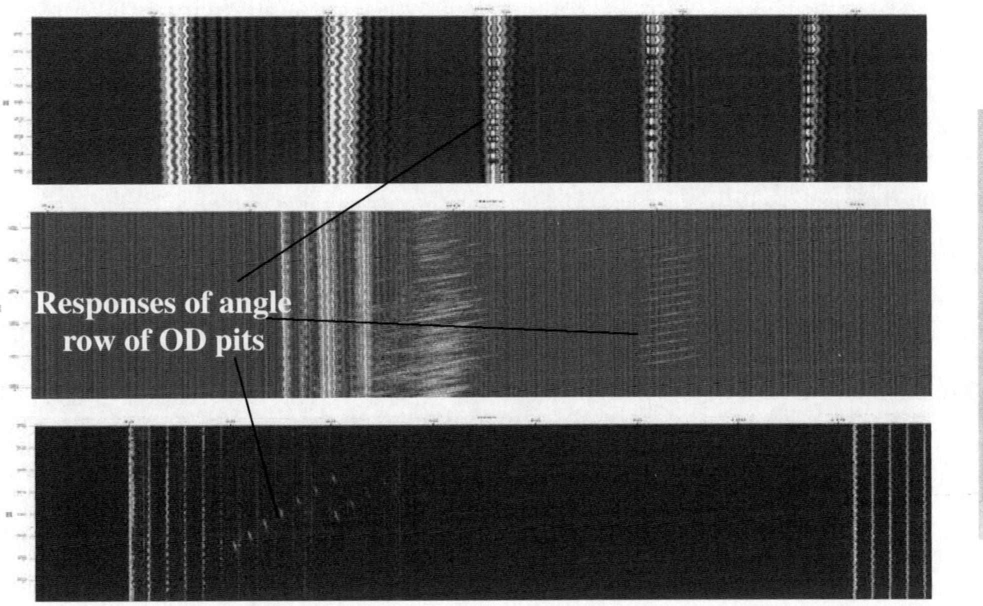

Figure 29. 2D PE axial B-scans of PT with angle row of OD pits 0.7mm wide and 0.7mm deep and picture of this PT. Probes: standard axially oriented, positioned concentrically and eccentrically with attached 45° and 35° conical mirrors, f=10MHz, FL=100mm, D=9.5mm. Color scale is shown in Fig. 1.

The circular probes give ability not only to detect various flaws (ID and OD pits, axial and circumferential notches), but they also allow sizing flaws. Figs. 27-29 show that flaw length and position in the axial direction can be easily measured. Flaw depth in the radial direction can be estimated by measuring time-of-flight of the respective response for the NB tube-probe and the response duration for the cone-probe (since duration of the UT response for angle probe always depends on the flaw depth). Flaw width and position in the circumferential direction can be also estimated by using some special techniques described in [7]: e.g. probes positioned eccentrically in relation to the tube, circular equiangular probes creating angle beam in the circumferential direction, and others.

Note that non-rotating technique can be also used as a fast screening detection technique; after which, if necessary, a detailed scanning of areas with detected flaws can be performed by employing standard techniques.

7. TANGENTIAL BEAM TECHNIQUE

One of the main problems in the UT inspection of tubes is the depth measurement of shallow flaws. Standard techniques allow measuring depth of flaws which are deeper than ~0.2mm only. Preliminary experiments show that tangential UT beam in the PE mode may give a possibility not only to detect but also measure depth of the very shallow ID and OD flaws: about ~15μm deep. This technique provides good (almost linear) correlation between measurement results and flaw actual depths. If one excites within tube wall the UT shear wave propagating tangentially to the tube ID, then even a shallow flaw (scratch, pit or crack), located at the ID, would be a good reflector for such UT beam, since this flaw blocks the beam to the maximum extent. Any other technique will give a worse reflection. To create such "tangential" acoustic beam one should use a large angle of incidence for the beam entering the tube. After refraction at the ID, the beam reflects from the OD and propagates almost tangentially to the ID (see Fig. 30). Any shallow flaw even slightly elongated in the radial direction will efficiently reflect the UT beam, since this flaw blocks its way of propagation. Time and angle ranges of the obtained PE response depend on the flaw depth.

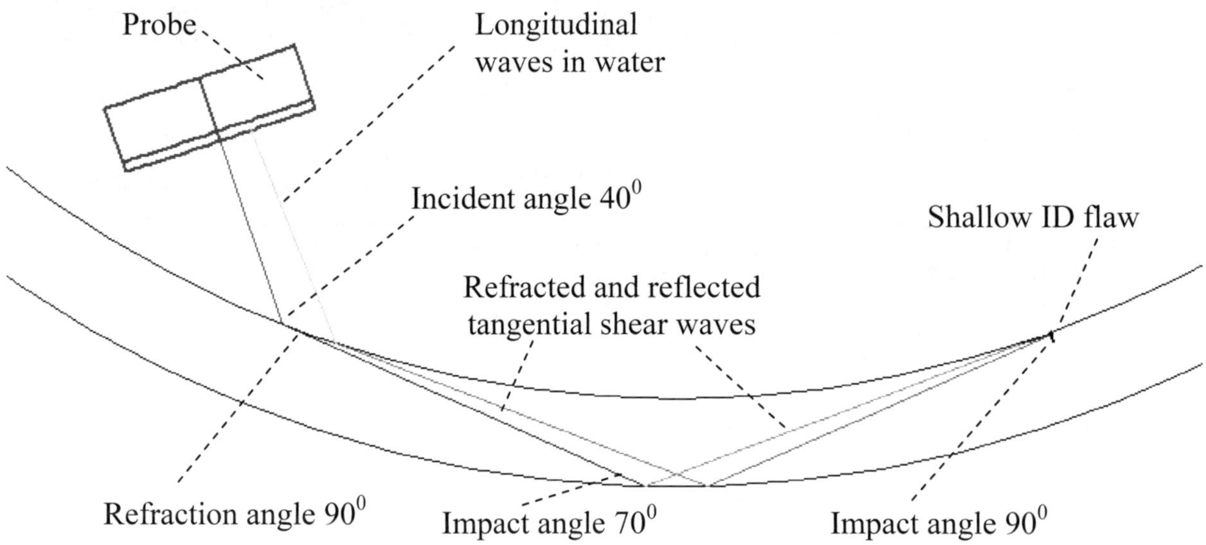

Figure 30. Tangential PE UT beam technique.

2D circumferential B-scans, performed by using a tangential beam technique on a few PT specimens containing various ID flaws, are presented below in Figs. 31-32. These scans clearly show that even the very shallow flaws (about 15-50μm) can be reliably detected.

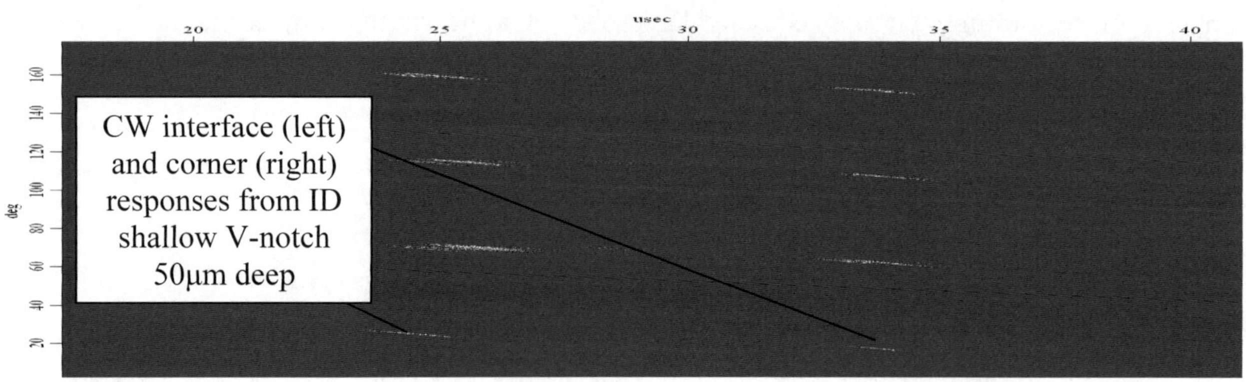

Figure 31. Circumferential 2D CW PE B-scan of four shallow axial ID V-notches (30° tip angle) with depths 0.05mm, 0.1mm, 0.15mm, and 0.2mm in the PT. Probe: f=20MHz, flat, D=8mm, WP=18mm, $\alpha=33^0$. Color scale is shown in Fig. 1.

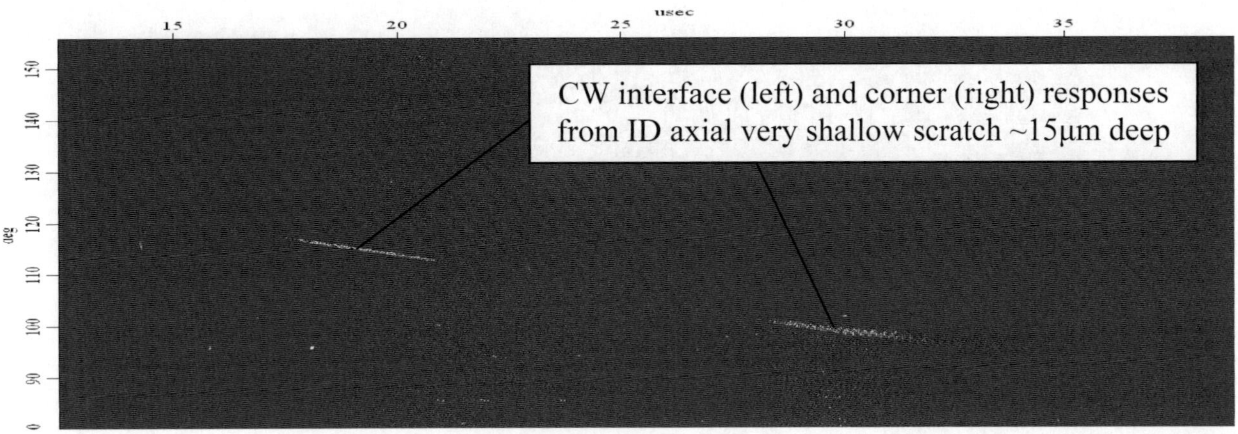

Figure 32. Circumferential 2D CW PE B-scan of axial very shallow ID scratch about ~15μm deep. Probe: f=20MHz, flat, D=8mm, WP=18mm, $\alpha=33^0$. Color scale is shown in Fig. 1.

Note that tangential beam technique also provides a strong interface reflection directly from extremely shallow surface-open flaws (down to ~5μm), if probe incident angle is large enough (about 40°). At the smaller incident angles the interface response amplitude quickly decreases. Recall that UT probes, used in the tangential beam technique for shallow flaws detection and sizing, should have a high center frequency (i.e. small wavelength), because otherwise a shallow flaw will not be detected.

Presented results clearly demonstrate that tangential beam technique allows reliably detecting very shallow flaws by using the interface and corner responses and also measuring depth of the shallow ID flaws down to ~0.015-0.02mm deep by using duration and/or length of the corner responses. Flaws about ~15μm deep can be detected and sized at the limit of sensitivity, while flaws about ~10μm deep cannot be detected by using the corner responses.

8. 3D-VISUALIZATION OF NB, CW, CCW, AND PC 3D B-SCANS

Using different 3D time-of-flight B-scans, the orientation, shape, and dimensions of flaws can be determined. Special software, that can automatically measure parameters of various responses and generate 3D image of the flaw, was developed. Obtained 3D images provide convenient and reliable information concerning flaw identification, characterization and sizing. Typical 3D time-of-flight images of various flaws are presented in Figs. 33-37 below.

Images in Figs. 33-37 demonstrate that software used for data processing can definitely lead to a decrease in the number of required replicas, which should be taken and analyzed to determine flaw shape, orientation, and dimensions. This provides savings in terms of valuable outage time and worker dose uptake. This software also provides convenient and reliable information on flaw orientation, geometry, and dimensions. It will lead to simplification, improved efficiency, reliability and decreased time of data analysis. Using this software, the old stored inspection data can be revisited, re-analysed and compared with the most recent inspection data. All these advantages will allow developing more accurate and reliable procedures, instructions and guidelines concerning flaw identification, characterization and sizing.

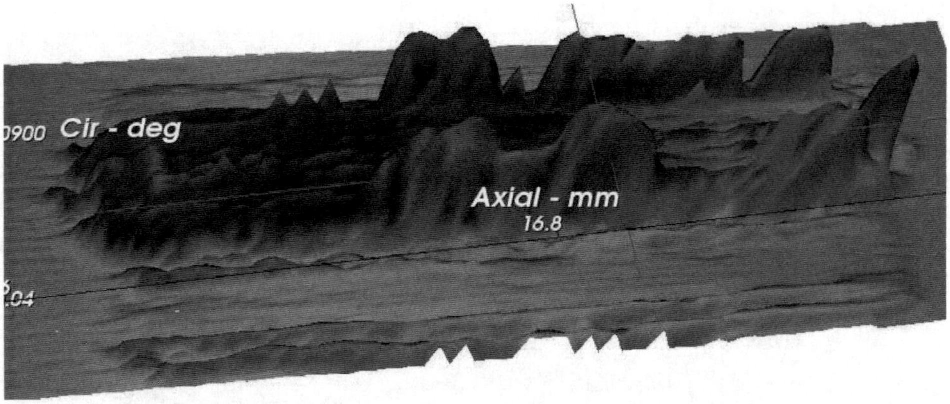

Figure 33. Time-of-flight 3D image derived of 3D NB PE B-scan of axial rectangular ID notch (25mm long, 2.5mm wide and 0.5mm deep) in the PT with two fatigue cracks ~1.7mm deep. Probe: NB, D=9.5mm, FL=45mm, f=15MHz, WP=12mm. Scale 5:1 in the depth direction.

Figure 34. Time-of-flight 3D image derived of NB PE 3D B-scan of two-step axial ID rectangular notch (primary notch 0.15mm deep, 2.5mm wide, and 25mm long; secondary notch 0.076mm deep, 0.15mm wide, and 3mm long in the centre of primary notch) in the PT. Probe: NB, FL=10mm, f=20MHz, D=6.35mm, WP=10mm. Scale 10:1 in depth direction.

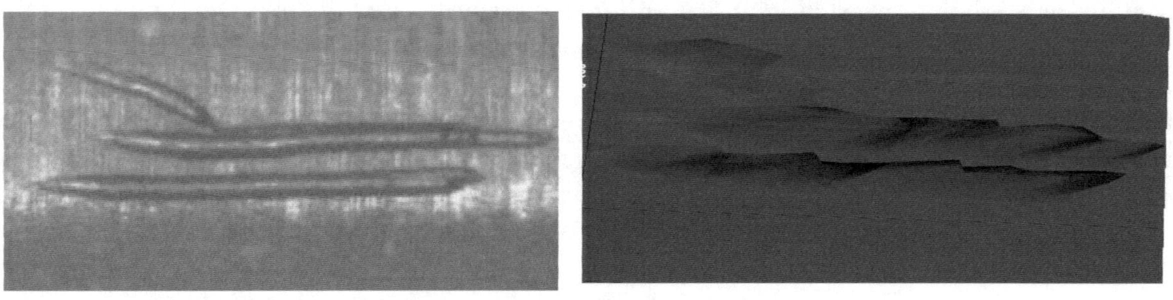

Figure 35. Optical image of replica of the ID flaw (~5.5mm long, ~2.5mm wide with maximum depth ~0.083mm) in the PT and time-of-flight 3D image derived of 3D NB PE B-scan of flaw. Probe: NB, D=6.35mm, f=20MHz, FL=10mm, WP=10mm. Scale 10:1 in the depth direction.

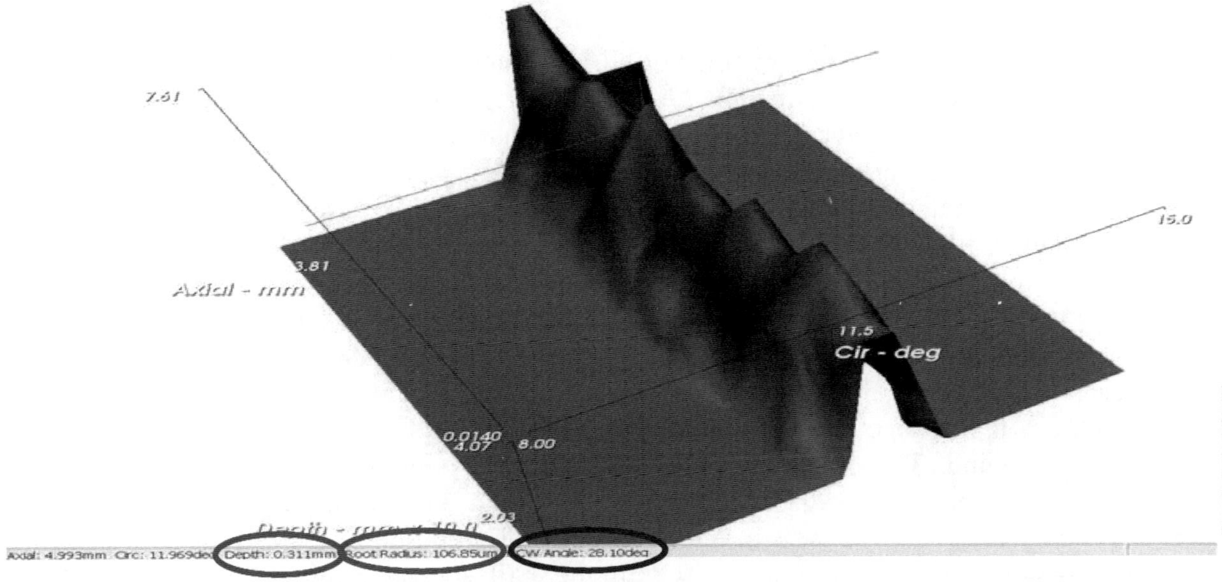

Figure 36. Time-of-flight 3D image derived of CW, CCW and CPC fused scans of ID axial V-notch (0.3mm deep with 60^0 root angle and root radius 100μm). Probes: CW and CCW, FL=33mm, f=10MHz, D=9.5mm, WP=20.6mm, $\alpha=25^0$. Scale 10:1 in depth direction.

Note that moving 3D-cursor in Fig. 36 to any point, the following flaw parameters at this point can be assessed: depth (circled in green at the bottom of the image), root radius (circled in red), and flaw side inclination angle (circled in blue). Recall that flaw root radius is a very important flaw parameter for risk assessment, because this radius determines the degree of the stress concentration and, respectively, the probability of crack growth.

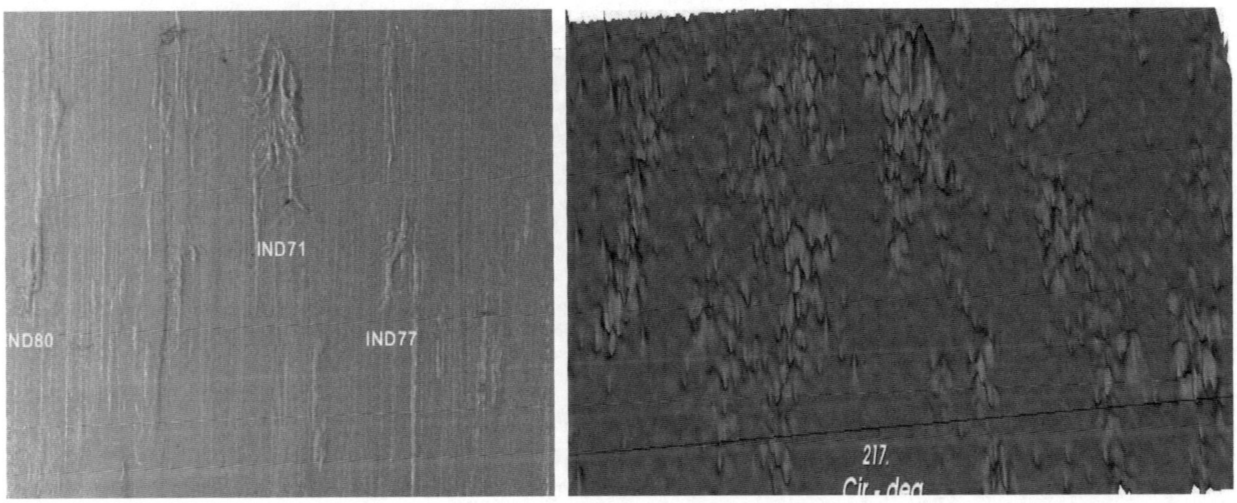

Figure 37. Optical image of replica of a few very shallow debris frets from reactor fuel channel (~3.9mm long, ~3.4^o wide with maximum depths ~0.11-0.13mm) and time-of-flight 3D image derived of 3D NB PE B-scan of these flaws. Probe: NB, D=6.35mm, f=20MHz, FL=10mm, WP=10mm. Scale 10:1 in the depth direction.

9. PROBE OFFSET INFLUENCE

Due to the PT deformations the positions of all UT probes tend to change and deviate from the nominal values. The total deformation (due to the irradiation, temperature and aging effects) consists of a few components: PT axial elongation, diametrical expansion, wall thinning, tube sag, and also off-centering of the Rotating Probe Module. All these factors lead to the variations in the probe water-paths. This, in turn, entails changes in the UT beam trajectory, reflection and refraction angles, focusing, and UT pulse time-of-flight. All these changes affect performance of the UT probes and ability of the inspection system to detect, characterize and size flaws. Therefore, these factors affect flaw probability of detection and sizing accuracy. To meet the inspection requirements, all UT probes should perform at various offsets. The ranges of offsets of the each probe were calculated [8] based on the effect of the each deformation component, code requirements, inspection specifications, and current procedures and practices for flaw detection and sizing. The obtained results allowed determining the ranges of offsets, within which each probe could meet the requirements. It turned out [8] that both pairs of shear wave probes (axial and circumferential) in the PC modes of operation and 20MHz NB probe do not meet the requirements approximately within the half of the required range of offsets. In addition, their sensitivity is extremely non-uniform within the required range of offsets.

For example, the 20MHz NB probe does not meet the requirements on resolution, sensitivity, and sizing accuracy within the offset ranges from -3 mm to -2 mm and from +3 mm to +8 mm (see Fig. 3). The whole required offset range for this probe lies from -3 mm to +8 mm at the end of PT life - 30 years. One can see in Fig. 38a that at zero offset the flaw can be reliably detected and accurately sized, while at the offsets equal to -2 mm and +3 mm, flaw is detected at the limit of sensitivity (Figs. 38b-38c) and its dimensions cannot be measured. If offset is <-2mm or >+3mm, flaw cannot be detected at all.

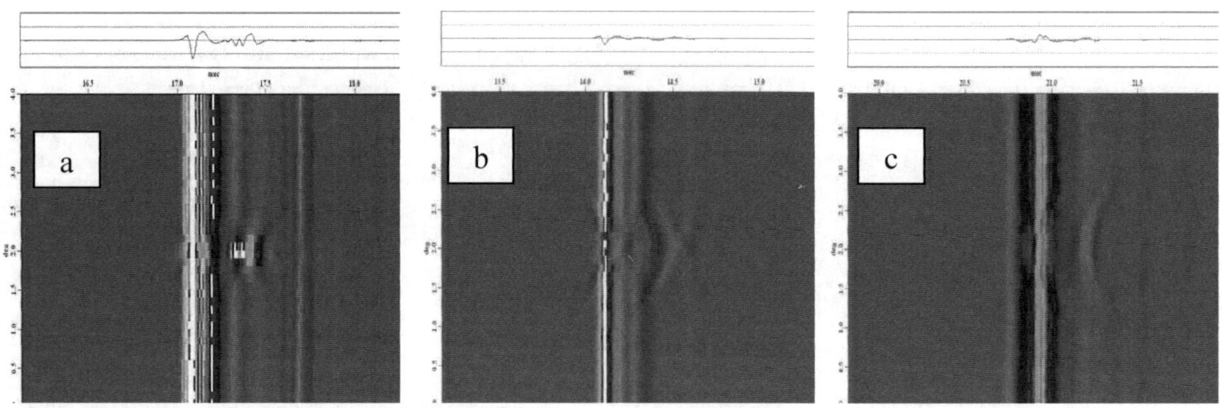

Figure 38. 2D NB PE B-scans at the middle of the notch derived from 3D NB PE B-scans of the ID axial calibration notch 6.35mm long, 0.15mm wide and 0.15mm deep. Probe: NB, f=20MHz, D=6.35mm, FL=10mm. a - offset is 0mm, b - offset is -2mm, c - offset is +3mm. Color scale is shown in Fig. 1.

In addition, sensitivity of this probe is non-uniform and highly variable within the required range of offsets: the response amplitudes vary about ~30 dB, within the specified range of the offsets (see Fig. 39).

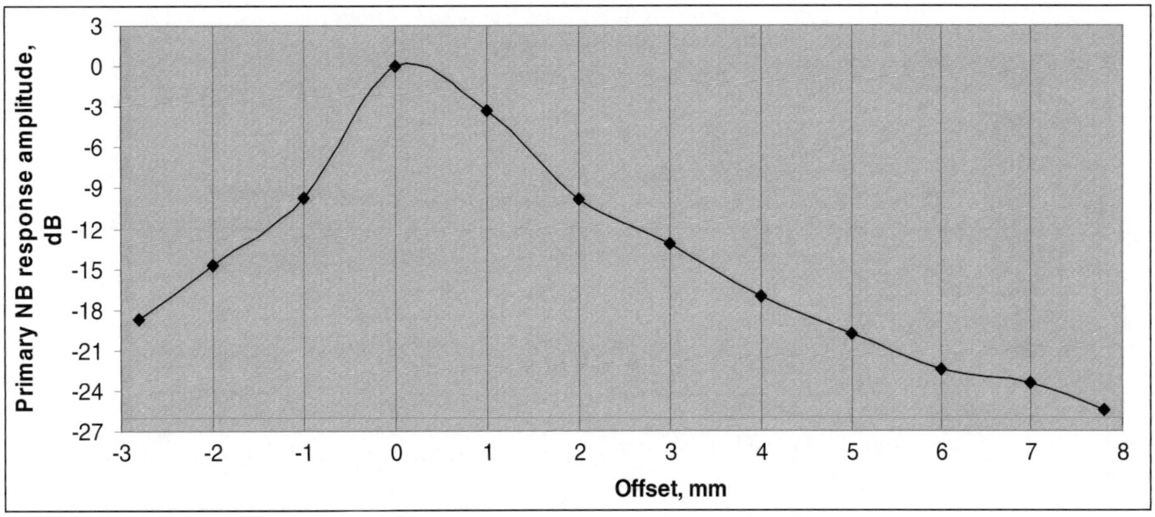

Figure 39. Response amplitude of 20MHz NB probe on clean tube vs. probe offset.

Based on the obtained results, the modification of both pairs of existing PC probes and 20MHz NB probe should be performed in order to meet the requirements and obtain uniform sensitivity within the required range of offsets. This modification can be based on usage of the novel probes with "logarithmic" acoustic lenses providing stretched focal zones and uniform sensitivity [9]. Such probes meet the requirements within the whole required range of offsets.

Any standard spherically focused probe provides a good frontal resolution in the focal point, which leads to high sensitivity and accuracy of measurement only in one narrow area in the axial direction – in the focal spot. However, before and after focal point the frontal resolution of a standard spherically focused probe quickly deteriorates, which leads to the low sensitivity and poor accuracy of measurement outside the focal area.

Probably, the best solution of this problem is to use probe with a stretched focal zone, which can provide good frontal resolution, high sensitivity and good accuracy of measurement within the whole required range. Such probe contains special "logarithmic" acoustic lens and forms narrow weakly diverging UT beam, i.e. this probe has a stretched focal zone. The concept of the "logarithmic" acoustic lens is presented below in Fig. 40. Such a lens forms a narrow weakly diverging acoustic beam throughout the required axial range. The central part of the lens focuses the UT waves close to the transducer (see blue rays in Fig. 40), while the peripheral parts of the lens focus the UT waves far from the transducer (see red rays in Fig. 40), and the medium parts of the lens focus the UT waves in the middle of the focal zone (see green rays in Fig. 40). In other words, a logarithmic lens is a focusing acoustic lens with a varying focal length: small for the central part of the lens and large for the peripheral parts of the lens. As result, the acoustic field created by such transducer has a stretched focal zone, where the UT beam is narrow and collimated. Such probe creates needle-like narrow and weakly diverging acoustic beam. Picture of the UT transducers with different diameters, various center frequencies, and different logarithmic acoustic lenses (spherical and cylindrical), i.e. focusing the UT beams within different required axial ranges, are shown in Fig. 41 below. Such probes are very useful when flaw position is unknown or variable.

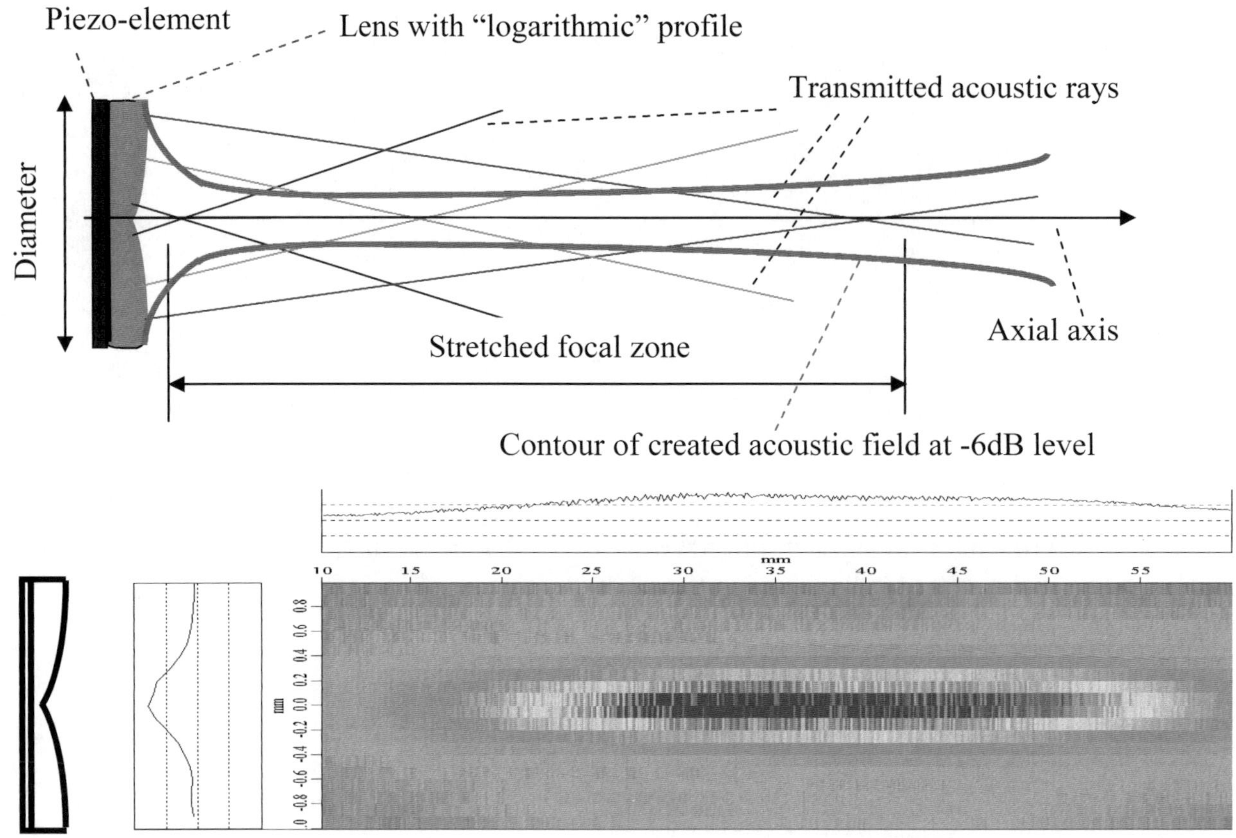

Figure 40. Schematic of the acoustic field and measured acoustic field formed by the probe with "logarithmic" acoustic lens. Color scale is shown in Fig. 1.

Figure 41. UT transducers with different diameters, various center frequencies, and different logarithmic acoustic lenses (spherical and cylindrical), i.e. and focusing the UT beams within different required axial ranges.

10. GUIDED WAVE TECHNIQUE

UT guided wave (GW) method can be a reliable and rapid technique for assessing the condition of tubes by detecting various flaws. The idea of the UT detection of flaws located far from a transducer is based on the excitation of the traveling acoustic GW, propagating along the tube in the axial or circumferential direction. The term "guided" is used, because the UT wave travels along the tube guided by its geometric boundaries. Different types of the UT waves can be used for inspection: longitudinal, shear, surface, torsional, and Lamb waves. Each of these waves has its advantages and limitations. Signals reflected from geometric irregularities, material imperfections and/or defects are detected in the PE mode. As a result, the location of the defect and sometimes its size can be determined. Because of the long-inspection range, the GW inspection technology is useful for quickly surveying a structure for defects, including areas that are difficult to access directly. Typically, the sensitivity of GW technique is lower than sensitivity of standard UT methods.

Fig. 42 shows beam-tracing simulations for surface and shear/longitudinal waves, respectively, propagating along the tube length.

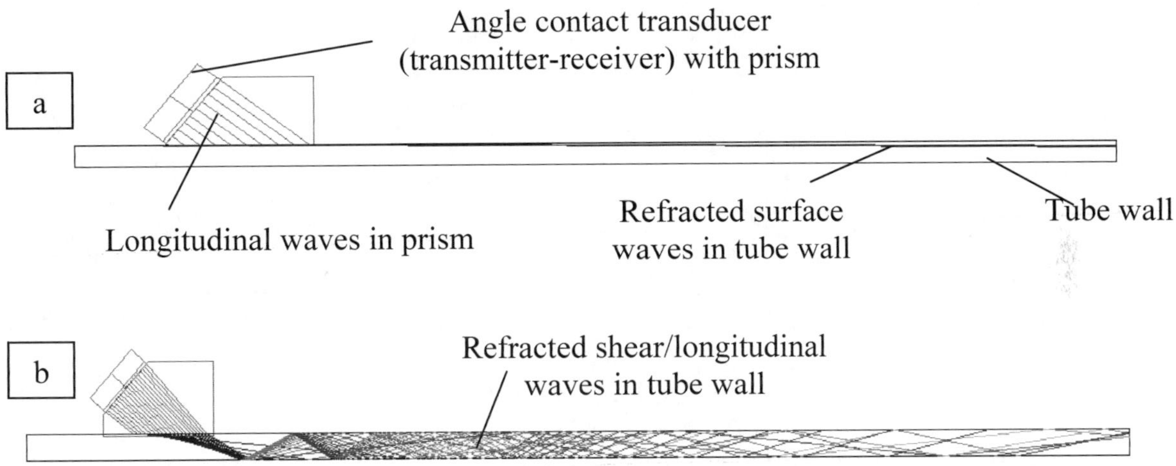

Figure 42. Beam tracing simulations of surface axial wave, propagating tangentially to tube surface (a), and of shear/longitudinal wave propagating within the tube wall (b).

Circumferential PE B-scans using surface waves, propagating in the axial direction, have been performed to detect circumferential ID notches (Fig. 43) using contact non-focused angle probe (f=2.5MHz and D=12.5mm) with prism fitting the ID of a large diameter tube.

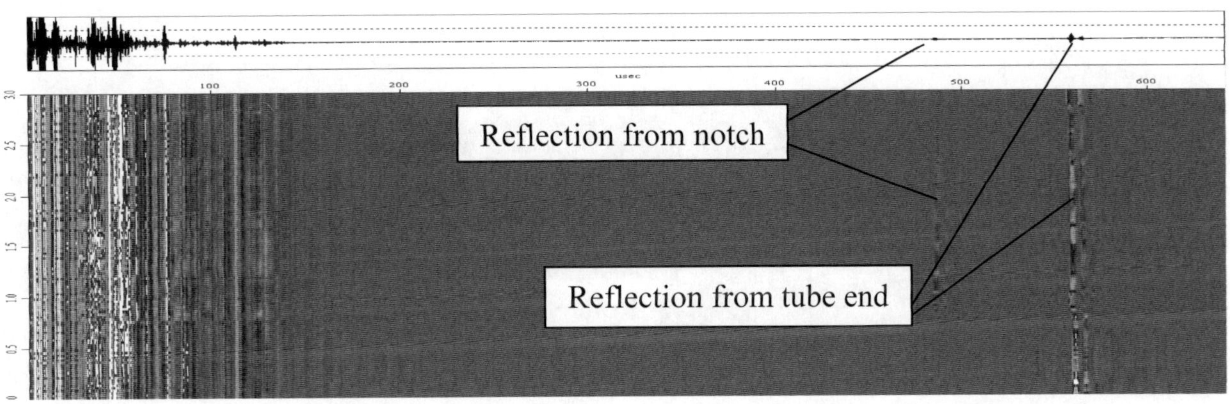

Figure 43. PE circumferential B-scan. Tube ID=18” and WT=10mm with ID circumferential notch 0.15mm wide and 0.3mm deep. Probe: contact angle surface wave containing plexiglass prism with curved surface fitting the tube ID, f=2.5MHz, D=12.5mm, flat, $\alpha=60^0$. Distance between probe and notch is ~800mm. Color scale is shown in Fig. 1

The GW technique, using acoustic wave propagating around the tube in the circumferential direction, does not need a rotation of the probe. In Fig. 44 the beam-tracing simulations of surface and shear waves, excited within the tube wall and propagating around the tube in the circumferential direction are shown.

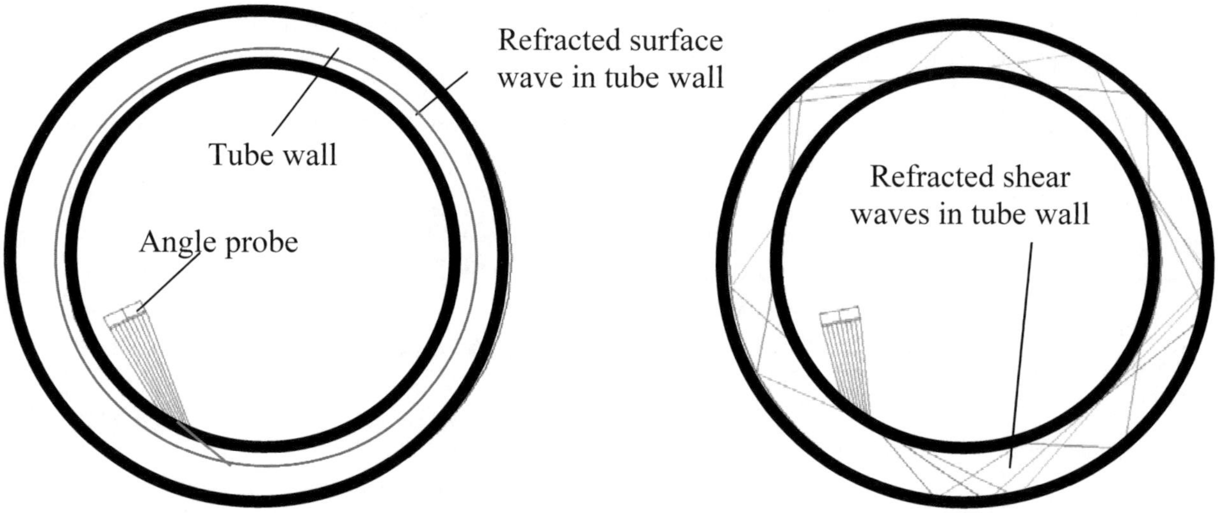

Figure 44. Beam tracing simulation of surface wave propagating tangentially to the tube ID surface in the circumferential direction (a) and of shear wave propagating within the tube wall in circumferential direction (b).

A few Axial PE B-scans, detecting axial ID notches and based on the technique shown in Fig. 44 and presented below in Figs. 45-47, show how large diameter tube (ID=18", WT=10mm) can be inspected from the ID by using the GW technique employing surface wave propagating within ~1m long inspection range in the circumferential direction. The angle probe, transmitting this surface wave, should be moved only axially; and at the same time it will inspect the whole surface of the tube. Time-of-flight of the received signal reflected from the ID axial flaw depends on the distance between probe and flaw in the circumferential direction. That is how circumferential position of the flaw can be measured.

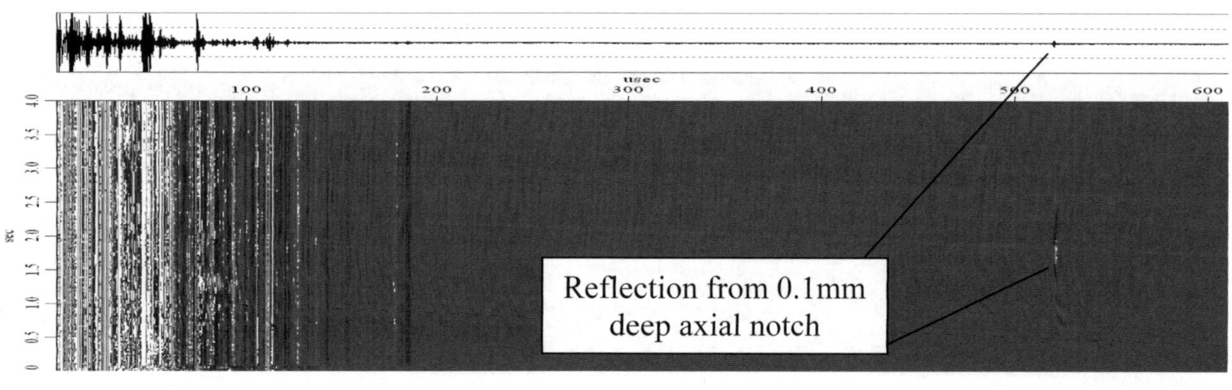

Figure 45. PE axial B-scan. Tube ID=18" and WT=10mm with ID axial notch 0.15mm wide and 0.1mm deep. Probe: contact angle surface wave with prism fitting the tube ID, f=2.5MHz, D=12.5mm, flat, $\alpha=60^0$, surface wave refraction angle $\beta=90^0$. Circumferential distance between probe and notch is $\sim180^0$ (i.e. ~720mm). Color scale is shown in Fig. 1.

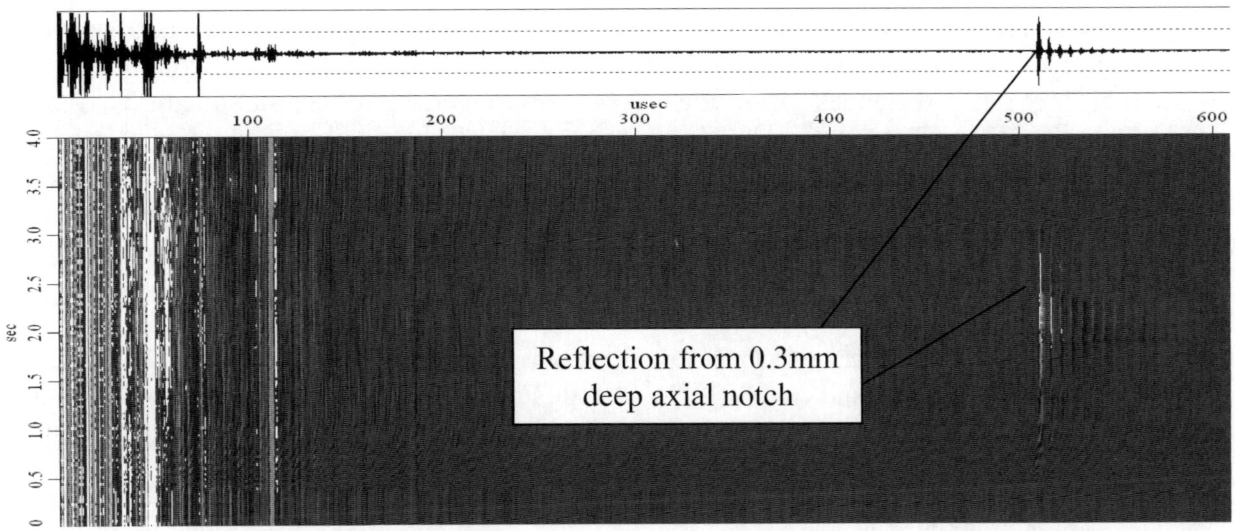

Figure 46. PE axial B-scan. Tube ID=18" and WT=10mm with ID axial notch 0.15mm wide and 0.3mm deep. Probe: contact angle surface wave with prism fitting the tube ID, f=2.5MHz, D=12.5mm, flat, $\alpha=60^0$, $\beta=90^0$. Distance between probe and notch ~180^0 (i.e. ~720mm). Color scale is shown in Fig. 1.

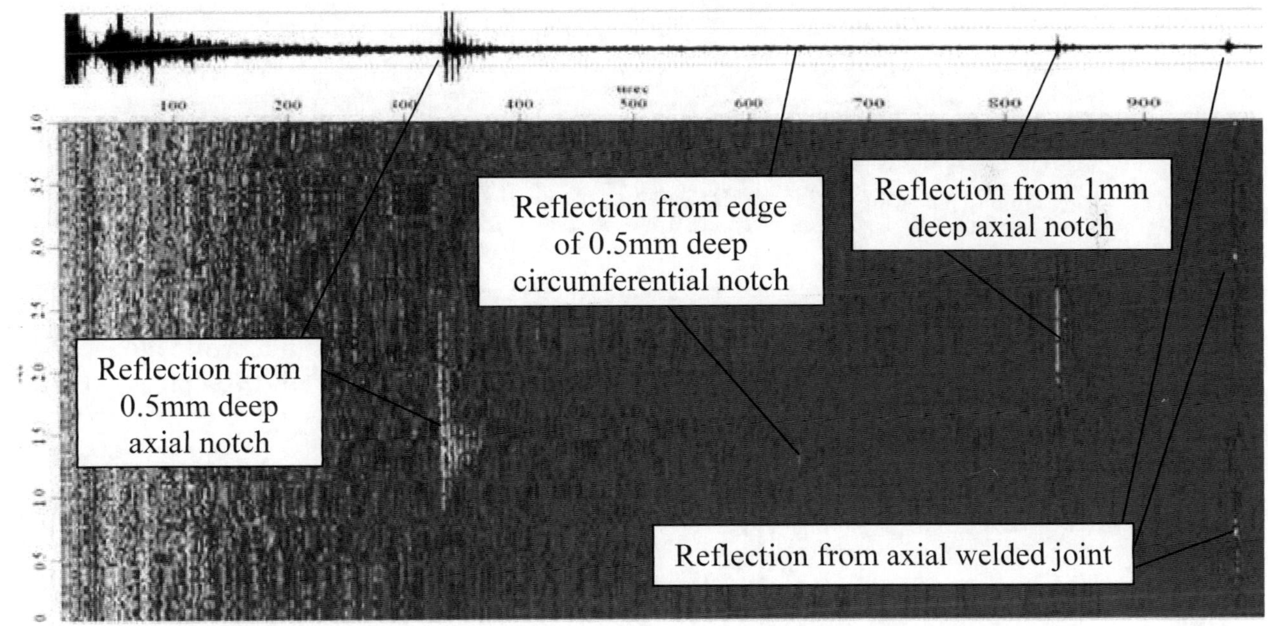

Figure 47. PE axial B-scan. Tube ID=18" and WT=10mm with one circumferential ID notch 0.15mm wide and 0.5mm deep and two axial ID notches 0.15mm wide and 0.5mm & 1mm deep located ~180^0 (i.e. ~720mm) apart. Probe: contact angle surface wave with prism fitting the tube ID, f=2.5MHz, flat, D=12.5mm, $\alpha=60^0$, $\beta=90^0$. Distance between probe and first 0.5mm deep axial notch is ~90^0 (i.e. ~360mm), distance between probe and 0.5mm deep circumferential notch is ~180^0 (i.e. ~720mm), distance between probe and 1mm deep axial notch is ~340^0 (i.e. ~1360mm), and distance between probe and axial welded joint is ~360^0 (i.e. ~1440mm). Color scale is shown in Fig. 1.

Images presented in Figs. 44-46 show that GW technique employing surface waves is promising. It allows detecting shallow (from 1% to 10% wall thickness) axial ID notches located at the large distances from the probe. Moreover, this technique can also detect circumferential ID notches using reflections from edges. However, this method cannot be used for OD flaws and flaw depth measurement, because surface wave propagates only along the ID surface. To detect the OD flaws and measure flaw depth, the shear GW technique employing shear waves (see Fig. 43), can be used. Recall that during field inspection, the GW reflections from elbows, valves and welds can be used as the reference points.

Another group of experiments was performed on a feeder pipe (OD=73mm and WT=5.2mm) ~3m long with two elbows and notch (Fig. 48a). Results are presented in Fig. 48b. They show that it is quite possible to detect signals reflected from both elbows and circumferential OD notch 0.5mm deep, 0.15mm wide and 10mm long.

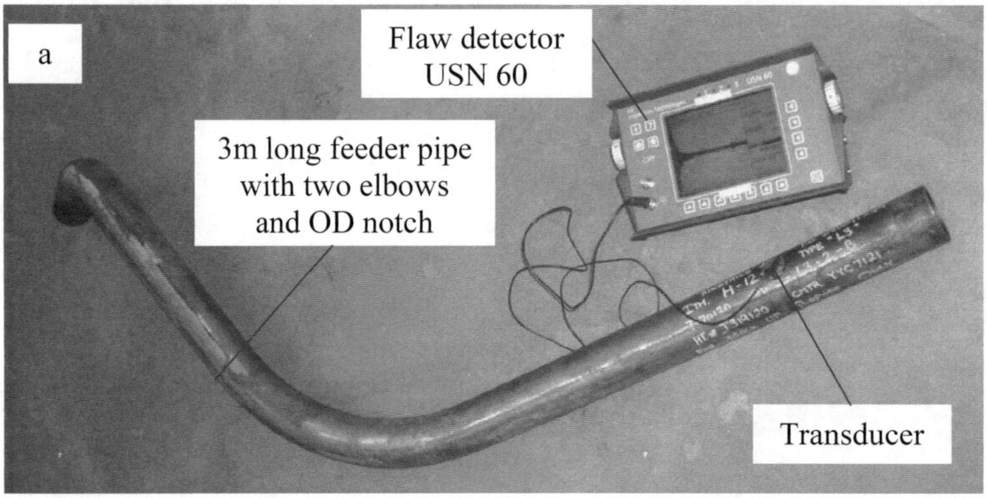

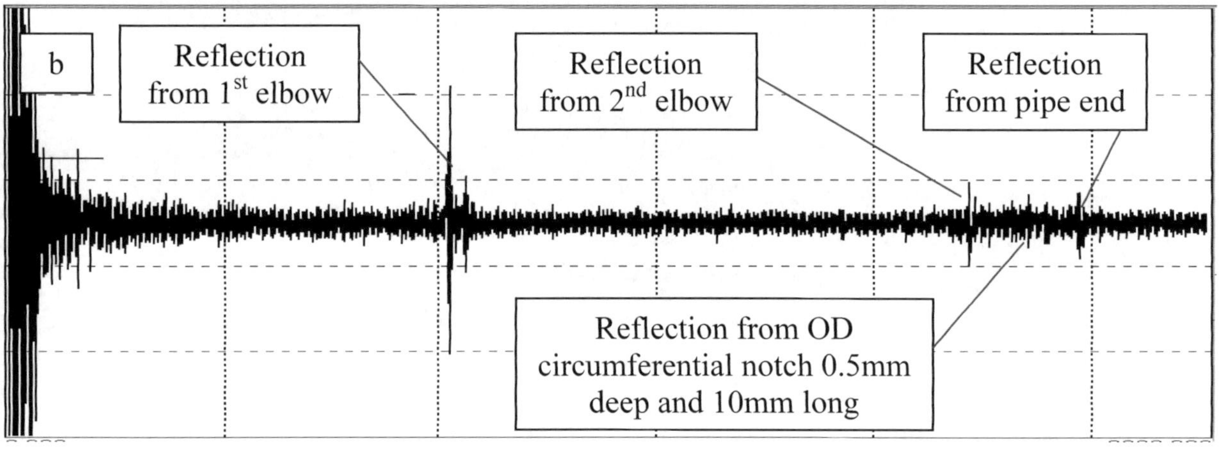

Figure 48. UT GW setup (a) for inspection of a long feeder pipe with two elbows and circumferential OD notch (0.5mm deep, 0.15mm wide and 10mm long) and result of testing (b) - A-scan. Probe: contact angle surface wave with prism, f=2.5MHz, incident angle 90^0. Distance between probe and 1st elbow ~1.4m, between probe and 2nd elbow ~2.7m, between probe and notch ~2.8m, and between probe and pipe end ~2.9m.

11. RESONANCE INSPECTION METHOD (RESONANCE SPECTROSCOPY)

Sometimes spectrum of the received UT signal can be very efficient for part inspection. As an example, some results, obtained using the UT contact method (resonance inspection method or acoustic resonance spectroscopy) for porcelain insulated cutouts testing are presented below. Using this technique one can detect the presence of crack and characterize the material property by analysing the acoustic response of the insulator. The idea of the contact resonance inspection method is based on tapping a cutout by impact hammer, "listening" to its vibration response using accelerometer, attached to the cutout, and performing a Fourier analysis of the received signals. The obtained "spectral signature" of a cutout (vibration behaviour) depends on the cutout size, material, internal imperfections (e.g. cracks), and other factors.

Development and testing of methodology of UT resonance technique was performed on the samples of new and used (with and without visible cracks) cutouts made by different manufacturers. Experimental setup consisted of impact hammer, accelerometer, charge amplifier, and real time spectrum analyzer (see Fig. 49). Accelerometer, connected to spectrum analyzer, was attached to the cutout by wax.

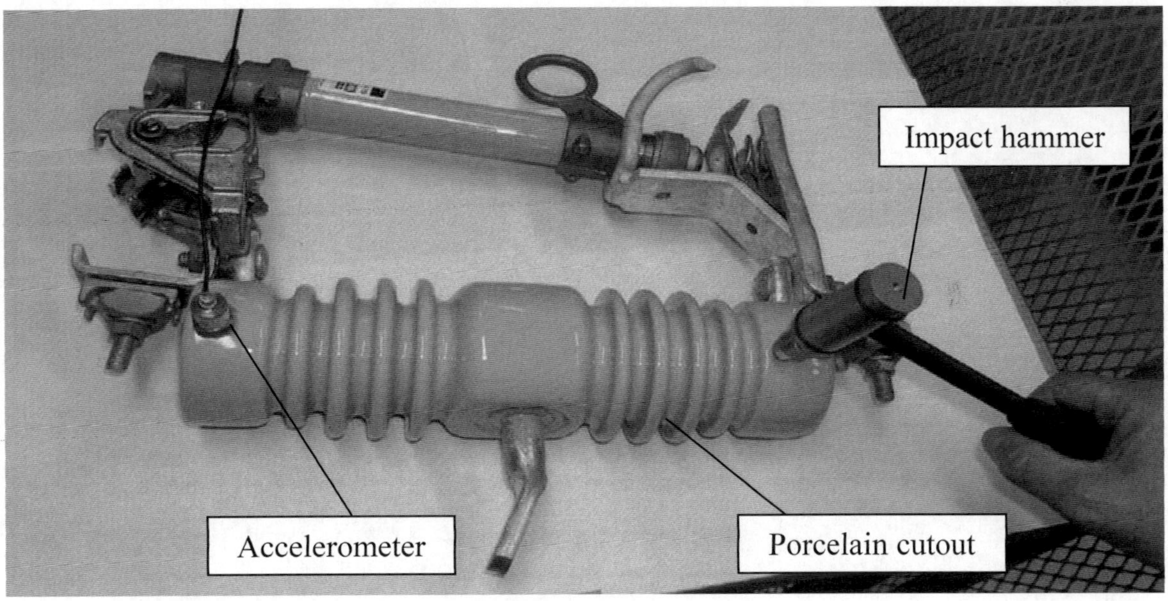

Figure 49. Accelerometer and impact hammer used for cutout testing.

After the impact excitation by impact hammer, the spectrum of the received acoustic signal from accelerometer was watched on the screen of spectrum analyzer connected to computer. The averaging procedure has been used as a statistical signal processing and allowed eliminating the most serious random errors related to the variations in accelerometer installation and impact hammer hitting force.

Typical spectra of the un-cracked and cracked cutouts are presented in Figs. 50-51. The obtained results were consistent and robust, i.e. "spectral signatures" of different cutouts of the same type even at various tapping forces and accelerometer attachments were similar. This consistency is very important, since it demonstrates that UT resonant method can be practically used for cutouts testing.

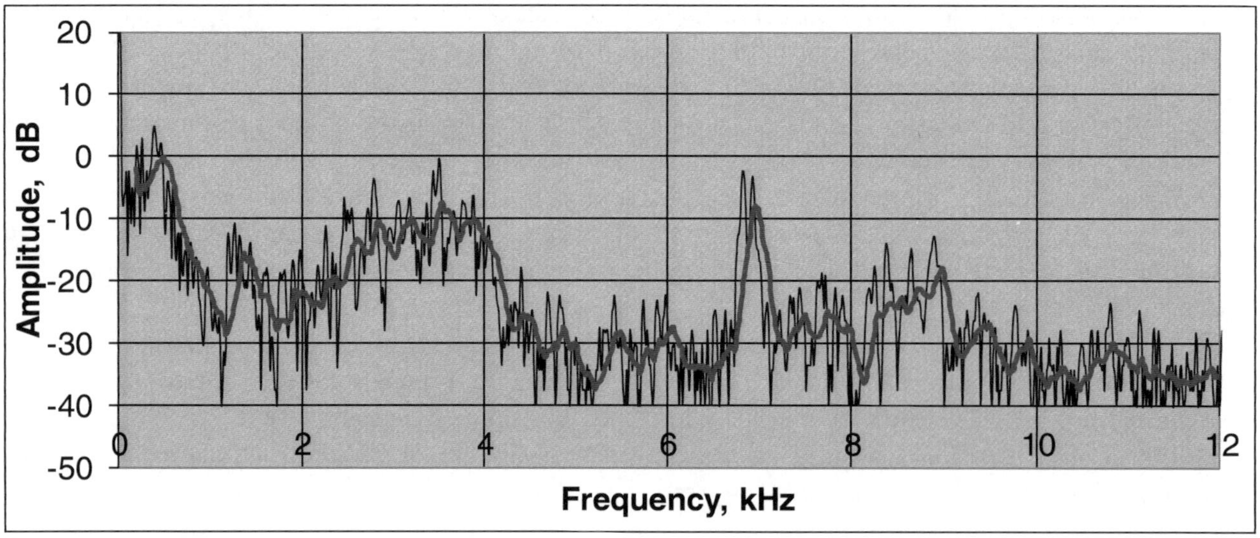

Figure 50. Averaged spectrum of the un-cracked cutout. Red curve is the trend-line.

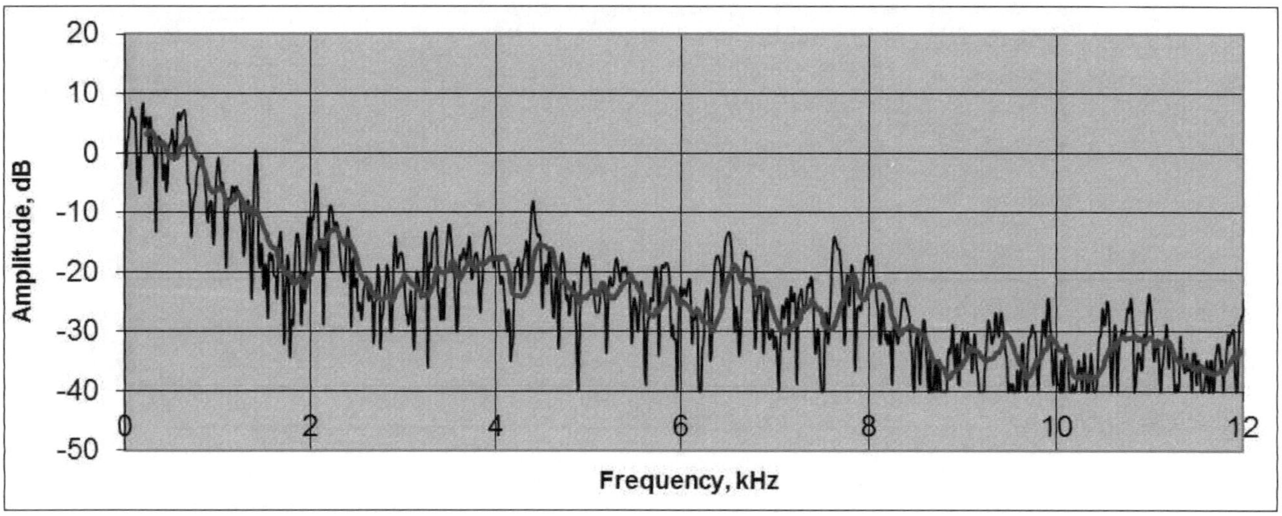

Figure 51. Averaged spectrum of cutout with large visible crack. Red curve is the trend-line.

The obtained results (compare Figs. 50 and 51) show that difference between spectra of the un-cracked and cracked cutouts is substantial. Besides the ability to detect cracked cutouts, the developed technique allows distinguishing new and old cutouts and also cutouts made by different manufactures. Moreover, using UT resonance inspection method based on the cutout spectrum analysis, it is possible not only to detect the large existing cracks, but also estimate the cutout condition and predict crack appearance (by detecting the incipient crack).

12. NON-CONTACT LASER-ULTRASONIC TESTING

Sometimes it is necessary to perform inspection of parts without direct mechanical contact between components of the inspection system and part under test. As an example, some results, obtained using laser-ultrasonic (laser-UT) technique and equipment for testing the porcelain insulators located on the high-voltage transmission lines towers, are presented below.

One of the most promising inspection techniques for insulator testing is the UT method (resonance inspection method or acoustic resonance spectroscopy), which can detect the presence of a crack ("puncture") by analysing the acoustic response of the insulator. This method involves hitting the test object and determining its vibration behaviour in the time domain and/or via a Fourier analysis in the frequency domain. Cracks reveals themselves as a peak-reflection signal in the time domain and as a change in the insulator vibration spectrum in the frequency domain.

Laser induced excitation and optical detection of acoustic waves in materials has been studied for many years. It is a completely non-contact method, and the inspection system can be utilized remote from the work-piece. Non-contact laser-UT inspection method, employing pulse laser excitation and optical interferometer reception, demonstrated sufficient sensitivity to distinguish between insulator samples with and without cracks in the shell. The non-contact excitation can be achieved by heating test object locally using a generation laser: short pulse and high peak power laser. Non-contact measurement (see Fig. 52) is done by using detection laser (long pulse laser with long coherence length) and optical interferometer, which measures spectrum and amplitude of the reflected optical signal.

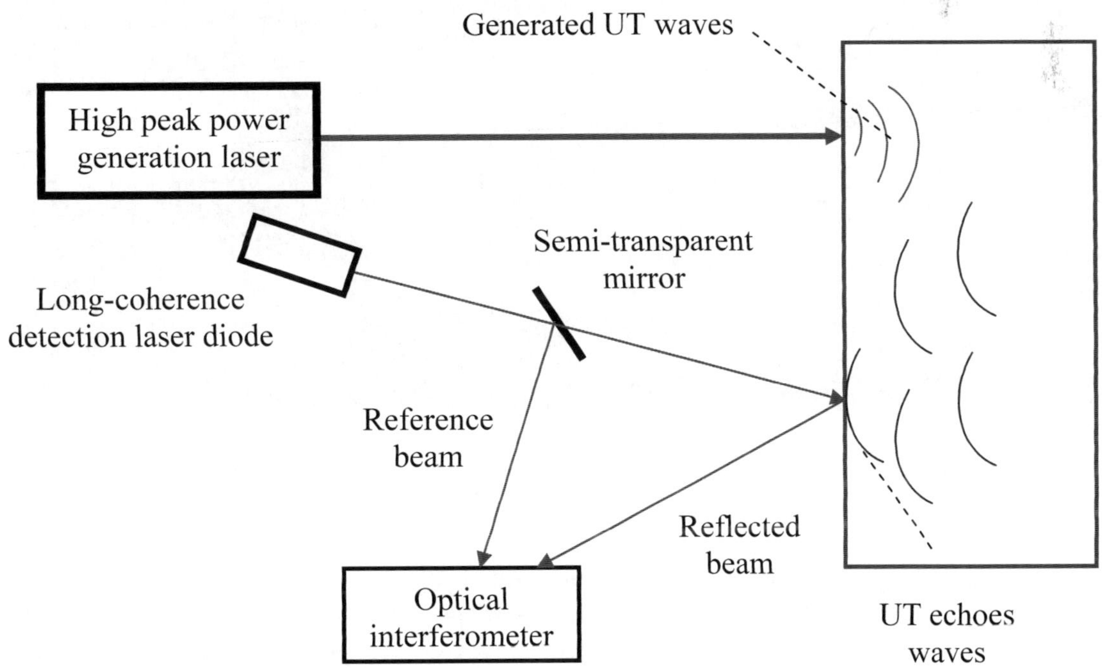

Figure 52. Schematic of laser-UT inspection technique.

The presence of an internal crack in the insulator affects the characteristics of the acoustic signature of insulator and respectively the parameters of the reflected optical response: phase, amplitude and spectrum. As result, acoustic responses of the cracked and un-cracked insulators to the generation laser pulse excitation are different, and these differences can be reliably detected by detection laser and optical interferometer in time and frequency domains.

The testing has been performed on the cracked and un-cracked insulators model 50-KIP. The cracked samples had one, two and three large visible cracks (about a few centimeters long). The testing system consisted of generation laser - Q-switched Nd:YAG oscillator model CFR200 (Quantel, USA) and optical interferometer (model Quartet, Bossa Nova Technologies, Culver City, CA, USA), see Fig. 53.

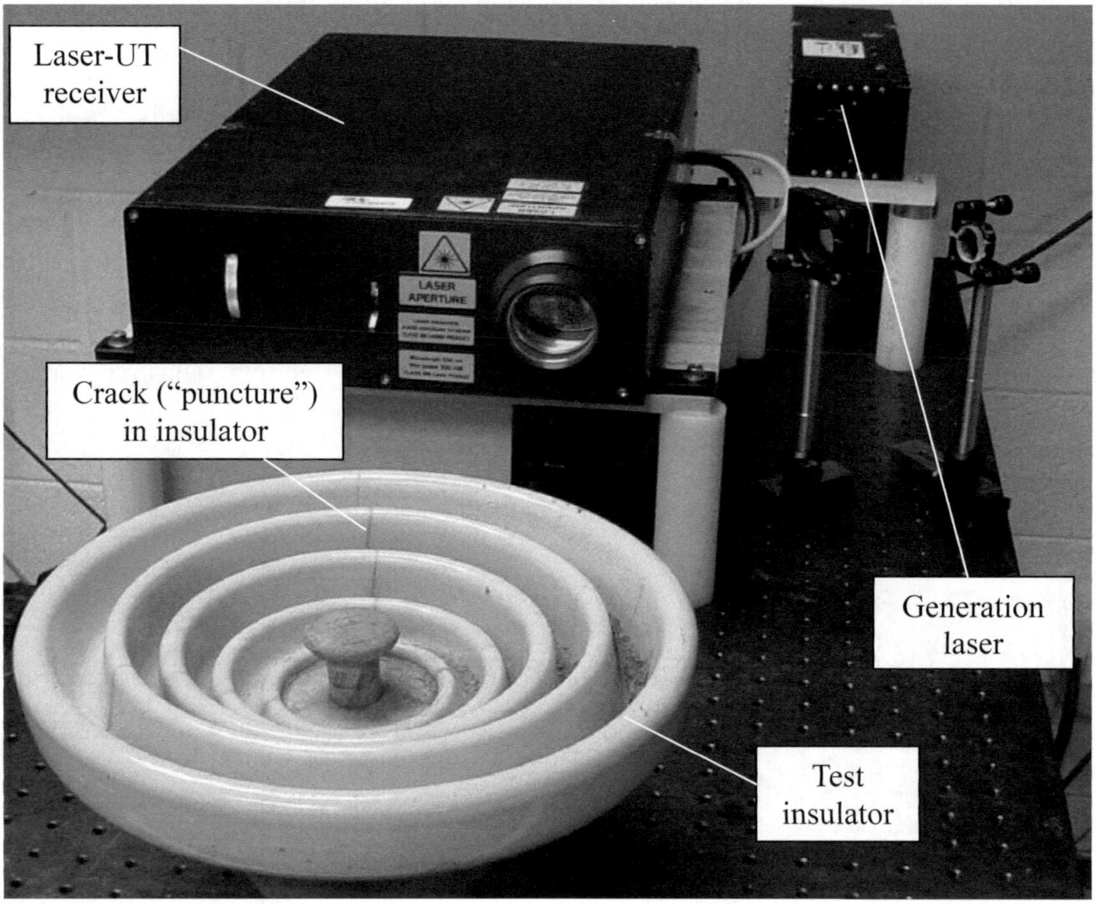

Figure 53. Setup of laser-UT inspection system for insulator testing.

Signals of the cracked insulators in the time domain have peaks (probably, crack reflections) positioned at 50-260μs, which are not present in the responses of the un-cracked insulators (see Fig. 54). Signals of the un-cracked insulators have peaks within the range 270-310μs, which are generated by the acoustic waves travelling around the insulator circumference. These peaks are not seen in the responses of the cracked insulators (see Fig. 54), because in such samples the UT waves cannot go around due to reflections from the cracks.

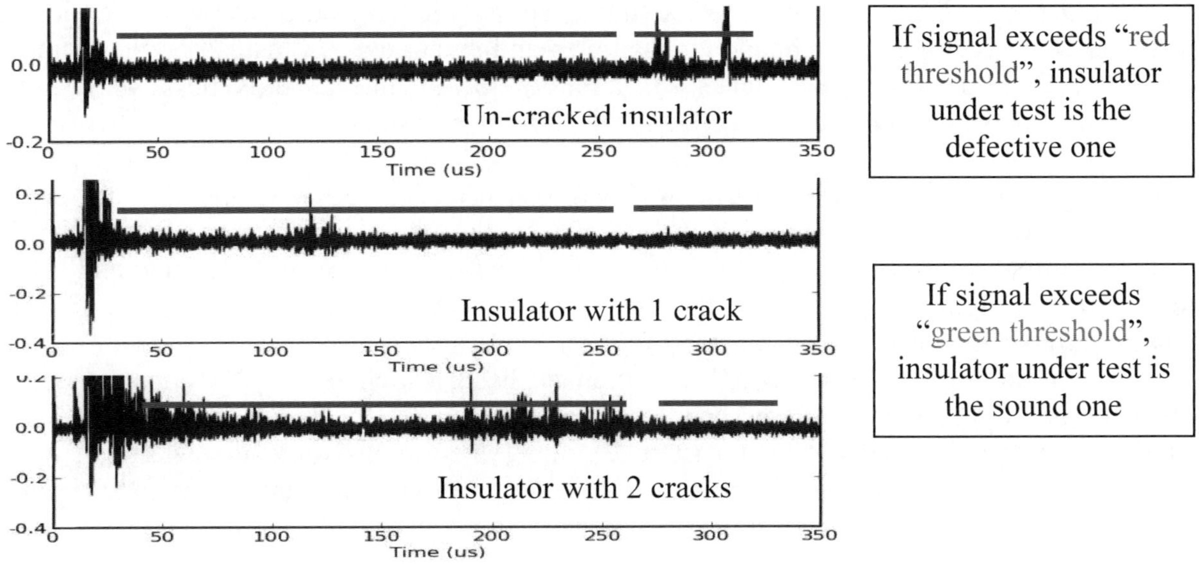

Figure 54. Time-domain signals of cracked and un-cracked insulators.

Spectra of the cracked insulators are concentrated at low frequencies (below 80kHz) and do not have the high-frequency components (at ~300kHz) and also duplets and triplets, which are typical for spectrum of the un-cracked sample (Fig. 55). Probably, the reason is the excitation of various types of waves and vibrations, which can exist only in an un-cracked insulator, because cracks reflect the waves and therefore block their propagation within the sample.

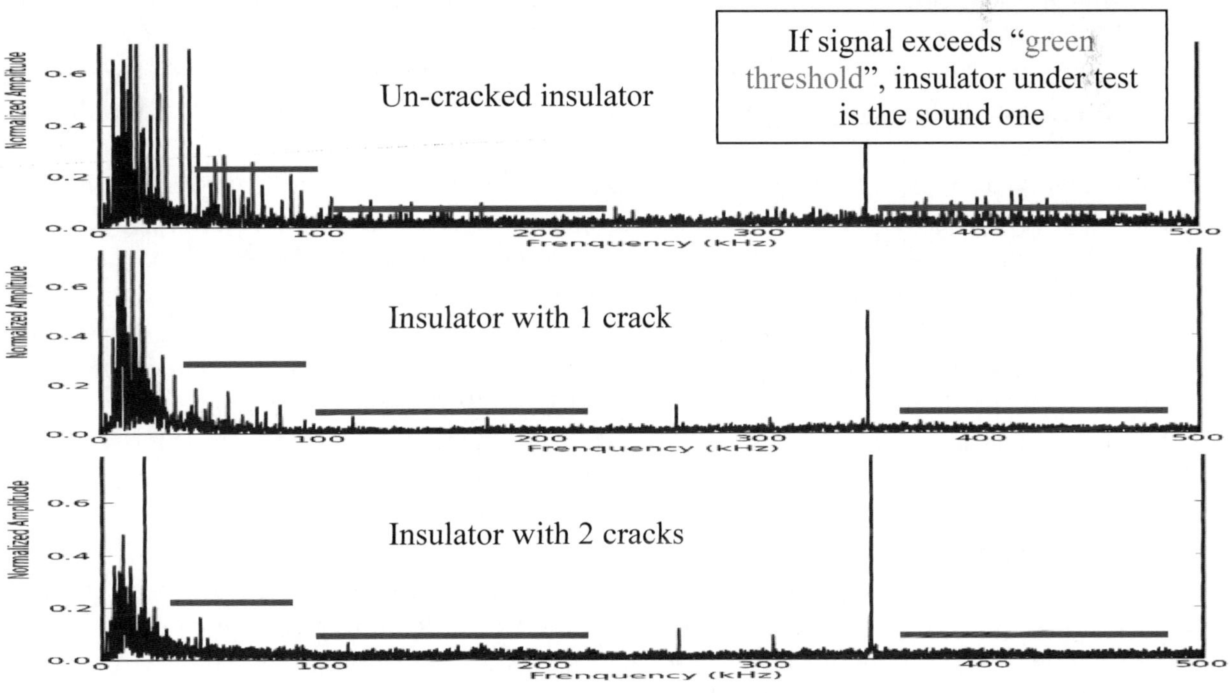

Figure 55. Spectra of cracked and un-cracked insulators.

Non-contact laser-UT inspection method demonstrated that it had a sufficient sensitivity to distinguish between defective and sound samples and even between insulators containing one, two, and three large visible cracks. Simple criteria of the defectiveness were used to differentiate between sound and defective insulators.

The non-contact examination of insulators was performed at rather small distance between inspection system and insulator: about ~50cm. At the same time, in practice rather large distance is required: about ~30m (this is the distance between truck on the ground and insulator located on the high-voltage tower). There are two options in developing a prototype of the non-contact laser-UT inspection system for practical application. The first option is to develop a system consisting of two parts: small measurement head, which can be positioned close to the insulator, and large processing unit, positioned far away and connected to measurement head by fiber-optic cable. The second option is to develop system, consisting of powerful generation laser, emission optics, reception laser, collection telescope, and other units, located at significant distance from the insulator under test.

13. FLAW ROOT RADIUS ASSESSMENT

Flaw root radius (RR) is a very important parameter for pressure tube risk assessment, because it determines degree of stress concentration and probability of crack growth. Unfortunately, the direct measurement of RR cannot be performed, because of a few reasons:

1. The typical wavelength λ of the UT wave, generated by shear wave angle probe with center frequency f=10MHz used for PT inspection ($\lambda = c/f \approx 0.235$mm, where $c \approx 2.35$mm/μs is the speed of propagation of shear UT wave in ZrNb). Respectively, the resolution which this wave can provide is about $0.5\lambda \approx 120$μm. Subsequently, if flaw RR is less than ~0.1mm, then transducer most probably will not “notice” it. It means, the small RR (about 30-50μm) cannot be assessed.
2. The other reason is the diameter of the acoustic beam, which is much larger than the flaw RR. It means that only a small portion of the transmitted UT wave impinges on the flaw root and reflects from it.
3. Even this weak wave, reflected from the small area of the flaw root, which usually has a complex shape, is very diverging. Therefore, only a small fraction of the reflected wave goes back in the direction of probe and impinges on it, while the major portion of the reflected wave misses the transducer.

The most reliable indirect method for RR assessment is probably the combined technique, which employs simultaneously the circumferential PC (CPC) response and both CW and CCW PE responses from left and right sides of the notch (see Section 3). Using it, the time interval Δt between center of the CPC notch response and intersection point of the CW/CCW responses can serve as criterion of RR assessment. Fig. 56 shows in the large scale how interval Δt between center of the PC response and intersection point of the CW/CCW responses can be measured. Fig. 57 below shows how these time intervals Δt are measured for notches with different RR. The physical reason, why interval Δt depends on the flaw RR, is related to analysis of the PC, CW and CCW responses based on the physical acoustics approximation. The CPC reflection always comes earlier than PE CW and PE CCW reflections are starting (they start from the root portions of the flaw), because the CPC time-of-flight is smaller than PE time-of-flight from flaw root. As result, the cross-over of CW and CCW responses is always positioned later in time than the CPC response. The CW and CCW responses for a flaw with large RR are separated by a wide angle range; therefore the crossover of these responses (in comparison with flaw with the same depth but small RR) is positioned closer to the PC response.

Emphasize that in order to measure time interval Δt, it is not necessary to use a novel technique – combined CW+CCW+CPC B-scan. The same images, as shown in Figs. 56-57, can be obtained using special software for data fusion and standard CW, CCW and CPC B-scans.

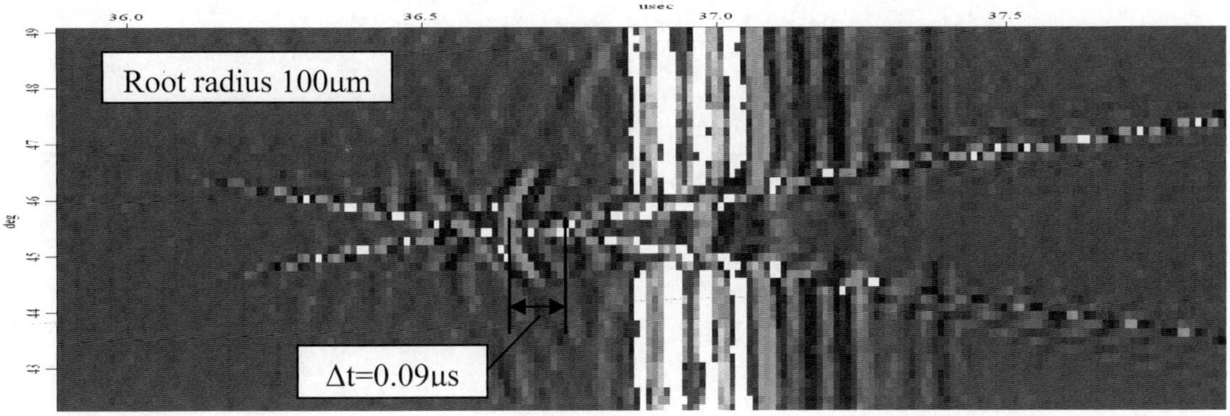

Figure 56. Circumferential combined PE CW+PE CCW+CPC B-scans of symmetric axial V-notch (depth d=0.3mm, width 0.5mm, $60^0/60^0$, RR=0.1mm. Probes: CW+CCW, f=15MHz, FL=39mm, D=9.5mm, WP=20.6mm, $\alpha=25^0$. Color scale is shown in Fig. 1.

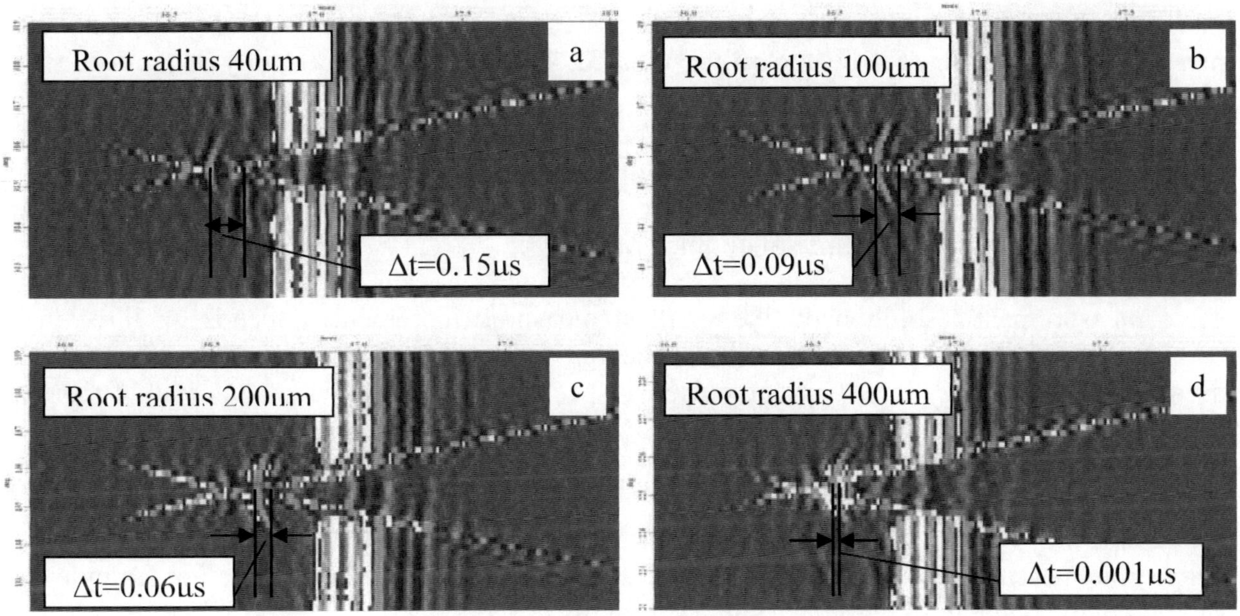

Figure 57. Circumferential combined PE CW+PE CCW+CPC B-scans of symmetric axial V-notches: depth d=0.3mm, width 0.5mm, $60^0/60^0$, root radii 0.04mm (a), 0.1mm (b), 0.2mm (c), and 0.4mm (d). Probes: CW+CCW, f=15MHz, FL=39mm, D=9.5mm, WP=20.6mm, $\alpha=25^0$. Color scale is shown in Fig. 1.

Based on the results, obtained on symmetric, asymmetric and off-axis V-notches with various RR, the correlation between Δt and RR for a wide range of RR was generated: see Fig. 58.

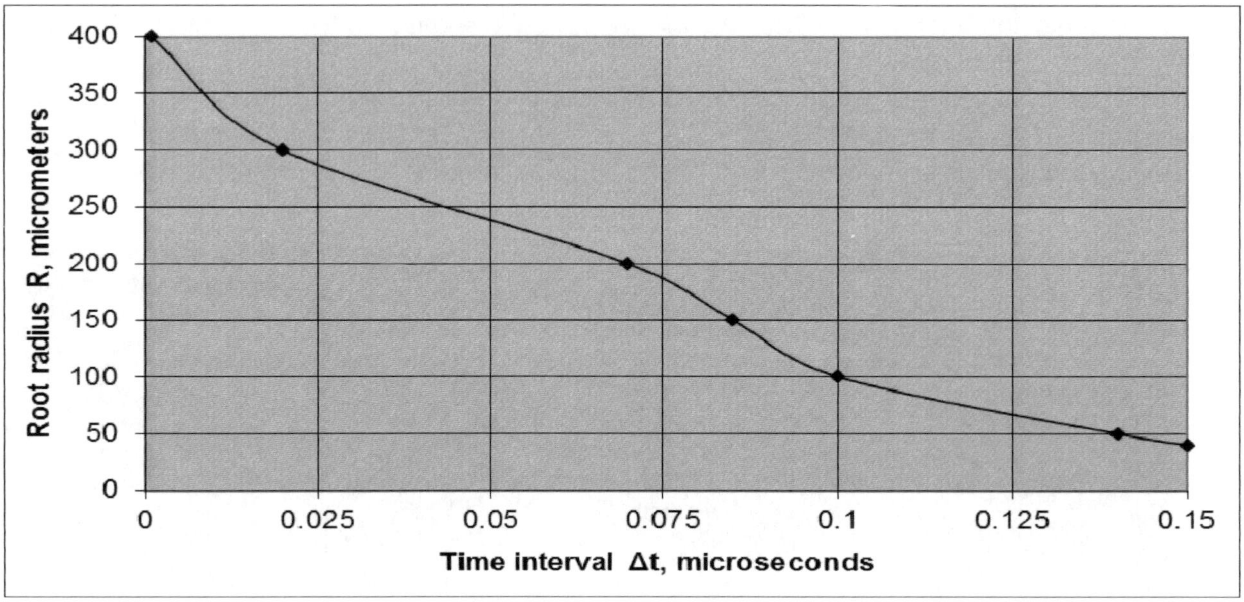

Figure 58. Correlation between Δt and RR.

The proposed combined technique is using the indirect method of RR estimation based on the measurement of time interval Δt, see Figs. 56-57. The combined uncertainty δt_{comb} of this measurement determines the accuracy of RR assessment. If time difference Δt is accurately measured, then flaw RR can be correctly estimated. The combined uncertainty δt_{comb} depends on many parameters (variations in flaw shape and position, temperature changes leading to variations in probe parameters and UT beam trajectory, changes in inspection system parameters, variations in probe parameters and position, human errors, and others), but the major factor, which usually determines the value of this uncertainty, is the measurement resolution δt of the UT pulse time-of-flight.

Time measurement resolution δt (the smallest measured increment) depends on three major parameters related to each other: probe center frequency f, probe bandwidth BW, and digitization (sampling) rate. Currently used inspection systems have probes with f=10MHz, BW≈100%, and acquisition cards with 125MHz digitization rate. These parameters allow providing time measurement resolution δt=0.008μs. Used acquisition cards allow, if necessary, increasing the digitization rate up to 500MHz. In addition, the up-sampling procedure can be also employed, which will lead to a further increase of the digitization rate and respectively to a better time measurement resolution δt. Recall that for this purpose, the digitization (sample) rate should be high enough to reproduce signals digitally.

The upper frequency of shear wave probes with center frequency 10MHz is ~20MHz at -20dB level. It means that Nyquist frequency, which should be at least twice greater than the upper frequency, is supposed to be greater than 40MHz for these probes. (The Nyquist frequency, i.e.

the highest frequency that can be accurately represented, is about one-half of the sampling rate). Subsequently, the sampling rate should be at least twice greater than the corresponding Nyquist frequency. Currently used digitization rate is more than sufficient to correctly reproduce the shape of UT responses. As result, the response positions and time interval between responses can be accurately measured. Note that obtained resolution $\delta t=0.008\mu s$ can be further improved by using wide-band probes with higher frequency and acquisition card with higher digitization rate, because at these conditions system will be able to transmit and receive short UT pulses without distortions, correctly reproduce them in the digital form, and accurately measure all time intervals.

High resolution in time δt can be easily recalculated in the resolution δs of spatial measurement $\delta s=c\cdot\delta t=(2.35\cdot 0.008)/2\approx 0.01mm=10\mu m$. The obtained result, spatial resolution $\delta s\approx 10\mu m$, means that all spatial measurements, including RR estimation, can be performed with high accuracy. It occurs because the proposed method of RR assessment is based on the combined indirect technique, where high measurement resolution, about $10\mu m$, is achieved in the direction of the UT beam trajectory. Recall that such high axial resolution of the test system is based on the wide-band high-frequency UT probes and high digitization rate of acquisition card.

However, recall that combined uncertainty δt_{comb} is determined not only by the time resolution δt (although this is a major component), but also by many other factors mentioned above. That is why the combined uncertainty δt_{comb} will be always greater than uncertainty δt.

14. DATA PROCESSING AND STRUCTURE INTEGRITY ASSESSMENT

Strictly speaking, all obtained data, if possible, should be statistically processed [10]. It means that in order to perform detection, the detection criterion, data probability distribution (e.g. discrete binomial distribution of detection measurements or continuous normal probability density distribution), probability of detection (POD), confidence level, and cumulative POD should be determined. For noise analysis such parameters as the noise probability density distribution and probability of false positive should be defined. The results of measurements, such as sizing criteria, mean values, confidence probabilities, minimum required number of measurements, and standard deviations of samples related to the flaw depth, width and length measurements should be calculated.

If necessary, flaws can be measured using some other independent techniques: mechanical measurements, replication, another NDE technique, or destructive measurements. As result, the true values of flaw dimensions will be defined. Using these true values and mean values obtained during UT examination, the measurement errors will be determined. Based on the obtained results, the UT inspection system should be calibrated using reference samples.

The measurement uncertainties, related to the impacts of various essential parameters affecting the detection, noise, and flaw depth, width and length measurements, should be determined.

Then structure condition assessment should be performed, based on information related to the flaw type, shape, dimensions, position and orientation, and also on estimation of probability of structure failure (tube rupture) and structure integrity evaluation. Flaw growth rate and deterioration rate of the tube should be also determined, in other words, the obtained results should be correlated to the remaining strength of the test tube and used, in accordance with the appropriate codes and standards, to calculate the remaining life time of the tube. If the UT testing is used for material characterization, then obtained results (size of the detected area with parameters which differ from "normal", its location, and "severity level" should be correlated to the material rate of deterioration and remaining life time of the structure under test.

15. CONCLUSIONS

1. Sometimes results of the UT inspection of tubes are not satisfactory due to inability to detect and characterize flaw (i.e. determine its type and shape) and size it. To improve the inspection performance, one should apply different methods and probes, which e.g. allow "seeing" flaw at various angles. The information, obtained by using various transducers and methods, should be combined.
2. Data processing and combination of different techniques and probes (combined technique, technique based on the multiple reflections of the large diameter NB probe, technique based on the cylindrically focused probe, tangential beam technique, and others) give a possibility to "reproduce" the flaw, determine its type and position, and rather accurately size the flaw.
3. Modified or novel UT techniques, probes, software, and data analysis procedures (non-rotating probes, GW technique, collimating probes, 3D visualization software package, and others) were developed to improve flaw identification, characterization and sizing.
4. Such novel techniques, probes and software packages lead to more accurate determination of the flaw type, orientation and shape, improved accuracy of flaw sizing and estimation of its root radius, decreased analysis time, and improved efficiency of analysis. All this, in turn, will lead to a decrease in the number of required replicas, and provide savings in terms of valuable outage time and worker dose uptake.
5. Based on the obtained results, the measurement data, if possible, should be statistically processed, and the respective structure condition assessment (calculation of remaining lifetime of the test tube) should be performed.

16. REREFENCES

1. Trelinski M., "Inspection of CANDU Reactor Pressure Tubes Using Ultrasonics", Proceedings of 17th World Conference on Nondestructive Testing, Shanghai, China, October 2008.
2. Karpelson A., "Quasi-Tomographic" Ultrasonic Technique for Tube Inspection", *The e-journal of NDT*, vol. 10, No. 6, June 2005.
3. Karpelson A., "Quasi-Tomographic Ultrasonic Inspection of Tubes", Proceedings of IV Pan-American Conference on NDT, Buenos-Aires, Argentina, October 2007.
4. Karpelson A., "Ultrasonic Detection of Cracks in Pressure Tubes", Proceedings of 32nd Canadian Nuclear Society Conference, Niagara Falls, Canada, June 2011.
5. Karpelson A., "Improved Flaw Characterization and Sizing in Pressure Tubes", *The e-journal of NDT*, vol. 17, No. 9, September 2012.
6. Karpelson A., "Ultrasonic Detection of Dry Contact between Metal Parts", *The e-journal of NDT*, vol. 17, No. 5, May 2012.

7. Karpelson A., "Ultrasonic Testing of Tubes without Probe Rotation", *The e-journal of NDT*, vol. 17, No. 3, March 2012.
8. Karpelson A., "Impact of Ultrasonic Probe Offset on Flaw Detection and Sizing during Inspection of CANDU Reactor Pressure Tubes", *The e-journal of NDT*, vol. 20, No. 7, July 2012.
9. Karpelson A., "Piezo-transducers with acoustic lenses forming narrow weakly diverging ultrasonic beams", Proceedings of 9th European Conference on NDT, Berlin, Germany, September, 2006.
10. Karpelson A., "Probability of Detection and Uncertainty of Measurement during Ultrasonic Inspection of CANDU Reactor Pressure Tubes", *The e-journal of NDT*, vol. 20, No. 7, July 2015.

Pulsed Eddy Current Measurements of Bulk Steel Pipe Wall Thickness

K. Faurschou[1,2],
J. Morelli[2], T.W. Krause[1] and P.R. Underhill[1]
[1] Dept. of Physics, Royal Military College of Canada, Kingston, ON
[2] Dept. of Physics, Eng. Physics and Astronomy, Queen's University, Kingston, ON

Queen's UNIVERSITY

Truth, Duty, Valour • Vérité, Devoir, Vaillance

Outline

- Experimental Work
  - Motivation
  - Experimental Setup
  - Probe Types
  - Data analysis
- Theoretical Work
  - Theory
  - Wall Thickness Model
  - Validation
  - Future Work

Motivation

- Medium term nuclear waste is stored in an array of steel pipes set in concrete, known as tile holes
- The steel is mainly there to contain the radiation
- Moisture in the earliest tile holes may have caused some corrosion in steel pipe, characterized as bulk wall loss

Test Piece

- 8.25" diameter 7.5 mm thick carbon steel
- 20% thinner and 20% smaller diameter than tile hole liner

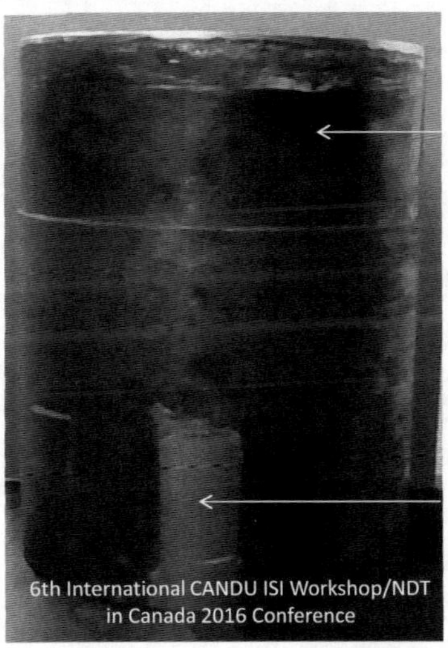

1.3 mm deep (17%) full circumference flaw (thin wall)

Unflawed (thick wall)

1 mm deep (13%) localized flaw

Experimental Apparatus

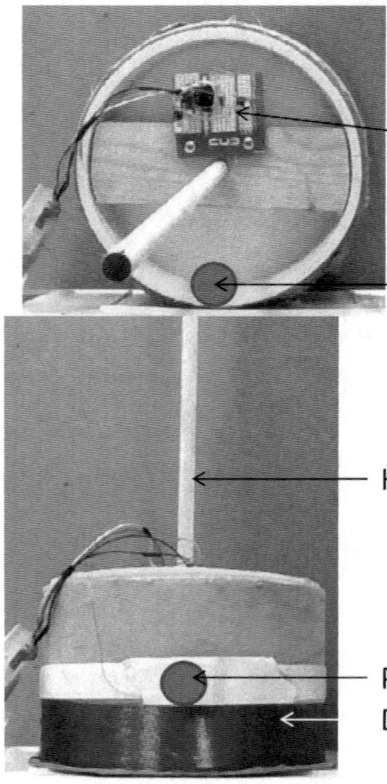

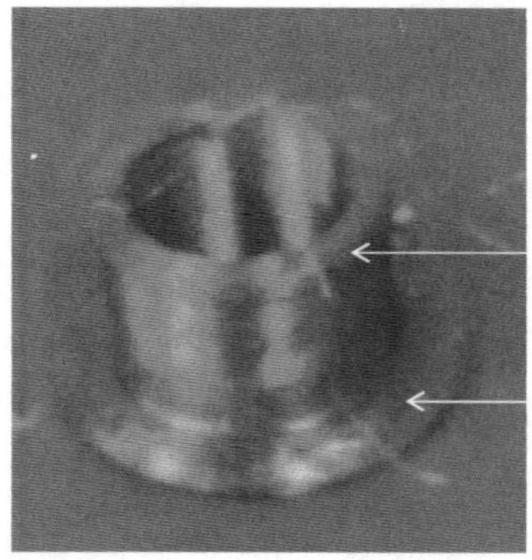

Signal Conditioning

- 10v drive pulse is filtered with a 3kHz low pass filter to eliminate ringing
- Pickup signal goes through two stages of amplification
- Each amplification stage is stitched together
- Further digital filtering

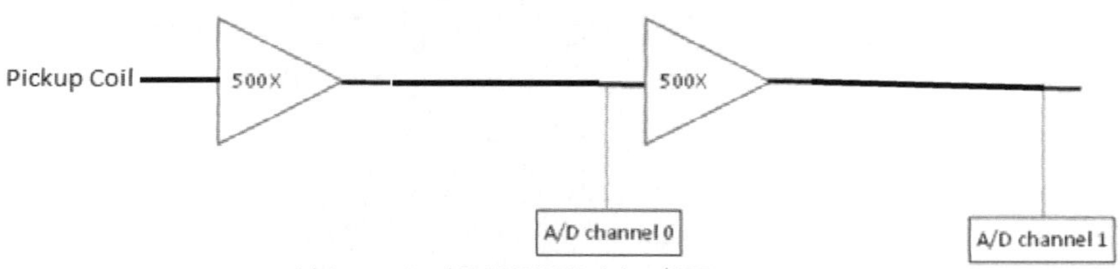

Measurements

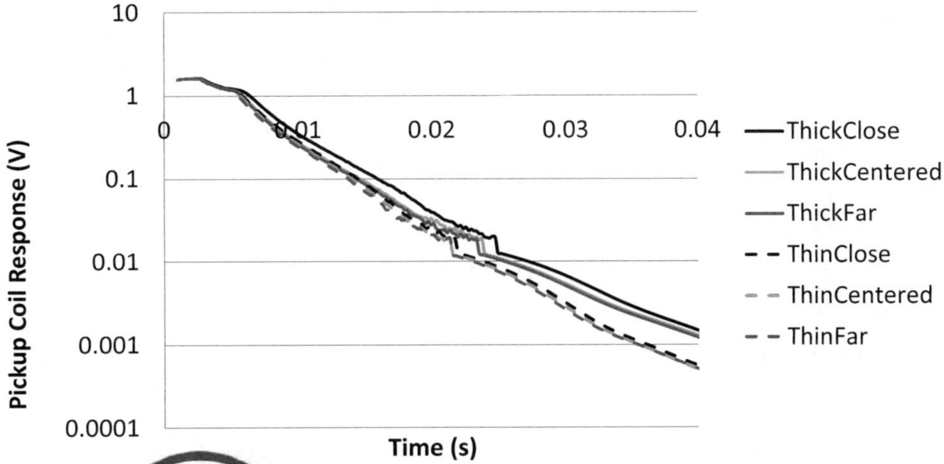

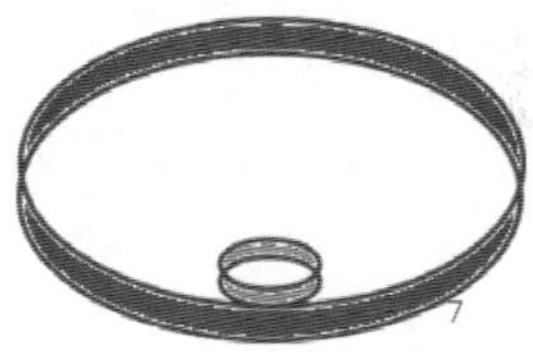

Measurements

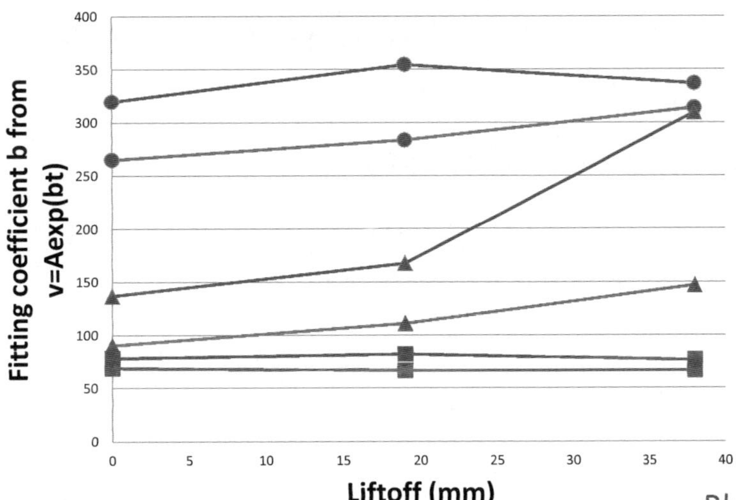

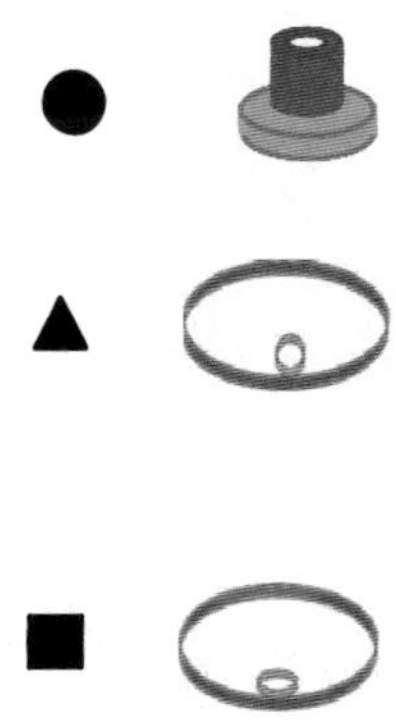

Blue is 7.5mm thick
Red is 6.2 mm thick

Probe Property Summary

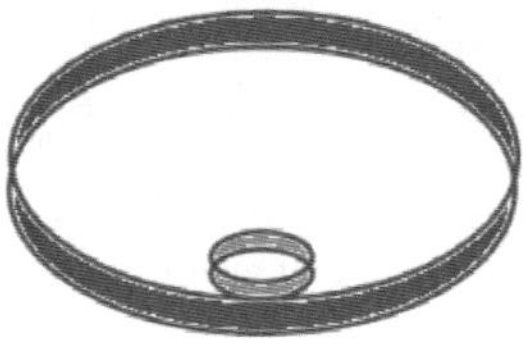

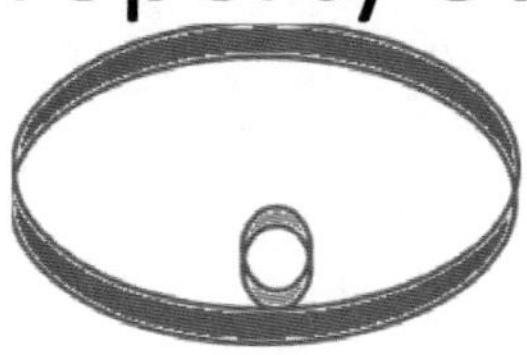

Both coils parallel to pipe

Pros
- Liftoff insensitive

Cons
- Lower sensitivity to wall loss
- Missed localized flaw

Drive coil parallel and pickup coil perpendicular to pipe

Pros
- Increased sensitivity to wall loss

Cons
- Increased sensitivity to liftoff
- Missed localized flaw

Both coils perpendicular to pipe

Pros
- Detected localized flaw
- increased sensitivity to wall loss

Cons
- Lower maximum liftoff threshold

Model Motivation

- Model allows sensitivity analysis of essential parameters
- Facilitates probe optimization
- Can be modified to work for similar projects (ferromagnetic pipe thickness)
- Possibility of using an inverse problem algorithm

General Theory

- For the time scales used in Pulsed Eddy Current (PEC) the diffusion equation is a valid simplification of Maxwell's equations
- $\nabla^2 J = \mu\sigma \frac{\partial J}{\partial t} + \mu\epsilon \frac{\partial^2 J}{\partial t^2}$, where A is magnetic vector potential
- Solution is a summation of exponentials multiplied with Bessel functions
- The first exponential eigenvalue (longest decay time) is $\tau \sim \mu\sigma L^2$ used for experimental analysis

Flat Plate Model

- Uses Desjardins et al. model[1]
- Solves the diffusion equation
- $\frac{d^2 J(r,z,t)}{dr^2} + \frac{dJ(r,z,t)}{rdr} - \frac{J(r,z,t)}{r^2} + \frac{d^2 J(r,z,t)}{dz^2} = \mu\sigma \frac{dJ}{dt}$
- Transformed to spatial and temporal frequency space
- $\frac{d^2 J(\gamma,z,\omega)}{dz^2} = (\gamma^2 + i\mu\sigma\omega) J(\gamma, z, \omega)$

[1] D. Desjardins, T.W. Krause and L. Clapham, 'Transient eddy current method for the characterization of magnetic permeability and conductivity', NDT&E International

Flat Plate Model

- Solved for impulse response then convoluted with input waveform
- Square wave input allows analytic inverse Fourier Transform
- Truncating infinite integral over spatial frequencies (γ) introduces high frequency noise at late times

Problem Geometry

Model

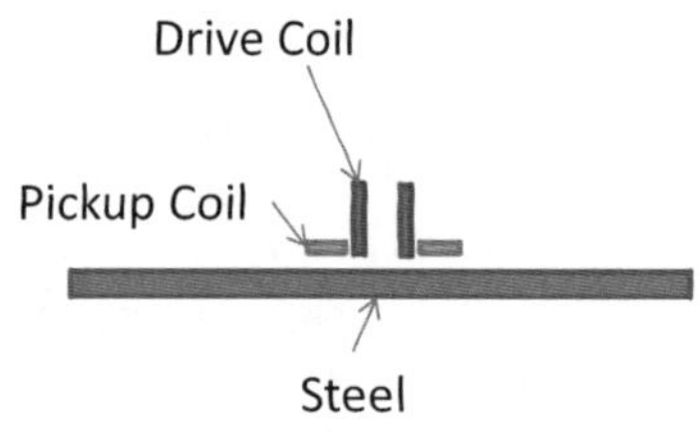

Experiment

Validation

Wall Thickness Sweep

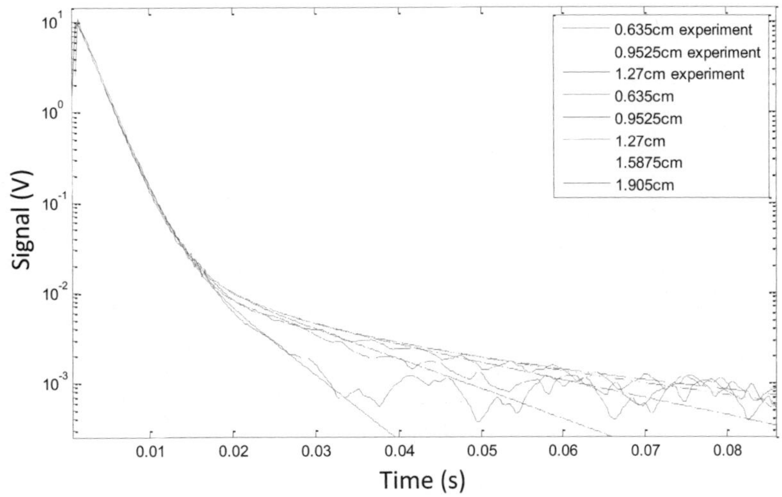

6th International CANDU ISI Workshop/NDT in Canada 2016 Conference

15

Isolated Liftoff Effects

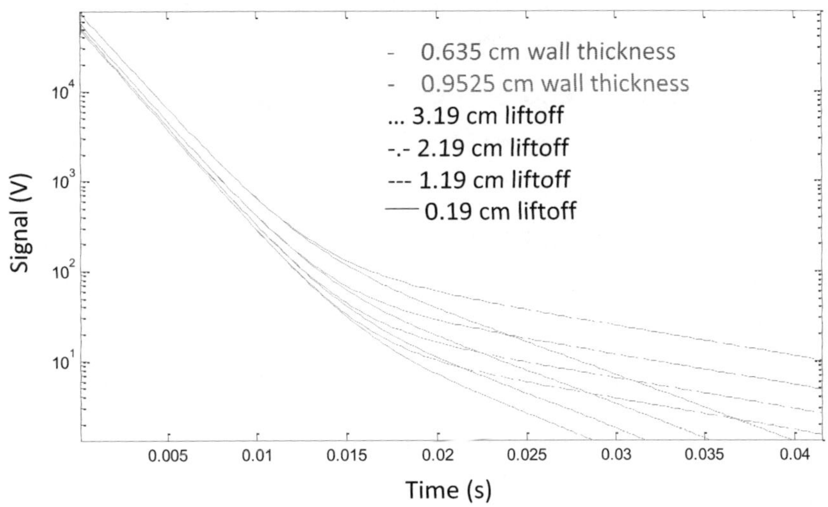

6th International CANDU ISI Workshop/NDT in Canada 2016 Conference

16

Wall Thickness Sweep

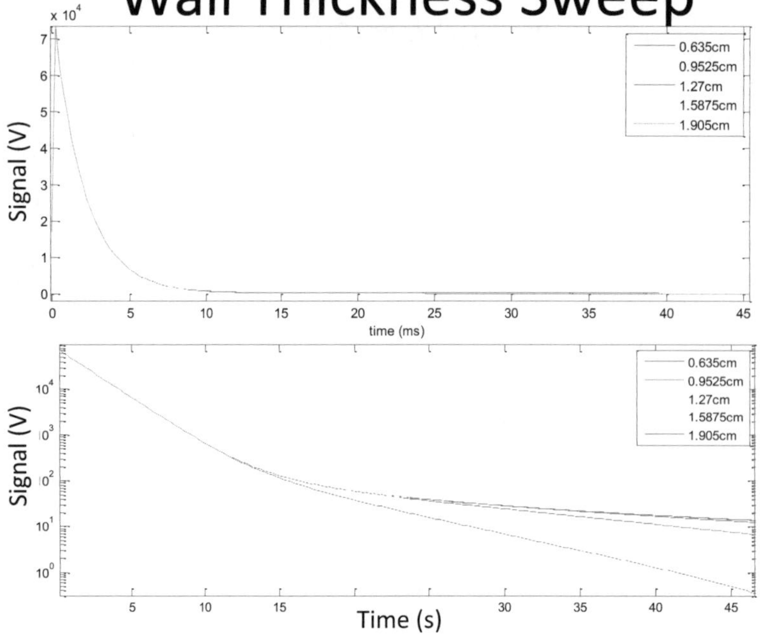

Principal Components Analysis

- Each eigenvector represents as much variance as possible under the constraint of eigenvector orthogonality
- $Y = \sum S_i V_i$ used to reconstruct original signal
- PCA allows small shape variations to be reflected in only a few scores
- When used directly on the model data, only peak shape (~ 1st 10 ms) affects the PCA
- Since all wall thickness effects appear after 15 ms PCA allows anything that affects the early signal to be compensated for

Separating Magnetic Permeability from Wall Thickness

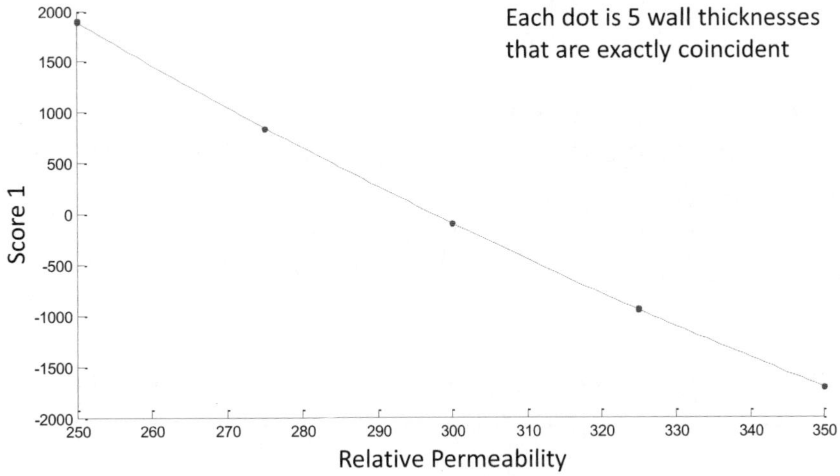

Changes in Wall Thickness sensitivity with Permeability changes

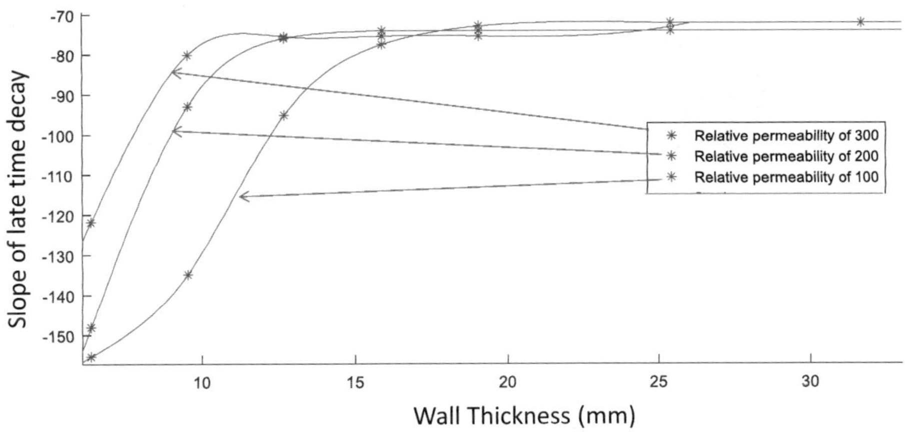

Summary

- Pickup coil axes perpendicular to the pipe axis increase sensitivity but increase liftoff effects
- Drive coil axes parallel to the pipe axis increase eddy current density at the expense of flaw localisation
- A flat plate analytical model has been produced
- It is possible to separate permeability and liftoff from wall thickness effects

Future Work

- Add more variables to PCA analysis such as conductivity, to see how many parameters can be extracted
- Evaluate how probe parameters affect the sensitivity response
- Measure the permeability of the validation samples experimentally

Acknowledgements

- Thanks to Canadian Nuclear Laboratories for collaboration on the tile hole project
- Thanks to University Network for Excellence in Nuclear Engineering and National Sciences and Engineering Research Council for funding

Questions?

Measurements

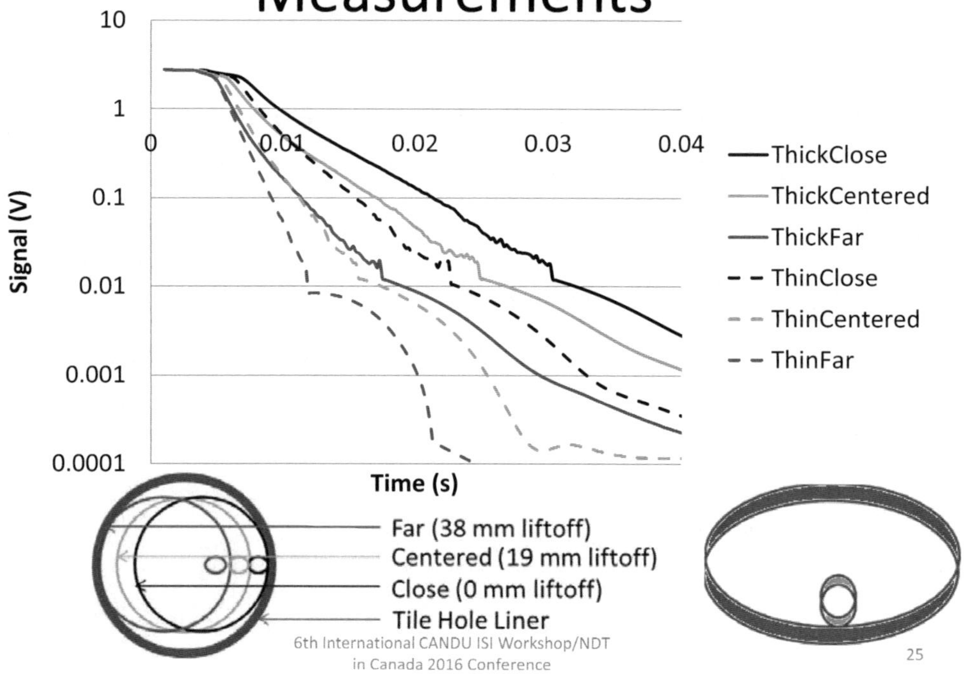

6th International CANDU ISI Workshop/NDT in Canada 2016 Conference

25

Measurements

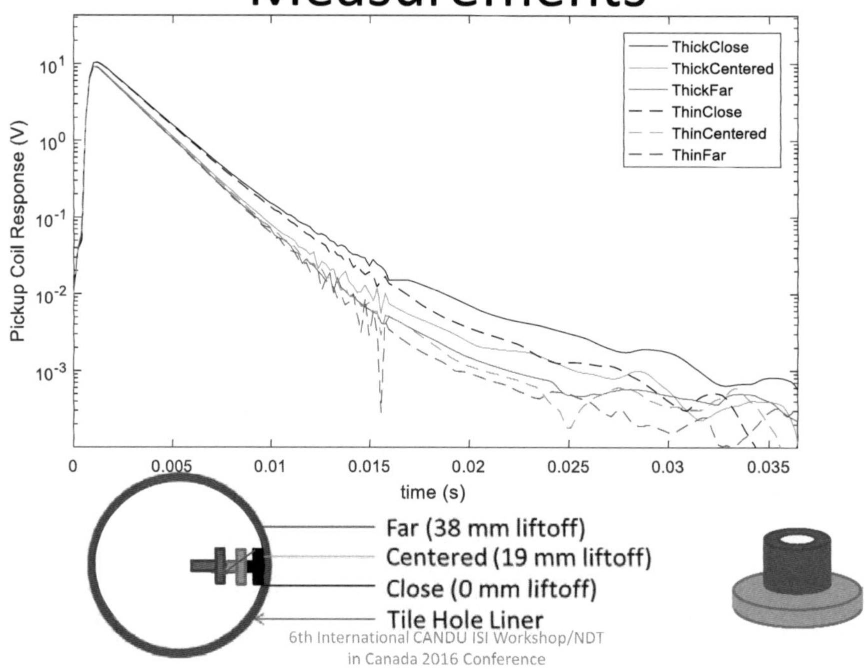

6th International CANDU ISI Workshop/NDT in Canada 2016 Conference

26

NDT in Canada 2016 Conference

CINDE in conjunction with CANDU

ECHO NDE

Ultrasonic Creep Wave Technique

Prepared by: Nick Marx

INTRODUCTION

The "creep wave" is the primary wave in a group of other waves that constitute the creep wave technique; the wave produced as the first critical angle is reached. In steel the first critical angle is approximately 31.5°.

This is the incident angle (the angle of the transducer relative to the test surface) at which longitudinal (compression waves) become no longer present. At this point the compression wave (now at a refracted ~76° travel just above the inside surface of the piece and are called "creep" waves.

In basic manual ultrasonic systems, they are produced by attaching the transducer to a CDS wedge, aka the "snail" wedge due to its design:

For automated systems the probe would like something like this:

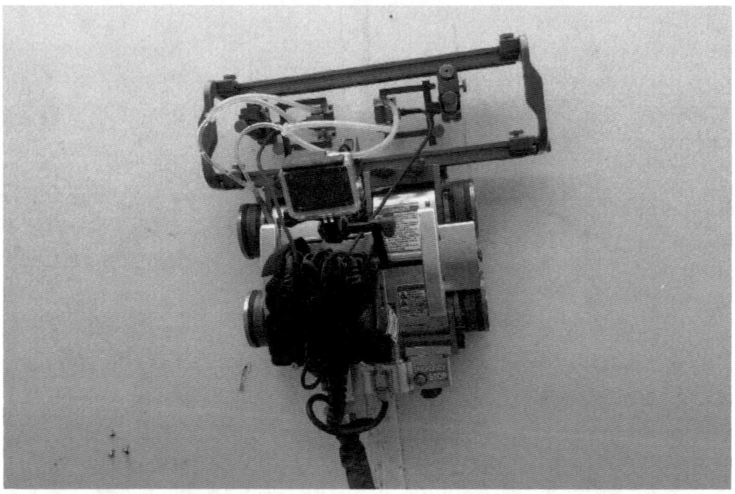

With a presentation that looks something like this:

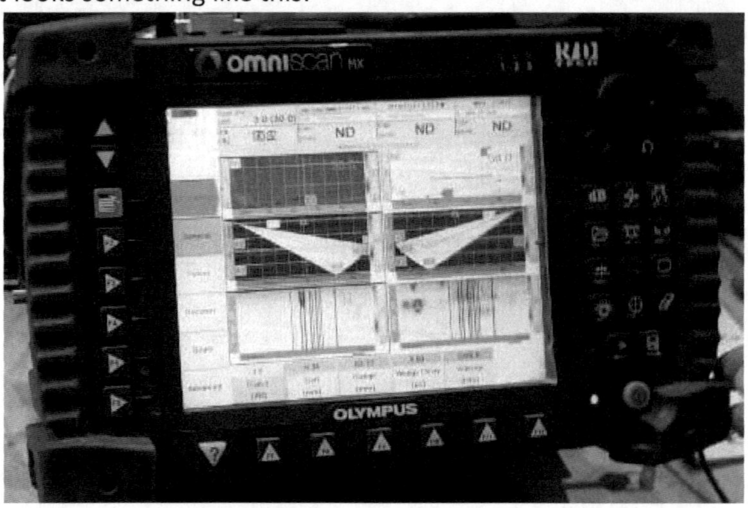

Instead of like this:

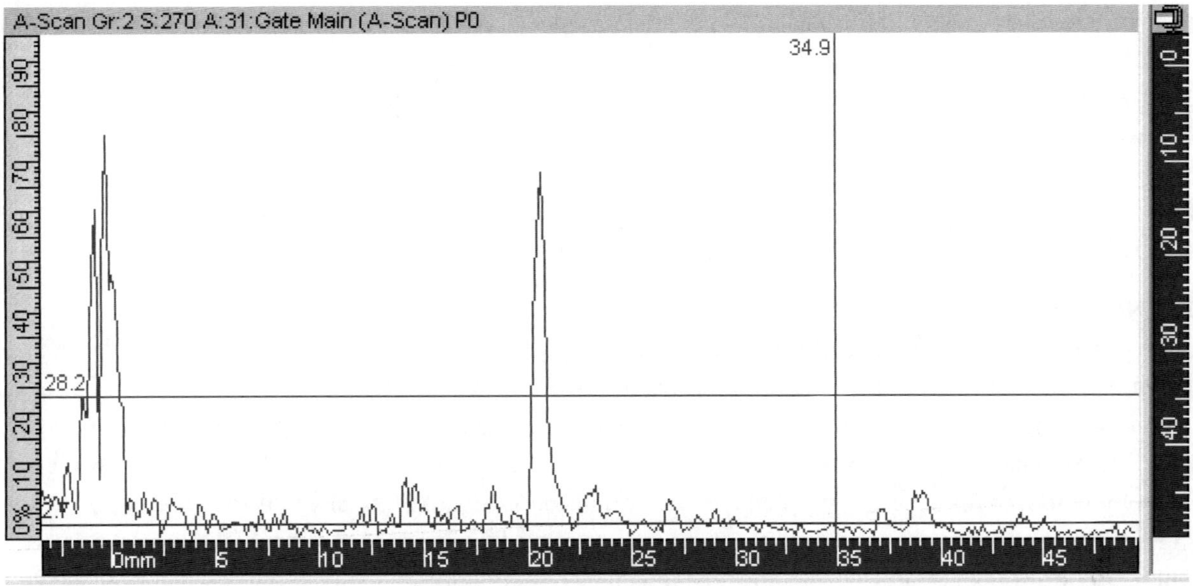

For the purposes of simplicity and ease of depiction this paper will deal with the manual equipment as the automated systems have much of the functionality we want to discuss built into the software. The same basic principles apply for either case.

PRINCIPLE

The technique uses:

- a refracted 76° compression wave which in turn travels parallel to the inside surface and just within its bounds called the creep wave.
- A 30° shear wave which mode converts to a 70° compression wave when it strikes the inner surface and will return as a reflected wave from internal surface breaking defects aka 30-70-70
- A refracted 70° compression wave that is produced from the tip of a surface breaking defect; this is a TOFD signal.

They are depicted as follows:

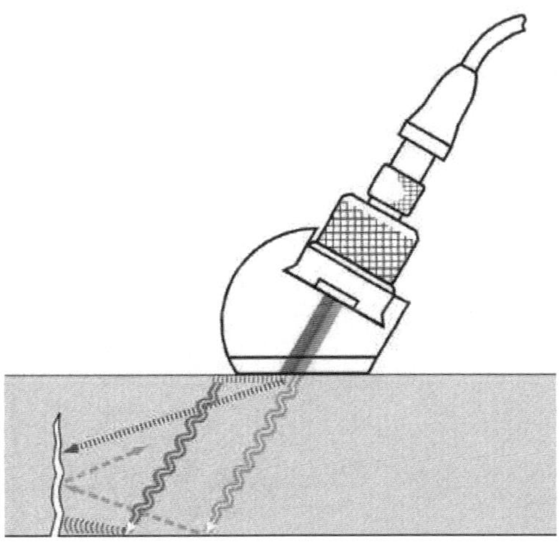

The creep wave is very sensitive to surface breaking linear defects. As it travels just under the inside surface of the part it is not affected by part surface condition such as rust or roughness. As discussed later, we will see the effectiveness of this wave in finding small linear defects.

The 30-70-70 shear wave picks up the "corner trap" signal of defects in the mid range of depth; between ~25 & 70% thru-wall with real effectiveness between 30& 60%.

The 70^0 compression wave hits the side of the defect in the upper cross section from ~50 to 100% thru-wall (of course at 100% thru-wall no NDE technique would be required!) with accurate assessment between 60 & 90%.

WAVE DYNAMICS

When thinking of ultrasonic testing, the normal premise of soundpath and screen range need to be put aside. Defect depth is measured quite differently when using the creep wave technique. The presence of one, two or three signals are the first broad interpretation. Defects where there is only one signal (the creep wave) are less than 25% thru-wall and it is the amplitude of the signal which is interpreted for depth. Care needs to be exercised when anomalies are picked up which have an amplitude ≤ to 50%FSH; due to the affinity for surface breaking defects, this wave can – and may – also pick up things like internal undercut, sharp root edge and other such discontinuities that are of no concern during this inspection.

As the amplitude of the creep wave approaches 100% FSH the defect is approaching 25% thru-wall and we need to look for some indication of the mode converted shear wave; even a small change in the baseline at the calibrated for location is enough to suggest 25% or more thru-wall. The signal characteristics are used to determine defect depth. Although there won't be much lateral movement, the deeper defects (35 to 50% range) will have some; as well, the increase, decrease and maximum amplitude says a lot about smaller defects (25 to 35% range). Once the amplitude of this signal approaches ~50% FSH we need to look for the 70^{0}TOFD signal. This signal is produced by the 70^0 compression wave striking the defect side surface and causing the tip of the defect to "ring" or echo; it is this TOFD signal that is picked up by the instrument. Its amplitude, as well as its location on the baseline, along with the wave dynamics of the shear wave are used to determine defect depth.

It is now clear that the conventional use of the time baseline is not feasible as the creep and shear wave signals do not move according to soundpath travel.

CALIBRATION

For accurate defect quantification the calibration piece must be of the same – or very close to – thickness of the part being examined. The transducer is first placed on the calibration piece so as to pick up a cut end (100% thru-wall). The sweep and delay are then used to place the signal from the creep wave at half screen width (5) and the shear wave signal is placed at relative increment 4. The calibration block has a number of EDM (Electrical Discharge Machining) notches; these can be whatever one is wanting to resolve but are usually cut at 20, 40, 60 & 80% thru-wall.

Calibration block:

Calibrated Screen:

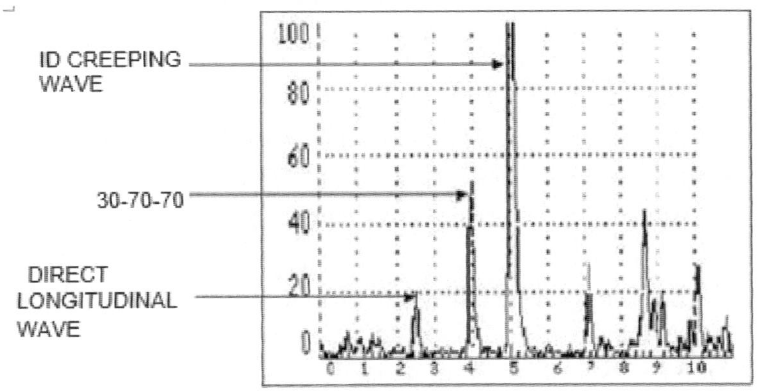

Calibration of amplitude (amount of sound) then occurs using the various EDM notches. The object is to get all three signals without having the creep wave signal go too far over 100%FHS. This normally occurs around the 40% EDM. That amplitude then becomes our reference setting and the EDM notch used is noted. Care needs to be taken when testing as the surface condition of the part is rarely as smooth and clean as the calibration block.

TECHNIQUE HISTORY

The technology was first developed and perfected by EPRI for use on critical piping welds in nuclear facilities where effecting repairs could be a very expensive proposition. With proper calibration and good experience with the technique, defect characteristics and extent (depth) can be accurately measured. Accuracy is very dependent on the operator, the calibration, ultrasonic equipment and the part being tested. In general accuracy to within ±5% of actual is normally achieved and sometimes even better down to 2% of actual. The quantified defect depth will then be compared to a matrix that takes into account the potential consequence of failure, cost of repair and likelihood of failure – along with any number of other factors – to determine the course of action. These can range from continuing to operate with a monitoring plan to immediate shutdown. In general, the services the equipment in our industry is not so critical nor are the repairs so costly.

We started off with this technology using only manual handheld equipment. We found a limited market for the technique where looking for ID related service cracks was an advantage. Thanks to the benevolence and foresight of our previous employer H. John Roth we were able to purchase appropriate equipment, build calibration blocks and take courses in the proper use of the equipment.

For approximately ten tears our focus was on small equipment where there was no other way to detect small ID cracks. Given the relative ease of repair, quantification was not as critical as with the EPRI work. We would be able to tell our clients that: they had no cracks, shallow cracks (≤10%), significant cracks (>10% to ≤40%) or serious cracks (<40%).

Approximately eight years ago we were approached by a client who knew of this technique and wanted to test it against a number propane bullets (4) they had at a gas plant. We decided to invest in some phased array equipment that would do what we needed and at a much faster pace. We also followed up on the creep wave inspection of these 4 bullets with internal WFMPI to prove up the feasibility and accuracy of our technique.

The creep wave inspections found 2 bullets with issues:

- V4031 indicated 3 suspect locations; traditional UT was used and it was determined that 2 of them were caused by weld profile. The 3rd was found to be ~10% thru-wall and 59mm long
- V4032 one indication was found with creep wave to be 1.69m long and 33% thru-wall.

WFMPI was carried out on all welds in both V4031 & V4032; the results corroborated not only the existence of the indications found with the creep wave techniques found with the creep wave technique but also that there were no other indications found; this corroborated the creep wave inspection.

Once a repair plan was devised all areas with indications from the creep wave in both vessels were examined with B&W MPI. This inspection concurred with the creep wave findings.
The indication in V4031 was determined to be a "missed edge" and was blend-grinded away. There was no recordable depth.
The 1.69m indication in V4032 was excavated to 5mm until all evidence of the indication was removed; the shell is 15.9mm thick.

These results gave us conclusive proof as to the viability of the creep wave technique in lieu of internal WFMPI.

Following are copies of the applicable NDE reports:

ECHO NDE

ULTRASONIC PHASED ARRAY INSPECTION REPORT

DATE: MARCH 04, 2009	**REPORT #: PAET-0009**	**PAGE 2 of 4**
CLIENT: [redacted]		
LOCATION: [redacted] GAS PLANT	**PROJECT: UT PHASED ARRAY CREEP WAVE ON WELDS**	
ITEMS EXAMINED: LPG BULLET [redacted] V-4031 ON ALL CIRCS, LONGSEAMS, NOZZLES (NO REPADS)		
PROCEDURE: UT-10	**CLIENT PO # / JOB #/ AFE:**	
TECHNIQUE: **CREEP 17mm**		
ACCEPTANCE CRITERIA: ASME VIII DIV I	**SPECIFICATION:** ASME V, ART. 4	

TEST RESULTS

PSV INFORMATION:

TYPE – 1905MC	SET PRESSURE – 185PSI
SN – [redacted]	TEMPERATURE – 160F
CRN – 01832.568312	CAPACITY – 14587 SCFM
TAG – 104	MANUFACTURER - CONSOLIDATED
4" INLET, 6" OUTLET	SERVICED BY SYNTECH, AUG 2003

NOZZLES NOT SCANNED (REPADS) – 20" MANWAY, 1 X 6", 2 X 4", 4 X 3"

NOZZLES SCANNED (NO REPADS) – 3 X 2"

RESULTS:

3 SUSPECT INDICATIONS NOTED AT TIME OF INSPECTION.
1) LS1 @ 1189mm TO 1248mm
2) LS3 @ 290mm
3) CIRC4 @ 4617mm TO 4789mm

A TRADITIONAL UT SHEARWAVE INSPECTION WAS CONDUCTED ON ALL SUSPECT AREAS.

- INDICATIONS ON LS3 AND CIRC4 WERE DETERMINED TO BE WELD PROFILE, NO DEPTH NOTED TO INDICATIONS.

- INDICATION ON LS1 DETERMINED TO BE CRACK LIKE LINEAR INDICATION FROM 1189mm TO 1248mm (59mm LONG). INDICATION HEIGHT 15mm TO ID SURFACE (2mm THROUGHWALL HEIGHT, ROUGHLY 10%).

NO OTHER INDICATIONS NOTED AT TIME OF INSPECTION. ALL OTHER WELDS ACCEPTABLE AS PER CODE.

Misc. Charges	STAMP	Regular Hours	**See**
	EDWIN TYMENSEN CGSB / SNT-TC-1A UT II / MT II / PT II #11424 API 510, CERT # 27479	Overtime Hours	**Time**
		Shift Premium	**Summary**
		Kilometers	**CET-0015, CET-0016,**
		Sub / Man Day	**CET-0017**
Client Representative:	Technician:	Assistant: DARRYL COLEMAN / TYLER BAITION	

347A – 28042 Hwy 11 / Red Deer County, AB / T4S 2L4

Ph: (403) 347-7042 Fax: (403) 347-7052 www.echonde.com

ECHO NDE

M - 4751

MAGNETIC PARTICLE/LIQUID PENETRANT INSPECTION REPORT

Date Mar 18/09 | Page | of

CLIENT: [redacted] | ECHO JOB #

CONTRACTOR:

LOCATION: [redacted] GAS PLANT | PROJECT

ITEMS EXAMINED: V-4031 NGL STORAGE BULLET | MAINTENANCE

PROCEDURE: | CLIENT P.O.#/JOB#:

ACCEPTANCE CRITERIA: | SPECIFICATION:

SURFACE CONDITION: ☐ Clean Base Metal ☐ As Ground ☐ Machined ☐ Shot Blast ☐ Painted ☐ Other

MINIMUM LIGHT INTENSITY ≥100 fc VISIBLE ≥1000μW/cm² FLUORESCENT

MPI METHOD

☐ AC ☐ DC ☐ Continuous ☐ Residual ☐ 12V ☐ 120V ☐ Other:

EQUIPMENT TYPE

☐ Yoke ☐ Coil Serial #: Last Calibration Date:

☐ Blacklight Serial #: Last Calibration Date:

MPI MEDIUM ☐ Dry Colour: ☐ Wet ☐ Fluorescent ☐ Black on White Particle Size:

EXCESS PARTICLE REMOVAL ☐ Flow ☐ Exhalation ☐ Other

LPI METHOD

Penetrant: S/N ☐ Vis ☐ Fluorescent ☐ Water Wash ☐ Solvent Removable

Developer: S/N ☐ Wet ☐ Dry ☐ Nonaqueous

Dwell Times Penetrant Min: Max: Developer Min: Max:

TEST RESULTS

A WFMPI WAS PERFORMED ON V-4031 NGL STORAGE BULLET INTERNALLY AS REQUESTED BY CLIENT. MPI WAS DONE ON ALL CIRCS, LONGSEAMS AND [illegible]

RESULTS: THE 3RD CIRC FROM THE WEST HAD A 1" LINEAR INDICATION BESIDE A [illegible]

OF THE WELD.

THERE WERE NO OTHER INDICATIONS FOUND AT THE TIME OF INSPECTION.

ALL OTHER AREAS WERE ACCEPTABLE TO CODE.

GENERATOR

STAMP

GORD BOS #11303

CGSB RTII / MTII

SNT-TC-1A RTII / MTII

Regular Hours

Overtime Hours

Client Representative: | Technician:

Echo NDE Inc. Red Deer, Alberta

Phone: (403) 347-7042 * Fax: (403) 347-7052 * www.echonde.com

White - *Client* Canary - *Office* Pink - *Invoice*

ECHO NDE

MAGNETIC PARTICLE INSPECTION REPORT

DATE:	MAR 26-2009	REPORT #	MDC-0002		PAGE 1 of 1
CLIENT:	[redacted]			ECHO JOB #:	
CONTRACTOR:	SAME				
LOCATION:	LSD [redacted]	PROJECT:		LPG BULLET [redacted]	
ITEMS EXAMINED:	SEE BELOW				
PROCEDURE:	MT-3A	CLIENT PO # / JOB #:			
ACCEPTANCE CRITERIA:	ASME VIII DIV 1	SPECIFICATION:	ASME V, ART 7		

METHOD

SURFACE CONDITION

☑ Clean Base Metal	☐ As Ground	☐ Machined	☐ Sand Blast	☐ Painted	☑ Other:	BUFFED
☐ AC	☐ DC	☑ Continuous	☐ Residual	☐ 12V	☑ 120V	☐ Other:

EQUIPMENT TYPE

☑ Yoke	Serial #: 9563	Last Calibration:	JUNE 01 - 2008	Prod Spacing: 3 - 8 inches
☐ Blacklight	Serial #: 26591	Last Calibration:		
☐ Light Meter	Serial #: 1225653	Last Calibration:		

MINIMUM LIGHT INTENSITY: ≥ 100 fc Visible, ≥ 1000 μW/cm2

MEDIUM TYPE (LIST CONTRAST AS WELL):	MAGNAFLUX	MEDIUM MANUFACTURER:	MAGNAFLUX
☐ Dry Colour:		☐ Fluorescent	☑ Black on White
MPI Particle Application Method:	☐ Powder Puffer	☑ Spray	Excess Particle Removed via Exhalation & / or Flow

TEST RESULTS

BLACK ON WHITE MPI EXAMINATION WAS CONDUCTED ON THE LPG BULLET V-4031 AT THE FOLLOWING AREAS AS REQUESTED BY CLIENT.

LONG SEAM #1 FROM 46" TO 49" (3" LONG)

_ MPI ON *"MISSED EDGE"* INDICATION AS PER ECHO NDE PHASED ARRAY REPORT BEFORE ANY BLEN GRINDING HAD TAKEN PLACE.

_ MPI ON *"MISSED EDGE"* INDICATION AFTER BLEND GRINDING WAS COMPLETE

RESULTS: THERE WERE NO INDICATIONS LOCATED AT TIME OF EXAMINATION, AREA WAS ACCEPTABLE TO CODE.

Misc Charges	STAMP	Regular Hours	SUMMARY
1 GALLON	DARRYL COLEMAN	Overtime Hours	
	CGSB / SNT-TC-1A RTII / UTII / MTII/PTII # 4369	Shift Premium	
		Kilometers	
Client Representative:	Technician Signature:	Assistant: DAN WONG	

347A - 28042 Hwy 11 / Red Deer County, AB / T4S 2L4

Ph: (403) 347-7042 Fax: (403) 347-7052 www.echonde.com

ECHO NDE

ULTRASONIC PHASED ARRAY INSPECTION REPORT

DATE: MARCH 04, 2009	**REPORT #: PAET-0013**	**PAGE 1 of 2**

CLIENT: [redacted]	
LOCATION: [redacted]AS PLANT	**PROJECT: UT PHASED ARRAY CREEP WAVE ON WELDS**
ITEMS EXAMINED: LPG BULLET [redacted] V-4032 [redacted]	
PROCEDURE: UT-10	**CLIENT PO # / JOB #/ AFE:**
TECHNIQUE: CREEP 17mm	
ACCEPTANCE CRITERIA: ASME VIII DIV 1	**SPECIFICATION:** ASME V, ART. 4

METHOD

EQUIPMENT MANUFACTURER: OLYMPUS	SERIAL #: OMNI-100493	CALIBRATION DATE: MAY 21, 2008
SCANNER / MANIPULATOR HXMT-103	UMBILICAL LENGTH:	5.8m (80" TRADITIONAL UT CABLE LENGTH)
CAL. BLOCK(S): IIW BLOCK, ASME BLOCK, CREEP BLK.	COUPLANT: WATER / UTX	PROBE CABLE LENGTH: 2.5m

TEST PIECE: ☒ CREEP BLK.	REF REFLECTOR:	Type: ☒NOTCH	Size:	Depth: 20%	Response Ht: 80%FSH
THICKNESS: ☒	SHEAR WAVE: ☒	LAMINATION: ☒			

	ANGLE	WAVE	FREQ	SIZE	MFG.	S/N	RANGE	Ref. db	Scan db	TL db
1	0 DEG	LONG	5 MHZ	0.250"	Stresstel	015NT3	1.000"	58	+6	64
2	70 DEG	SHEAR	2.3 MHZ	0.375"	Sonatest	T024854	200mm	46	+6	52
3	60 DEG	SHEAR	2.3 MHZ	0.375"	Sonatest	T024840	200mm	45	+6	51
4	45 DEG	SHEAR	2.3 MHZ	0.375"	Sonatest	T024849	200mm	44	+6	50

SURFACE CONDITION: Clean Base Metal: ☒ As Ground: ☐ Painted: ☒ Other:

TRANSVERSE SCAN COMPLETED:	**Manual Creep Wave Transverse Scan along welds and around Tees – ACCEPTABLE**
NORMAL BEAM SCAN COMPLETED:	**ACCEPTABLE**

TEST RESULTS

A UT PHASED ARRAY CREEP WAVE INSPECTION WAS COMPLETED ON SWEET LPG BULLET [redacted] V-4032 ON CIRC 5 (NUMBERED EAST TO WEST) FROM 3980mm TO 20mm AS REQUESTED BY CLIENT. (0mm AT TOP DEAD CENTER – SCANNED C.W. LOOKING EAST). 17mm WALL THICKNESS.

RESULTS:

- INDICATION ON CIRC5 SKEW 90 FROM 4696mm TO 6277mm DETERMINED TO BE CRACK LIKE LINEAR INDICATION. INDICATION SIZED WITH TRADITIONAL UT SHEARWAVE FROM 4672mm TO 6358mm (1686mm LONG), 11mm TO ID SURFACE (6mm THROUGHWALL HEIGHT, ROUGHLY 33%). INDICATION STOPS AT 6358mm WHICH IS WHERE IT RUNS UP INTO A LONGSEAM. AS THE INDICATION MOVES UP THE WELD, IT APPEARS TO MOVE FROM THE WELD INTO THE SHELL MATERIAL.

Misc. Charges	STAMP		
	EDWIN TYMENSEN CGSB / SNT-TC-1A UT II / MT II / PT II #11424 API 510, CERT # 27479	Regular Hours	**See**
		Overtime Hours	**Time**
		Shift Premium	**Summary**
		Kilometers	**CET-0015, CET-0016,**
		Sub / Man Day	**CET-0017**
Client Representative:	Technician:	Assistant: DARRYL COLEMAN / TYLER BAITON	

347A – 28042 Hwy 11 / Red Deer County, AB / T4S 2L4
Ph: (403) 347-7042 Fax: (403) 347-7052 www.echonde.com

ECHO NDE

M- 4752

MAGNETIC PARTICLE/LIQUID PENETRANT INSPECTION REPORT

Date MAR. 19/09		Page 1	of 1
CLIENT: [redacted]		ECHO JOB #	
CONTRACTOR: AS ABOVE		WO# 9-1265	
LOCATION: [redacted] GAS PLANT		PROJECT CC# 81012	
ITEMS EXAMINED: V4032 NGL STORAGE BULLET		MAINTENANCE	
PROCEDURE: MT-2A		CLIENT P.O.#/JOB#:	
ACCEPTANCE CRITERIA: ASME SEC. VIII, DIV. 1, APPENDIX 6		SPECIFICATION:	

SURFACE CONDITION: ☑ Clean Base Metal ☐ As Ground ☐ Machined ☐ Shot Blast ☐ Painted ☐ Other

MINIMUM LIGHT INTENSITY ≥100 fc VISIBLE ≥1000μW/cm² FLUORESCENT

MPI METHOD

☑ AC ☐ DC ☑ Continuous ☐ Residual ☐ 12V ☑ 120V ☐ Other:

EQUIPMENT TYPE PROD SPACING 3"-8"

☑ Yoke ☐ Coil Serial #: ECHO-001 Last Calibration Date: APR. 1/08

☑ Blacklight Serial #: Last Calibration Date: MAR. 18/09

MPI MEDIUM ☐ Dry Colour: ☑ Wet ☑ Fluorescent ☐ Black on White Particle Size:

EXCESS PARTICLE REMOVAL ☑ Flow ☑ Exhalation ☐ Other

LPI METHOD

Penetrant: S/N ☐ Vis ☐ Fluorescent ☐ Water Wash ☐ Solvent Removable

Developer: S/N ☐ Wet ☐ Dry ☐ Nonaqueous

Dwell Times Penetrant Min: Max: Developer Min: Max:

TEST RESULTS

A WFMPI WAS PERFORMED INTERNALLY ON V4032 NGL STORAGE BULLET AS REQUESTED BY CLIENT. MPI WAS DONE ON ALL CIRCS, LONGSEAMS, AND CAT-D NOZZLE WELDS AND A 1"-2" AREA TO EACH SIDE.

RESULTS: THERE WAS A 62" LONG CRACKLIKE INDICATION FOUND RUNNING PARALLEL WITH THE 3RD CIRC FROM THE WEST.

THERE WERE NO OTHER INDICATIONS FOUND AT THE TIME OF INSPECTION. ALL OTHER AREAS WERE ACCEPTABLE TO CODE.

1 GALLON MPI CONSUMABLES	STAMP	Regular Hours	8
GENERATOR	GORD BOS #11303	Overtime Hours	2
2 GAS MONITORS	CGSB RTII / MTII	Sub/Man Day	—
BUFFING CONSUMABLES.	SNT-TC-1A RTII / MTII	Kilometers	315
Client Representative:	Technician:	[redacted]	

Echo NDE Inc. Red Deer, Alberta

Phone: (403) 347-7042 * Fax: (403) 347-7052 *

White - *Client* Canary - *Office* Pink - *Invoice*

ECHO NDE

MAGNETIC PARTICLE INSPECTION REPORT

DATE:	MARCH 25/26-2	REPORT # MDC-0001		PAGE 1 of 1
CLIENT:	[redacted]		ECHO JOB #:	
CONTRACTOR:	SAME			
LOCATION:	LSD [redacted]	PROJECT:	LPG BULLET [redacted]	
ITEMS EXAMINED:	SEE BELOW			
PROCEDURE:	MT-3A	CLIENT PO # / JOB #:		
ACCEPTANCE CRITERIA:	ASME VIII DIV 1	SPECIFICATION:	ASME V, ART 7	

METHOD

SURFACE CONDITION

☑ Clean Base Metal ☐ As Ground ☐ Machined ☐ Sand Blast ☐ Painted ☑ Other: BUFFED

☐ AC ☐ DC ☑ Continuous ☐ Residual ☐ 12V ☑ 120V ☐ Other:

EQUIPMENT TYPE

☑ Yoke	Serial #: 9563	Last Calibration: JUNE 01 - 2008	Prod Spacing: 3 - 8 inches
☐ Blacklight	Serial #: 26591	Last Calibration:	
☐ Light Meter	Serial #: 1225653	Last Calibration:	

MINIMUM LIGHT INTENSITY: ≥ 100 fc Visible, ≥ 1000 μW/cm2

MEDIUM TYPE (LIST CONTRAST AS WELL): MAGNAFLUX **MEDIUM MANUFACTURER:** MAGNAFLUX

☐ Dry Colour: ☐ Fluorescent ☑ Black on White

MPI Particle Application Method: ☐ Powder Puffer ☑ Spray Excess Particle Removed via Exhalation & / or Flow

TEST RESULTS

BLACK ON WHITE MPI EXAMINATION WAS CONDUCTED ON THE LPG BULLET V-4032 AT THE FOLLOWING AREAS AS REQUESTED BY CLIENT.

CIRC SEAM #5 FROM 7 O'CLOCK TO 9 O'CLOCK (62" LONG)

_ REPAIR AREA AT DIFFERENT INTERVALS THROUGH-OUT THE INDICATION REMOVAL PROCESS TO A MAXIMUM DEPTH OF 5MM.

_ REPAIR AREA AFTER FIRST & FINAL PASS

RESULTS: THERE WERE NO INDICATIONS LOCATED AT TIME OF EXAMINATION REPAIRED WELD AREA WAS ACCEPTABLE TO CODE.

Misc Charges	STAMP	Regular Hours	SUMMARY
1 GALLON	DARRYL COLEMAN	Overtime Hours	
	CGSB / SNT-TC-1A RTII / UTII / MTII/PTII # 4369	Shift Premium	
		Kilometers	
Client Representative:	Technician Signature:	Assistant: DAN WONG	

347A - 28042 Hwy 11 / Red Deer County, AB / T4S 2L4
Ph: (403) 347-7042 Fax: (403) 347-7052 www.echonde.com

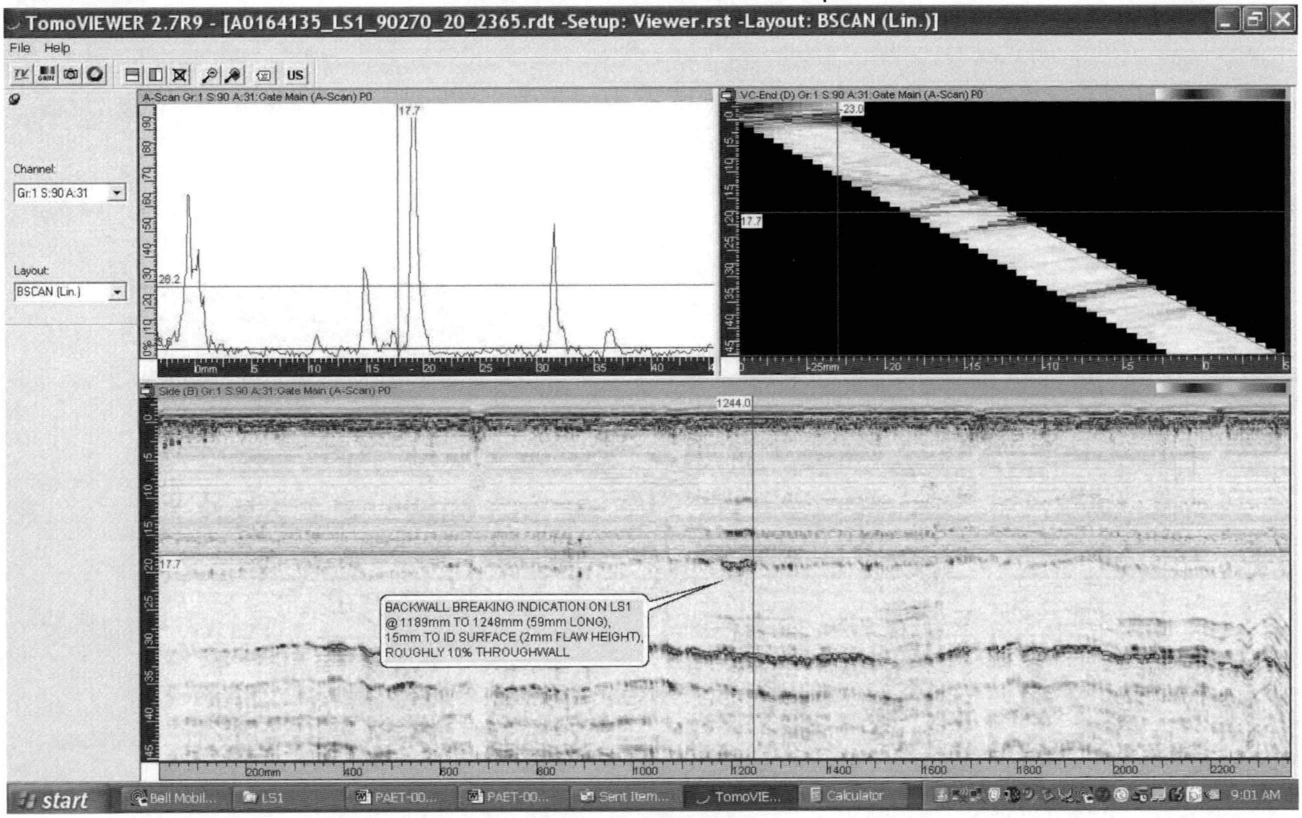

Scan result for V4031

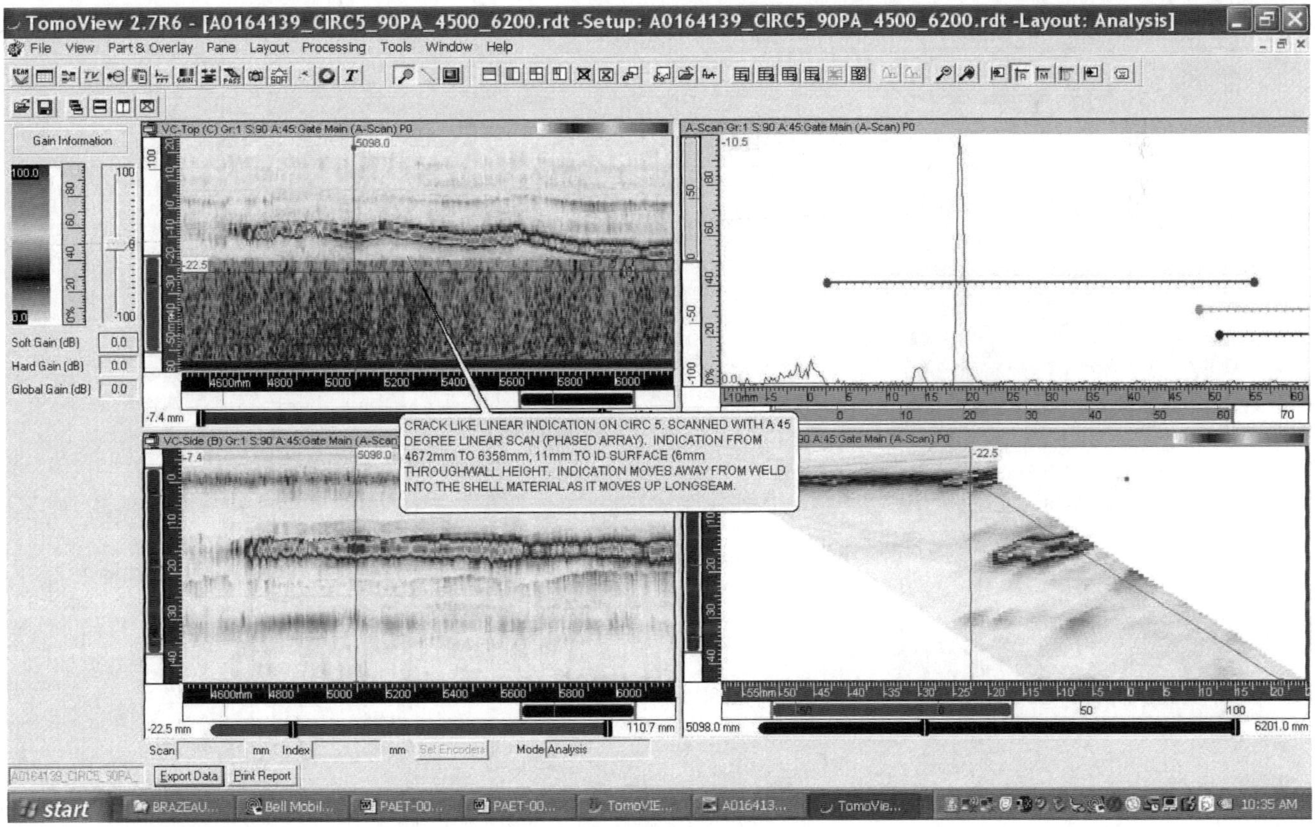

Scan result for V4032

MORE RECENT HISTORY

A project was awarded to complete the same creep wave technique on a number (16) of propane bullets for a large propane supplier. As this type of inspection would be a real departure from the norm (and might be questioned by ABSA), the Inspector determined that a more demonstrable method be developed to prove the validity of the technique.

To this end a number of welded plates were built with EDM notches of varying depth.

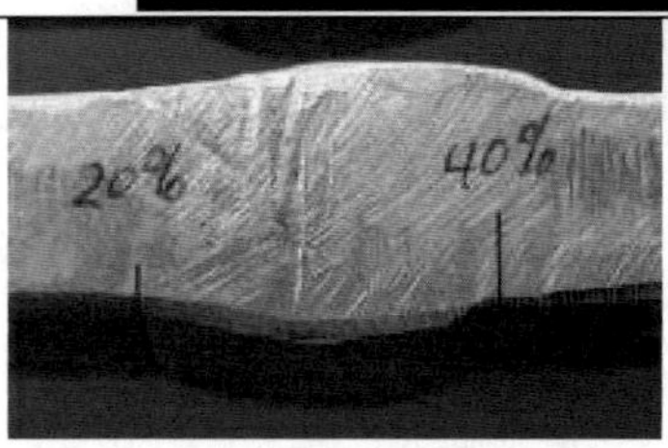

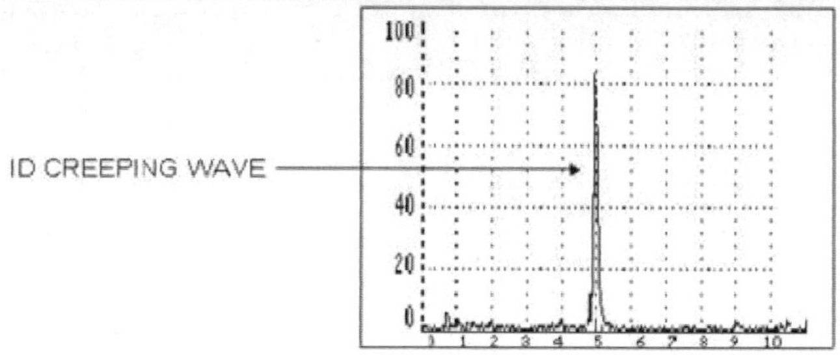

20% test piece & accompanying 'A' Scan of same

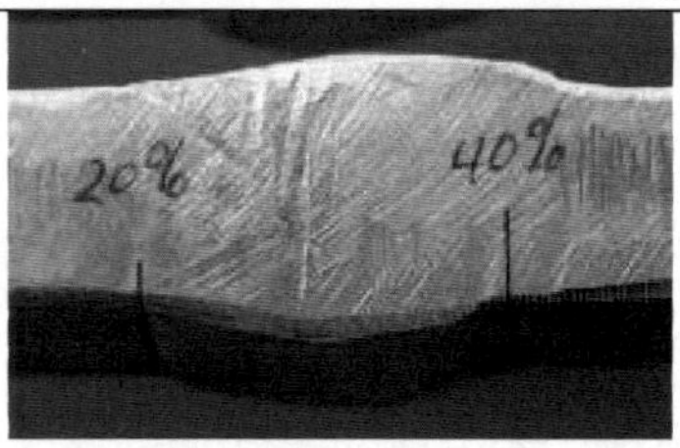

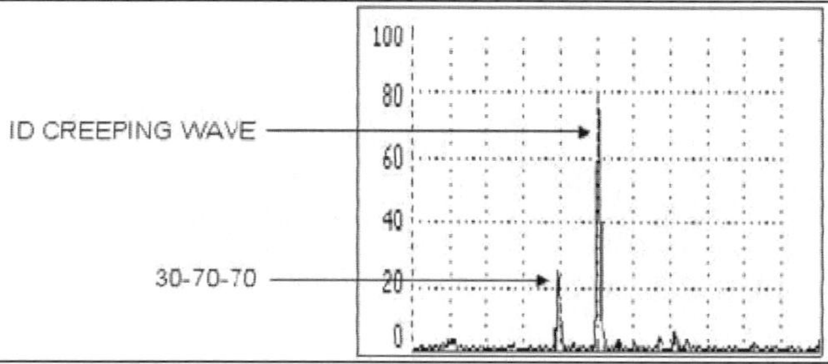

40% test piece & accompanying 'A' Scan of same

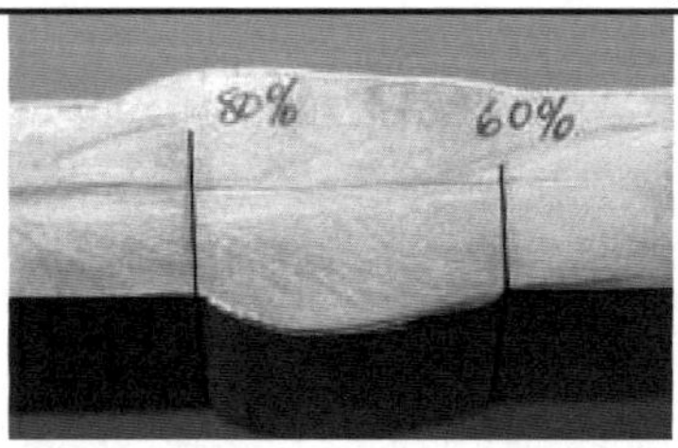

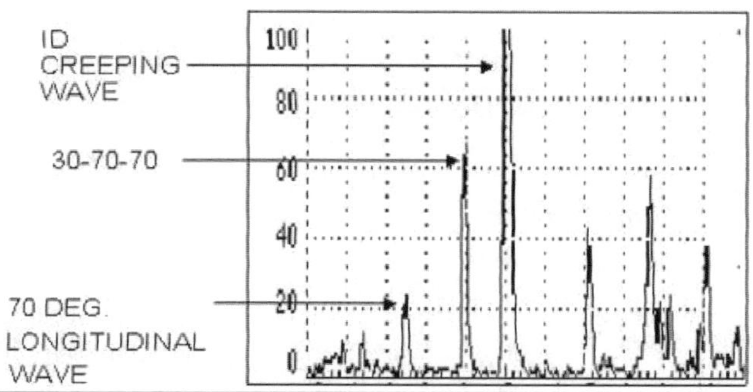

60% test piece & accompanying 'A' Scan of same

PRESENT DAY - APPLICATION

Once the propane bullet inspections were reported to ABSA – without there being an internal, we had to work hard to prove that the technique we were using was of an equivalent level sensitivity as the WFMPI. The latest edition of the AB-506 which covers inspections has actually addressed propane (and other LPG storage units) as being acceptable to inspect with some form of ultrasonic testing.

The issue is that any other method that may achieve similar levels of identity and quantification will also find any inherent defects present at fabrication. As one NDE tech from another firm stated: *"We do one day of initial inspection and then three days of repairs"*. One major advantage for the service conditions for which we apply this technique is that defects internal to the part (slag, porosity…) are effectively non-detectable. What makes this such an advantage is that we are inspecting equipment that has already been accepted as meeting Code and has been in service. This method is used in lieu of internal WFMPI for searching for and finding cracks in the HAZ of welds while disregarding any other potential defects.

The other main advantage is the facility of the creep wave to react to potential defects. This allows for the scanning of large vessels in a relatively short period of time. On page 16 is a comparison of a conventional shear wave inspection using a 60° transducer and a creep wave 'snail' (CDS) wedge on a 2% thru-wall EDM notch. The calibration used for amplitude was the ASME Sec V set-up for shear wave.

TODAY'S EQUIPMENT & PROCESS

Gone are days of using ladders, scaffolds, platforms and other devices to gain access to all parts of the welds. So too is the need to have someone moving the probe along. The new system now requires only a team of 2 versus the older setups where 3 people were required.

The UT probes are contained in a motorized carriage with magnetic wheels; the carriage has a laser beam which is used to track the equipment as it crawls along. There is also a camera mounted to the front that sees both the weld and the laser beam; this information is relayed to a Tablet where the operator can control the unit with a joy stick.

For winter conditions, the couplant is a mixture of water and glycol. Depending on conditions, the vessel may need to be swept of snow and ice to effect a full inspection.

One limiting factor in an otherwise exemplary system is that the length of the creep wave is no more than ~45mm and usually closer to ~25mm. Given the standoff of the wedge nose to point of sound entry it is possible that proper inspection of thicker parts with single bevel welds may not be possible.

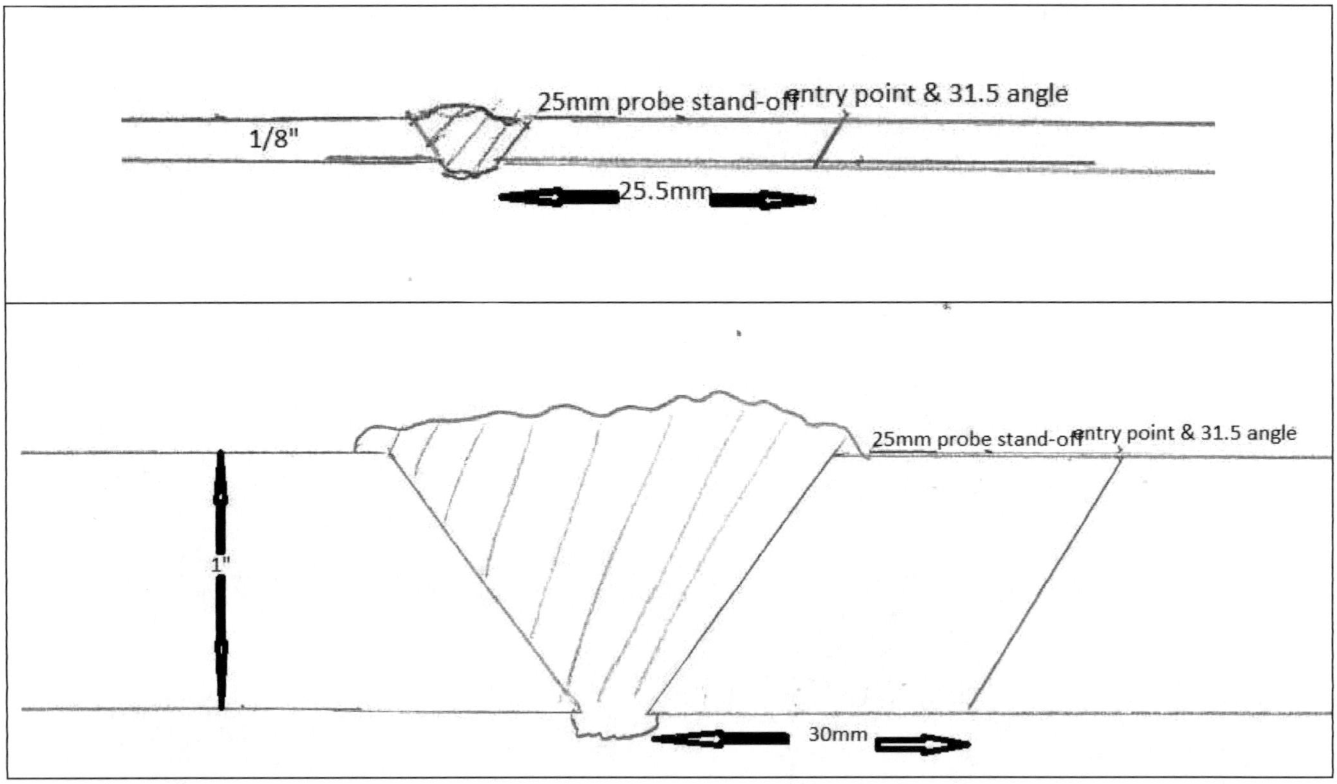

From the illustration above we can see that single bevel welds 1" thick are at the practical limit for the creep wave. While it is possible that under optimum circumstances that thickness could be exceeded the reliability of any test would only be valid if defects were found.

Double welds or critical welds (such as those EPRI deals with) have their cover passes ground flush to remove that as a hindrance to the inspection.

Below are the comparisons discussed on page 14. As well there are 2 scans shown side by side of a shear wave reflection from the 80% notch and a creep wave reflection from a 20% notch – both with the same amplitude setting:

1" thick carbon steel block with a 2% & a 5% thru-wall EDM notch	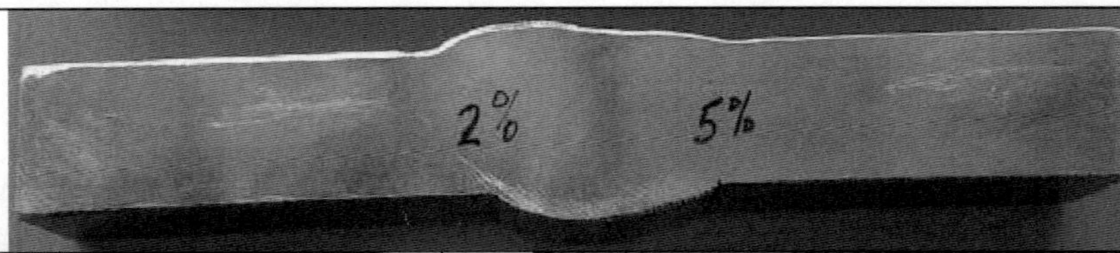
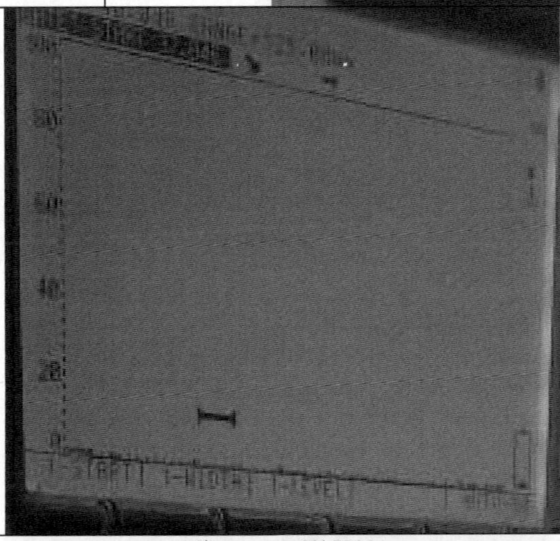	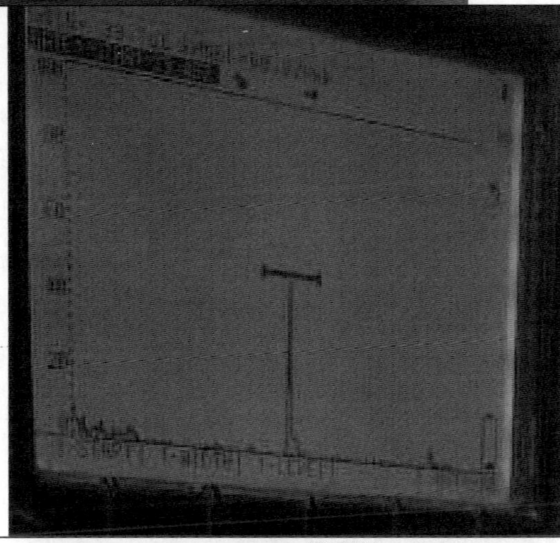
Shear wave 2% EDM	*creep wave 2% EDM*
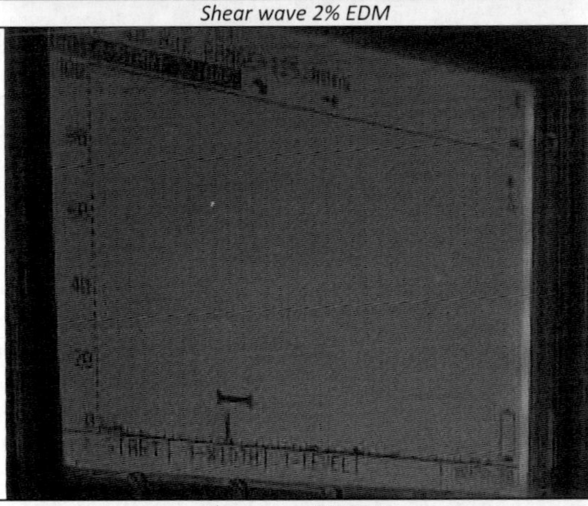	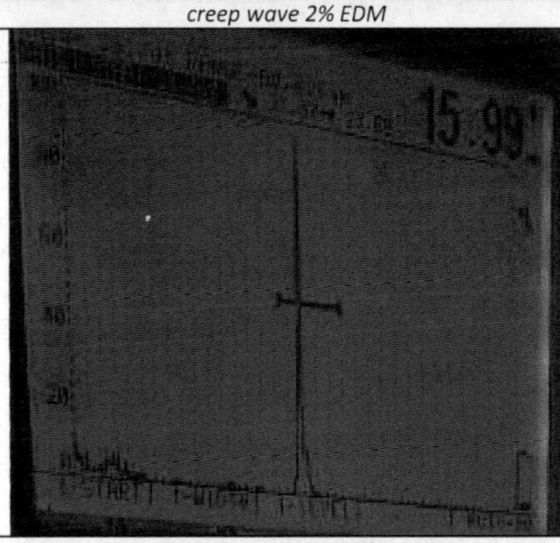
Shear wave 5% EDM	*creep wave 5% EDM*
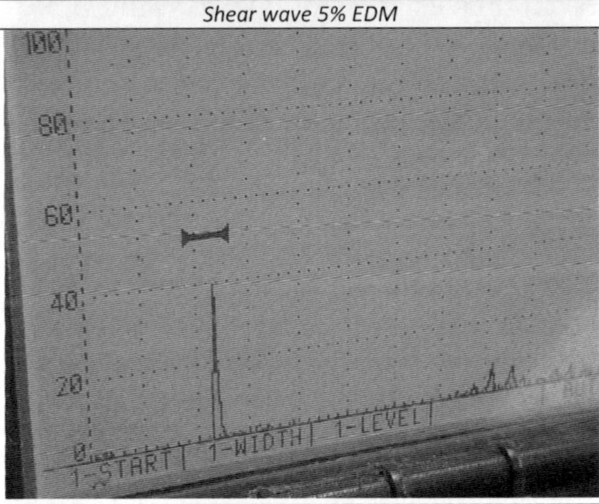	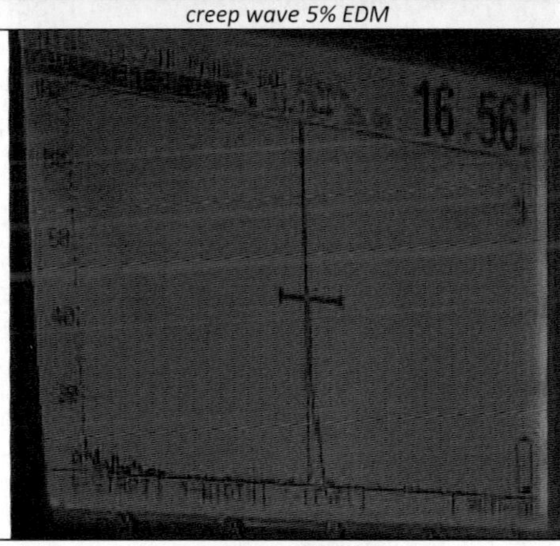
*Shear wave **80%** EDM*	*creep wave **20%** EDM*

A Method for Quantitative Analysis of Transient Eddy Current Testing

D. HORN, B.A. LEPINE, J. LEI

Canadian Nuclear Laboratories
Chalk River, ON, K0J 1J0, Canada

Phone: +1 613 584 3311; e-mail: dag.horn@cnl.ca, brian.lepine@cnl.ca, jia.lei@cnl.ca

Abstract
Transient eddy current (TEC) testing, driven by square waves, is also known as pulsed eddy current (PEC) testing and offers practical advantages over conventional, sinusoidally-driven eddy current testing in certain applications. By effectively sampling a continuous and very broad range of frequencies, the TEC technique generates data useful for materials of low electrical resistivity or high magnetic permeability, both of which may require testing at very low frequency. TEC also has an advantage for complicated configurations, which may have characteristic responses at a variety of frequencies and are therefore best analyzed by multi-frequency techniques. While quantitative analysis techniques have evolved over half a century for discrete-frequency eddy current testing, these techniques do not apply directly to TEC testing. A method for quantitative analysis of signals obtained in TEC testing is presented here. The method transforms time-domain signal response to the frequency domain; established response functions from traditional eddy current testing can then be used with a trial set of initial geometrical and material properties to calculate signals for comparison. A multi-frequency optimization of the geometrical and material properties is performed to minimize the difference between the calculated and the observed signal, and the optimized values are proposed as solutions. Applications to high-permeability materials like carbon steel, to low-resistivity materials like aluminum, and to complicated geometries, are discussed, as is integration of the technique with commercial instruments and software.

Keywords: transient eddy current (TEC), pulsed eddy current (PEC), multi-frequency analysis

1. Introduction

Traditional eddy current testing may be challenged by materials that are thick, highly conductive, ferromagnetic, or at variable distances from the test coil, which can limit electromagnetic skin depth or cause large signals from lift-off. In these cases, transient eddy current or pulsed eddy current (TEC or PEC) testing may confer practical advantages. Since the decay of a sharp pulse can be decomposed into contributions from a broad range of frequencies, the very low frequencies needed to increase skin depth are accessible. The relationship between time-dependent and time-harmonic responses is recognized in the scientific literature dealing with TEC or PEC testing; see, for example, Safizadeh et al. [1]. Comparisons have been made of modeled and observed signals transformed from the time domain to the frequency domain, and discrete-frequency analysis of the Fourier-transformed signals at specific frequencies, optimized for response to effects of interest, has been proposed (Pavo [2]). These methods have been used, for example, as a rationale for application of Principal Component Analysis (Pan [3]).

Such calibrated techniques for geometrical reconstruction do not perform a simultaneous calculation of signals from the full set of frequencies available from transformed TEC measurements at every inspection point. Within large-scale (e.g. finite element or analytical) models, this is computationally cumbersome and would not be practicable in common NDE applications. As a consequence, quantitative methods for analysis of TEC are of great interest. Several analysis methods are surveyed in a contribution to this conference from J.A. Buck et al. on carbon steel inspection [4].

Simple, but physically realistic, approximations of phase and amplitude response at discrete frequencies, such as those provided by skin depth considerations, have evolved over half a century of practice in conventional eddy current testing. See for example, Cecco et al. [5]. These approximations are computationally tractable when applied to many frequencies at once and have been incorporated into a suite of techniques referred to as Multi-Frequency Analysis, or MFA [6,7]. The sum, over a large number of frequencies, of the residuals of measured signals relative to calculated signals, can be obtained and minimized for varying parameters for which the frequency-dependent behaviour is understood. Parameters of interest include defect extent, remaining conductor thickness, and lift-off of the sensor from the surface. The combination of transformation to the frequency domain, analytical approximations to frequency dependence, and residual minimization to obtain material and geometrical properties is presented here as self-consistent means of TEC signal analysis.

2. TEC signals and their transformation

TEC pulses are typically driven as square-wave voltage inputs. The voltage is held at a constant value for only a few milliseconds and then cut to zero; to minimize coil heating, the duty cycle is often less than 50%. The decay of the field is measured at either the rise or the fall of the pulse, with the region of interest digitized and the rest ignored. Figure 1 shows schematically the time relationship between the driving pulse and the acquired data.

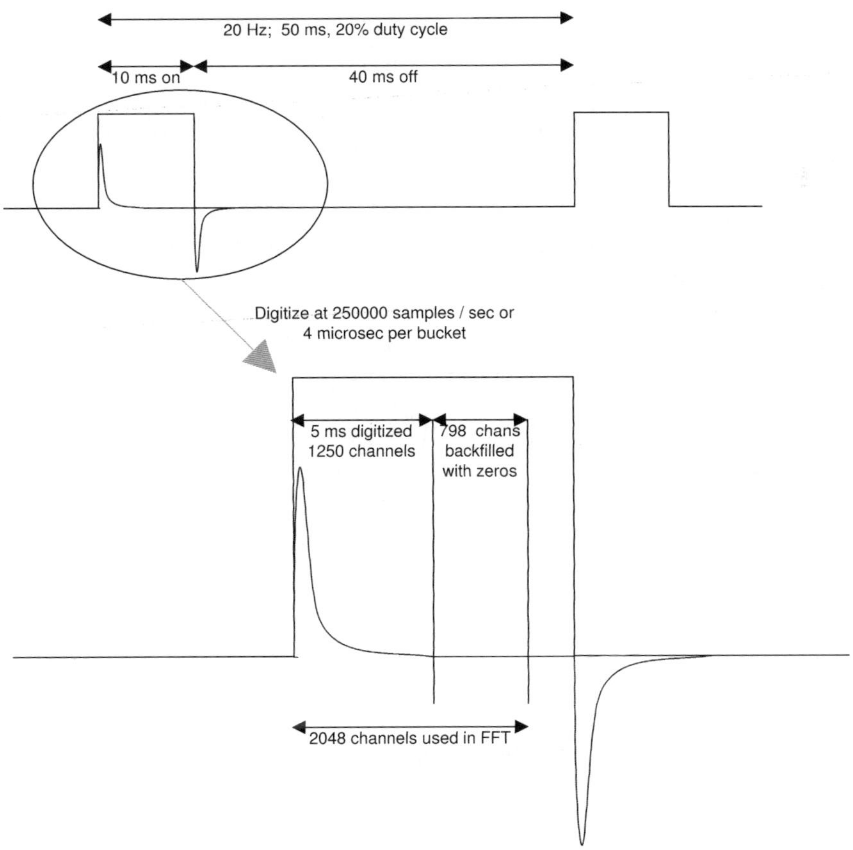

Figure 1. Time structure of TEC pulses and acquired data. The solid black line is the driving pulse, and the solid blue line is the digitized data from a sensing coil.

For high conductivity, high magnetic permeability, or thick samples, there is particular interest in the later times of the pulse decay, as these correspond to the lower frequencies and the larger skin depths; this can influence choice of cycle times. Since later times have lower amplitude, signal-to-noise ratio in this region is also important.

The transformation from time domain to frequency domain is accomplished by a fast Fourier transform (FFT). During early development, this was done in Microsoft EXCEL; subsequently, a feature of InspectionWare[1] was used for direct extraction of frequency-based data.

Analysis requires a minimum of two signals: a reference point of the defect-free full-thickness material and a test point where one or more unknown parameters are to be evaluated. In addition, use of phase angle conventions from classical eddy current testing require knowing the phase angle corresponding to lift-off at each frequency, obtained by lifting the probe off the sample at the reference point. The rotational offset in the complex plane at each frequency is deduced by performing trigonometry on the real and imaginary components of the complex division result. In Figure 2a, below, the phase angle offset increases (in the clockwise direction) with increasing frequency, ranging from approximately 90 to 180 degrees over many different frequencies. A quantitative lift-off distance calibration, from data taken at different values of lift-off, may also be performed if the probe-to-test-piece distance is to be extracted from the subsequent analysis. Lift-off phase angles can be obtained by the complex division of lift-off signal by reference-point signal. Test-point data are shown in Figure 2b, normalized by the complex division of the test point signal by the reference-point signal, and with phase angles referenced to the lift-off phase at each frequency. See the upper left-hand side of Figure 3 for a generalized view of the data acquisition and transformation process.

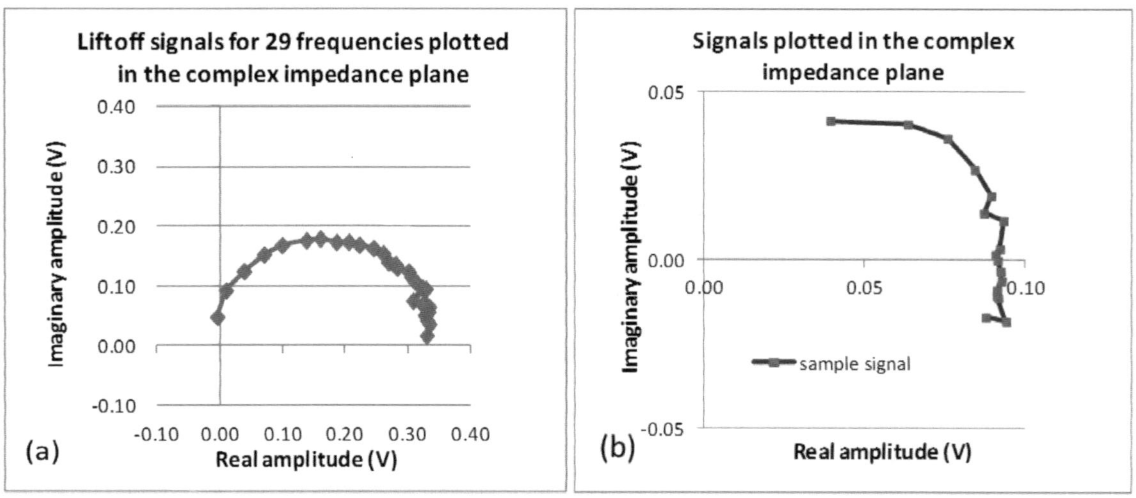

Figure 2. Transformed signals for discrete frequencies plotted in a complex impedance plane representation: (a) Signal for 1 mm lift-off, relative to reference, plotted in the complex plane. (b) Signal at test point, relative to lift-off, after rotation by lift-off phase.

[1] InspectionWare is a software platform from UTEX Scientific.

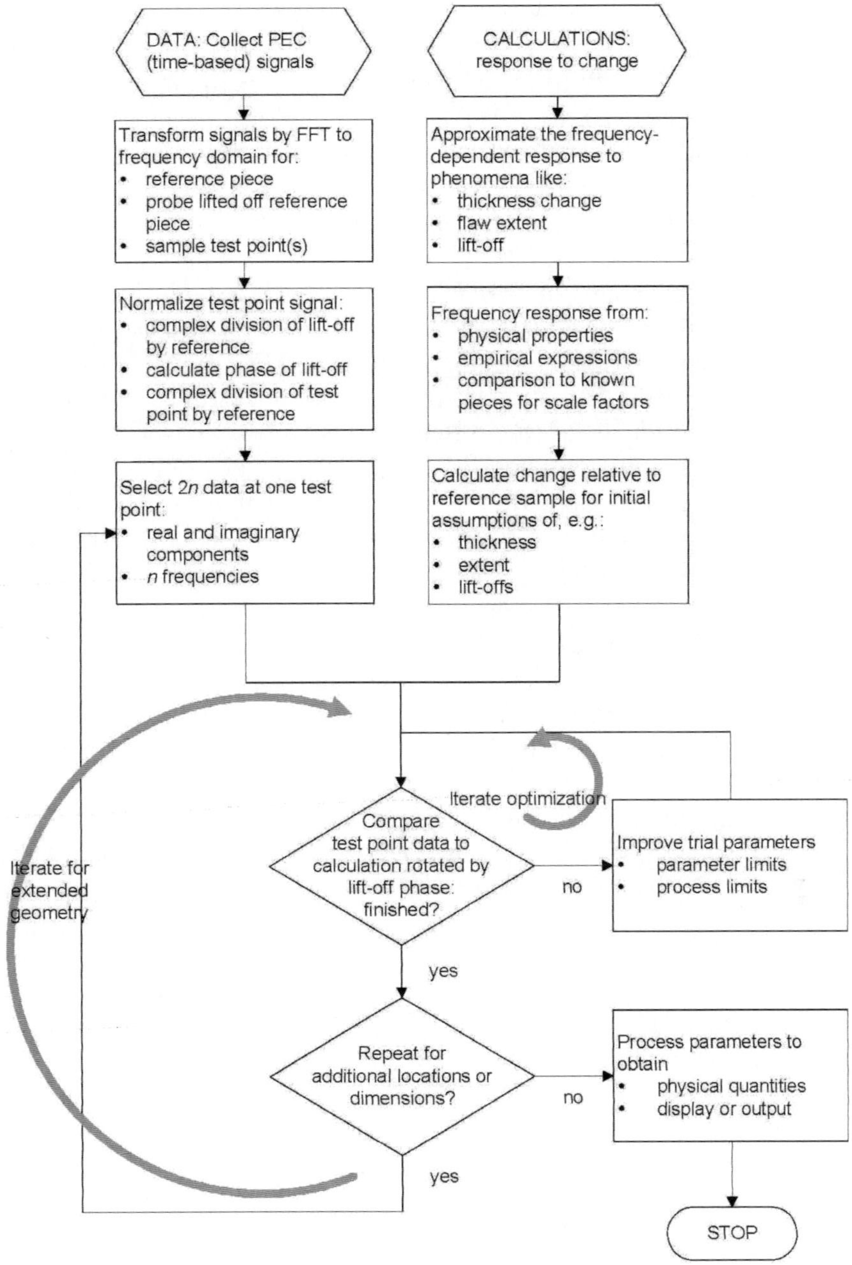

Figure 3. Process for transforming transient eddy current signal responses into geometrical measurement quantities.

3. Frequency domain analysis

Frequency-dependent estimates of signal response to wall loss, the areal extent of a flaw, or variations in material-to-sensor separation (lift-off) can often be approximated by expressions containing the skin depth, δ, which is defined as

$$\delta = 50\sqrt{resistivity/(frequency * relative\ permeability)},$$

where the skin depth in mm is obtained from resistivity in µΩ-cm, frequency in inverse seconds and relative permeability (dimensionless). The signal amplitude of an infinitesimal void at depth x is proportional to $e^{-2x/\delta}$, and the phase, φ, attributable to the signal is $\varphi = 2x/\delta$. Weighted integration of these quantities, over the range of the feature's depth, can provide a convenient analytical expression for frequency dependence [7]. Alternatively, an empirical expression or more elaborate calculations may be used.

The method uses initial estimates of the parameters of interest in analytical expressions to obtain the resulting signal phase and amplitude at multiple frequencies (upper right-hand side of Figure 3). These are compared to the measured TEC signals transformed into the frequency domain in Figure 4a. Iterative optimization of the estimated parameters minimizes the summed (squared) residuals of the differences between data and calculation until the required convergence is reached (Figure 4b). Wall loss, areal extent of the flaw, lift-off, or other parameters are then saved as output. If data from a sequence of positions have been collected, the process can continue to analyse a one-dimensional strip or a two-dimensional surface; see the lower part of Figure 3.

4. High-conductivity examples

The National Research Universal (NRU) Reactor at Canadian Nuclear Laboratories has a calandria vessel surrounded by an aluminium-walled light-water shield, with regions where wall thickness inspection is impeded by corrosion product on the near surface. This presents the problem of inspecting a high-conductivity material through an unknown lift-off. Samples relevant to this geometry, both of large-area wall loss (scalloped regions) and small-area features (flat-bottom holes) were inspected by TEC and the results analyzed.

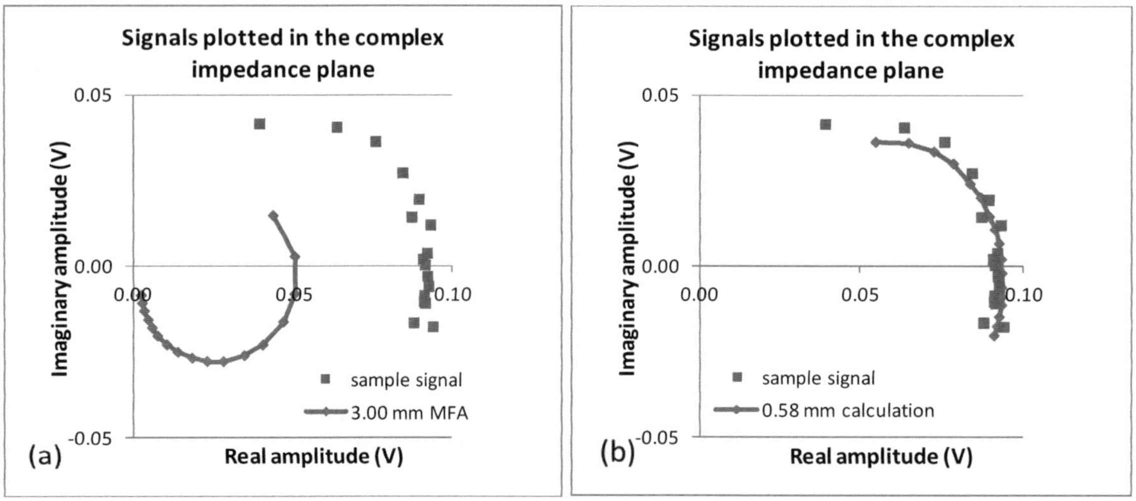

Figure 4. (a) Initial and (b) final calculated frequency trajectories relative to measured signal at a flat-bottom hole of 0.5 mm remaining wall (same data as Figure 2b). The initial arbitrary values of depth and diameter were adjusted to minimize the sum of the squared residuals between the calculation and data.

Localized defects: In this example, the surface of an aluminum sample, 8 mm thick (Figure 5a), containing three holes of 4 mm diameter (0.0, 0.5, and 1.0 mm remaining wall) and one of 1.7 mm diameter (0.0 mm remaining wall), was scanned. The depths obtained by the technique of Figure 4 were within 0.1 mm of the nominal sample values, but due to the spatial response of the probe (field spread, coil diameter, and coil separation) the scan image gave surface extents much larger than the hole diameters (see Figure 5b). In this technique the analysis also extracts an amplitude related to the areal extent of the flaw. A calibrated scaling factor was applied to obtain a reasonable diameter for a realistic 3D reconstruction of the sample, as in Figure 5c.

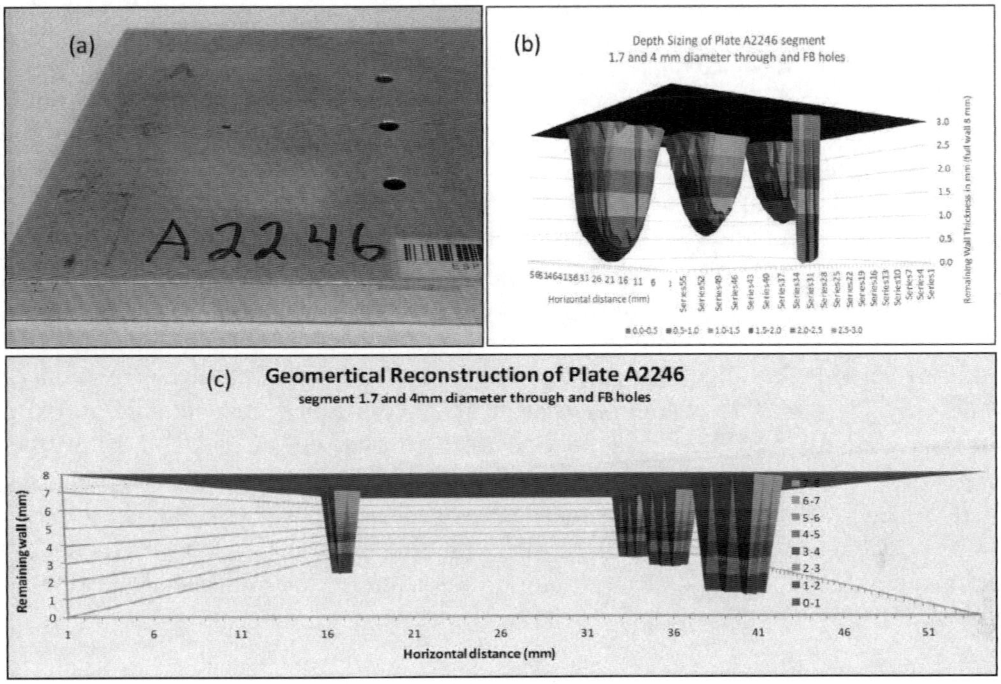

Figure 5. (a) 8-mm Al sample with three 4-mm flat bottom holes (0.0, 0.5, and 1.0 mm remaining wall) and one 1.7-mm diameter hole (0.0 mm remaining wall). (b) correct depth sizing but misleading surface extent due to field spread. (c) 3D reconstruction of sample by TEC/MFA including calibrated extent.

Generalized wall loss: In contrast to the test piece above, the scalloped sample contains gradually changing wall thickness. Since the sensing area is then much smaller than the feature, localization is less important, and the amplitude parameter related to areal extent, used in the previous example, need not be extracted; see Figure 6. The ability to ignore a parameter can be useful, in that it frees the analyst to make different use of the information collected at different frequencies. For example, in the NRU reactor application mentioned above, an additional parameter representing the thickness of the corrosion layer could be extracted from the measurements.

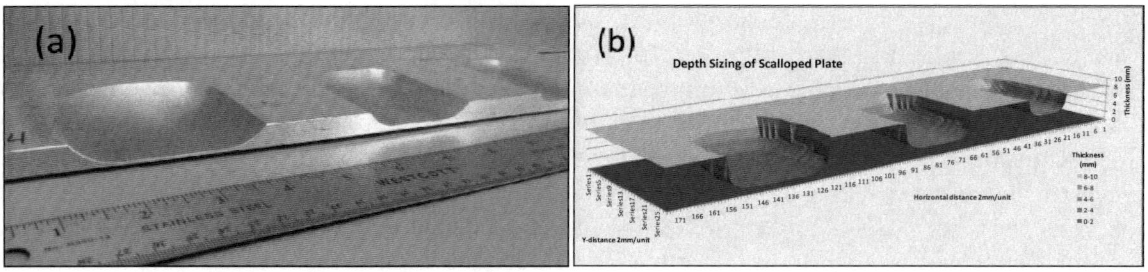

Figure 6. (a) Photo and (b) reconstructed image of scalloped test piece.

5. Ferromagnetic example

Carbon steel is a common material in waste management, pipeline, and other large-scale applications. Unlike aluminium, zirconium, stainless steel, Inconel and other materials commonly encountered in the nuclear and aerospace industries, carbon steel has relative permeability greater than one. This reduces skin depth, making electromagnetic inspection of unsaturated material difficult. As with highly conductive samples, TEC testing with sensitivity to late times in the signal decay tail can overcome this difficulty.

Inspection of carbon steel pipes in waste management "tile hole" applications is reported in this conference [4]. Wall thickness scans from that work along eight parallel strips of pipe wall are analyzed by the TEC/MFA method with resolution sufficient to show the spiral structure of the piping (typically 0.5 mm oscillatory thickness variation). Figure 7 shows an "unrolled" image of the wall thickness map for one of the pipes inspected. In this particular case it was possible to confirm the TEC results with ultrasonic testing measurements. In other applications, e.g. for pipelines under insulation, confirmation with ultrasonic testing may not be practical.

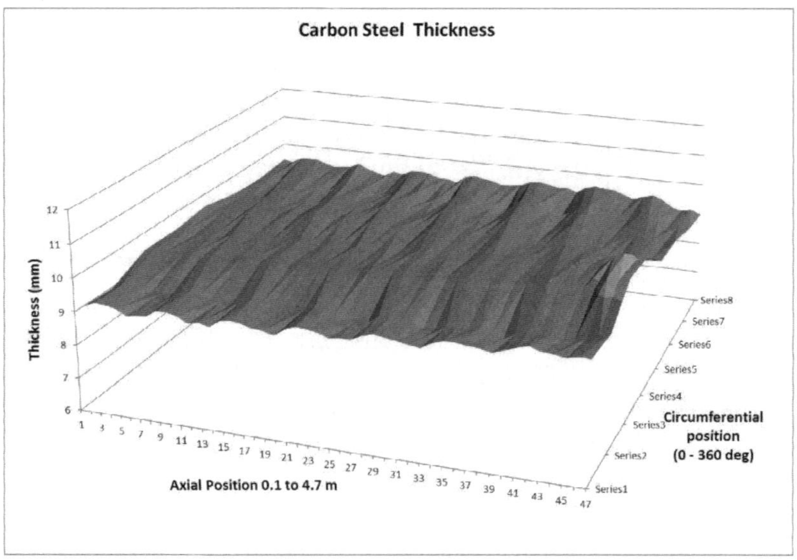

Figure 7. Thickness of carbon steel pipe from TEC analyzed by a multi-frequency technique after transformation into the frequency domain.

6. Future exploitation

Confounding effects from lift-off, feature geometry, and thickness of remaining material can be disentangled and quantified when TEC measurements are transformed to be processed by multi-frequency techniques. This extends electromagnetic inspection capability to otherwise challenging materials, such as those with high magnetic permeability or high conductivity. In addition to nuclear power and waste management applications, both carbon steel inspection in pipelines and infrastructure, and aluminium alloy inspection in the aerospace industry are possible uses of the TEC/MFA technique.

The method is the subject of a recent patent disclosure [8]. Potential future exploitation ranges from in-house use developing dedicated inspection techniques to the licensing of technology for commercial systems. Intermediate between these extremes is collaboration with developers of inspection instruments and software in joint ventures.

References

[1] M.S. Safizadeh, B.A. Lepine, D.S. Forsyth, and A. Fahr, Time-Frequency Analysis of Pulsed Eddy Current Signals, Journal of Nondestructive Evaluation 20 (2001) 73.

[2] József Pávó, Numerical Calculation Method for Pulsed Eddy-Current Testing, IEEE Transactions on Magnetics, 38 (2002) 1169.

[3] Mengchun Pan, Yunze He, Guiyun Tian, Dixiang Chen, and Feilu Luo, PEC Frequency Band Selection for Locating Defects in Two-Layer Aircraft Structures With Air Gap Variations, IEEE Transactions on Instrumentation and Measurement, 62 (2013) 2849.

[4] J.A. Buck, C. Kramer, J. Lei, and B.A. Lepine, Field Implementation of a Transient Eddy Current System for Carbon Steel Pipe Thickness Measurements, contribution to this conference.

[5] V.S. Cecco, G. Van Drunen, and F.L. Sharp, "Eddy Current Manual", Volume 1, AECL Report, AECL-7523 Rev 1, (1983).

[6] D. Horn and R. Lakhan, Multi-Frequency Analysis of Bobbin Probe Data, AECL Report, CS-ET-155 (1998 November).

[7] D. Horn and R. Roiha, Multifrequency Analysis of Eddy Current Data, Proceedings 16th World Conference on Nondestructive Testing, Montreal, August 30-September 3, 2004, NDT.net, Vol. 9 No.11, Nov 2004, ISBN 0-9736577-2-3.

[8] D. Horn, B.A. Lepine, and J. Lei, United States Provisional Patent Application No. 62/289,515, Method for Non-Destructive Analysis of Multiple Structural Parameters, filed 2016 February 1.

Low Frequency Eddy Current for Sub-Surface Crack Sizing

P.R. Underhill
T.W. Krause

1

Outline

- Motivation
- Probe Design
- Results
- Next Steps

2

Why Low Frequency EC?

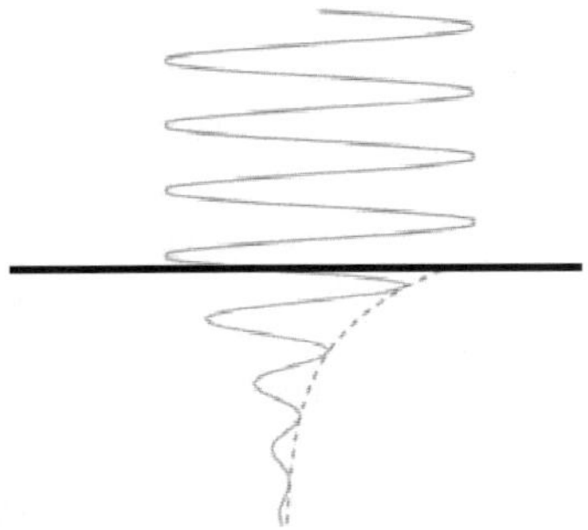

Aluminum	Resistivity (μΩ•cm)
7050-T76	4.36
6061-T4	4.30
7075-T6	5.39
2024-T3	5.33

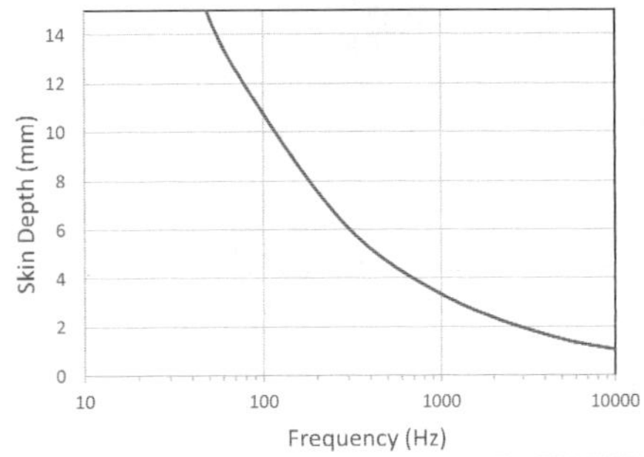

$$\delta(\text{mm}) = 50.3\sqrt{\frac{\rho}{\mu_r f}}$$

3

Problem

4

Problem

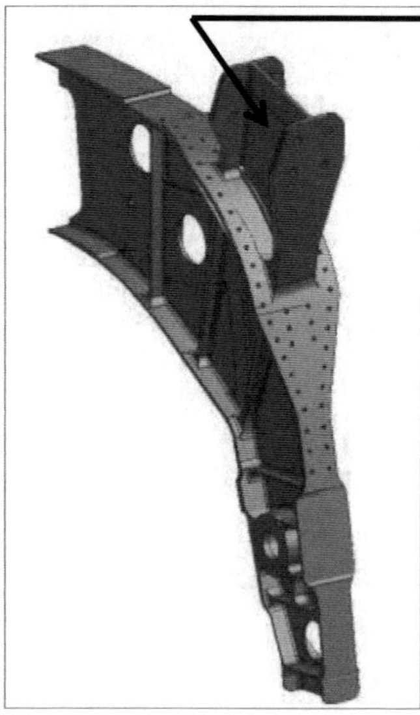

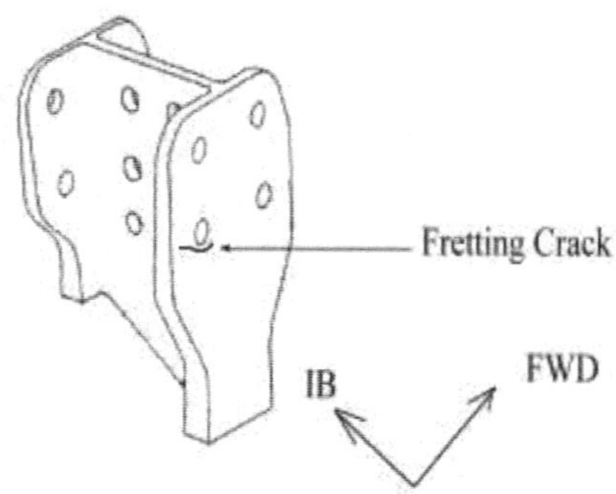

[5]

QETE Standard

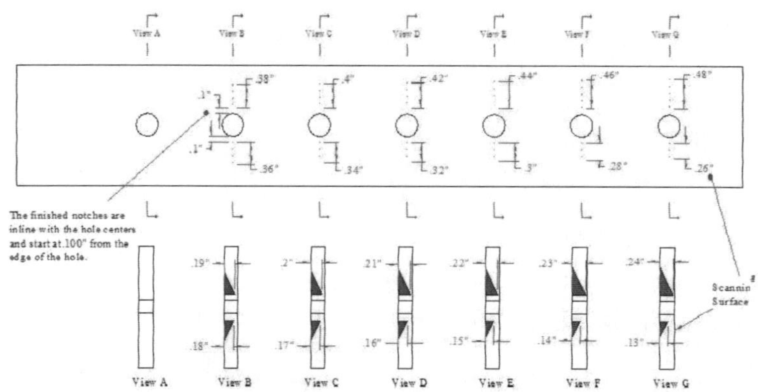

QETE Project A014812, F188 Stub Flange LFEC Calibration Block
Material = Aluminim 40% IACS, +/- 2%
Surface Finish =
Block Tolerance = +/- .005"
Hole Tolerance = +.000 / -.002
Notch Width = .005" +/- .002", Length to Depth Ratio =- 2/1
Die-Sink Electrode Angle = 26.6 deg
31 July 2014

EDM Notches

[6]

Probe Configurations

Concentric

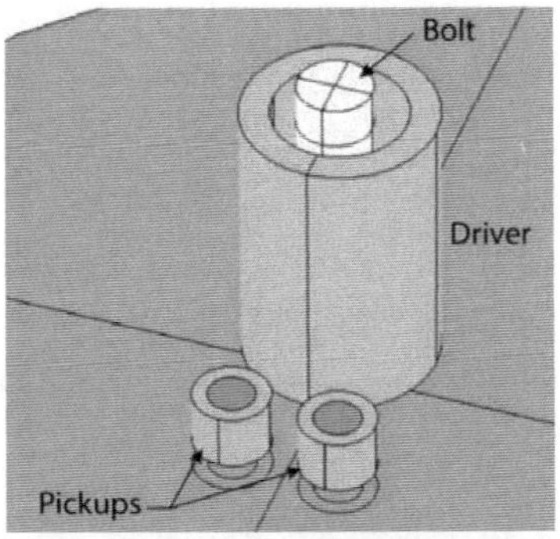

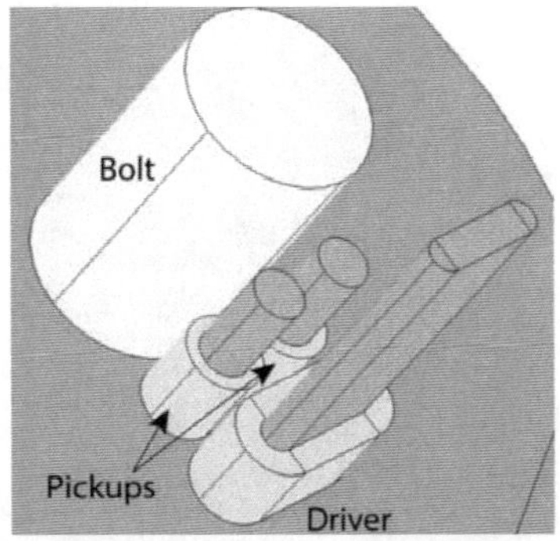

Beside

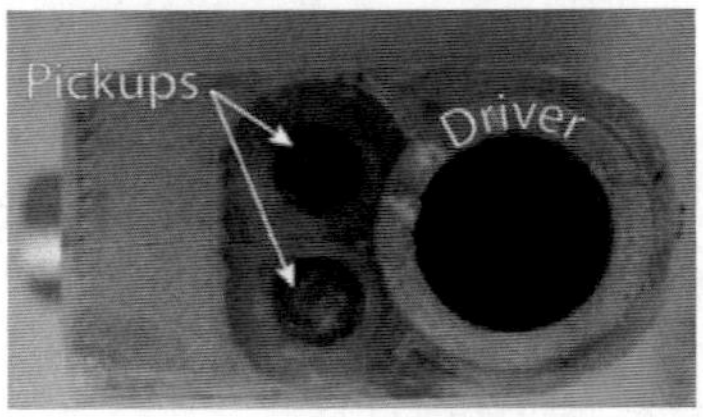

7

Modelling (Comsol)

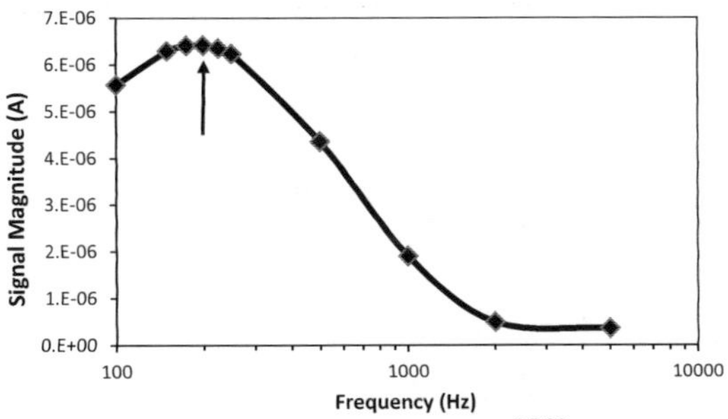

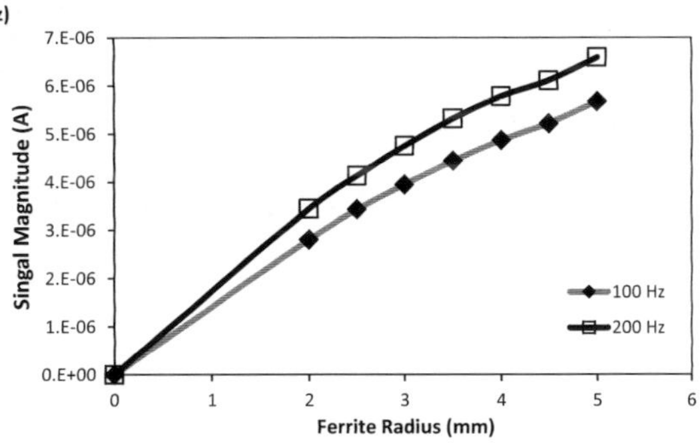

8

Concentric Probe

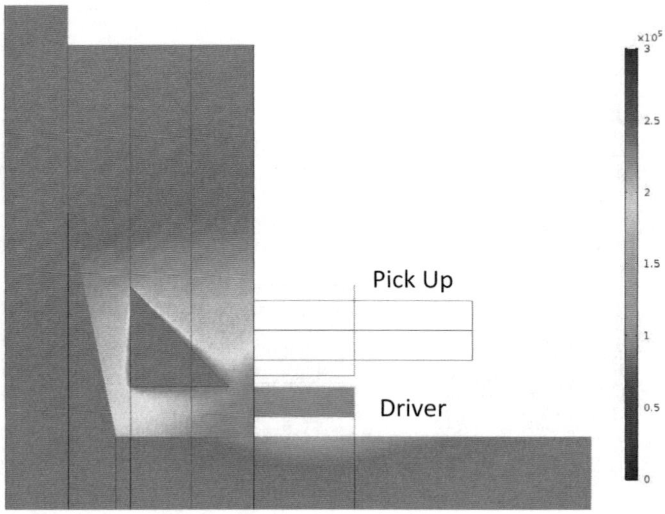

9

Beside Configuration

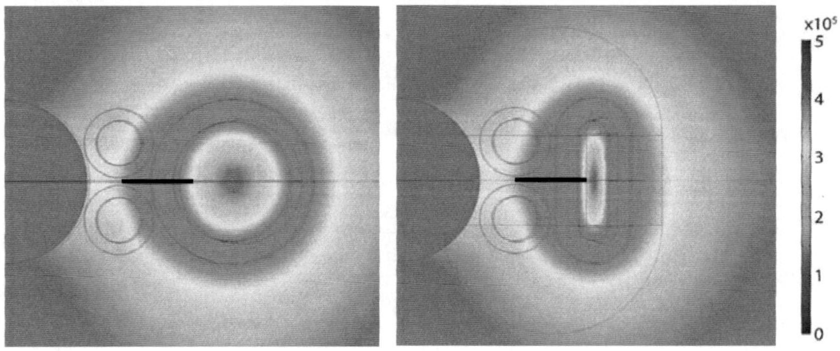

10

Concentric Configuration

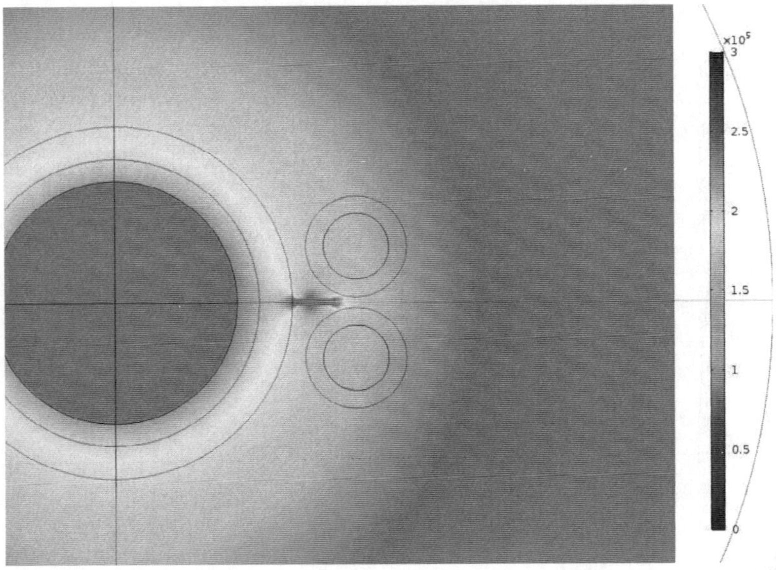

11

Testing

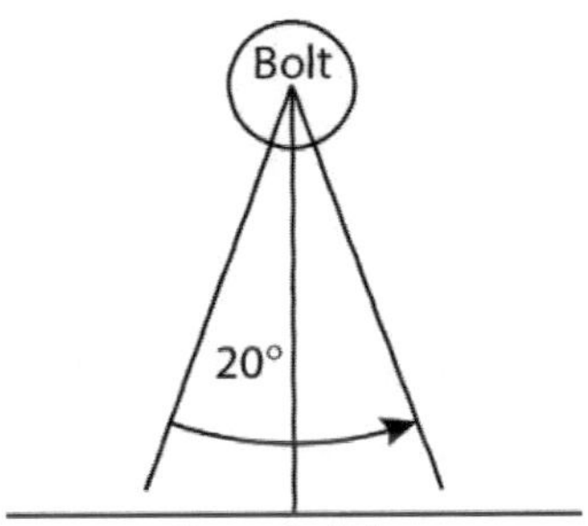

12

Results (Concentric)

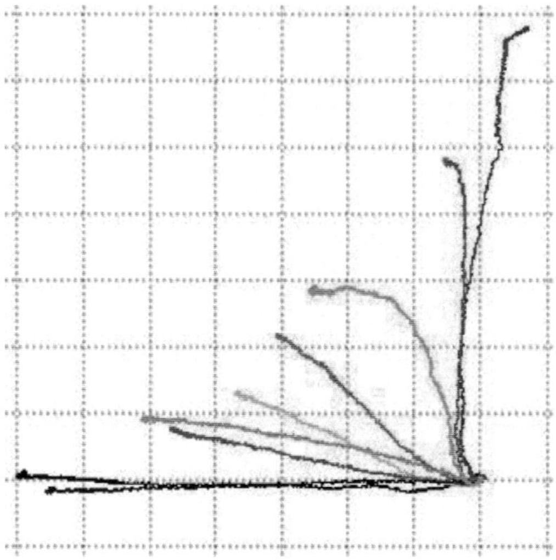

200 Hz

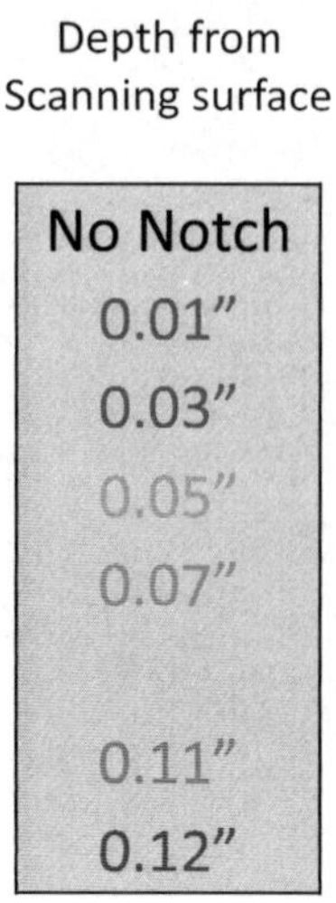

13

Results (Concentric)

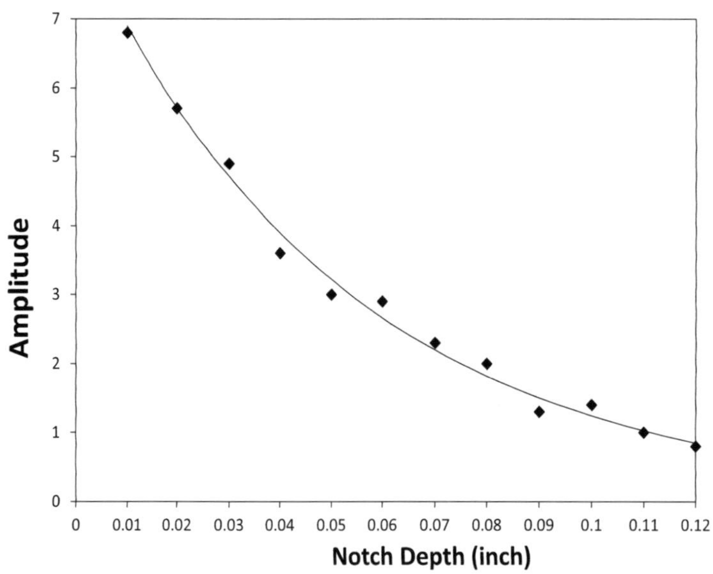

14

Results (Beside)

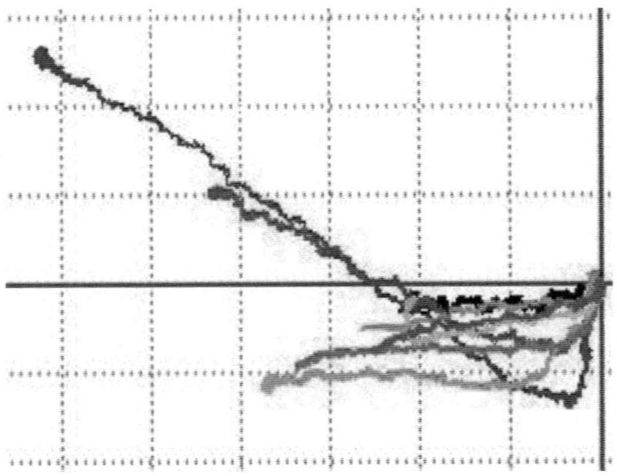

Depth from
Scanning surface

No Notch
0.01"
0.03"
0.05"
0.07"

0.11"
0.12"

500 Hz

15

Results (Beside)

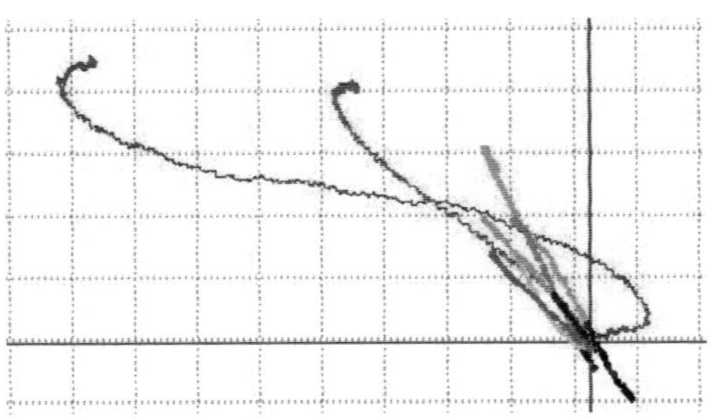

Depth from
Scanning surface

No Notch
0.01"
0.03"
0.05"
0.07"

0.11"
0.12"

560 Hz and 5 kHZ

16

Next Steps

- "Real" piece
    - Changing edge distance
    - Bolt not perpendicular to scanning surface
    - Inspection surface not parallel to scanning surface

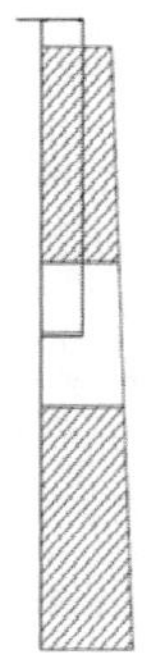

Section B-B

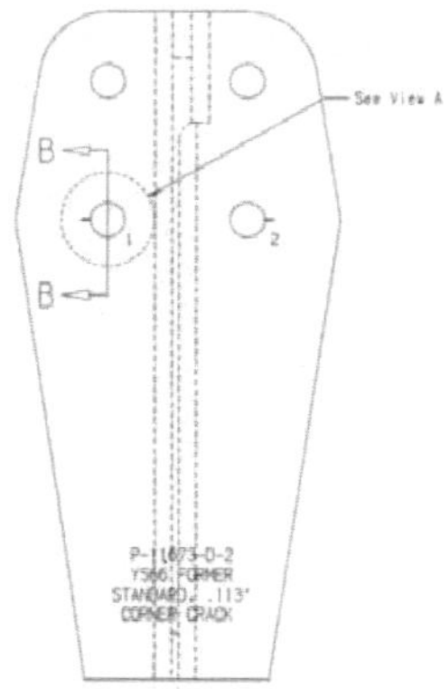

17

Questions

18

Next Steps

- Issues with uneven scanning surface
- Do we really need a different probe for each bolt size?

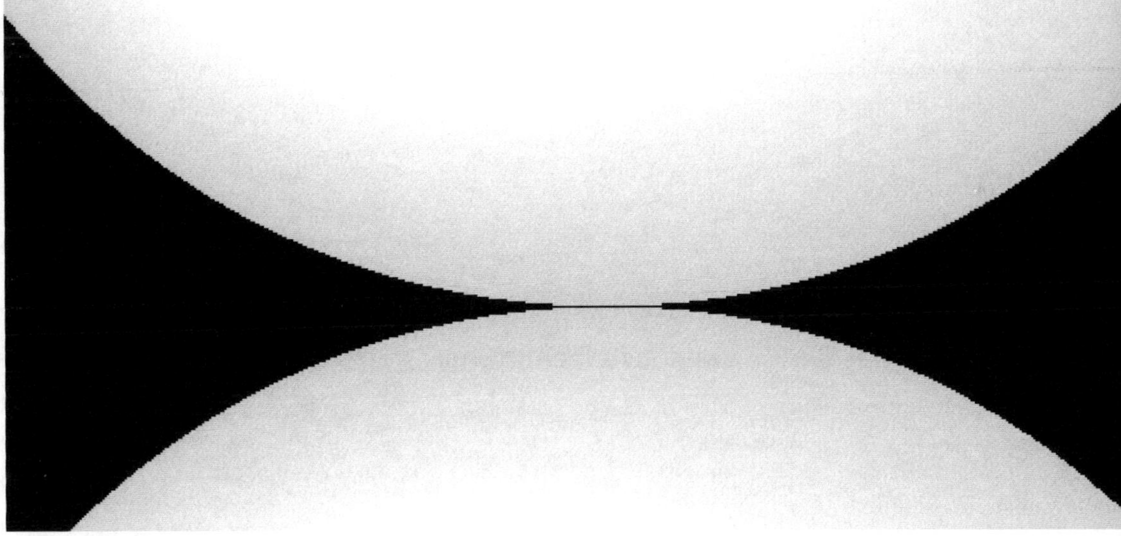

19

Low temperature compliance testing of wind turbine applications for the 'cold climate' market

pieterjan.jordaens@owi-lab.be

sirris
driving industry by technology

Vrije Universiteit Brussel

www.owi-lab.be

Introduction

- **Sirris - collective technology centre in Belgium**
  - Supporting companies with implementing technology innovations
  - Multidisciplinary R&D and innovation projects in technology industry
  - Different technology sectors: Automotive, Energy, Aerospace, ICT, ...
  - Different key expertise: ICT, Manufacturing, Mechatronics, Materials
  - High-tech test and R&D infrastructure

sirris
driving industry by technology

- **OWI-Lab - RD&I center for wind energy in Belgium**
  - Set-up in 2010 as a new application lab at Sirris to support wind energy R&D
  - Scope: wind energy in general - focus on '**offshore wind**' and '**cold climate**'
  - Range of new and unique test & monitoring infrastructures
  - Partnership with 3 Belgian universities for wind energy research (VUB, KU Leuven, UGent)
  - Member of IEA Wind Task19 Wind Energy in Cold Climates

Task 19

Unique test infrastructure – large climate chamber

- **Public large climatic test chamber for wind turbine applications** (+60°C to -60°C / Humidity / IR-Solar load)
- **Focus: climatic validation tests of wind turbine equipment (cold, hot-tropical and offshore climates)**

XANT

Full size small & mid-range wind turbine nacelle (or assembly) tests

Power electronics tests
Pitch & Yaw cold starts
Hydraulic brake tests
Generator tests

SIEMENS

Functional component testing with or without wind turbine auxiliaries (forced cooling, pumps, heating, expansion tank, lubrication unit,…)

Electrical, mechanical and hydraulic components as gearboxes, transformers, switch gears, power electronics, anti & de-icing systems,…

R&D tests on the behavior of fluids, oils and hydraulics in a full functional set-up

Driver for the 'testing topic' within the wind power business

$$COE = \frac{CAPEX + OPEX}{AEP}$$

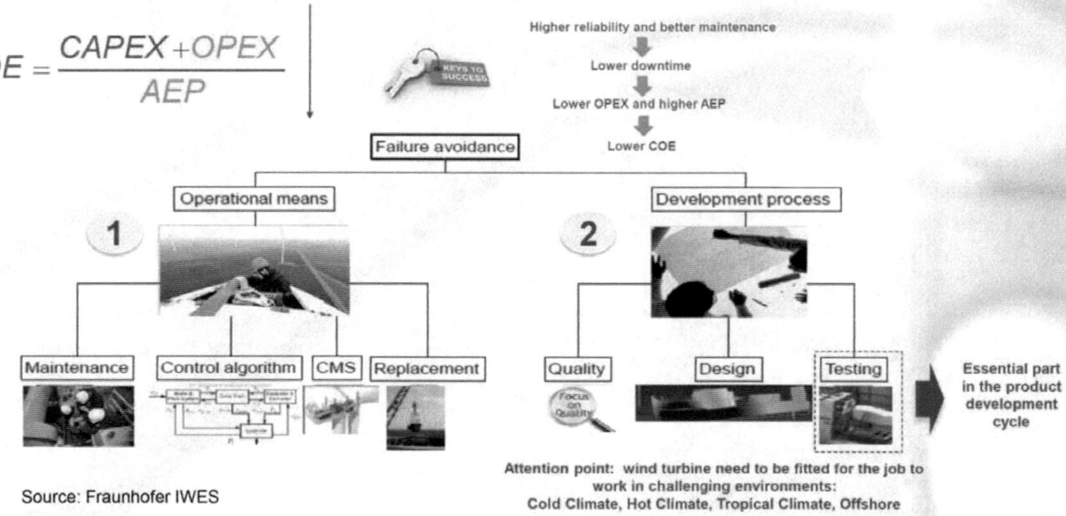

Source: Fraunhofer IWES

Attention point: wind turbine need to be fitted for the job to work in challenging environments: Cold Climate, Hot Climate, Tropical Climate, Offshore

Different public test centers have been set-up to support R&D

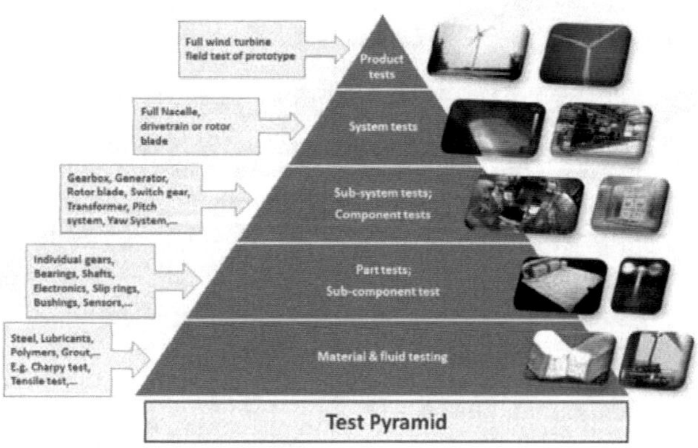

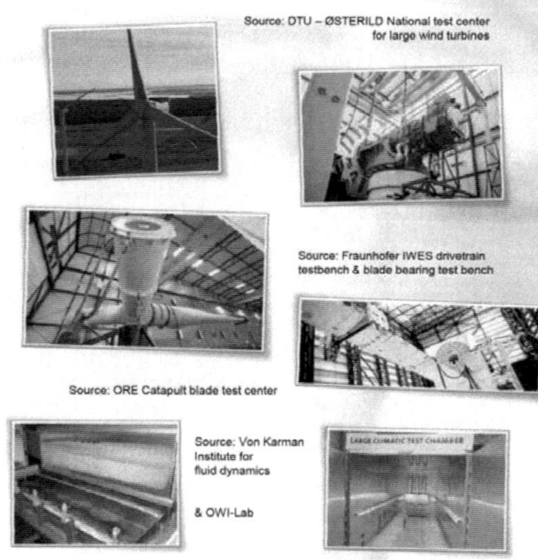

Purpose of prototype testing in the climate chamber

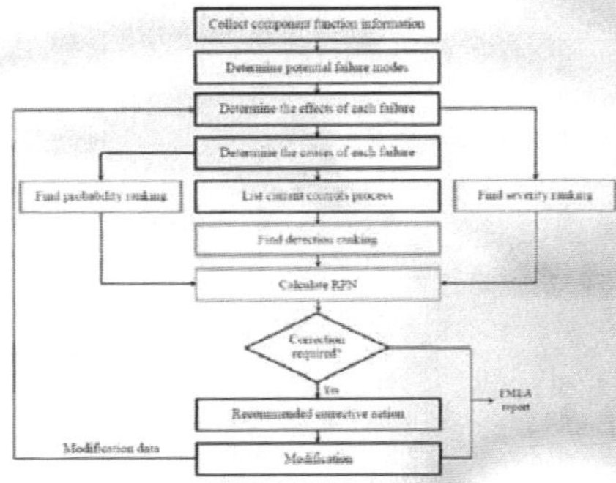

FMEA Process – Pillay and Wang 2003

Design validation

Checking if simulation models are valid (model validation)

Risk mitigation

Certification

Deliver reliable machinery + shorten the time-to-market

Cold temperatures influence materials, lubrication and performance of systems → Some influences are difficult or time consuming to simulate

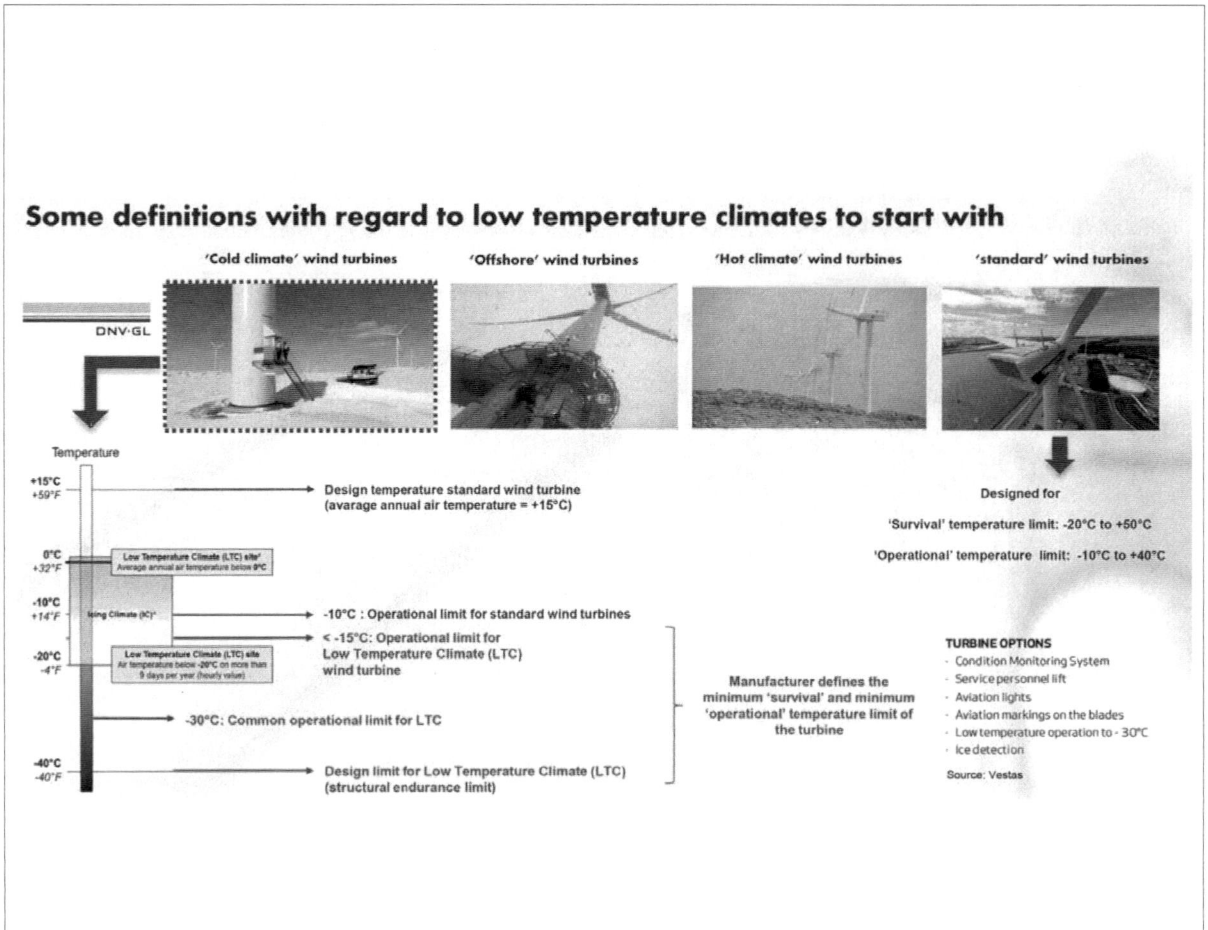
Some definitions with regard to low temperature climates to start with
'Cold climate' wind turbines
'Offshore' wind turbines
'Hot climate' wind turbines
'standard' wind turbines
DNV·GL
Temperature
+15°C
+59°F
0°C
+32°F
-10°C
+14°F
-20°C
-4°F
-40°C
-40°F
Design temperature standard wind turbine
(avarage annual air temperature = +15°C)
Low Temperature Climate (LTC) site'
Average annual air temperature below 0°C
Icing Climate (IC)'
-10°C : Operational limit for standard wind turbines
< -15°C: Operational limit for
Low Temperature Climate (LTC)
wind turbine
Low Temperature Climate (LTC) site
Air temperature below -20°C on more than
9 days per year (hourly value)
-30°C: Common operational limit for LTC
Design limit for Low Temperature Climate (LTC)
(structural endurance limit)
Manufacturer defines the
minimum 'survival' and minimum
'operational' temperature limit of
the turbine
Designed for
'Survival' temperature limit: -20°C to +50°C
'Operational' temperature limit: -10°C to +40°C
TURBINE OPTIONS
· Condition Monitoring System
· Service personnel lift
· Aviation lights
· Aviation markings on the blades
· Low temperature operation to - 30°C
· Ice detection
Source: Vestas

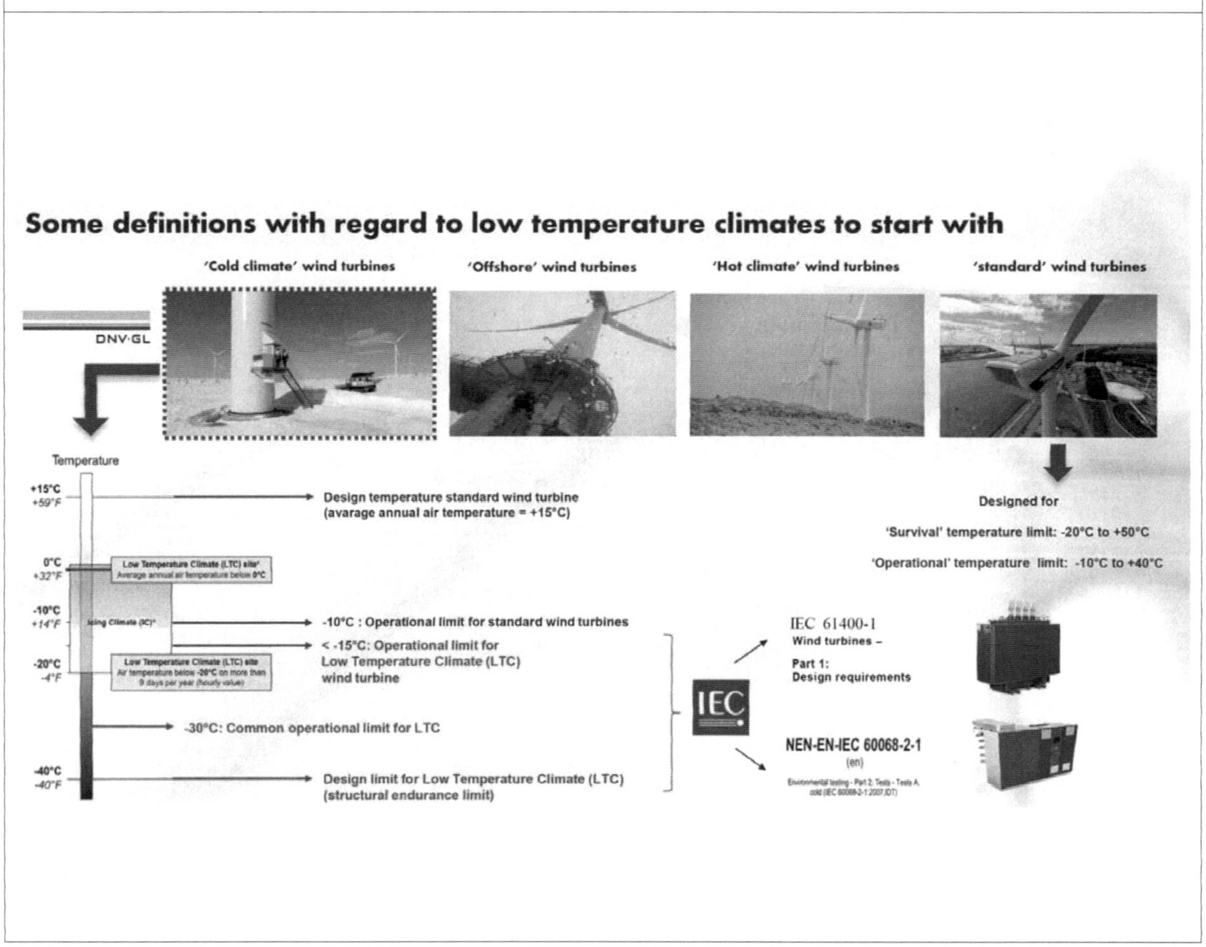
Some definitions with regard to low temperature climates to start with
'Cold climate' wind turbines
'Offshore' wind turbines
'Hot climate' wind turbines
'standard' wind turbines
DNV·GL
Temperature
+15°C
+59°F
0°C
+32°F
-10°C
+14°F
-20°C
-4°F
-40°C
-40°F
Design temperature standard wind turbine
(avarage annual air temperature = +15°C)
Low Temperature Climate (LTC) site'
Average annual air temperature below 0°C
Icing Climate (IC)'
-10°C : Operational limit for standard wind turbines
< -15°C: Operational limit for
Low Temperature Climate (LTC)
wind turbine
Low Temperature Climate (LTC) site
Air temperature below -20°C on more than
9 days per year (hourly value)
-30°C: Common operational limit for LTC
Design limit for Low Temperature Climate (LTC)
(structural endurance limit)
Designed for
'Survival' temperature limit: -20°C to +50°C
'Operational' temperature limit: -10°C to +40°C
IEC
IEC 61400-1
Wind turbines –
Part 1:
Design requirements
NEN-EN-IEC 60068-2-1
(en)

New trend: Cold climate & offshore meet in the Baltic sea

FINLAND

Designing to withstand Arctic conditions

29 February 2016 by Gerard O'Dwyer , Be the first to comment

FINLAND: Much is at stake for Finland's first commercial offshore wind farm, the €120-million 40MW Tahkoluoto project in the Gulf of Bothnia, in the north of the Baltic Sea, west of Finland.

Frozen... The 2.3MW pilot turbine (right) is sited off Finland, testing the arctic weather conditions on the turbine and steel cassion foundations (left) ahead of installing the country's first commercial offshore project (top)

29/02/2016 Source: WIND POWER MONTHLY

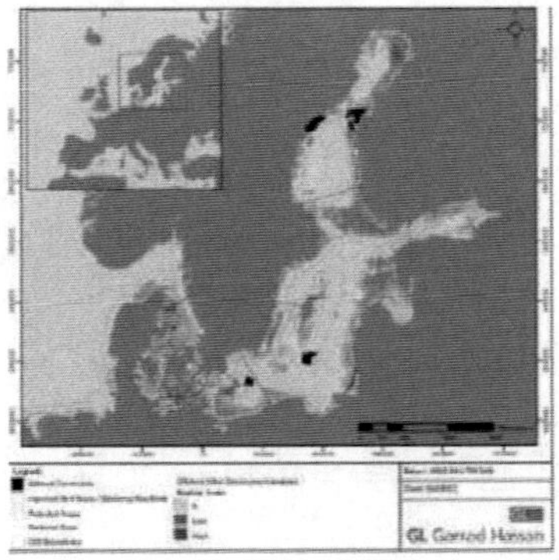

Source: Baltic Sea Region Energy Cooperation

6.2M126 SENVION

Design data

Nominal power	6,150 kW (MV-side)
Cut-in wind speed	3.5 m/s
Nominal wind speed	14 m/s 13.5 m/s
Cut-out wind speed	25 m/s 30 m/s
Restart cut-in wind speed	20 m/s 25 m/s
Operating temperature range	-20 – +35 °C

Example of temperature specification

Largest "non standard" sectors for wind energy today !

Dedicated solutions and low temperature adaptations are developed to cope with the challenges of cold climates wind farms

- Overview study of the 'Available technology' made by IEA Wind Task 19
- Study made public recently – July 2016 (see website)
- The study summarizes existing technologies and solutions from weather modelling, to ice detectors and turbine manufacturers that deal with wind energy in cold and icing conditions
- Low temperature adaptations chapter included

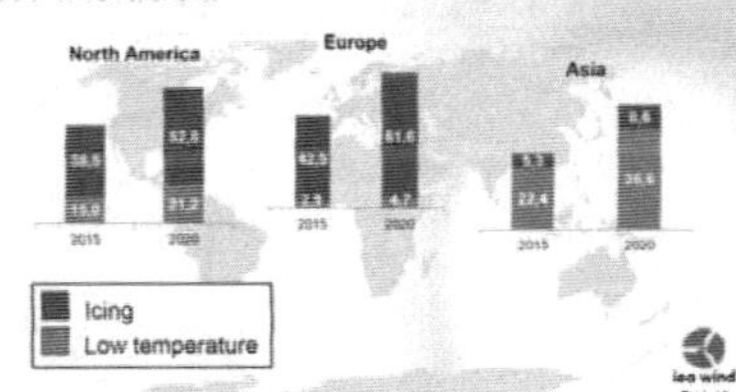

http://www.ieawind.org/task_19.html

Recent article 29/07/2016: http://www.windpowermonthly.com/article/1403504/emerging-cold

Overview of most popular measures in "cold climate package"

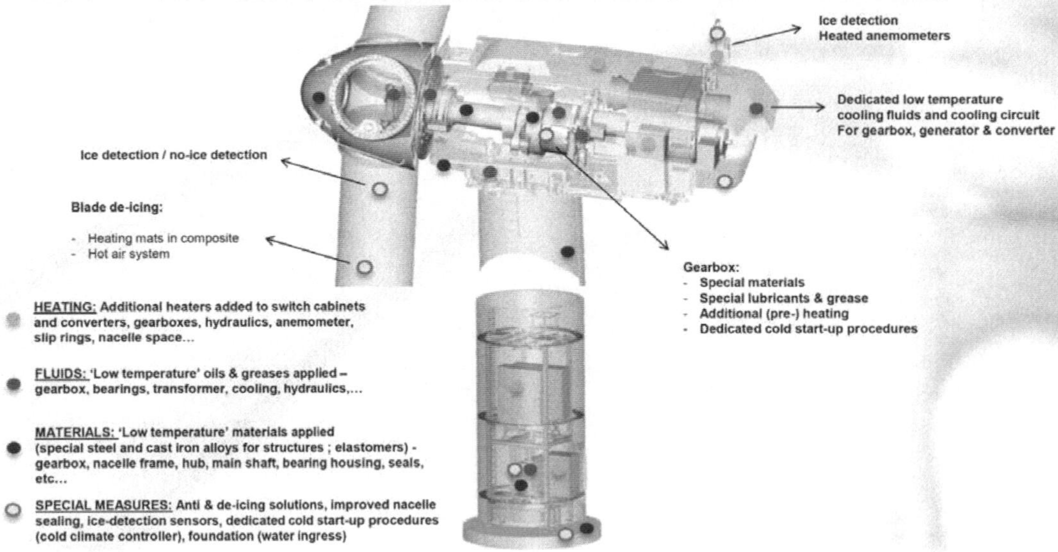

Insights in field performance in low temperatures

10 minute interval sample turbine output (Jan 2016)

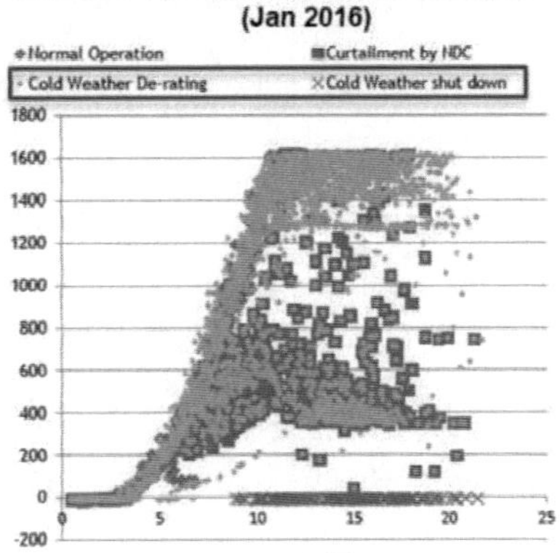

Source: Azure international / Clean Energy – Winterwind 2016

Location wind farm: Mongolia (cold climate)

- Dry air conditions at this location in Mongolia – almost no influence of icing on power curve, only effect of the low temperatures
- Power loss due to:
  - **Cold weather shut-downs**:

    The control system will shut the turbine down when the average temperature measured in the nacelle drops below -30°C.

  - **Cold weather de-rating**:

    When T< -15°C and >-30°C the turbine will be operating below the optimal power curve. Reasons include oil heating and limiting stress forces

Other industries: low temperatures in grid infrastructure

- Example of a 15MVA power transformer test campaign
- Test program at -40°C and **-50°C** for applications used in Canada, Russia, China, Mongolia and Scandinavia.
- The transformer filled with environmentally friendly, but highly viscous synthetic ester fluid.
- High viscosity at low temperatures impedes natural fluid flow and thus leads to a change of thermal performance.
- Especially relevant in case of cooling by external radiators, which is standard for power transformers. Fluid flow in the small radiator ducts is completely different from flow at the tank wall, even more so in case of high viscosity.
- Results presented at: cigre Symposium Dublin

SIEMENS

Lessons learned when testing wind turbine machinery

1.

3.

2.

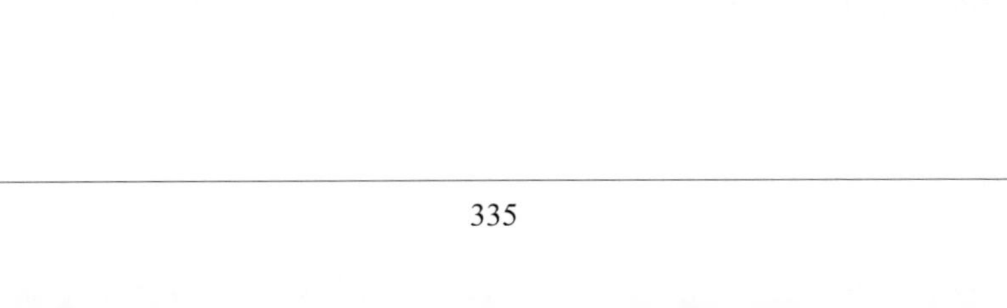

Liquid filled wind turbine transformers

- **Risks during low temperatures:**
  - Under load – no risk due to transformer losses
  - Stand-still in low temperature > 24h
    - Risk of tank underpressure and related air suction due to shrinking of the transformer oil
    - Risk of overpressure during cold start-up and warm-up in general as air can not evacuate fast enough
    - Risk of accelerated temperature rise in windings as natural cooling convection is limited at low temperature (stiff fluids)
    - Failure of auxiliaries and leakage (pumps, cooling circuits, radiators, tripping sensors, seals, etc.)

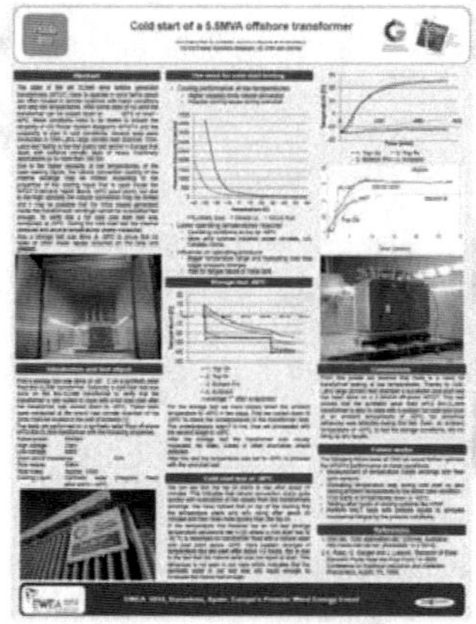

Poster and Paper available
EWEA 2014: Cold start of a 5.5MVA offshore transformer

Liquid filled wind turbine transformers

- **Lessons learned during climate chamber tests:**
  - Design of the transformer tank flexibility should take low temperature cool down and start-up into account (under- and overpressure)
  - Usage of correct type of pressure control in tank (hermetically sealed, gas cushion, breather, expansion tank)
  - The importance of seals (eliminate air suction / leakage)
  - Usage of the right auxiliaries that can work in cold climate
- **Continuous optimization and R&D effort:**
  - New fatigue tank test bench – topic under & overpressure and effect on lifetime of transformer tank
  - Field monitoring campaign of instrumented WTG transformer in cold climate

Wind turbine gearboxes – cold start-up tests

- **Risks during low temperatures:**
  - Ensure sufficient lubrication of gears and bearings during cold start-up procedure (CSP) as the oil is stiff.
  - Cold-start-up time must be within a certain allowable time limit (based on customer requirements)
  - Failure of auxiliaries and leakage (pumps, cooling circuits, filters, sensors, seals, etc.) → ensure survival limit of -40°C of all parts, even during a grid-disconnection.
  - Ensure sufficient oil flow in pumps to mitigate risk of cavitation
  - Heaters needed for pre-heating, but surface temperature of heaters should be low enough in order not to burn gearbox oil.

Wind turbine gearboxes – cold start-up tests

- **General procedure (OEM / Supplier dependent):**

Lubrication oil is heated up in separate oil tank and the gearbox oil sump from -40°C to +10°C. Friction losses from gearbox are than used to heat-up the oil quicker, next rotor speed is increased in a pre-determined procedure until partial load and full load can be applied.

- **Lessons learned during climate chamber tests:**
  - Generic lessons linked to the associated risks – example oil pumps, cavitation, seals,...
  - Climate chamber test = Standard test in validation trajectory (Cf. Automotive or off-highway vehicle tests)
  - Average cold-start-up time for gearbox: 5h – with wind
- **Continuous optimization and R&D:**
  - Testing of new lubrication oils and their performance during cold-start

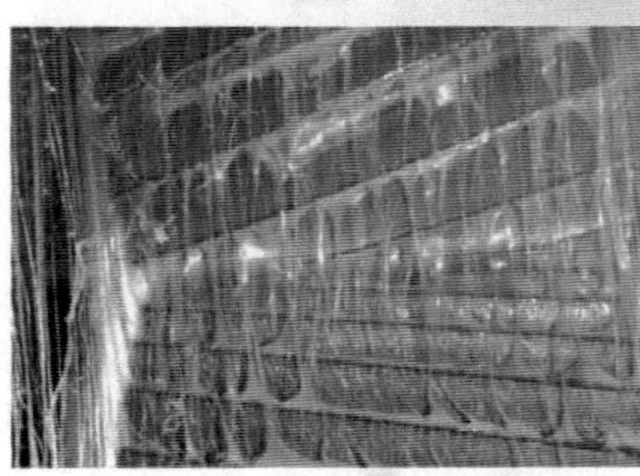

Other components: pitch / yaw / brakes / converters

- Time-to-market essential
- Full scale test to proof cold climate compliance to customer
- Cold start-up procedure (CSP) of pitch ; yaw and brake systems
- Converter / generator test

Statement DNV-GL note on Engineering Details: Wind Turbine operation down to –20°C

Requirements for certification: Evidence on safe operation of the main mechanical and electrical components. At least:

1. Components of the safety system and braking systems including the pitch system and the hydraulic system
2. Blade bearing, Main bearing (s)
3. Gearbox
4. Generator
5. Yaw bearing
6. Electrical Cabinets in Hub, Nacelle, Tower
7. Medium voltage switchgear and transformer

Contact person & more information

pieterjan.jordaens@owi-lab.be

be.linkedin.com/in/pieterjanjordaens/en

@OWI_lab

Knowledge sharing – LinkedIn Group:
Offshore Wind Infrastructure
Application Lab (OWI-Lab)

www.owi-lab.be

Pulsed eddy current detection of loose parts in nuclear reactor steam generators

T.W. Krause[1], D.P. Johnston[1,2], J.A. Buck[3], P.R. Underhill[1] and J. Morelli[2]

[1] Dept. of Physics, Royal Military College of Canada, Kingston, ON
[2] Dept. of Physics, Eng. Physics and Astronomy, Queen's University, Kingston, ON
[3] Canadian Nuclear Laboratories, Chalk River Laboratories, ON

Outline

- Motivation
- Experimental Set up
- Signal Processing
- Results
- Summary

Motivation

Steam Generators (SGs) consist of thousands of tubes that are supported by plates at various locations along the SG.

Metallic loose objects can become wedged between the tube sheet and the tube. Vibration can cause fretting wear of the steam generator tube

Previous work[1] suggests PEC works well in the presence of ferromagnetic materials

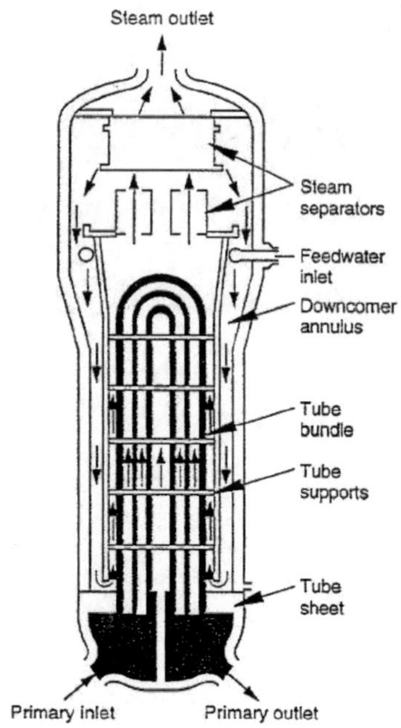

Recirculating steam generator

[1]J.A. Buck, P.R. Underhill, S.G. Mokros, J. Morelli, V.K. Babbar, B. Lepine, J. Renaud and T.W. Krause, "Pulsed eddy current inspection of support structures in steam generators", IEEE Sensors Journal, 15, No. 8, August 2015, Pgs. 4305-4312.

[3]

Experimental Set Up

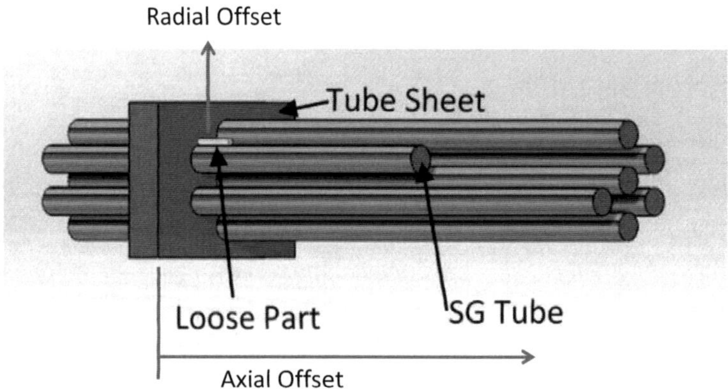

Radial offset used up to four 3.2 mm thick plastic shims
Two orientations

- Parallel: parallel to SG tube
- Perpendicular: parallel to Tube Sheet

6th International CANDU ISI Workshop/NDT in Canada 2016 Conference

[4]

Loose Parts

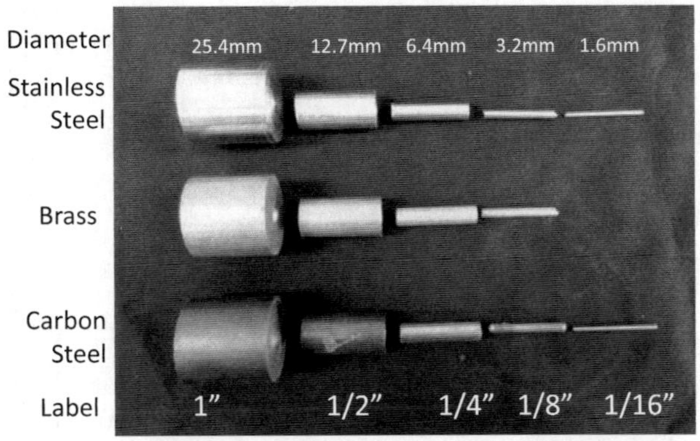

Cylinders above 25.4 mm (1") long.

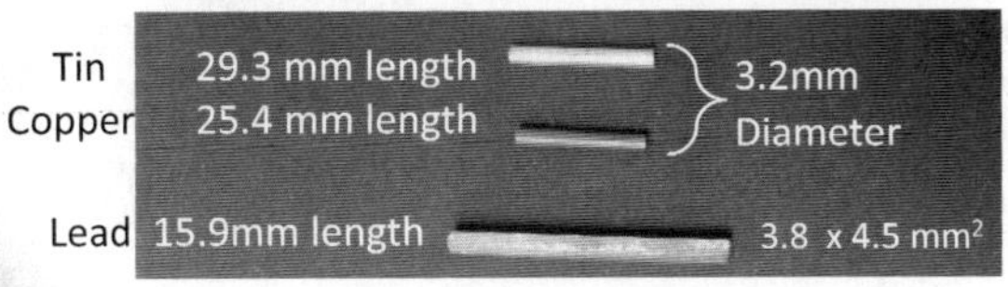

Material	Conductivity
	(MS/m)
Stainless Steel	1.4
Brass	14.3
Carbon Steel	5.9
Tin	7.9
Copper	58.1
Lead	4.9

6th International CANDU ISI Workshop/NDT in Canada 2016 Conference

Experimental Set Up

Parallel (||) Orientation

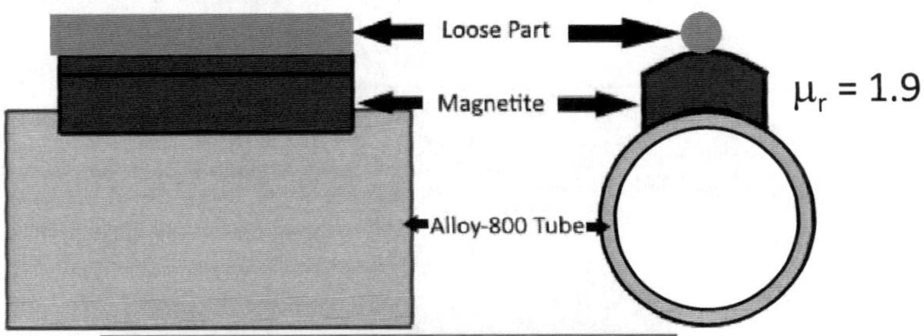

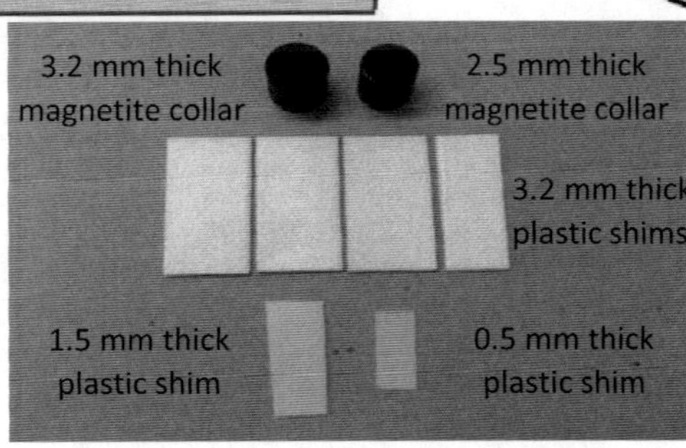

6th International CANDU ISI Workshop/NDT in Canada 2016 Conference

6

Pulsed Eddy Current Testing

- Pulsed Eddy Current (PEC) testing uses a square pulse excitation of a drive coil, which
  - Contains numerous discrete frequencies
  - Provides near DC magnetization of ferrous structures
- PEC takes advantage of long diffusion time of electromagnetic fields

$$\tau = \sigma \mu \ell^2$$

where σ is the conductivity, μ the permeability and ℓ the characteristic length

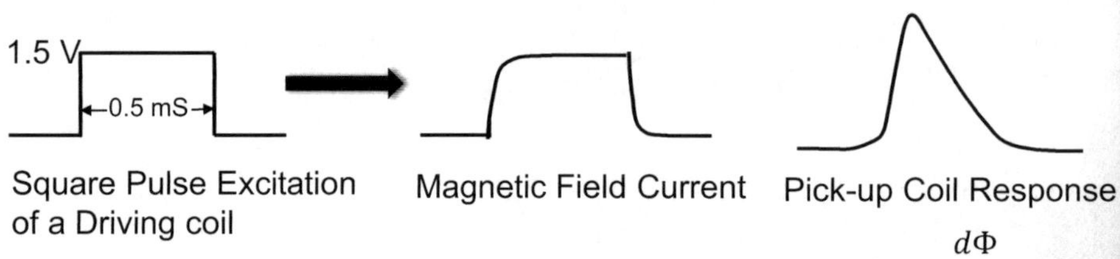

$$\varepsilon = -\frac{d\Phi}{dt}$$

Faraday's Law

PEC Probe

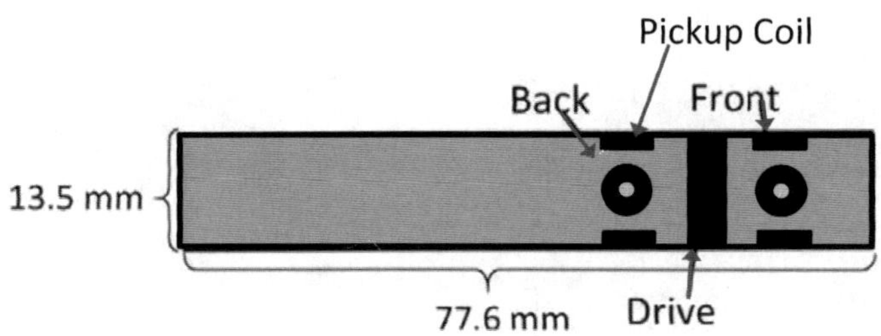

- Readings every 2 mm over 130 mm range
- Only used pickup coils aligned with loose part
- A/D using NI 6356 USB DAQ @ 1 MHz/channel
- Input pulse 1.5V square wave @ 1 kHz 50% duty cycle

Signal

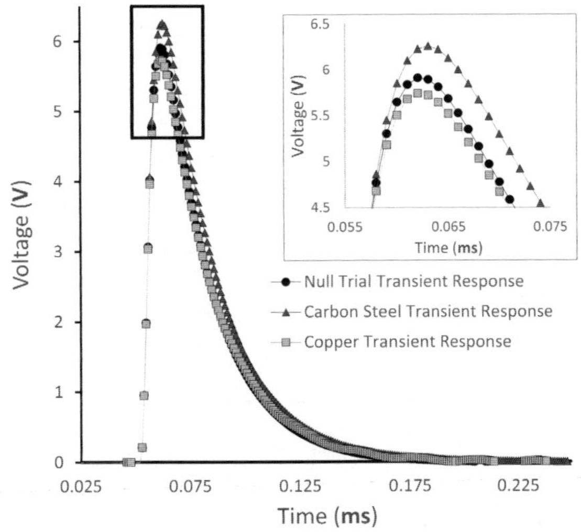

6th International CANDU ISI Workshop/NDT in Canada 2016 Conference

9

Modified Principal Components Analysis

- Modified principal components analysis[2] (MPCA) is a least squares method used to reduce PEC signals to a series of vectors (V_i) and scores (s_i) that best represent the data in a least squares sense
- MPCA differs from PCA in that the mean is not subtracted from the data points
- Using a small number of vectors (3-5), the original signal can be reproduced to a high level of accuracy
- This reduces the data variation to a manageable number of scores

10

[2] P. Horan, P.R. Underhill and T.W. Krause, 'Pulsed Eddy Current Detection of Cracks in F/A-18 Inner Wing Spar without Wing Skin Removal using Modified Principal Component Analysis', Nondestructive Testing & Evaluation International, Vol. 55, Apr. (2013), 21-27.

Signal Processing

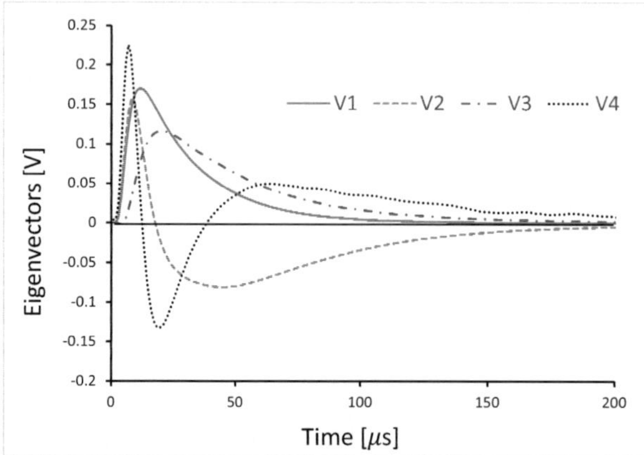

Used four eigenvectors (V_i) from Modified Principal Component Analysis resulting in a 1.0% difference between original and reconstructed signal

11

6th International CANDU ISI Workshop/NDT in Canada 2016 Conference

Results for Carbon Steel (CS)

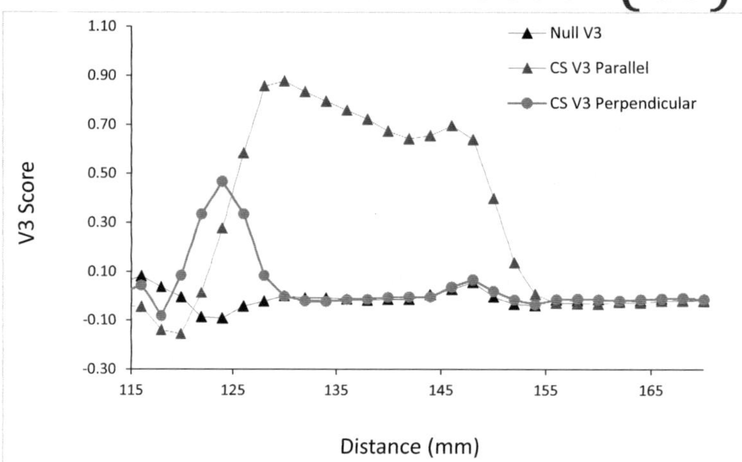

1/16" (1.6 mm) diameter carbon steel loose part in both parallel and perpendicular orientations with no radial or axial offset (OS)

12

6th International CANDU ISI Workshop/NDT in Canada 2016 Conference

Cluster Analysis

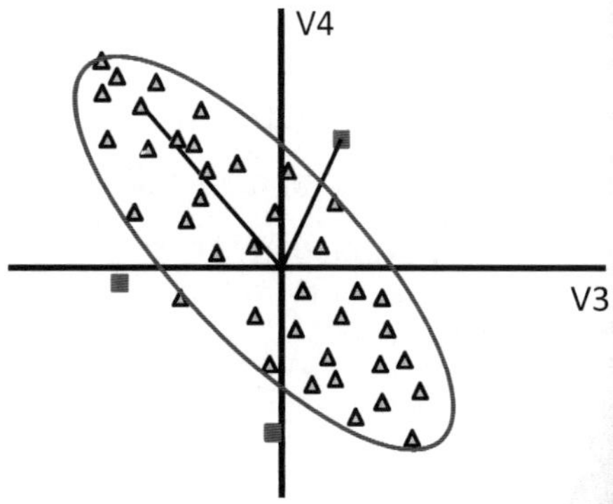

Mahalanobis Distance (MD)

13

Cluster Analysis

- Use 5 points closest to tube sheet
- Plot 3 scores (2-4) as a function of position for null trials

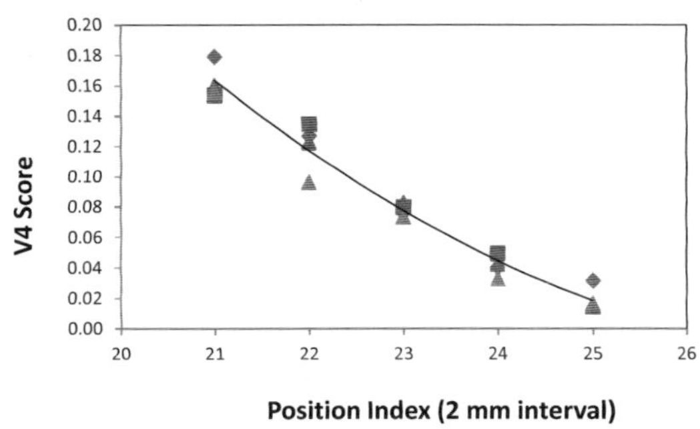

- Use fit to shift scores to zero centre
- 5 positions x 4 scores = 20 points per cluster

14

Cluster Analysis

- Shift all data using functions obtained from null trials
- Calculate covariance matrix for null trials
- Calculate Mahalanobis distance for unknown tubes
- If one point is greater than cut-off register a hit

15

6th International CANDU ISI Workshop/NDT
in Canada 2016 Conference

Results (by Material)

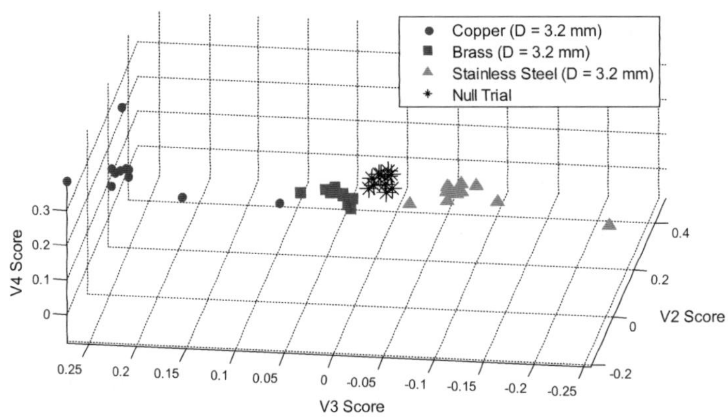

1/8" (3.2 mm) diameter loose parts of varying composition in parallel orientation with no radial or axial offset (OS)

16

6th International CANDU ISI Workshop/NDT
in Canada 2016 Conference

Results (by Size)

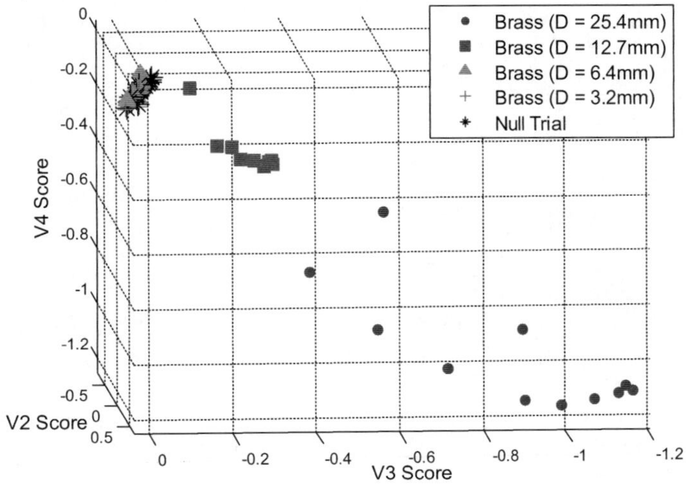

Brass loose parts varying in diameter from 1/8" (3.2 mm) to 1" (25.4 mm) with no radial or axial offset (OS)

6th International CANDU ISI Workshop/NDT in Canada 2016 Conference

17

Results (by radial offset)

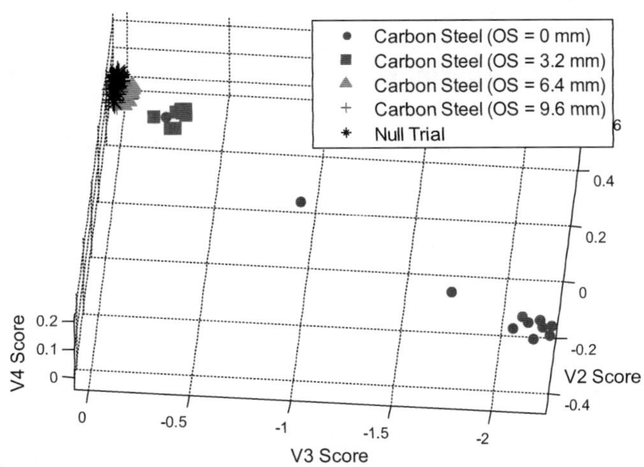

1/8" (3.2 mm) diameter carbon steel loose parts on tube sheet with varying radial offset (OS) from steam generator tube

6th International CANDU ISI Workshop/NDT in Canada 2016 Conference

18

Results: Detectable Distance

Loose Part (diameter)	Maximum radial offset detectable	Maximum radial offset detectable[1] (magnetite present)
1/8"Carbon Steel (3.2 mm)	9.6 mm	3.2 mm
1/16" Carbon Steel (1.6 mm)	3.2 mm	3.2 mm
1/8"Brass (3.2 mm)	3.2 mm	3.2 mm
1/8" Stainless Steel (3.2 mm)	3.2 mm	ND (< 2.5 mm)
1/16" Stainless Steel (1.6 mm)	1.5 mm	ND (< 2.5 mm)
1/8" Copper (3.2 mm)	3.2 mm	2.5 mm
1/8" Tin (3.2 mm)	1.5 mm	ND (< 2.5 mm)
1/6" Lead (4.2 mm)	0.5 mm	ND (< 2.5 mm)

Various Screw Type Diameters

#6	3.3 mm
#8	4.1 mm
#10	4.8 mm

[1]ND- Not-Detectable with minimum thickness (2.5 mm) magnetite collar present

19

Summary

- Loose parts on tube sheet next to SG tube detectable (in contrast to conventional Eddy Current)
- Relative score clustering locations change with material
- Magnitude of score variation larger in higher conductivity materials
- Scores switch sign between non-magnetic and ferromagnetic materials
- Score variations proportional to physical dimensions of part, with larger parts producing larger score variations
- Score variations drop off quickly with radial offset
- Effect of orientation (|| or ⊥) on score variation small; magnitude of variation approximately equal but occurred for fewer number of data points
- Analyzed signal variations attributed to electromagnetic interaction (size of part and distance) and relaxation time

20

Acknowledgements

- First author would like to thank Vijay Babbar and Brian Lepine of Canadian Nuclear Laboratories and Sean Sullivan of Ontario Power Generation for useful discussions.
- University Network of Excellence in Nuclear Engineering (UNENE)
- Natural Sciences and Engineering Research Council of Canada (NSERC)

Questions

?

6th International CANDU ISI Workshop/NDT
in Canada 2016 Conference

New Methodology for NDT Assessment of In-service Wood Poles

María-José RODRÍGUEZ-ROBLERO [1], Fernando J. TALLAVÓ [1], Giovanni CASCANTE [1], Mahesh D. PANDEY [1]

[1] Department of Civil and Environmental Engineering, University of Waterloo; Ontario, Canada;
Phone: +1 519-888-4567, Fax: +1 519-888-4349; e-mail: mjrodrig@uwaterloo.ca, fjtallav@uwaterloo.ca, gcascant@uwaterloo.ca, mdpandey@uwaterloo.ca

Abstract
Wood poles are extensively used in North America to support electrical lines. To ensure the reliability of the electrical network, the condition of the poles is assessed usually through visual and sounding inspection. These methods have the disadvantage that they are subjective, they provide a local evaluation of the condition, and are unable of detecting early decay.

This paper summarizes the results of a research project that included the development of a new methodology to assess the condition of wood poles based on ultrasonic waves, the construction and calibration of the testing device, numerical modelling, and laboratory and field tests. The proposed methodology considers wood as an orthotropic material, takes into account the variability in the elastic and mechanical properties, and the effect of temperature and moisture content in these properties. Also provides a quantification of the damage and the remaining strength through statistical indexes. The method has the advantage that is nonintrusive, provides quantitative measurements of the internal condition of wood poles and allows detecting early decay in wood poles.

Keywords: ultrasonic testing, reliability, wood poles, maintenance, condition assessment

1. Introduction

Wood poles are usually used in both, transmission and distribution lines; and are preferred over other materials because of their lower price, high strength to weight ratio, low electrical conductivity, easiness of storage, handle and climb, wide availability and lower installation cost *[1,3,5]*. However, as wood is an organic material, environmental factors such as temperature, moisture conditions, bacteria or fungi may induce internal decay, which reduces the strength of the poles. In order to avoid sudden failures and to optimize the schedule of pole replacement and rehabilitation, the assessment of the actual condition of in-service poles is needed.

Design and safety specifications for electrical networks require that the poles should be replaced or rehabilitated when the strength of the pole had reduced to 60% of the required design strength *[13]*. To satisfy this requirement, the network owners implement periodic inspections using different testing methods to determine the actual condition of the poles. The most common methods are visual inspection, sounding inspection and evaluation using a resistograph. Visual inspection is an economical method, but it is limited to the external condition of the pole. The sounding inspection consists on the identification of void areas or defects by the sound perceived when the pole is hit with a hammer. Likewise to visual inspection, sounding has lower reliability because it is subjective, since it relies on the experience of the inspector and it is a qualitative evaluation. Another method of inspection consists of measuring the local penetration resistance with a resistograph, which is a patented instrument that correlates the drilling resistance to the density of the wood. This method has a lower statistical significance; given that the measurements are punctual and do not necessarily represent the general condition of the pole. Figure 1 shows two examples where the internal decay may not be detected by the conventional methods. If the poles were inspected through visual inspection the internal decay cannot be noticed because there are no signs of deterioration in the surface of the pole. If the assessment was based on resistograph measurements, depending on the section and number of measurements done the decayed areas may not be identified.

Figure 1: Example of wood poles for which the traditional methods of inspection may not detect the internal deterioration

This paper presents the results of a research project which main objective was to develop an assessment method for in-service wood poles based on ultrasonic and statistical measurements. The proposed solution provides a quantitative estimation of the internal decay in the cross-section and an estimation of the remaining strength. The method gives a global assessment of the cross section and it is statistically significant. A new field testing device and software was developed and the proposed methodology also innovates in the consideration of the orthotropic nature of wood and the effect of moisture content and temperature on the mechanical properties.

2. Theoretical background

2.1 Nondestructive ultrasonic testing of wood

A method is considered nondestructive when the characteristics of the element being inspected are not affected by the tests. It is referred as ultrasonic testing the application of sound waves above human hearing (greater than 16 kHz) for material evaluation. The usual ultrasonic testing makes use of a transmitter and a receiver to measure the first arrival of compressional waves (P-waves) *[2]*. Generally, for ultrasonic testing are used piezoelectric transducers as transmitter and receiver. A piezoelectric transducer is a device that converts electrical energy in ultrasonic energy and vice versa *[14]*. These transducers need to be in contact with the object tested, and also require to be calibrated for the specific material under study. One of the improvements in the methodology presented herein is the use of an array of ultrasonic transducers, rather than one transmitter and one receiver. Arrays have the advantage that make possible to take different measurements from a single location and can be used to produce images for the visualization of internal structure of a component *[10]*.

2.2 Wave propagation in wood

Wood is an orthotropic material, since its properties are different in three orthogonal directions: longitudinal, radial and transverse. To fully describe the wave propagation in wood under plain strain conditions are required eight parameters: mass density (ρ), three elastic moduli (the elastic modulus in the longitudinal direction E_l, elastic modulus in the radial direction E_r, elastic modulus in the tangential direction E_t), three Poisson's ratios (ν_{tr}, ν_{tl}, ν_{rl})

and the shear modulus G_{rt}. These parameters vary between species, within the same wood species and depend on the temperature and moisture content *[4]*. Consequently, to obtain reliable ultrasonic measurements it is needed to take into account the intrinsic variability of these parameters and its influence on the wave velocity. For this, the velocity should be considered as a random variable and its probability distribution needs to be determined. A detailed explanation on how the probability distribution was determined may be found on *[2, 4]* where the methodology is demonstrated for red pine poles. For the purpose of this paper only a brief description of the methodology is presented.
The derivation of the probability distribution of wave velocity was based on laboratory tests, the consideration of uncertainty in the mechanical properties of wood and a simplified model of wave propagation in a cylindrical orthotropic material, validated with the results of 60,000 finite element numerical simulations *[2]*. The values of P-wave velocity (V_p) were estimated using the simplified method of analysis for wave propagation in wood given by the relationship between E_l (the elastic modulus in the longitudinal direction) and ρ is the mass density:

$$V_p = \sqrt{\frac{E_l}{\rho}} \tag{1}$$

Figure 2 shows the array of transducers and the receiver locations considered. Due to the anisotropy of wood, the ray paths between the transmitter and the receivers are curved, except for the receiver located at the opposite face of the transmitter. This relation is not linear because wood is not a homogeneous material, the cross-sections are not perfectly circular and the pith is not always located at the center *[3]*.

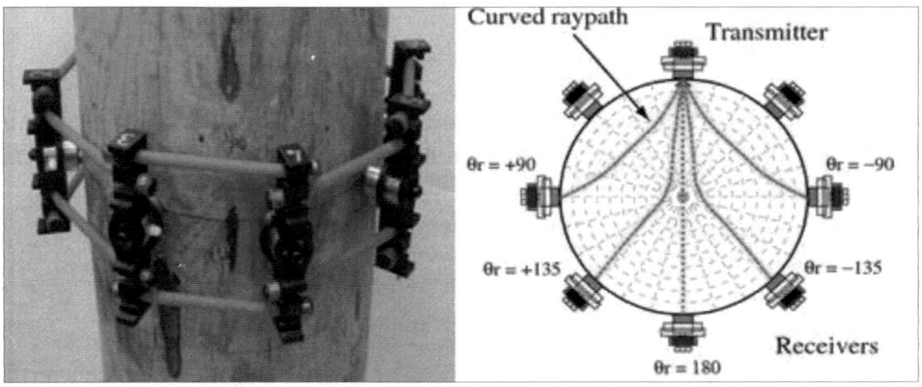

Figure 2: Ultrasonic transducer array used in the testing

Monte Carlo simulations (MCS) can be used to define the probability distribution. The wave velocity was studied with respect to the receiver location angle. For each receiver location, the probability distribution that describes better V_p was defined through the method of the Probability paper plots (PPP). The distributions studied were the normal, lognormal, gamma, Gumbel and Weibull distribution. Probability paper plots allow determining whether the data follows the assumed distribution. According to the results, the normal distribution provides the best-fitting of the data for wave velocity for the general case (all the variables are considered as random variables).

2.3 *Consideration of effect of temperature and moisture content*

The material parameters (elastic moduli, mass density, Poisson's ratios, shear modulus) depend on the temperature and moisture content, since water has an important effect on the mechanical properties of wood *[4]*. The proposed method includes a model to predict the moisture content based on the records of relative humidity and temperature.

The influence of the temperature and moisture content may be analyzed from their effect on the moduli of elasticity and mass density. Given that the square of the wave velocity is directly proportional to the modulus of elasticity and inversely proportional to mass density (equation (1); the variations on the mass density induced by the changes in moisture content have an effect on wave velocity. The moisture content for the wood poles was studied during a full year to evaluate the effect of the moisture content and temperature on wave velocity. The poles were exposed to long-term (seasonal) and short-term (daily) variations in temperature and humidity. According to the results, the minimum values of moisture content occur between spring and summer, while the maximum values in winter. The variability in the moisture content and the temperature was considered using finite element numerical simulations. Monte Carlo simulations were done to evaluate the probability distribution of wave velocity when the mass density, the elastic moduli, Poisson's ratios and the moisture content are modeled as random variables. There were considered two situations: spring/summer and fall/winter. According to the results from 60,000 simulations and using the PPP method, the normal distribution gives the best fit of the data. From the simulations, a moisture-temperature factor to correct the velocity (R_{V_p}) for different moisture content (MC) and temperature (T) was derived and it is represented as:

$$R_{V_p}(MC,T) = (-0.000045T - 0.0111)MC - 0.0004T + 1.152 \quad \textbf{(2)}$$

The previous expression allows adapting the probability distributions of wave velocity for a given moisture content and temperature.

2.4 *Definition of new statistical indices for condition assessment*

One of the main disadvantages of the sounding and visual inspection is that they do not provide a quantification of the existing damage or an estimation of the remaining strength. The proposed method uses ultrasonic measurements (wave velocity, transmission factor, elastic moduli in the radial and tangential directions) to compute two statistical indexes: the Overall dissimilarity index (ODI) and the Condition rating index (CRI). These two parameters allow quantifying the damage existing in the cross section and estimate the remaining strength of the pole.

The ODI represents the number of standard deviations that the test results differ from the expected value for a new pole. Values of ODI in the range of [-2,-3] indicate that the pole has 98% probability of being decayed. The ODI is computed at each receiver location according to the next expressions:

$$\mathrm{ODI}_{\theta_r} = \mathrm{DIV}_{\theta_r} \cdot \mathrm{W} + \mathrm{DIA}_{\theta_r} \cdot (1 - \mathrm{W}) \quad \textbf{(3)}$$

where DIV_{θ_r} is the dissimilarity index for the wave velocity (equation (4)), DIA_{θ_r} the dissimilarity index for the transmission factor (equation (5)) and W is a weight factor calculated from equation (6).

$$DIV_{\theta_r} = \frac{V_p - \mu_{V_p}}{\sigma_{V_p}} \quad (4)$$

In the previous equation V_p is the wave velocity computed, μ_{V_p} is the expected value and σ_{V_p} the standard deviation for the wave velocity for a sound pole of the same wood species at the same receiver location.

Considering A_f the measured transmission factor (reciprocal of the attenuation factor) computed in the frequency domain, μ_{A_f} the expected value and σ_{A_f} the standard deviation for a sound pole, the dissimilarity index for the transmission factor is defined as:

$$DIA_{\theta_r} = \frac{A_f - \mu_{A_f}}{\sigma_{A_f}} \quad (5)$$

The weight factor is defined as a function of CV_V (the coefficient of variation for the wave velocity) and CV_A (coefficient of variation for the transmission factor).

$$W = \frac{1}{1 + \frac{CV_V}{CV_A}} \quad (6)$$

The condition rating index CRI represents the strength relation of the expected modulus of elasticity in the radial direction of the inspected pole, and the expected modulus of elasticity of a sound pole. A value of CRI equal to one indicates that the pole has a similar strength to that of a new one. This parameter allows verifying the requirement for the strength for wood poles indicated in the specifications *[13, 15],* since CRI equal to 0.60 corresponds to the end of the service life of a pole. More detailed information about the determination of the index may be found in *[8]*.

3. Methodology for the field testing

As indicated before, the condition assessment is based on three ultrasonic parameters: wave velocity, wave attenuation and the elastic moduli in the radial and tangential direction. These parameters are determined from the measurements of the first arrival of the compression waves propagating through the pole and the full waveform recorded at each receiver. The field testing is performed using a calibrated array of eight ultrasonic transducers evenly spaced around the pole cross section (Figure 2). The array is a customized belt that can be adjusted to test poles of different diameters and improves the coupling between the transducers and the surface. In each test, one of the transducers functions as a transmitter and the other seven as receivers. The test is repeated until each transducer has been used as a transmitter. The information registered is processed using a software develop specifically for the testing device. The program performs the required signal processing, computes the arrival times, and evaluates the correlation function between moisture content of the pole, and the in-situ relative humidity and temperature. Finally, the software generates an assessment report for each cross section tested and includes the information of the test in a data base. The assessment report summarizes the main information that may be used for maintenance decisions, such as ODI, CRI and a tomographic image of the cross section. The tomographic image shows the comparison of the measured wave velocity with the expected value of a sound pole, and allows identifying visually the decayed areas of the cross section. This tomographic image is generated by discretizing the cross section in the 25 regions and the 28 ray paths shown in Figure 3. For each region it is computed the wave velocity solving the inverse problem from the travel time measurements of the compression waves using the Least squares method.

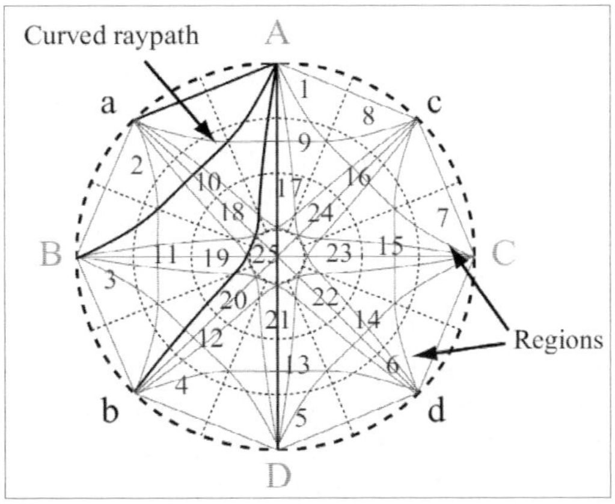

Figure 3: Discretization and ray paths considered for the tomographic method

The proposed methodology has been applied in the laboratory to thirty three new and aged red pine poles collected from Perth and Guelph in Ontario. There had also been tested five in-service wood poles at the Kleinburg Hydro One training centre. The results obtained in the laboratory and in the field demonstrate that the proposed methodology is able of detecting early decay and to provide a quantification of the internal condition of the poles and its remaining strength. Figure 4 shows an example of the tomographic image and the actual damage observed in the pole.

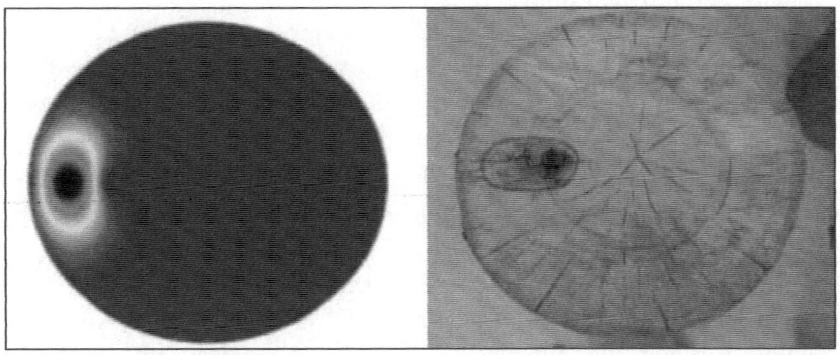

Figure 4: Example of the results from the tomographic image and the actual damage

Currently, a field testing prototype consisting of the transducers array and the portable electronic equipment built is being tested. The field testing program includes the assessment of 400 in-service wood poles in the Waterloo region. The sample includes poles of different ages (from newly installed to more than 60 years in service), of two species (red pine and cedar), located in urban and rural areas. The testing program includes measurements during winter and spring; and also the comparison of the results for cross sections located above the ground line and sections tested after excavating the soil surrounding the poles.

4. Conclusions

The main objective of this research project was to develop a new methodology for the condition assessment of in-service wood poles based on ultrasonic and statistical testing. The project included theoretical, experimental and field work, which demonstrated the applicability and the advantages of the methodology.

Ultrasonic waves have been used before for the assessment of wood poles, but it has not been taken into account the orthotropic nature of wood and the variability in the mechanical properties that affect the propagation of sound waves. The neglect of these inherent characteristics of the wood affects the reliability of the testing, and it impedes the detection of early decay. In this project the anisotropic nature of wood was considered by using an orthotropic model for wave propagation, and the variability of the parameters that affect wave propagation was studied and included through the probability distribution of wave velocity (V_p). This probability distribution was determined for each receiver location, and it was also taken into account the effect of the temperature and the moisture content in the velocity. The assumed wave propagation model and the derived probability distribution were verified through numerical simulations and laboratory testing. According to the results, more accurate results are obtained for the general case, where the parameters needed to study the wave propagation in wood (ρ, E_l, E_r, E_t, ν_{tr}, ν_{tl}, ν_{rl} and G_{rt}) are considered as random variables characterized by a probability distribution. Also from the study of the variation of the moisture content and temperature in one year, it is suggested to consider two scenarios: spring/summer and fall/winter.

An important limitation of the traditional methods of condition assessment of wood poles is their inability to quantify the damage. The proposed methodology overcomes this limitation by using two statistical indexes, the Overall dissimilarity index ODI and the Condition rating index CRI. These indexes quantify the extent of the decayed areas and the remaining strength of the pole. ODI is employed to construct a tomographic image of the cross section being tested, which can be used to identify visually the location and extent of the damaged areas. The CRI provides a quantification of the remaining strength and may be used to verify the strength requirements established in the specifications for overhead systems.

The novel features of the proposed methodology include also the use of an array of transducers and the development of a simple and efficient tool for nondestructive testing in the field. In addition, it was also developed software to process the inspection information and to generate a data base that may be used for management decisions.

The proposed NDT ultrasonic method has the advantage that is nonintrusive and provides quantitative measurements of the internal condition of wood poles. Also it is able of detecting early decay, which is an important improvement, given that the typical methods of inspection of in-service poles are not capable of detecting early stages of decay *[3]*. As indicated in *[12]* wood poles that are decayed, weathered or present significant checking may present shear failures that are undesirable failures because they occur suddenly. For the network owners it is essential to identify damage in due time to implement the pertinent actions and prevent unexpected failures. The proposed methodology substantially reduces the time required to perform the inspections, which may reduce the cost of maintenance of the electrical network by contributing to optimize the pole replacement and rehabilitation schedule.

5. Acknowledgements

This project received the financial support from Ontario Centers of Excellence (OCE) and the Natural Sciences and Engineering Research Council (NSERC). The authors also acknowledge the support received from Hydro One Networks Inc. and Waterloo North Hydro Inc., through the Waterloo Institute for Sustainable Energy (WISE) of the University of Waterloo.

References

1. Tallavó, F. (2009). "New methodology for the assessment of decayed utility Wood poles using ultrasonic testing." Ph.D. thesis, University of Waterloo, Ontario, Canada.
2. Tallavó, F. et al. (2011). "Estimation of the Probability Distribution of Wave Velocity in Wood Poles". Journal of Materials in Civil Engineering: 23(9): 1-10.
3. Tallavó, F. et al. (2012). "A novel methodology for condition assessment of wood poles using ultrasonic testing". NDT&E International 52: 149-156.
4. Tallavó, F. et al. (2012). "Probabilistic characterization of ultrasonic wave propagation in wood poles". Canadian Journal Civil Engineering 39: 484-493
5. Tallavó, F. et al. (2013). "Experimental and Numerical Methods for Detection of Voids in Wood Poles Using Ultrasonic Testing". Journal of Materials in Civil Engineering: 25(6): 772-780.
6. Tallavó, F. et al. "Statistical characterization of wave velocity in utility wood poles". Technical Report prepared for Hydro-One. July 13, 2011.
7. Tallavó, F. et al. "A condition rating index for the assessment of utility wood poles based on ultrasonic testing". Technical Report prepared for Hydro-One. July 13, 2011.
8. Tallavó, F. et al. "Calibration of the UPole-Testing: an innovative nondestructive evaluation method for condition assessment of wood poles". Technical Report prepared for Hydro-One. July 13, 2011.
9. Tallavó, F. et al. "A tomographic imaging method based on ultrasonic waves for the detection of decayed areas in wood poles". Technical Report prepared for Hydro-One. July 25, 2011.
10. Drinkwater, B. et al. (2006). "Ultrasonic arrays for non-destructive evaluation: A review." NDT&E International 39: 525-541.
11. Canadian Standards Association, CAN/CSA-C22.3 No. 60826-10 Design criteria of overhead transmission lines, Mississauga: Canadian Standards Association, 2010.
12. Daigle, O. (2013). "The effect of woodpecker damage on the reliability of wood utility poles". MSc thesis, University of Waterloo, Ontario, Canada.
13. Canadian Standards Association, Overhead systems, Mississauga, Ontario: Canadian Standards Association, 2010.
14. Blitz, J. and Simpson, G. Ultrasonic methods of non-destructive testing. London: Chapman & Hall, 1996.
15. N.E.S.C. Handbook, NESC Strength Requirements, ch.26, pp 293-308. McGraw Hill, 2002.

Three-dimensional nondestructive evaluation of cylindrical objects (pipe) using an infrared camera coupled to a 3D scanner

F. B. Djupkep Dizeu, S. Hesabi, D. Laurendeau, A. Bendada

Computer Vision and Systems Laboratory, Laval University
1065, avenue de la Médecine, Québec City (QC), G1V 0A6, Canada
Tel : (418) 656-2979, Fax: (418) 656-3159
Email: dizeubilly@yahoo.fr, s.hesabi12@gmail.com,{denis.laurendeau, hakim.bendada}@gel.ulaval.ca

Abstract

On the basis of its thermal behavior, the state of an object can be evaluated, without damages and without contact, using infrared thermography. The result of the nondestructive evaluation of an object can be qualitative (location of areas with defects) or quantitative (determination of the defects depth and size). In this paper, we use infrared thermography for the quantitative nondestructive evaluation of a long cylindrical object. In order to improve the results accuracy and for visual rendering, a three-dimensional formulation of the thermal nondestructive evaluation is adopted. The acquisition system is comprised of an infrared camera and a 3D scanner. Both instruments are registered to each other after a calibration procedure. The infrared camera records the time-dependent temperature of the inspected object while it is subjected to an external thermal stimulus, and the 3D scanner provides a set of 3D points representing the surface of the object. These experimental data (temperature and point cloud) are used to reconstruct the internal geometry of the object, i.e., to locate the defects and determine their depth and size, after solving an inverse geometry problem. Because the infrared camera cannot capture the whole object once, acquisition system is moved along the object. All the partial inspection results are put together to form the defects map. An efficient visual rendering is then obtained after that the mesh of the object is textured by the defects map.

Keywords: Infrared thermography, nondestructive testing, surface reconstruction, mapping, temporal tracking of the thermal front, 3D scanner.

1/ Introduction

Infrared thermography (IRT) has been widely used for qualitative nondestructive testing (NDT) of objects (metallic, non-metallic, composite, etc.) [1]. Qualitative NDT using IRT aims to locate regions of the inspected object where defects may exist. Quantitative NDT is performed when quantitative information such as the size and depth of defects are needed. Current thermal methods implemented for quantitative NDT use a one-dimensional model which make them accurate enough only for planar and almost planar objects with defects of simple geometry [2]. For object with a rather complex geometry, a three-dimensional formulation is necessary to satisfy the requirement of accuracy during quantitative NDT; an inverse geometry problem can then be solved to reconstruct the internal geometry of the inspected object [3-4]. Such an approach is very convenient to monitor the time-evolution of the degradation, due to corrosion for example, of the state of the objects.

In this paper we present the results of the quantitative inspection of a cylindrical metallic object (pipe). We use a new algorithm, the rear surface reconstruction by temporal tracking of the thermal front (RSR3TF) to determine the actual geometry of the object using the temperature history and the 3D point cloud of its frontal surface [4]. The thermographic data are recorded by an infrared camera, whereas, the point cloud is captured by a 3D scanner. These two devices are calibrated and for a visual rendering; the temperature map as well as the defect map can be used to texture the 3D point cloud. This is particularly appealing for the inspection of a long pipe because the inspecting system (IR camera and 3D scanner) can be moved along the pipe and the inspection results registered in a unique coordinate system.

In section 2 we describe the 3D thermal model which forms the direct problem of the inverse geometry problem considered. This inverse problem is solved using the RSR3TF algorithm which is described in section 3. In section 4 we present an overview of the process used to calibrate the IR camera and the 3D scanner. In section 5 the results of the quantitative inspection of a pipe by IRT are presented. We end the paper with a conclusion.

2/ Three-dimensional thermal model for NDT using IRT

Figure 1 shows the experimental setup for the three-dimensional NDT using IRT. The internal domain Ω which is actually the object itself is separated from the external domain by two boundaries: the frontal boundary Ω_f which is observed by the IR camera while it is submitted to an external stimulus and the rear boundary Ω_r which is unknowns and has to be reconstructed. For an isotropic object the temperature measured on the frontal surface using the IR camera can be numerically computed by solving the heat equation (1) associated to Neumann's boundary conditions (2) and (3) on boundary Ω_f and Ω_r respectively. The initial condition is given by equation (4).

$$\frac{\partial T}{\partial t} = \alpha\left(\frac{\partial^2 T}{\partial x^2} + \frac{\partial^2 T}{\partial y^2} + \frac{\partial^2 T}{\partial z^2}\right), X \in \Omega \quad (1)$$

$$\vec{n} \cdot \lambda \nabla T = q(X,t) + h_f(T_{amb} - T) + \sigma\varepsilon_f(T_{rad}^4 - T^4), X \in \Omega_f \quad (2)$$

$$\vec{n} \cdot \lambda \nabla T = h_r(T_{amb} - T) + \sigma\varepsilon_r(T_{rad}^4 - T^4), X \in \Omega_r \quad (3)$$

$$T(X, t = 0) = T_0, X \in \Omega \cup \Omega_f \cup \Omega_r \quad (4)$$

In equations (1) to (4), $X = [x, y, z]^{tr}$ is the vector of Cartesian coordinates, $T = T(X,t)$ is the temperature at point X at time t, $\vec{n} = [n_x, n_y, n_z]^{tr}$ is the outward-pointing normal vector at point X on the boundary (superscript tr is the transpose operator), α is the thermal diffusivity of the material, h is the convective heat transfer coefficient, ε is the emissivity, λ is the thermal conductivity, T_{amb} is the ambient temperature, T_{rad} is the mean radiant temperature, T_0 is the initial temperature and $q(X,t)$ is the heat flux density resulting from the external stimulus which can be a pulsed heating, a step- heating or a periodic heating.

Due to its ease of coding and implementing, its flexibility, its small computational cost and its great capability to handle complex and changing geometries, a meshless solver is used to find the numerical solution of the 3D thermal model (1)-(4) after the discretization of the domain $\Omega \cup \Omega_f \cup \Omega_r$ into a set of N nodes $\{X_j\}_{j=1}^{j=N}$ generated using the 3D point cloud of frontal surface and the corresponding normal directions (refer to [5] for more details).

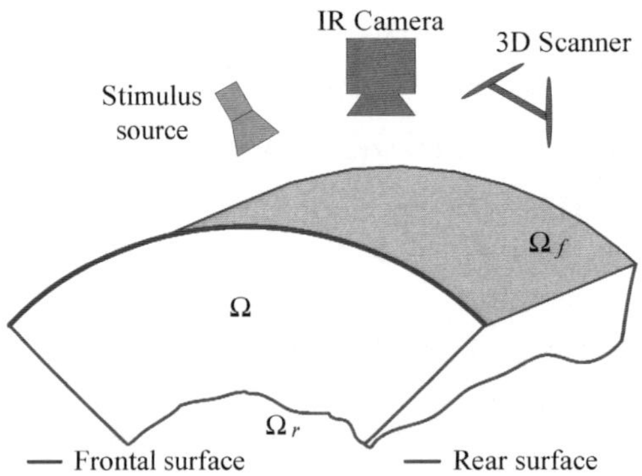

Figure 1. Setup for the three-dimensional NDT using IRT.

3/ Rear surface reconstruction by temporal tracking of the thermal front

Recently a new algorithm, the temporal tracking of the thermal front, has been proposed for defect characterization in NDT using IRT [4]. The basic idea is to consider NDT using IRT as an inverse geometry problem which consists in determining the internal geometry and the rear surface of an object only from the 3D point cloud and the time-dependent temperature of its frontal surface. To achieve this task two steps are necessary: the parametrization of the unknown rear surface and the update of internal geometry while the temporal tracking of the thermal front is performed.

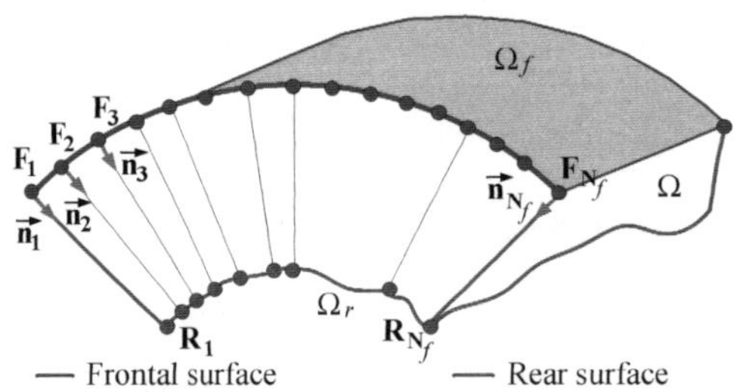

Figure 2. Parametrization of the unknown rear surface.

Figure 2 shows the parameters allowing the reconstruction of the rear surface. First of all, the frontal surface is represented by its discretized version as a set of N_f nodes $F_i, i = 1,2, \dots, N_f$ obtained after resampling its 3D point cloud with given spatial step sizes. Using the discretized version of the frontal surface, the unknown rear surface is parametrized by a set of N_f depths $d_i, i = 1,2, \dots, N_f$ such that:

$$\overrightarrow{F_i R_i} = -d_i \cdot \vec{n}_i \qquad (5)$$

In (5) $\vec{n}_i, i = 1,2,\dots,N_f$ is the outward-pointing normal vector at node F_i and $R_i, i = 1,2,\dots,N_f$ are the nodes of the rear surface.

The thermal front resulting from the external thermal stimulus has a propagation speed which depends on the thermal diffusivity of the material and a principal propagation direction which is the same as the normal direction to the frontal surface. In an isotropic object, each instant, the penetration depth of the thermal front is the same along the normal direction at each node of the frontal surface. This is summarized in Figure 3a. For each node $F_i, i = 1,2,\dots,N_f$, at time t_j, the thermal front reaches depth d_j measured from node F_i in the direction $-\vec{n}_i$. For $t < t_j$, there is no temperature change at depth d_j inside the object. The temperature at depth d_j starts to increase at time t_j. All points situated at the same depth along the normal directions are reached at the same time by the thermal front. For example, Figure 3b shows the typical temperature changes at different depths. At time t_1, the thermal front reaches the points a_1, b_1 and c_1 situated at depth d_1. Points a_2, b_2 and c_2 situated at depth $d_2 > d_1$ are reached at time $t_2 > t_1$, etc.

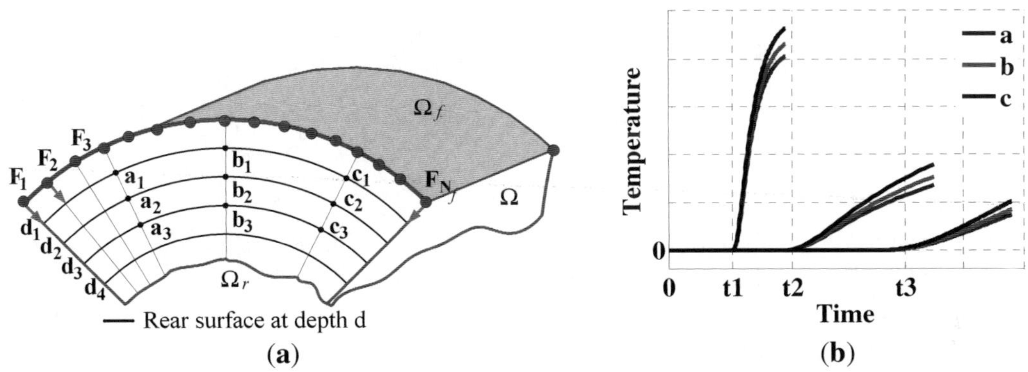

Figure 3. Propagation of the thermal front. (a) Possible positions of the rear surface. (b) Temperature history at different depths

Let us consider the generic case presented in Figure 4a. At pixel P the rear surface is situated at depth d_3. The tracking of the thermal front consists in following the thermal front while it propagates inside the object. In order to move from time to space and reversely from space to time, we use the following relation between time and penetration depth: $d = \sqrt{\pi \alpha t}$ [6,7]. In the RSR3TF algorithm we first assume that the rear surface is situated at depth d_1 and we verify that the partial error E_p, which is the root mean square of the difference between the theoretical temperature of the frontal surface provided by the meshless solver and its real temperature recorded by the IR camera for $0 < t < t_1$, is minimal. If it is the case, i.e., E_p is minimal, the size of the rear surface is determined by selecting among the pixels of the direct neighborhood of pixel P those for which E_p is also minimal. If for pixel P E_p is not minimal, the rear surface is situated at a depth greater than d_1; the depth $d_2 > d_1$ is considered and the previous steps is repeated. The typical temperature curves at pixel P for different positions of rear surface are presented in Figure

4b for a pulsed heating. The corresponding partial error is presented in Figure 4c. It takes a minimal value when the rear surface is assume to be at depth d_3.

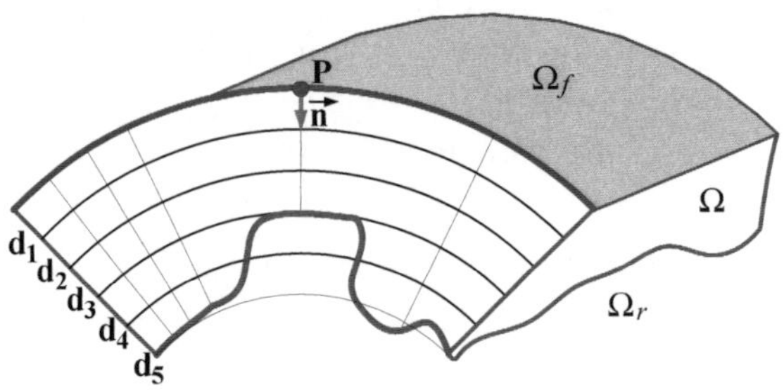

(a)

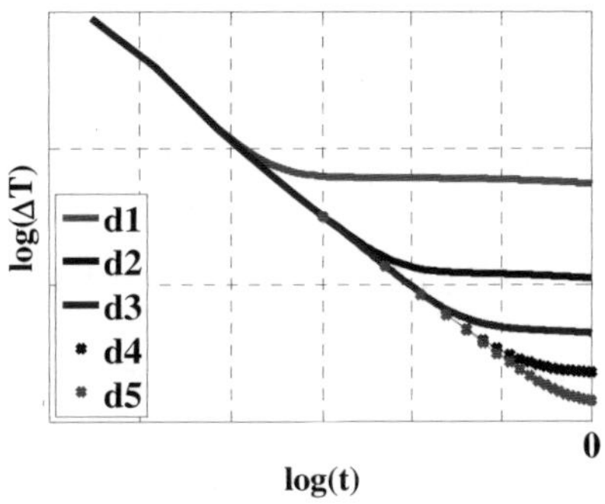

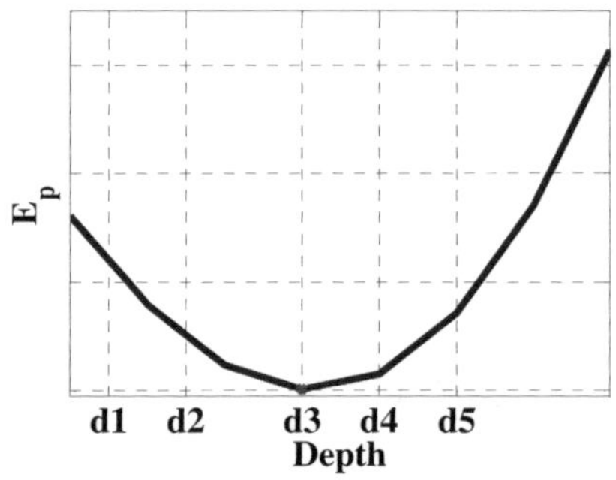

(b) (c)

Figure 4. Estimation of the depth of rear surface. (a) Temperature history at pixel P in case of pulsed thermography. (b) Partial error.

4/ Mapping IR Information on 3D data

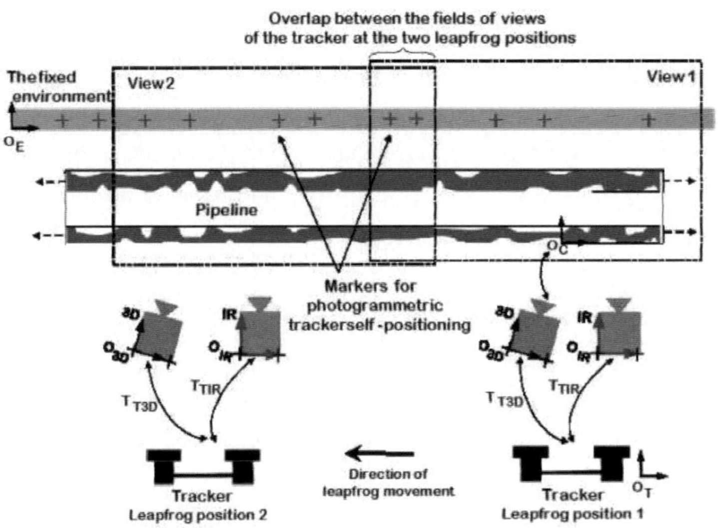

Figure 5. Acquiring IR data at leapfrog positions of the photogrammetric tracking system

Figure 5 shows the different steps of the inspection. The acquisition system (3D scanner and IR camera) is moved along the pipe and the IR images recorded at the successive positions are such that there is an overlapping between them. The procedure used to map the IR information (location, size and depth of subsurface defects) onto the 3D point cloud of the inspected object consists in defining the relationship between the coordinate systems of the sensors (3D scanner and IR sensor) after performing the intrinsic calibration of the IR camera followed by the extrinsic calibration of the two sensors.

The intrinsic parameters allow building the relationship between the 2D coordinates of a point in the IR image and the coordinates of that point in the sensor 3D coordinates system. We use a 2D planar target (Figure 6) with circular markers to accurately determine the intrinsic parameters. When the 2D planar target is placed close to a large object having a temperature greater/smaller than ambient temperature, the markers becomes visible to the IR camera as they reflect the radiation coming from that object. The coordinates of the centers of the markers in the IR image and in the camera 3D coordinates system are used to determine the intrinsic parameter of the camera [8].

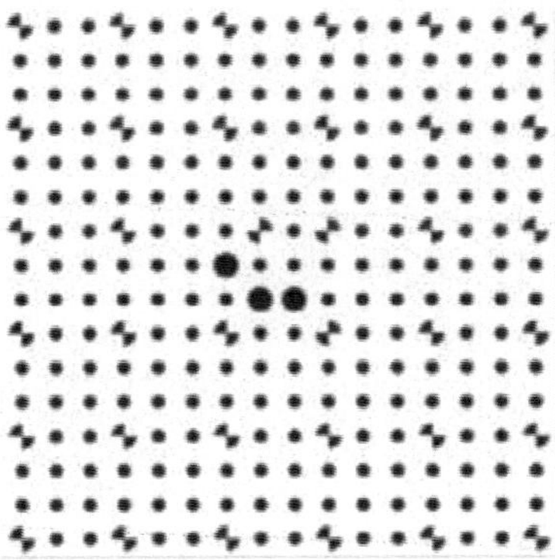

Figure 6. 2D Calibration target for finding intrinsic parameters

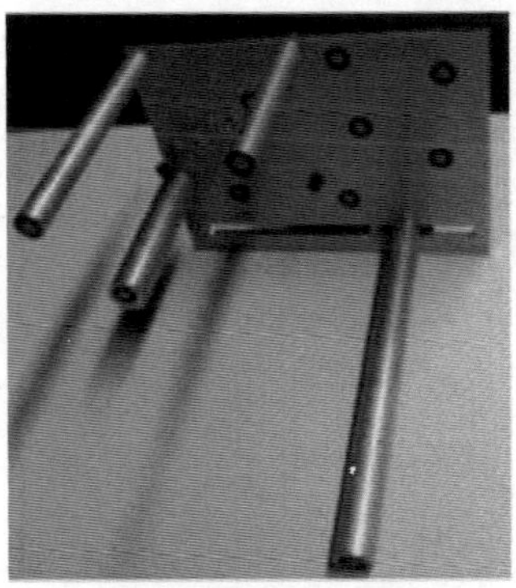

Figure 7. Non-coplanar tracking model

Figure 8. 3D calibration target for finding the calibration matrix M.

The extrinsic parameters represent the camera pose in the scene which is a rigid transformation consisting in a rotation and a translation. It is a transformation from the points in the global reference 3D coordinates system of the calibration target to the local reference 3D coordinates system of the camera. By installing retro-reflective markers on the camera that are visible from the photogrammetric tracker, we are able to track the camera and define the transformation between the camera case and the coordinate system of the tracker (see Figure 5). Since the configuration of markers (called the tracking model) has a significant effect on the accuracy of tracking and projection, we built a tracking model as presented in Figure 7 which consists of a planar plate with posts of different lengths in order to improve the accuracy of 3D tracking. However, the transformation matrix form the tracking model attached to the camera case to the camera coordinate system is still unknown. The problem of estimating this calibration matrix is called the tracker-camera calibration problem. Using a precise 3D target (Figure 8) with a known geometry in 3D space that is visible in both sensors, we can close the loop of transformation matrices

$AMB = C$, as illustrated in Figure 9. The matrix A represents the markers' transformation from tracker and is directly available from the tracker. The matrix B represents the extrinsic parameters of camera with respect to the 3D target. It is determined by detecting the white circular points at the top of the posts in a given IR image of the 3D target. The matrix C represents the 3D target's transformation from the tracker. It is determined by scanning the surface of the posts and defining a 3D coordinate system on the 3D target. The calibration matrix is then $M = A^{-1}CB^{-1}$.

Figure 9. Tracker-Camera (2D/IR) calibration. The setup to calibrate the tracker (C-Track) with the camera. The setup consists of the tracker (C-Track), the IR camera and the 3D target.

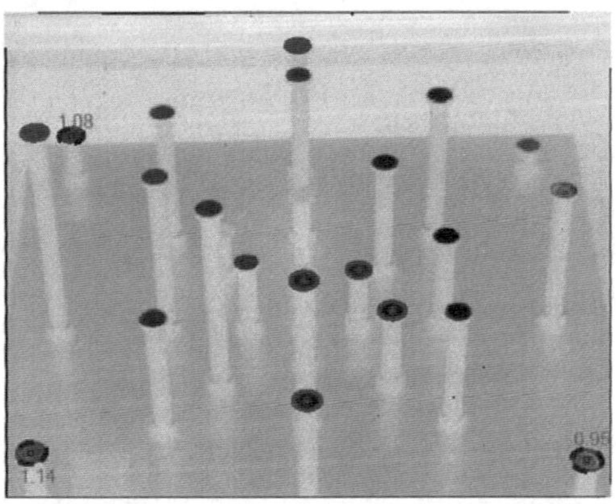

Figure 10. The reprojection error of calibration matrix M. The ellipses detected in the image are presented in red. The circles fitted to the 3D points acquired from the surface of posts are projected on the image which are illustrated in blue for these posts. The error is measured in pixels.

The accuracy of mapping the IR information on 3D point cloud is highly dependent on the accuracy of the estimation of the calibration matrix M. This accuracy can be evaluated using the reprojection error which can be smaller than $1.06pixel$ or $1.17mm$ (Figure 10). These results indicate that the calibration procedure is efficient and accurate for pipeline inspection.

5/ Experimental results

Figure 11 shows the aluminum cylindrical (external radius: $217mm$, thickness: $8mm$, Heigth: $300mm$) object inspected. A thin layer of high emissivity paint has been applied on its frontal surface for accurate temperature measurement with the IR camera (Figure 11a). Several defects have been created on its rear surface (Figure 11b); among them we have two flat-bottom defects D_{f1}(width: $13.5mm$; length: $174mm$; depth: $3mm$) and D_{f2}(width: $13.5mm$; length: $80mm$; depth: $2.5mm$), and four triangular defects D_{s1}(maximal width: $9mm$; Length: $180mm$; minimal depth: $4\ mm$), D_{s2} (maximal width : $3mm$; Length: $140mm$; minimal depth: $7.5\ mm$), D_{s3} (maximal width : $7mm$; length: $130mm$; minimal depth: $6\ mm$) and D_{s4} (maximal width: $3mm$; Length: $70mm$; minimal depth: $7mm$). The 3D point cloud of the frontal surface and that of the rear surface are presented in Figure 11c and Figure 11d respectively.

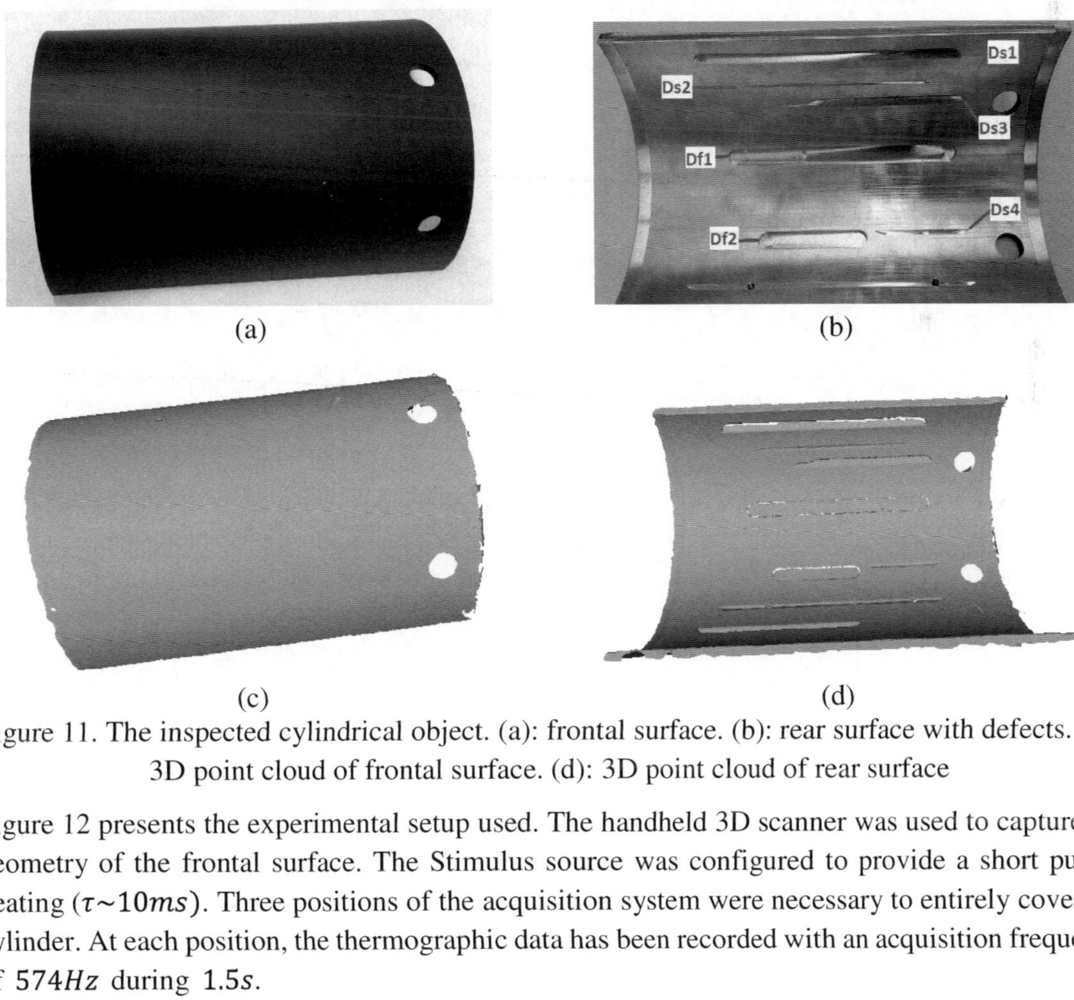

Figure 11. The inspected cylindrical object. (a): frontal surface. (b): rear surface with defects. (c): 3D point cloud of frontal surface. (d): 3D point cloud of rear surface

Figure 12 presents the experimental setup used. The handheld 3D scanner was used to capture the geometry of the frontal surface. The Stimulus source was configured to provide a short pulsed heating ($\tau \sim 10ms$). Three positions of the acquisition system were necessary to entirely cover the cylinder. At each position, the thermographic data has been recorded with an acquisition frequency of $574Hz$ during $1.5s$.

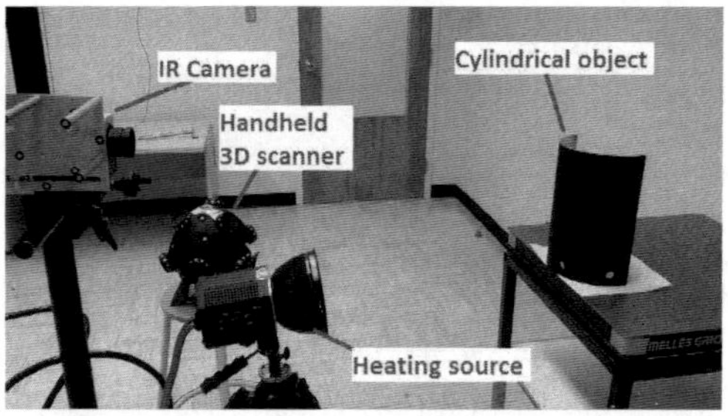

Figure 12. Experimental setup used

Figure 13 shows the results of the mapping of temperature on frontal surface.

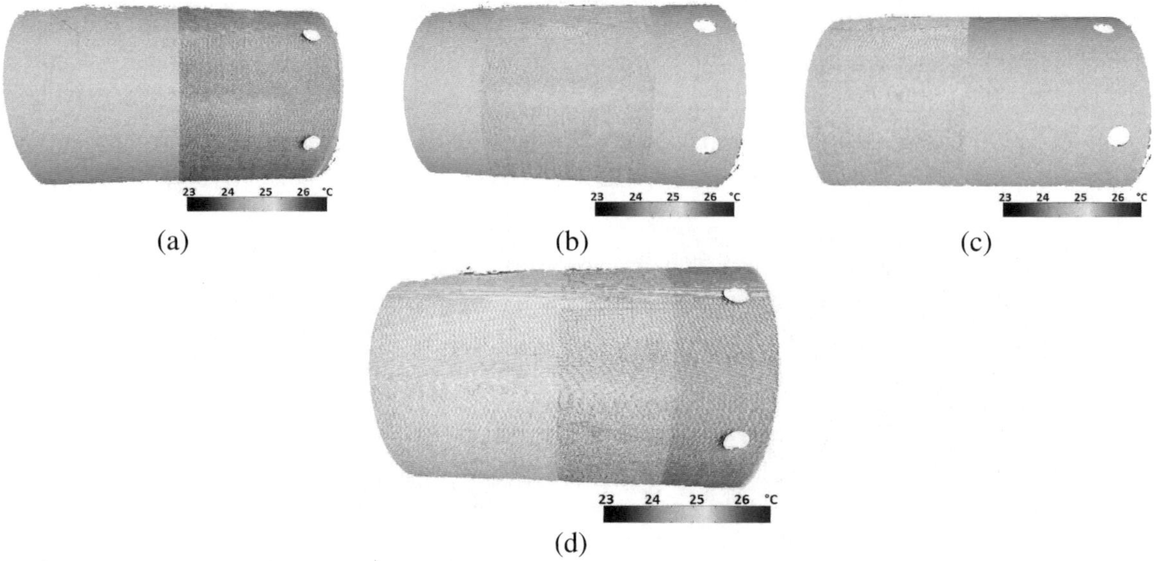

Figure 13. Results of the mapping of the temperature on the 3D point cloud of frontal surface. (a): first position of the acquisition system. (b): second position of the acquisition system. (c): third position of the acquisition system. (d): final mapping.

The measured temperature history on frontal surface is used to reconstruct the rear surface using the RSR3TF algorithm. The results of the mapping of defects on the frontal surface are presented in Figure 14. Figure 15 shows the same results but with an emphasis on the three-dimensional geometry of the reconstructed rear surface.

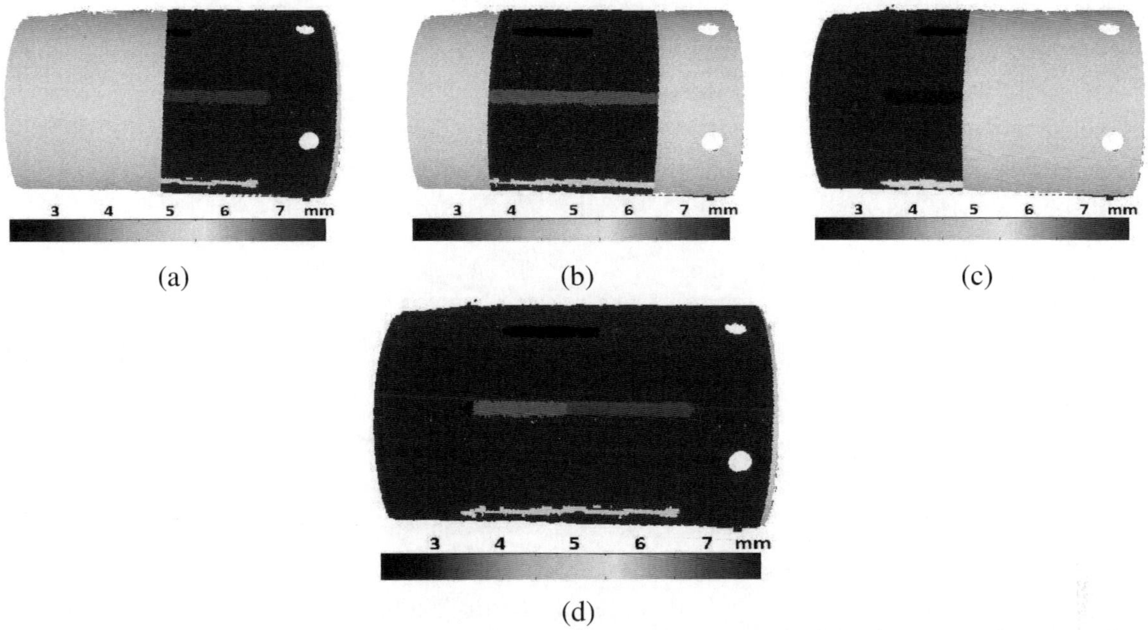

Figure 14. Results of the mapping of defects on the 3D point cloud of the frontal surface. (a): first position of the acquisition system. (b): second position of the acquisition system. (c): third position of the acquisition system. (d): final mapping.

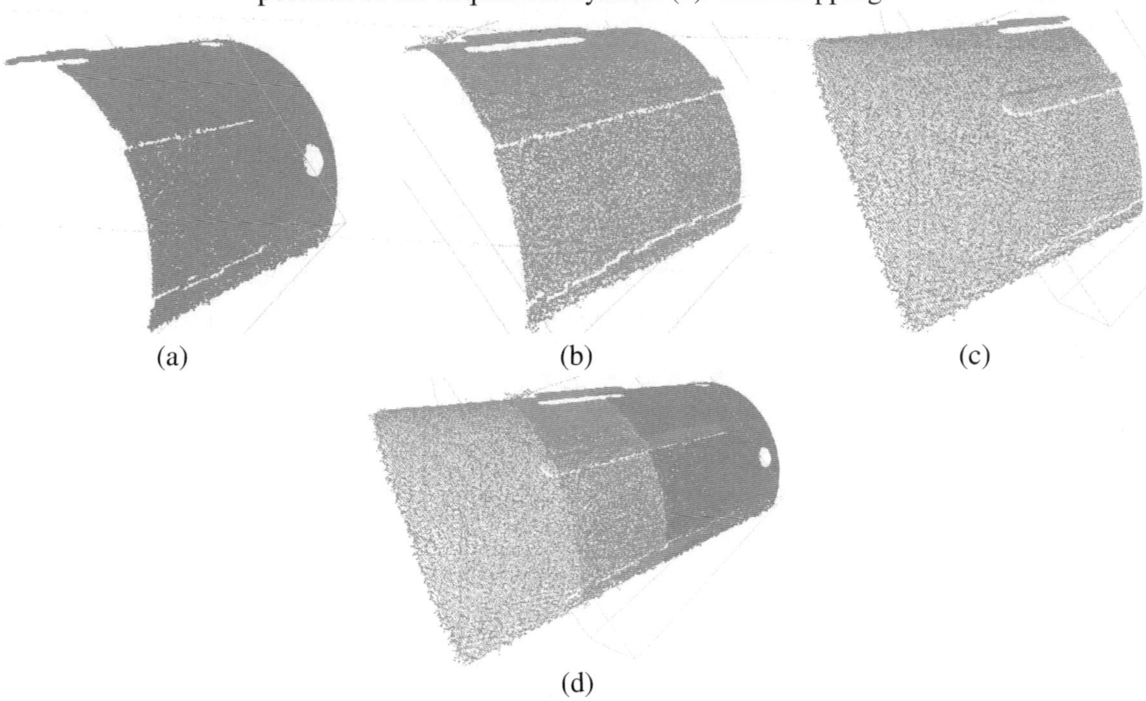

Figure 15. 3D point cloud of the reconstructed rear surface. (a): first position of the acquisition system. (b): second position of the acquisition system. (c): third position of the acquisition system. (d): final mapping.

6/ Conclusion

We have presented the results of the three-dimensional NDT of a cylindrical object (pipe) using an IR camera coupled to a 3D scanner. The results presented as the mapping of defects on the 3D point cloud of frontal surface are very interesting in terms of reconstruction of the internal geometry of the inspected object as well as for visual rendering of the inspections results.

References

1 X. Maldague, 'Theory and Practice of Infrared Technology for Non Destructive Testing', John-Wiley & Sons, 2001.

2 Y. Liu, X. Guo and G. Guo, 'A comparison between 3D and 1D numerical simulation models for infrared thermographic', NDT Proc. 10th European Conference on Non-Destructive Testing Vol 2, pp 1484-89 , june 2010.

3 F. B. Djupkep Dizeu, D. Laurendeau, A. Bendada, 'An approach for the 3D characterization of internal defects of objects by pulsed infrared thermography', Proc. 2015 ASNT annual conference, Salt Lake City, USA, pp 28-33, October 2015.

4 F. B. Djupkep Dizeu, D. Laurendeau, A. Bendada, 'Nondestructive testing of objects of complex shape using infrared thermography: rear surface reconstruction by temporal tracking of the thermal front', Inverse Problems (In press).

5F. B. Djupkep Dizeu, D. Laurendeau, A. Bendada, 'A localized radial basis function meshless method for modeling the nondestructive testing of complex shape objects using infrared thermography', Internal report, Computer Vision and Systems Laboratory, Laval University.

6 S. M. Shepard, J.R. Lhota, B. A. Rubadeux, D. Wang and T. Ahmed, 'Reconstruction and enhancement of active thermographic image sequences Optical Engineering 42 1337-42, 2003.

7 J. G. Sun, 'Analysis of pulsed thermography methods for defect depth prediction Journal of Heat Transfer' 128 329-338, 2006.

8 Z. Zhang, 'A flexible new technique for camera calibration', IEEE Trans. Anal. Mach. Intell. Vol 22, pp 1330–1334, 2000.

Application of a FMC/TFM Ultrasonic System to Inspection of Austenitic Welds

Raymond TEN GROTENHUIS [1], Andrew CHEN [2], Andrew HONG [1], Yadav VERMA [1]
[1] Ontario Power Generation, Toronto, Canada
[2] Kinectrics Inc., Toronto, Canada
Contact e-mail: ray.tengrotenhuis@ opg.com

Abstract. Ontario Power Generation (OPG) has successfully developed and deployed an ultrasonic inspection system based upon the combination of the Full Matrix Capture (FMC) UT acquisition technique with the Total Focusing Method (TFM) beamformer. The system, called Matrix Inspection Technique or MIT, was applied to selected carbon steel components that comprise the Primary Heat Transport System. The purpose of the inspection is the mapping of highly localized Flow Accelerated Corrosion (FAC) occurring at and adjacent the weld root of fitting to fitting welds. The MIT system has yielded outstanding results in multiple field inspection campaigns from 2010 to present. Results from these inspection campaigns also identified welds containing minor welding defects such as; lack of fusion, isolated porosity and slag inclusions. A programme was initiated to advance the capabilities of the MIT system beyond that of carbon steel to stainless steel components. The required changes to the MIT system include a relatively straight forward substitution of the transducer and a significant modification to the TFM beamformer strategy to address the anisotropic nature of the weld material. This paper describes the work conducted in the modification of the MIT beamformer and the results obtained on sections of 304 stainless steel pipe welded with 308L fill material.

1. Introduction

Effective inspection of anisotropic inhomogeneous materials is often a challenge due to the deleterious effects of the material properties on the bulk wave transmission of ultrasound. Reliable inspection capability for these materials must provide high probability of detection whilst minimizing false calls. Upon detection of an indication, the sizing and plotting of the indication within the search volume must be sufficiently accurate to support disposition in accordance with applicable codes. Common metallic members of this group of materials include austenitic stainless steel welds, coarse grain cast stainless steels and dissimilar metal welds. The task of development of inspection capability for this category of materials usually includes many if not all of the following; modelling, simulation, fabrication of mock-ups and samples containing realistic defects, procedure development, validation and subsequent training. Thus development of inspection capacity represents substantial commitments of time, personnel, capital, and other resources.

The FMC data acquisition technique combined with the TFM beamformer offer a new approach to the inspection of anisotropic inhomogeneous materials. The chief advantage of the FMC technique is that for a given transducer placed at a position and orientation with respect to the search volume, the maximum amount of information is obtained [7]. One advantage of the TFM beamformer is given a suitable model of the search volume, both material anisotropy and inhomogeneity may be addressed.

OPG has developed an inspection system based on FMC data acquisition operated in conjunction with a two stage TFM beamforming strategy, see reference [1] and [2]. This combination is referred to as the Matrix Inspection Technique or MIT. The system was originally developed to address localized FAC at and adjacent weld roots of carbon steel fitting to fitting welds. An example of a weld region profile is found in Figure 1a, a 3D reconstruction of an inspected fitting in Figure 1b. Encouraged by the MIT results, further development was initiated to extend the system capabilities to the inspection of austenitic weld material.

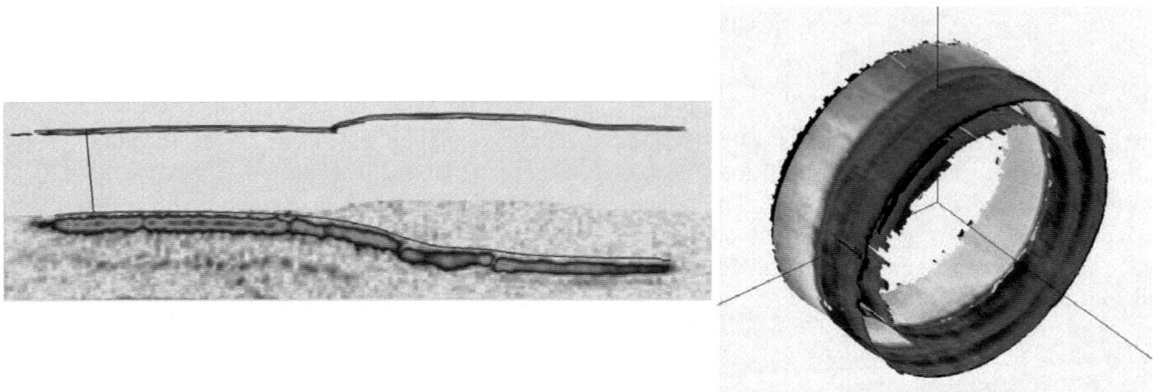

Figure 1 a) left - typical profile of a fitting to fitting weld. b) right - 3D reconstruction of a fitting to fitting joint aggregated from multiple profiles.

2. Proposed Solution

The solution adopted in this work follows the convention of dividing the problem into two separate issues; one of the inhomogeneous material, the other of the anisotropic characteristic. Widely accepted approaches to the problem of the inhomogeneous characteristic of austenitic welds is addressed either via numerical solution for the grain orientation, or by partitioning the volume into regions that may be approximated as homogeneous. An extensive body of work has been published regarding numerical modelling of austenitic weld structures, see references [3], [4] and [5]. Models exist for configurations including Vee and double Vee welds. Corresponding models for U groove welds have not been developed, in part due to the variability in the design of the weld preparation. In the case of U groove weld, derivation from macrograph is the preferred method for determining the orientation of the face centred cubic grains.

A macrograph of the subject weld was not available from the manufacturer. In lieu, a macrograph of a similar configuration was obtained from published literature found in [6], see Figure 2a. Separate regions in the macrograph were evaluated with respect to grain orientation. Regions were differentiated based upon criteria of an angle range of +/- 4 degrees defining a unique region. Epitaxial grain growth is observed from the fusion line of the weld preparation. Grain growth then turns towards the vertical direction following the path of heat transfer during pool solidification. Symmetry about the weld axis was introduced to simplify coding of the region, see Figure 2b.

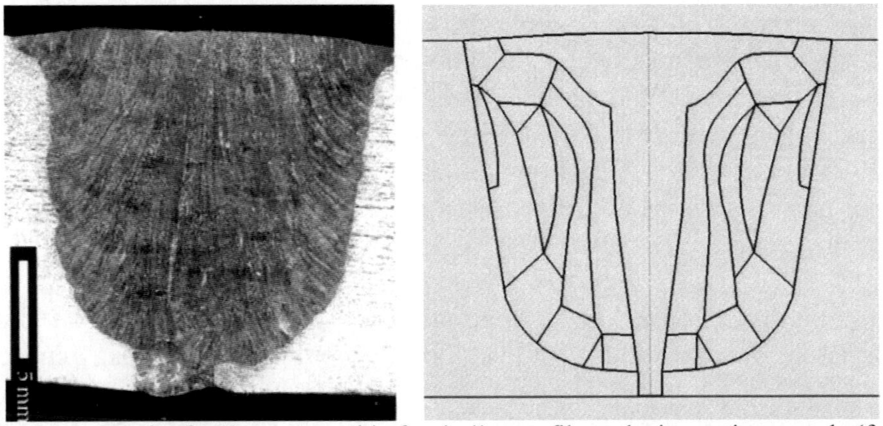

Figure 2 a) left - Macrograph of a U groove weld of a similar profile to the inspection sample (from Wirdelius)
b) right - corresponding weld model segregated into regions of homogeneous grain orientations

The anisotropic characteristics of the austenitic weld material are addressed by solving for the directionally dependant velocities of the group and phase components of the wave. While it is beyond the scope of this paper to provide the complete derivation, the velocities of the applicable wave modes with respect to direction may be obtained by solving the Christoffel equation for the corresponding eigenvalues and eigenvectors. The slowness surfaces for the quasi longitudinal, quasi transverse wave modes can be further derived. A plot of the phase slowness surface is found in Figure 3.

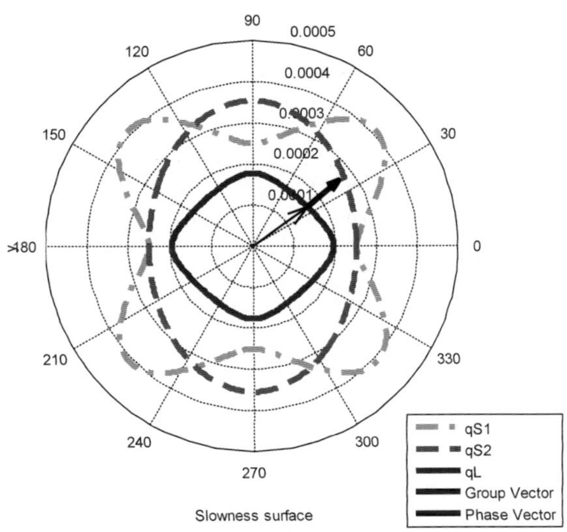

Figure 3 Slowness plots generated with elastic constants corresponding to 308L weld material. Quasi-longitudinal wave in blue, quasi-shear wave in red and green.

Within the anisotropic material the phase velocity and group velocity will nominally differ in the direction of the group vector. Recall that when performing signal summation with the TFM beamformer, an analytic representation of the original waveform is employed. The analytical waveform provides greater sensitivity to phase information compared to the time domain waveform. With TFM it is implied that the phase of the analytic waveform remains coherent with the group at all times within the reconstructed volume. In contrast, anisotropic materials with the difference between the phase and group velocity will exhibit a lack of coherence between the phase and group. The TFM beamformer will perform poorly under these conditions. It is necessary to calculate the displacement of the phase relative to the group for each point within the reconstruction volume. The net phase displacement for a given transmit-receive pair can then be compensated such that the phase is returned to a coherent state relative to the group.

Assumptions were applied in order to simplify the model, aid in code integration into the existing structure, and to reduce unduly long computation intervals. These assumptions included:

- Weld material properties are isotropic in the direction along the axis of the weld.
- Imaging functions are performed in L wave mode thus both S_H, S_V waves and mode conversion may be neglected.
- The properties of the parent material are isotropic.
- Weld material modelling is adequately addressed by defining 27 separate regions of uniform orientation with anisotropic properties.
- Solutions resulting in evanescent wave modes may be neglected.
- Elastic constants used to derive the slowness surfaces are appropriate for the weld fill material.

Portions of the existing code related to the imaging and edge detection of the exterior surface were retained without modification. Code segments were added to the interior imaging code. These segments address mapping of the grain orientation model into the interior search region [8], calculation of the Fermat solution for flight time into or transiting through the anisotropic regions and calculation of the phase displacement.

The adaptations to the existing MIT analysis code are summarized as follows:

- Extract corner features from detected exterior surface - define coordinates of weld toe
- Register model to detected corners.
- Map the weld model to interior volume to match weld area thickness as determined via cross correlation.

For every element in the array:

- Solve for flight times to isotropic regions excluding transit through anisotropic regions.
- Solve for flight times into anisotropic regions from anisotropic/isotropic boundary points applying group velocity vector.
- Calculate phase displacement for each point in anisotropic regions using phase velocity in the direction of the group vector.
- Solve for flight times in isotropic regions transiting through anisotropic regions.

The code modification required to address the anisotropic region departs from the conventional approach used in the TFM beamformer. The transit time for incoming ray is solved in sequential steps. Thus for the anisotropic region the process is akin to the methods used for ray tracing.

3. Code Evaluation Trials

A series of trials were conducted to evaluate the imaging capabilities of the modified analysis code. Improvement in the imaging of the weld discontinuities is taken as an indication of the validity of the modified analysis code.

3.1 Sample, Instrumentation and Test Apparatus

The inspection sample is a 6 inch NPS schedule 80 304 SS pipe segment containing a U groove weld, see Figure 4. The weld was fabricated using 308L fill material. The weld crown has been ground flat, but is not flush with the parent material. The manufacturer performed dimensional checks pre and post fabrication. Significant shrinkage of the sample occurred during fabrication resulting in a change in weld profile. A total of five manufactured defects have been inserted into the sample, three of which are appropriate targets for the goal of the test. The manufactured defects are; side wall lack of fusion, slag inclusion and root crack, see Figure 5.

A Peak NDT, Micropulse FMC operated with OPG's Neovision software was used for data acquisition. A five axis scanning apparatus was used for the tests, see Figure 4. The apparatus has a configuration of 3 linear axis; X, Y and Z with an independent rotary axis, theta. The fifth axis is typically used for transducer index and was not required for this application. Scanning speeds were set to 1 degree per sec with an acquisition index of 0.67 degrees per frame. Acquisition repetition frequency of 3 Hz is sufficient to minimize motion blur in the FMC frame.

Figure 4 Inspection sample mounted in scanning apparatus - left. 6 inch NPS schedule 80 sample - right.

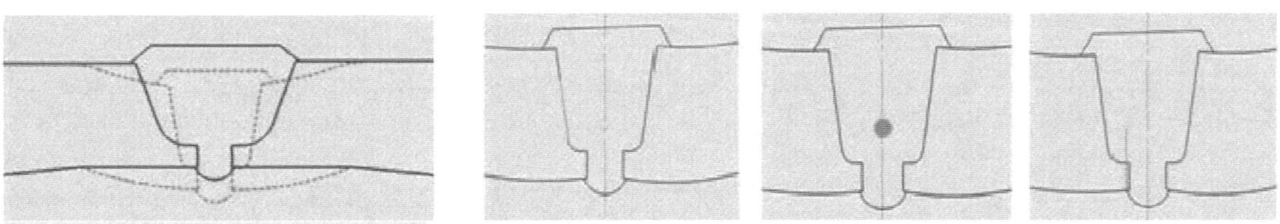

Figure 5 Sketch of the weld sample; solid - as designed, dashed - as manufactured - left. Lack of fusion, slag inclusion and root crack implanted into sample - right.

3.2 Transducers

Three transducers were utilized for this work. All are 128 element linear arrays. Properties of each transducer are provided in Table 1. The 3, 5 and 7.5 MHz transducer along with the waveform characteristics are found in Figure 6, Figure 7 and Figure 8 respectively.

Table 1 Characteristics of transducers used for application development.

Transducer	f_{NOM} (MHz)	f_{ACT} (MHz)	Pitch (mm)	Elevation (mm)	Pulse Duration	Bandwidth
A	3.0	2.9	0.47	8.0	1245 ns	42%
B	5	5.6	0.30	5.0	445 ns	75%
C	7.5	7.7	0.27	5.0	320 ns	71%

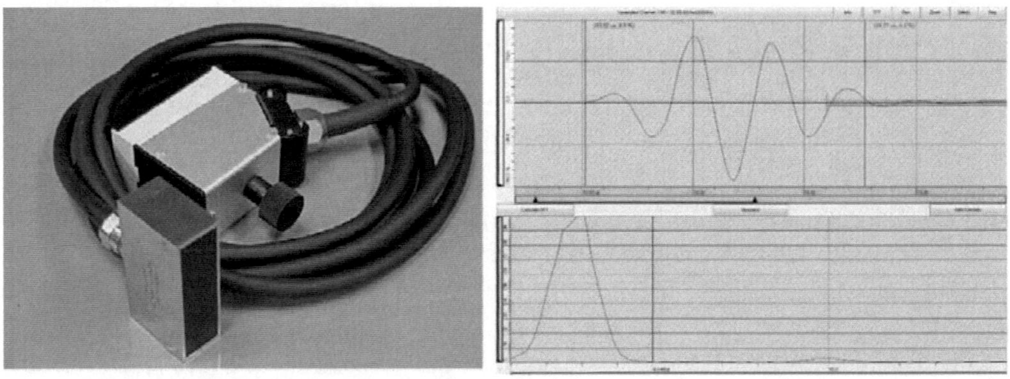

Figure 6 a) left - 3 MHz, 128 element transducer, b) right - waveform and frequency spectra plots.

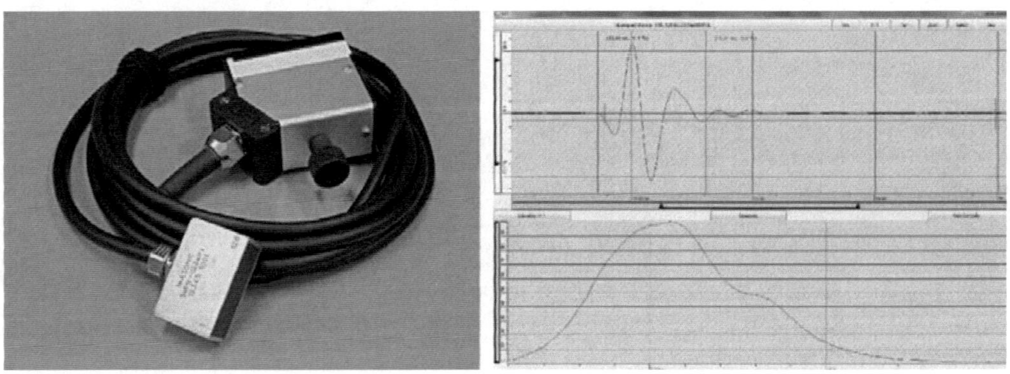

Figure 7 a) left - 5 MHz, 128 element transducer, b) right - waveform and frequency spectra plots.

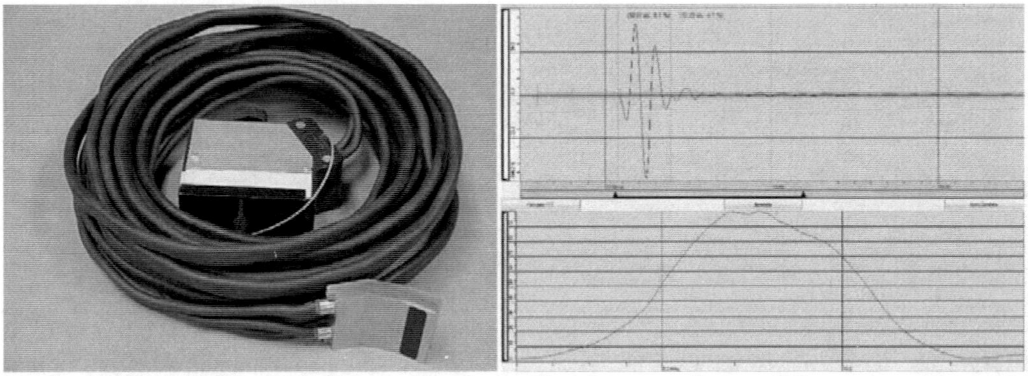

Figure 8 a) left - 7.5 MHz, 128 element transducer, b) right - waveform and frequency spectra plots.

4. Results

The results of the application of the model on scans conducted with the 3, 5 and 7.5 MHz transducers are presented in Figure 9, Figure 10 and Figure 11 respectively. In each figure, the upper row of images are results obtained assuming homogeneous, isotropic material properties while the lower row are the results obtained with the application of the model.

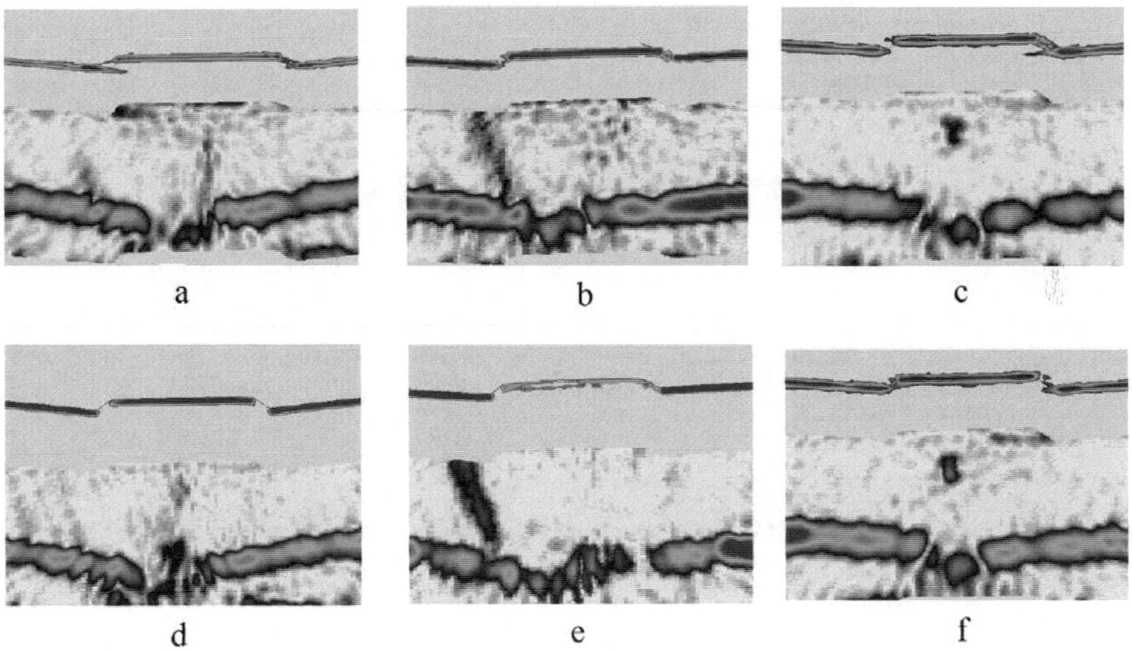

Figure 9 Result obtained with the 3 MHz transducer. Upper row - without code modification, lower row with code modification; a) and d) root crack, b) and e) lack of fusion, c) and f) slag inclusion.

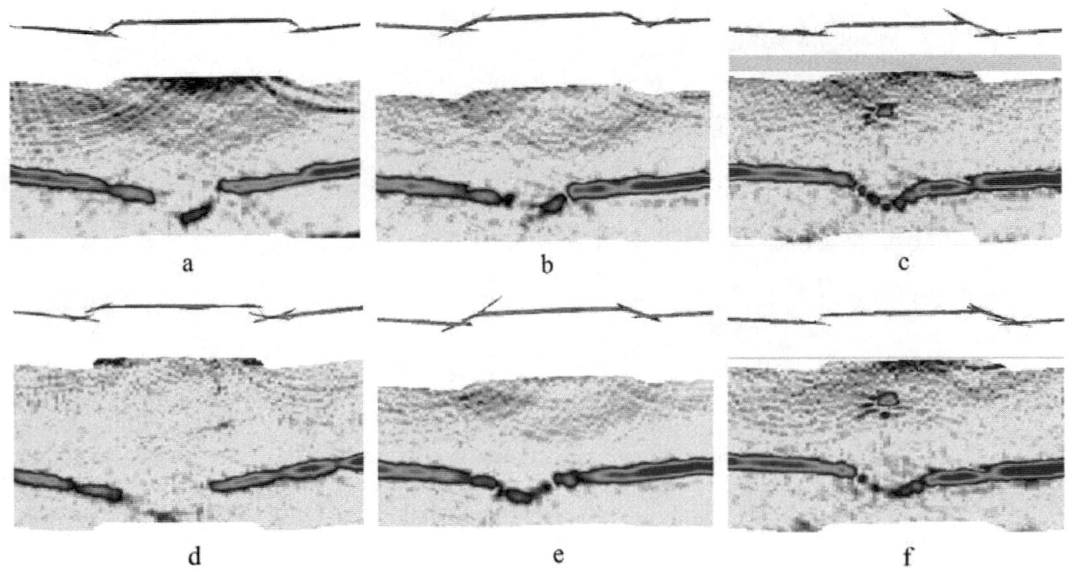

Figure 10 Results obtained with the 5 MHz transducer, upper row without code modification, lower row with code modification; a) and d) root crack, b) and e) lack of fusion, c) and f) slag inclusion. Minimal imaging improvement obtained.

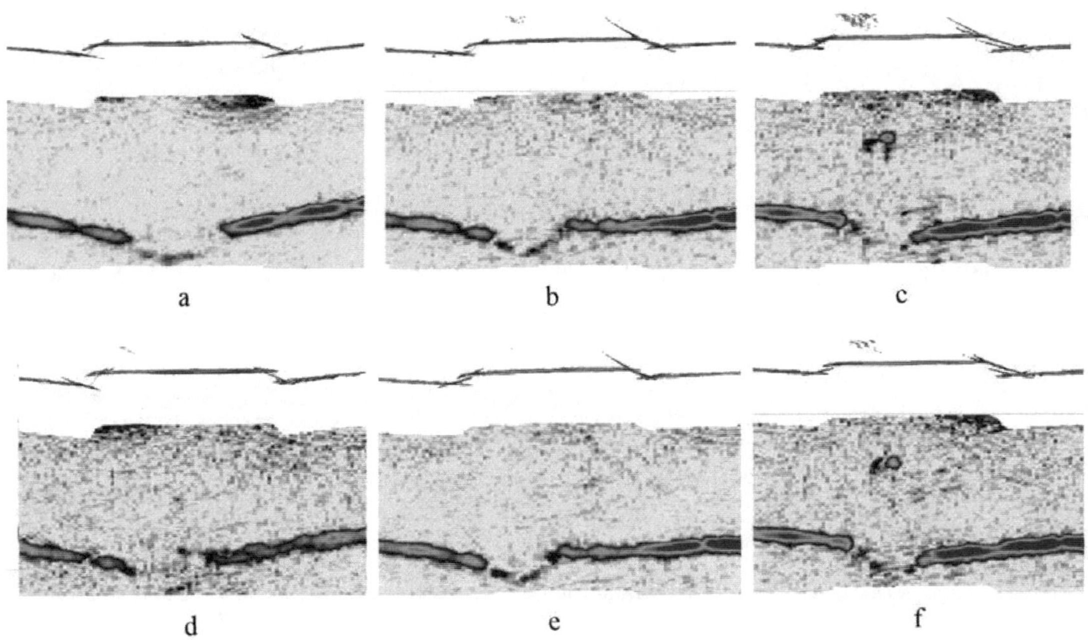

Figure 11 Results obtained with the 7.5 MHz transducer, upper row without code modification, lower row with code modification; a) and d) root crack, b) and e) lack of fusion, c) and f) slag inclusion. No appreciable imaging improvement obtained.

5. Discussion

The model is observed to improve the imaging of the implanted defects with data obtained using the 3MHz transducer. In contrast however, the model has a marginal influence on the scans obtained with the 5 MHz transducer and effectively no improvement to the results obtained with the 7.5 MHz transducer. It is speculated the difference in performance related to the relative sensitivity of the transducer frequency with respect to phase error. Additionally, the 7.5 MHz transducer may be inappropriate for this application given the scattering that would occur upon interaction with the coarse grains of the weld fill material.

The process of implanting controlled weld flaws in a sample involves some degree of deviation from the conventional weld procedure depending upon the nature of the defect and the means by which it is introduced. The deviation may be significant enough to invalidate the proposed model of the grain orientation within the weld. In this event, an adaptive approach to the weld model may be appropriate [9]. Further assessment of the modified code is required to determine if the error is a result of deviation between the grain orientation model and the actual sample or incorrect elastic constants used in the determination of the slowness surfaces

6. Conclusions

A weld model has been developed to separately address the inhomogeneous nature and anisotropic characteristics of the 308L weld material. The model has improved the imaging performance of the 3 MHz array for the three defects implanted in the sample, however the imaging performance of the 5 and 7.5 MHz array did not appreciably improve. Further work is needed to determine which specific aspects of the model depart from the characteristics of the weld sample.

7. Acknowledgements

The authors would like to thank EPRI for their kind assistance in the preparation of this work, the loan of the weld sample, and the insights provided by Dr. G. Connolly of EPRI.

8. References

[1] ten Grotenhuis R., Hong A., and Sakuta A., "Inspection of complex geometries using the full matrix capture (FMC) method", *9th Int Conf NDE,* Seattle WA, USA, (2012).

[2] ten Grotenhuis R., Hong A., "Imaging the weld volume via the total focus method", *4th International; CANDU ISI Workshop and NDT in Canada Conference*, Toronto (2012)

[3] Ogilvy J.A., "Ultrasonic beam profiles and beam propagation in an austenitic weld using a theoretical ray tracing model", *Ultrasonics* 24(6): 337-347 (1986)

[4] Ogilvy J.A., "Computerized ultrasonic ray tracing in austenitic steel', *NDT International* 18(2): 67-77 (1985)

[5] Connolly, G. D., "Modelling of the propagation of ultrasound through austenitic welds', Ph.D. Thesis Imperial College London (2009).

[6] Wirdelius H., Perrson, G., Hamberg, K., Hogberg, K., "Experimental validation of a software that models ultrasonic wave propagation through an anisotropic weld" *ULIAS Task 4, Swedish Nuclear Power Inspectorate (SKI)* 2008:05

[7] Holmes C, Drinkwater B and Wilcox P, "Post-processing of the full matrix of ultrasonic transmit-receive array data for non-destructive evaluation", NDT & E Int, 2005 38(8), 701-711.

[8] Althof R.J., Wind M.G.J., Dobbins J.T., "A rapid and automatic image registration algorithm with subpixel accuracy", *IEEE Transactions on Medical Imaging* 17 (1998) 957-966

[9] Fan, Z. and Lowe, M.J.S., 'Characterization of inhomogeneous and anisotropic steel welds by ultrasonic array measurements', *Progress in Quantitative Nondestructive Evaluation Proc.* pp.849-856, (2013)

NDT in Canada 2016 Conference

November 15 – 17, Burlington, ON, Canada

Non-Destructive Measurement of Residual Stresses by Ultrasonic Method

Yuri Kudryavtsev and Jacob Kleiman

Structural Integrity Technologies Inc.
Markham, ON, Canada
www.sintec.ca

OUTLINE OF THE PRESENTATION

- INTRODUCTION
  - Importance of Residual Stresses
  - Residual Stress Management
- MEASUREMENT OF RESIDUAL STRESSES BY ULTRASONIC METHOD (UM)
  - Ultrasonic Method of Residual Stress Measurement
  - Ultrasonic Equipment and Software for Residual Stress Measurement
  - Results of Residual Stress Measurements in Welded Elements of Structures
- SUMMARY

Importance of Residual Stresses

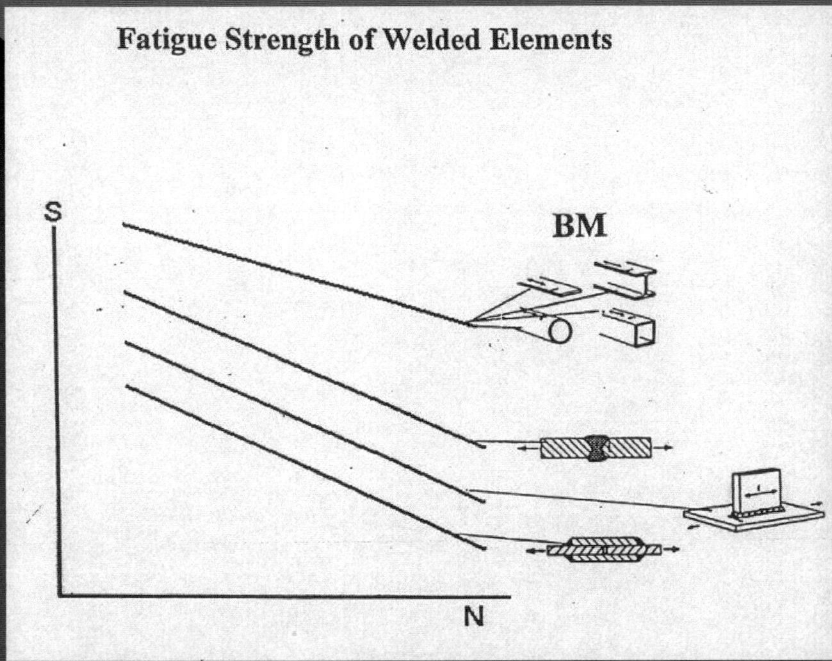

Effect of Residual Stresses

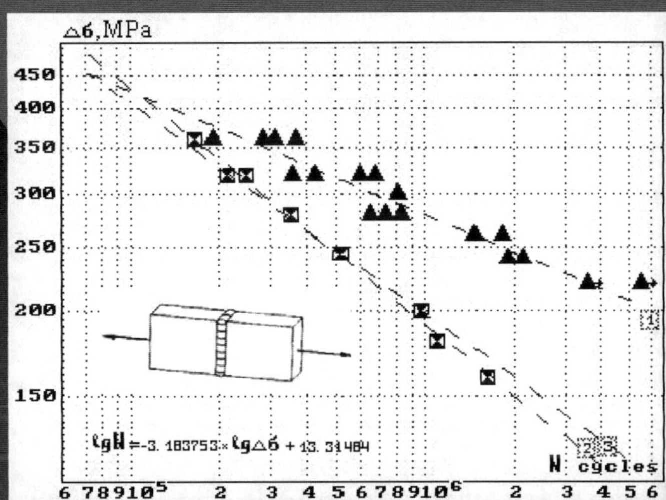

Fatigue curves of butt welded joint in low-carbon steel:
1 - without residual stresses;
2,3 - with high tensile residual stresses (fatigue testing and computation)

Effect of Residual Stresses

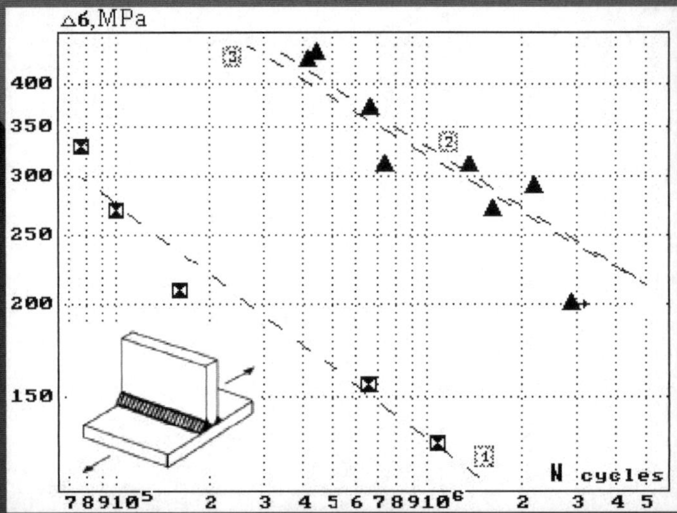

Fatigue curves of non-load caring fillet welded joint in high strength steel:
1 -in as welded condition;
2,3- after application of Ultrasonic Peening (fatigue testing and computation)

SINTEC

RESIDUAL STRESS MANAGEMENT

is an important component of the process of

DESIGN, MANUFACTURING, MAINTENANCE and REPAIR

of WELDED ELEMENTS and STRUCTURES

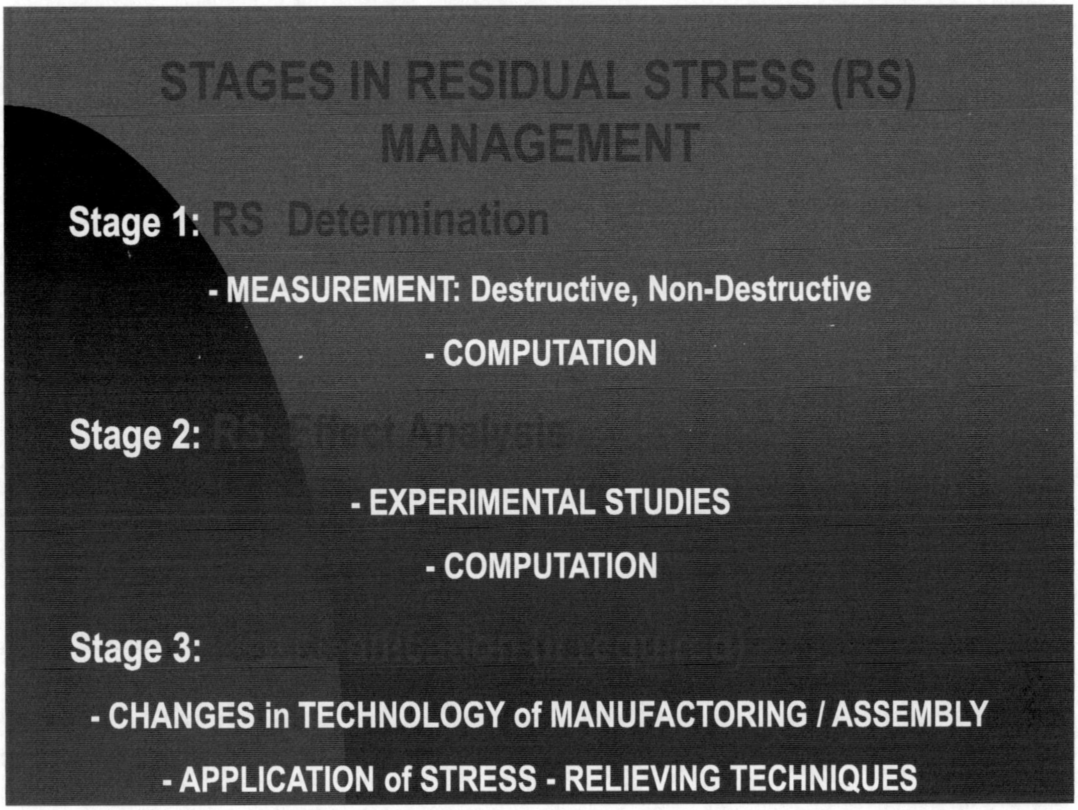
STAGES IN RESIDUAL STRESS (RS) MANAGEMENT
Stage 1: RS Determination
- MEASUREMENT: Destructive, Non-Destructive
- COMPUTATION
Stage 2: RS Effect Analysis
- EXPERIMENTAL STUDIES
- COMPUTATION
Stage 3:
- CHANGES in TECHNOLOGY of MANUFACTORING / ASSEMBLY
- APPLICATION of STRESS - RELIEVING TECHNIQUES

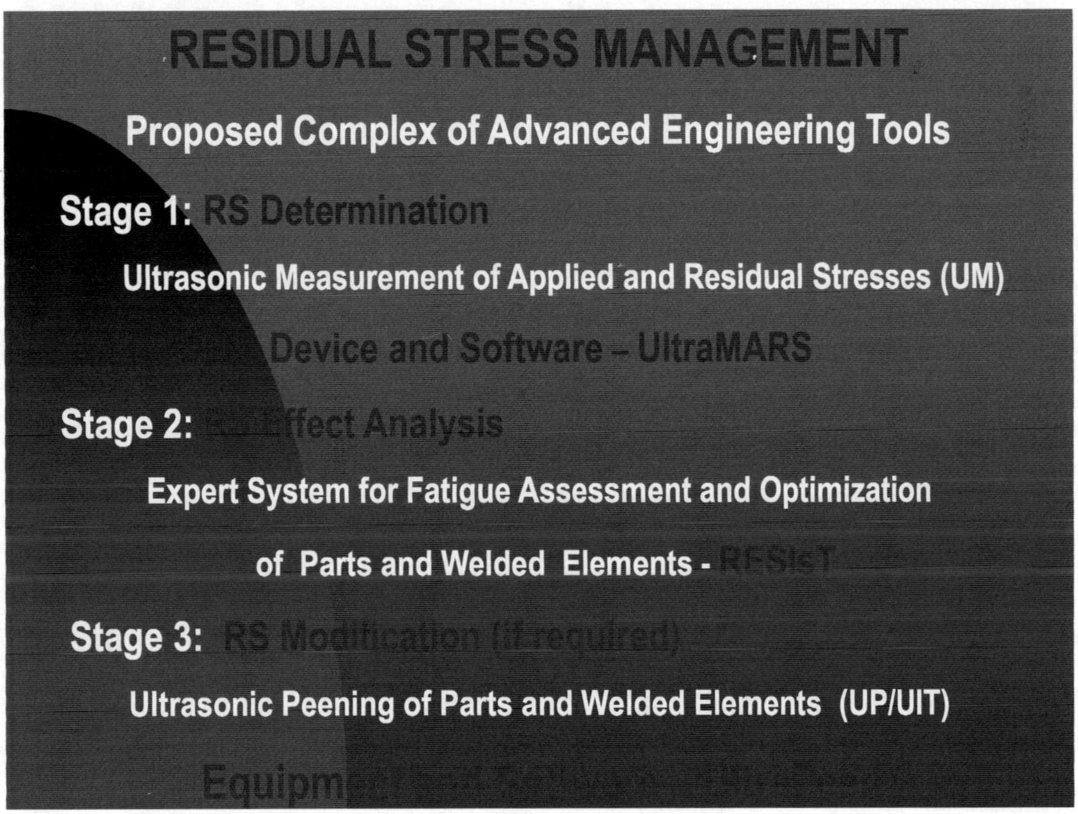
RESIDUAL STRESS MANAGEMENT
Proposed Complex of Advanced Engineering Tools
Stage 1: RS Determination
Ultrasonic Measurement of Applied and Residual Stresses (UM)
Device and Software – UltraMARS
Stage 2: Effect Analysis
Expert System for Fatigue Assessment and Optimization
of Parts and Welded Elements - RESIsT
Stage 3: RS Modification (if required)
Ultrasonic Peening of Parts and Welded Elements (UP/UIT)

RESIDUAL STRESS MANAGEMENT

Stage 1: RS Determination

Ultrasonic Measurement of Applied and Residual Stresses

Computerized complex - UltraMARS

Mark I

Mark II

Mark III

Measurement Unit

- Measurement Unit
- Transducers
- Transducer Holders
  - magnetic,
  - electromagnetic
- Supporting Software

Laptop

- Expert System for Residual Stress Analysis. Advanced database and analysis of the influence of residual stresses on the fatigue life of parts and welded elements

RESIDUAL STRESS MANAGEMENT

Stage 2: RS Effect Analysis

Expert System for Fatigue Assessment of Welded Elements

The RESIsT is an Expert System for optimization of welded elements and structures by fatigue criterion. The following important parameters of welded structures with the goal to enhance the fatigue performance are analyzed:

- preferred design of parts and welded elements;
- material selection;
- welding and other technological processes;
- residual stresses;
- application of improvement treatments
- influence of possible repair technologies;
- realistic service conditions.

RESIDUAL STRESS MANAGEMENT

Residual Stress in Fatigue Analysis

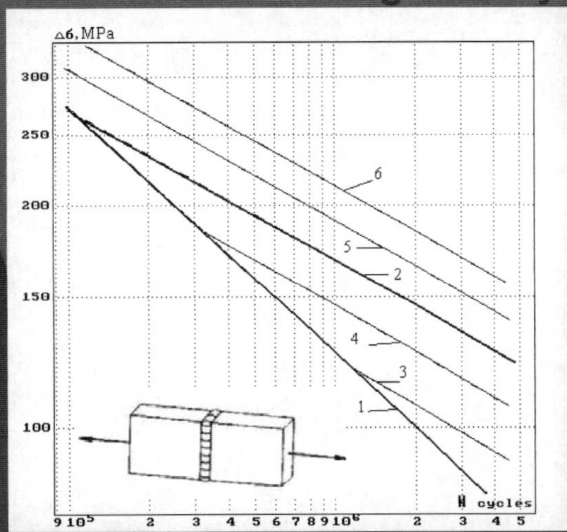

Fatigue curves of transverse loaded butt weld at R=0:

1 - with high tensile residual stresses;

2, 3, 4, 5 and 6 - with residual stresses

equals to 0 MPa, 200 MPa, 100 MPa, -100 MPa and -200 MPa

RESIDUAL STRESS MANAGEMENT

Stage 3: RS Modification

Ultrasonic Peening of Parts and Welded Elements UltraPeen

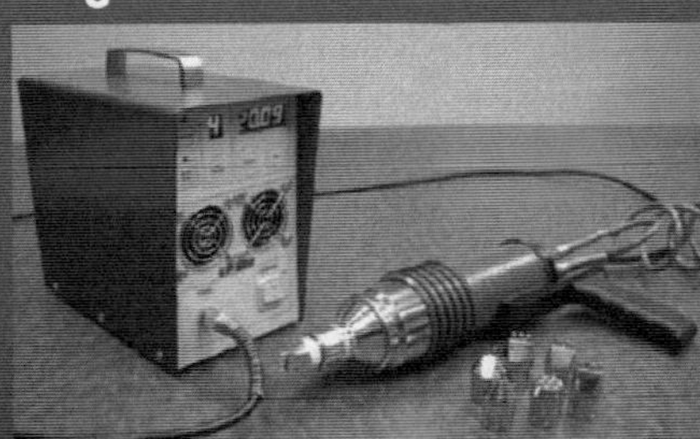

Ultrasonic transducer and Generator

- Piezoelectric transducer with the set of heads for different applications: weight– 2 kg, length – 420 mm
- Ultrasonic generator: weight – 4.5 kg dimensions – 350x250x90 mm
- Output frequency – 20 kHz
- Power supply – 250 watts

Software

- Optimization of the UP application parameters: maximum possible increase in fatigue life of parts and welded elements with minimum cost, labor- and power-consumption
- Quality monitoring of UP process
- Fatigue assessment of welded elements after UP

Ultrasonic Computerized Complex for Measurement of Residual and Applied Stresses UltraMARS-7

Ultrasonic Method of Residual Stress Measurement

- Ultrasonic stress measurement techniques are based on the acoustic-elasticity effect, according to which the velocity of elastic wave propagation in solids is dependent on the mechanical stress

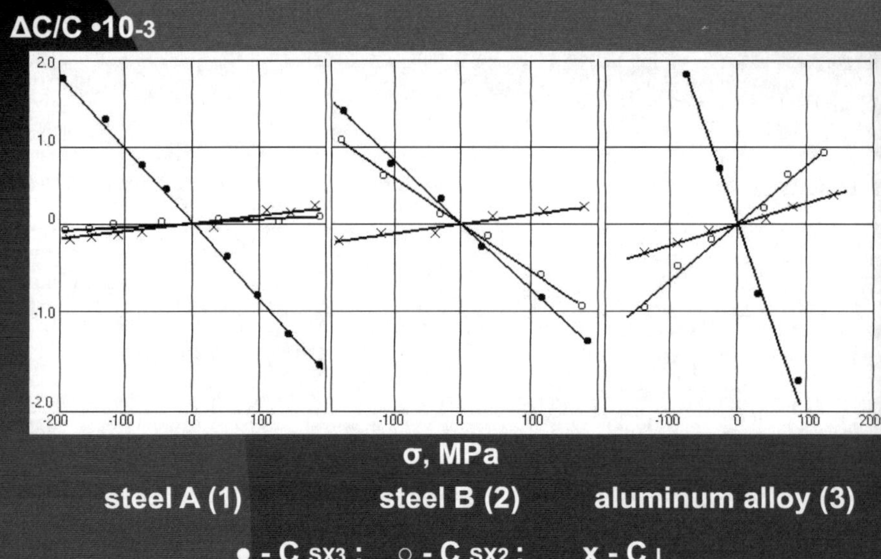

Acoustic-Elastic Relationships

$$C_{lx1}^2 = C_{lx10}^2 + C_{lx1o}^2 \left[2\frac{(a+b)\sigma_{kk}^2}{3K_0(<\lambda>+2<\mu>)} + \left(\frac{2(<\lambda.+<\mu>)\sigma_{11}^0}{<\mu>3K_0(<\lambda>+2<\mu>)} - \frac{<\lambda>(\sigma_{22}^0+\sigma_{33}^0)}{3K_0(<\lambda>+2<\mu>)} \right)(2b+c+4<\mu>+\frac{\sigma_{11}^0}{<\lambda>+2<\mu>} \right]$$

$$C_{Sx2}^2 = C_{Sx20}^2 + C_{Sx20}^2 \left\{ b\frac{\sigma_{kk}^0}{3K_0<\mu>} + \left[\frac{<\lambda>+2<\mu>}{3K_0<\mu>}(\sigma_{11}^0+\sigma_{22}^0) - \frac{2<\lambda>}{3K_0<\mu>}\sigma_{33}^0 \right]\left(1+\frac{c}{4<\mu>}\right) + \sigma_{11}^0 \right\}$$

$$C_{Sx3}^2 = C_{Sx30}^2 + C_{Sx30}^2 \left\{ b\frac{\sigma_{kk}^0}{3K_0<\mu>} + \left[\frac{<\lambda>+2<\mu>}{3K_0<\mu>}(\sigma_{11}^0+\sigma_{33}^0) - \frac{2<\mu>}{3K_0<\mu>}\sigma_{22}^0 \right]\left(1+\frac{c}{4<\mu>}\right) + \sigma_{11}^0 \right\}$$

- C_{lx1} – is the longitudinal wave propagation velocity,
- C_{sx2} and C_{sx3} – are the polarized shear wave propagation velocities,
- σ^0_{11}, σ^0_{22} and σ^0_{33} – are the residual stresses in the corresponding directions,
- ρ, λ, μ, a, b, c and K_0 – are material constants

DETERMINATION OF THE ACOUSTIC-ELASTIC COEFFICIENTS OF CONSIDERED MATERIALS

General view of the small sample loading set-up

Technique and Equipment
for Residual Stress Measurement UltraMARS®

Ultrasonic Transducers

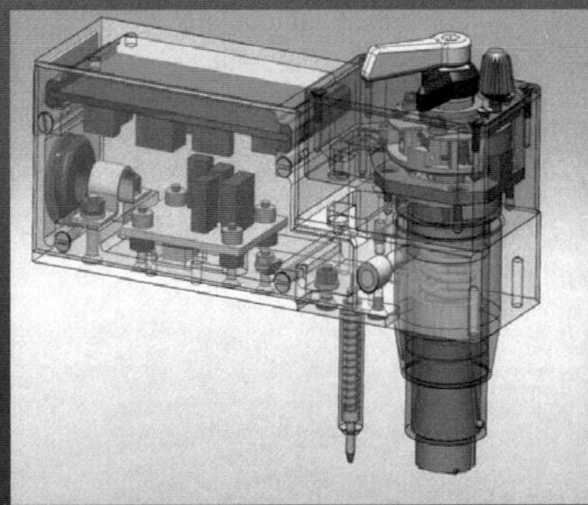

General and cross-sectioned views of one of the transducers mounted on the transducer holder and attached to the pre-amplifier

Ultrasonic Transducers

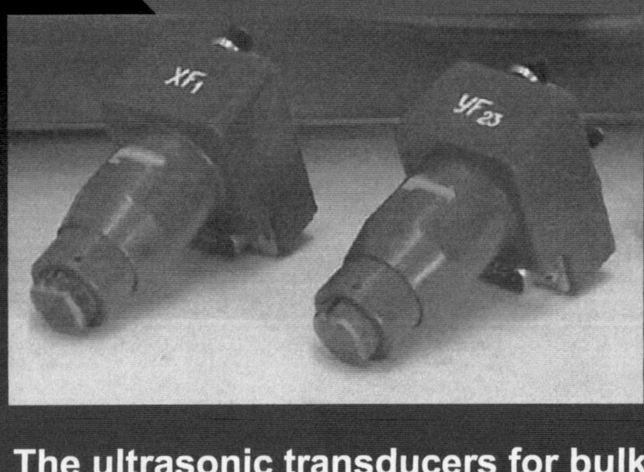

The ultrasonic transducers for bulk ultrasonic waves: color of UTs: red or orange, size of sender/receiver: 7x7mm or 10x10mm

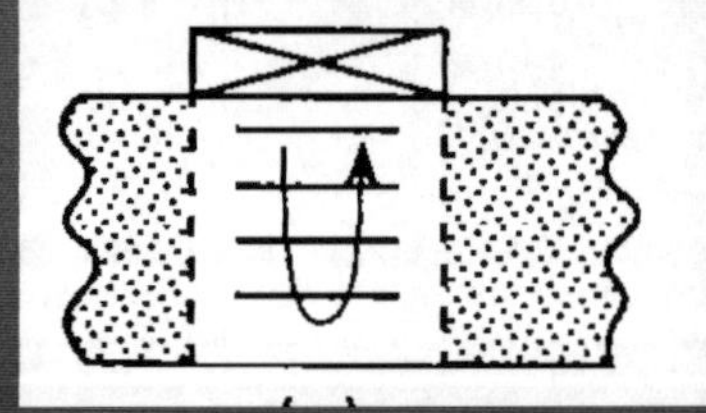

Schematic view of ultrasonic through-thickness pulse-echo measurement

Ultrasonic Transducers

Ultrasonic transducers for measurement of surface/subsurface residual stresses

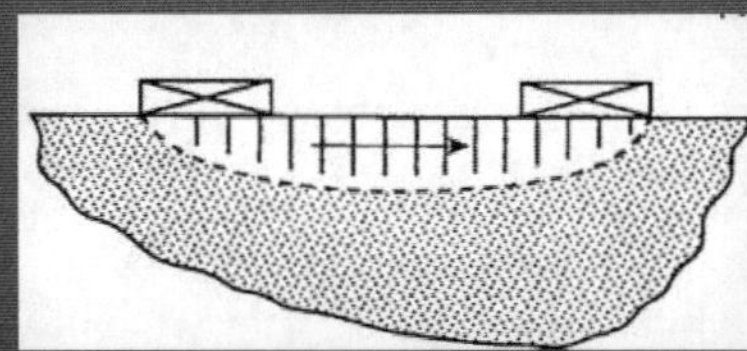

Schematic view of ultrasonic surface/subsurface pitch-catch measurement method

Surface measurement: color of UT – blue, measurement base 5mm, depth of penetration ~0.4mm. Subsurface measurement: color of UT – grey, measurement base 7mm, depth of penetration ~1.5mm.

The main technical characteristics of UltraMARS measurement system

- stress can be measured in materials with thickness 2 - 200 mm;
- error of stress determination (from external load): 5 - 10 MPa;
- error of residual stress determination: 20-30 MPa;
- stress measurement in fasteners (pins) 25-1000 mm;
- independent power supply (accumulator battery (12 V);
- overall dimensions of measurement device: 300x200x150 mm;
- weight of unit with sensors: 8 kg.

Distribution of RS near the end of stiffener

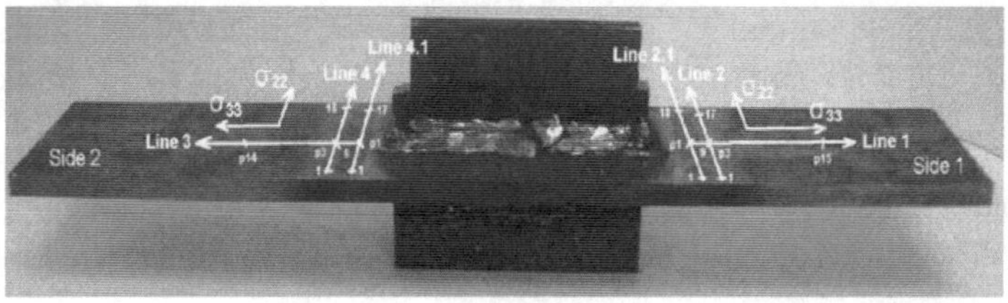

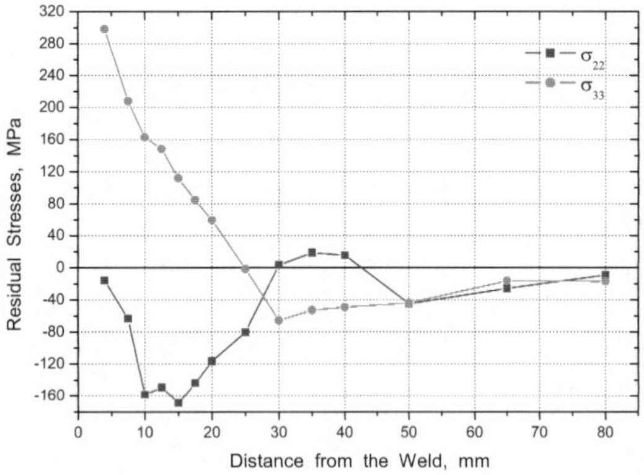

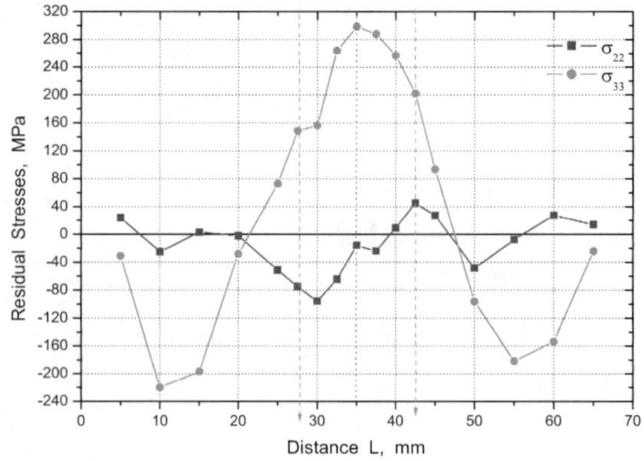

Distribution of RS in welded ample along lines 1 and 2.1

RS measurement at different locations (Line 2, 3, 4, 5 and 6) of welded panel after 1, 2, 10 and 2010 cycles of loading

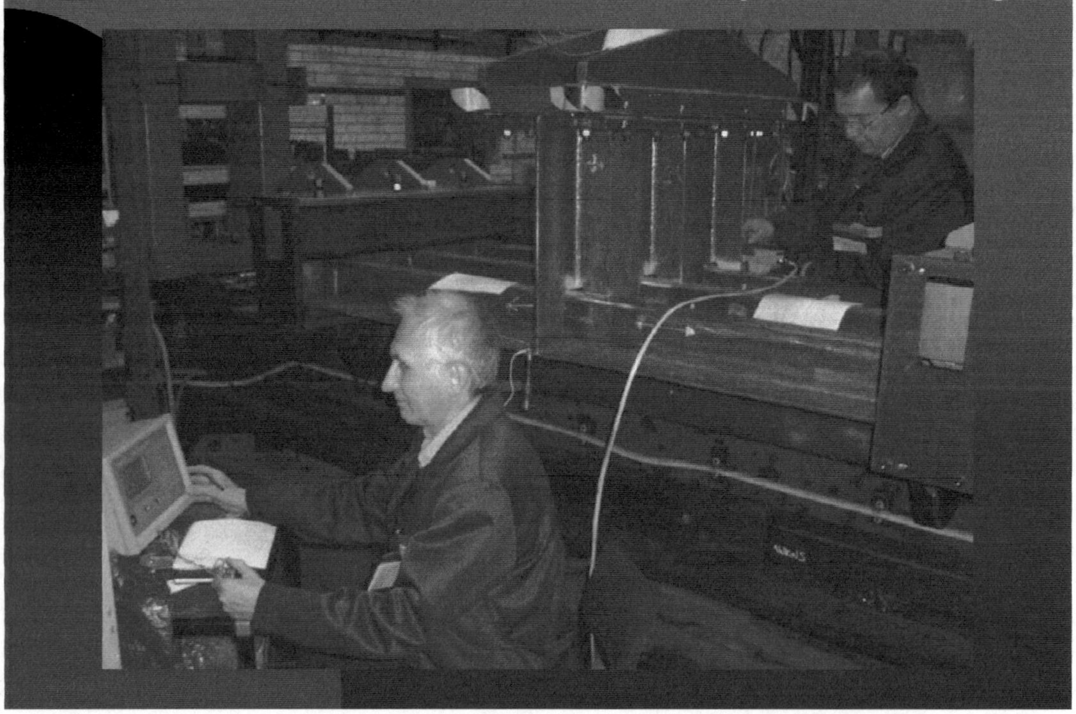

Distributions of RS along with the Line 6 in welded panel in as-welded condition (red and green circles) and after 2000 cycles of loading (black and blue circles)

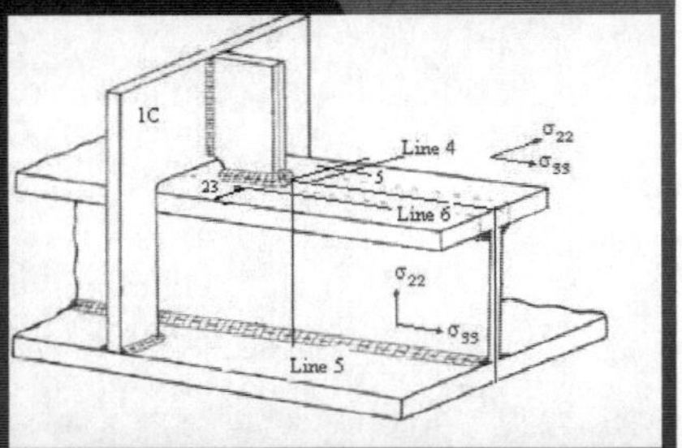

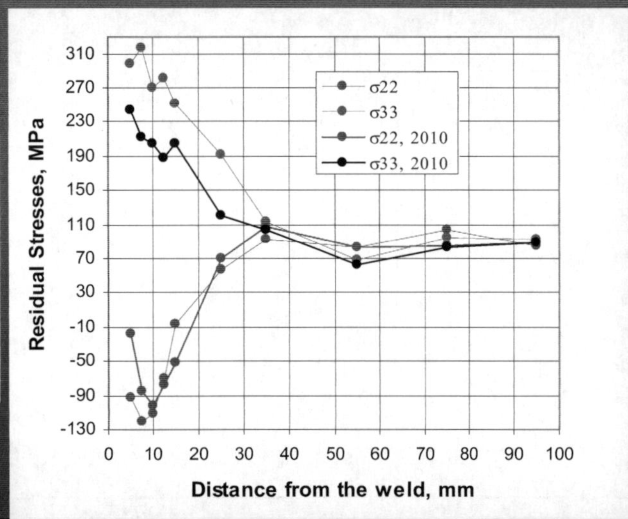

Sigma 33 – component of RS that is perpendicular to the weld,

Sigma 22 – component of RS that is parallel to the weld,

The relationship between the level of residual stresses and number of cycles of loading (red line) and max/min stresses during second cycle of loading (green line)

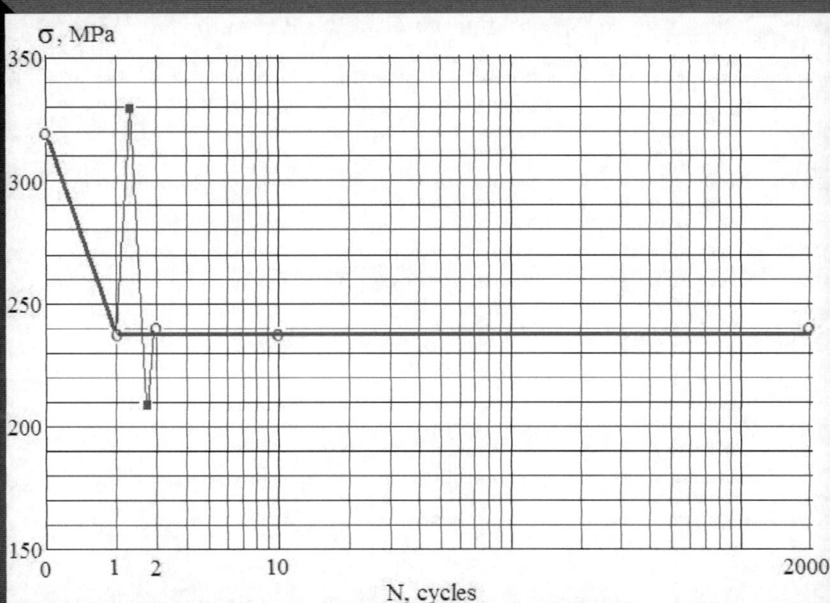

Residual Stress Management

Measurement of Residual Stresses, Fatigue Assessment and Improvement Treatment of Welded Elements

Examples of Industrial Applications

Axle Box of Large Mining Truck

Stage 1. Measurement of Residual Stresses in Axle Box Before and After PWHT

A general view of the axle box with the marking (in red) showing the zones A, B and C selected for RS measurement before and after heat treatment

Stage 1. Measurement of Residual Stresses in Axle Box Before and After PWHT

Results of RS measurement in zone C

before heat treatment / **after heat treatment**

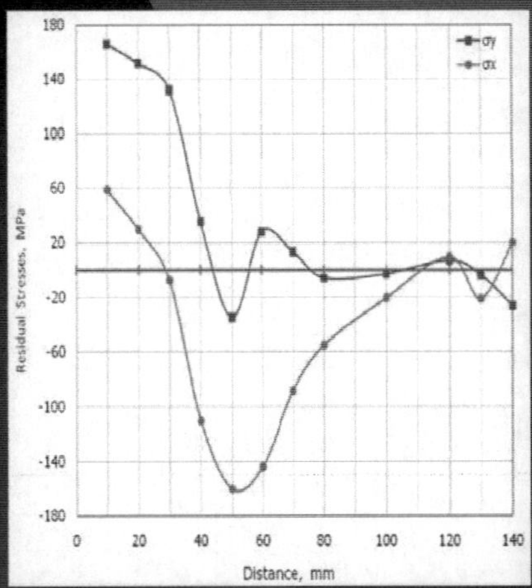

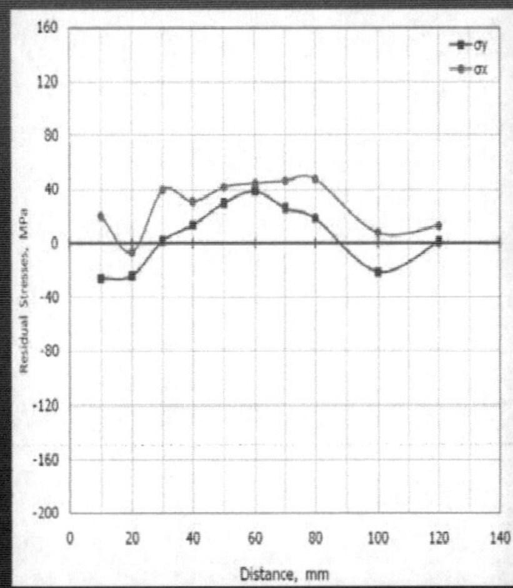

Stage 2. Fatigue Assessment of Welded Elements Before and After PWHT and UIT/UP

Fatigue curves of Fatigue Class (FAT) 100 welded elements with different levels of residual stresses (RS)

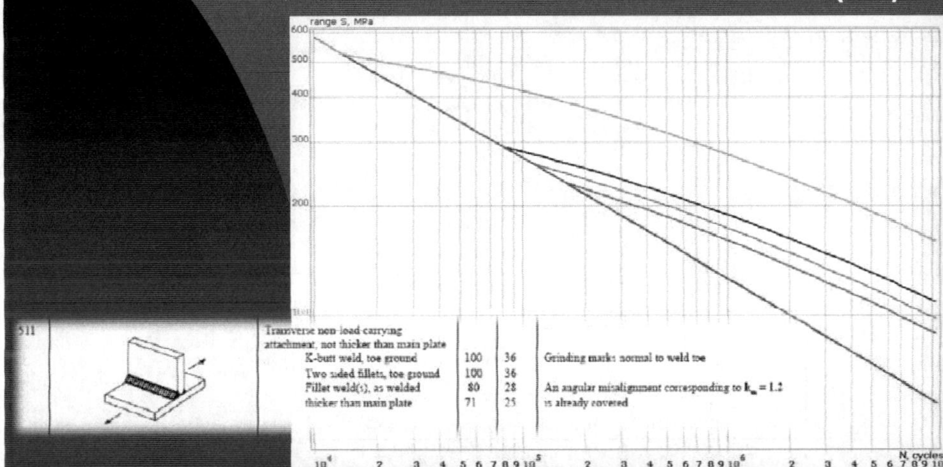

red line - in as welded condition, grey and blue lines – after heat treatment with the RS level equal to 0.1 and 0.2 of the yield strength of material, black line – zero level of RS, green line - after UIT

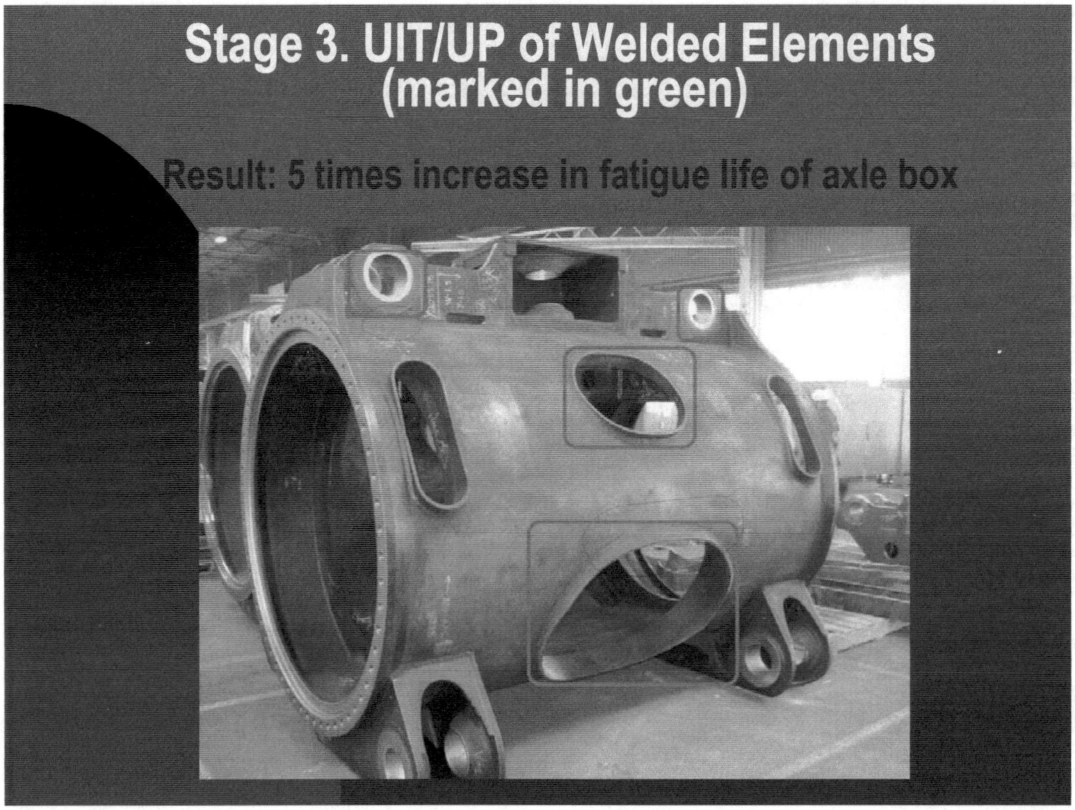
Stage 3. UIT/UP of Welded Elements
(marked in green)
Result: 5 times increase in fatigue life of axle box

Measurement of Residual Stresses
in a Welded Bridge

Measurement of Residual Stresses near the Welds in Pressure Hull of Submarine

SUMMARY

The developed advanced ultrasonic method, based on it portable instrument and the supporting software can be used for non-destructive measurement of residual stresses in laboratory samples and real parts and structural elements in many applications for a wide range of materials. The developed ultrasonic technique was successfully applied in construction industry, shipbuilding, railway and highway bridges, aerospace industry, oil and gas engineering and in other areas during manufacturing, in service inspection and repair of welded elements and structures.

Thank you
for your attention
www.sintec.ca

Finite Element Model of Eddy Current Measurement of Pressure Tube to Calandria Tube Gap

G. Klein[1,2], M. Luloff[1,2], T.W Krause[1], J. Morelli[2]

[1]Royal Military College of Canada, Kingston Ontario, Canada
[2]Queen's University, Kingston Ontario, Canada

Overview

- Introduction:
  - CANDU® Nuclear reactors;
  - Motivation;
- Background information:
  - Eddy currents
- Finite Element Method (FEM) models
  - Flat-Plate vs Curved geometry
- Experimental Measurements
  - PT wall thickness variation
- Future Work

CANDU® Reactor Fuel Channel

Image from:
http://www.civil.uwaterloo.ca/watrisk/research.html

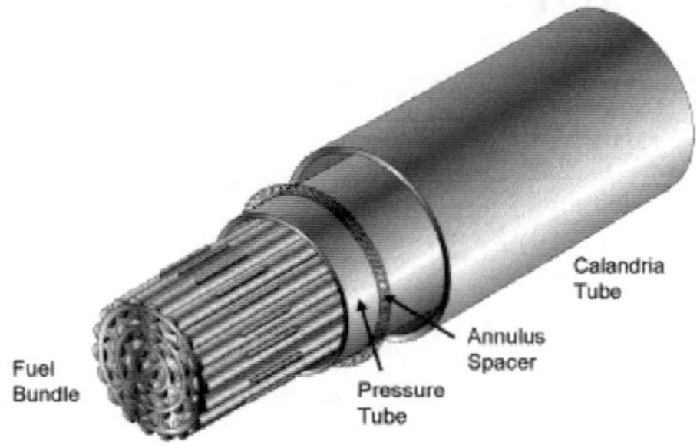

Image from: http://www.nuclearfaq.ca/cnf_sectionA.htm

Figure 1: Left: A CANDU® reactor core face showing the hundreds of fuel channels. Right: Components of a fuel channel.

Introduction: CANDU® Fuel Channel

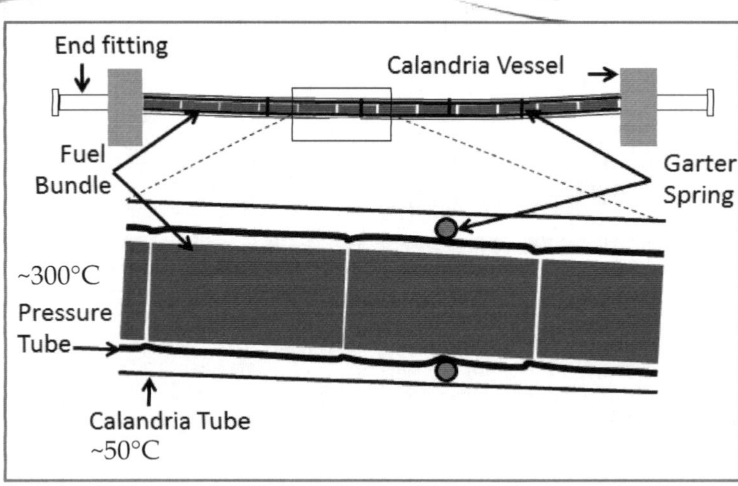

Figure 2: A schematic of a CANDU® fuel channel. Image was modified from [1].

[1] S. Shokralla, T.W. Krause, J. Morelli, 'Surface Profiling with High Density Eddy Current Non-Destructive Examination Data', NDT&E International, 62, March 2014, Pgs. 153-159.

PT-CT Gap Motivation

- Due to radiation, the pressure tube (PT) deforms in geometry
  - Length, wall thickness, radius
- PT can sag and contact the calandria tube (CT)
- Hydride blisters can then form on the PT
  - Causing cracking and failure of the PT
- Need to measure separation between PT and CT
  - PT-CT gap
- Measurements are done using an eddy current based probe

Modelling Motivation

- The PT-CT gap probe and associated measurement procedure requires a qualification process prior to it application
- Part of this process is understanding how essential parameters affect gap measurements
- This is very expensive and time consuming to do experimentally
- Modelling the PT-CT gap probe response under varying parameter conditions followed by validation can provide a less expensive option for the qualification process

Project Scope

- Current models use an approximation for the geometry of the measurement probe and approximate the PT-CT gap as separation between two flat plates
- Design a comprehensive model of the PT-CT gap measurement probe
  - Curved geometry model
- Compare flat-plate vs. curved geometry models
- Understand the effects of varying essential parameters:
  - Probe Lift-off variation (LO);
  - PT Resistivity variation;
  - PT Curvature variation;
  - PT Wall thickness (WT) variation;
- Validate model with experimental measurements

Shokralla et al./Dodd and Deeds Model

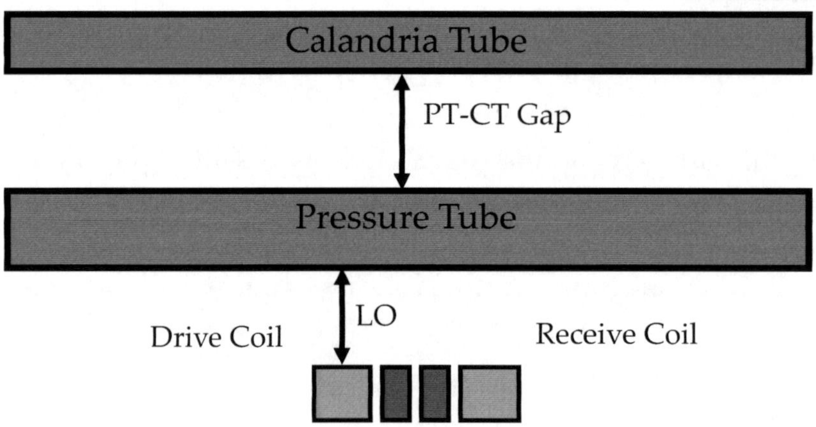

Figure 3: Shokralla et. al [2] application of Dodd and Deeds [3] model.

[2] S. Shokralla, S. Sullivan, J. Morelli and T. W. Krause, "Modelling and validation of Eddy current response to changes in factors affecting pressure tube to calandria tube gap meansurement," NDT&E International , no. 73, pp. 15-21, 2015.
[3] C. V. Dodd, W. E. Deeds, Analytical solutions to eddy-current probe-coil problems, Journal of Applied Physics 39 (6) (1968) 2829–2839.

PT-CT Gap Eddy Current Probe

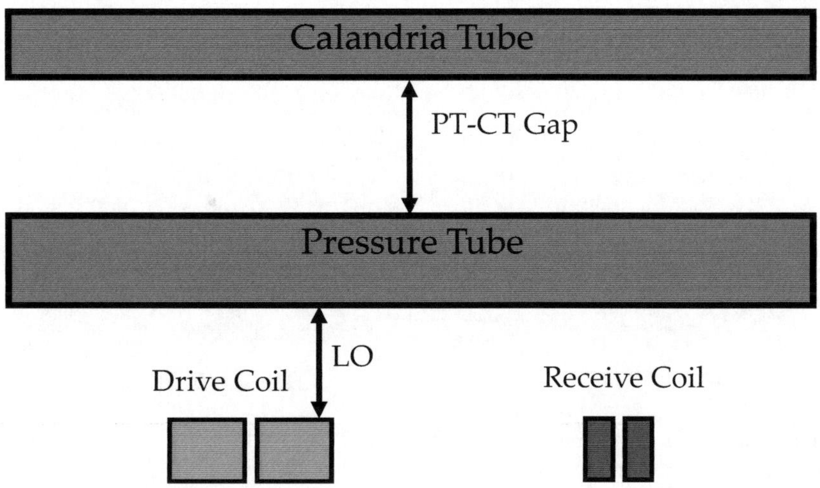

Figure 4: A representation of the experimental probe showing its basic components.

Eddy Currents Theory

- The eddy current based measurements rely on Faraday's law
  - Can be simplified due to low frequencies

$$\nabla \times \boldsymbol{H} = \boldsymbol{J} + \cancel{\sigma\epsilon_0 \frac{\partial J}{\partial t}}$$

- Excitation of the drive coil causes currents to be induced in the PT, CT and receive coil
  - eddy currents

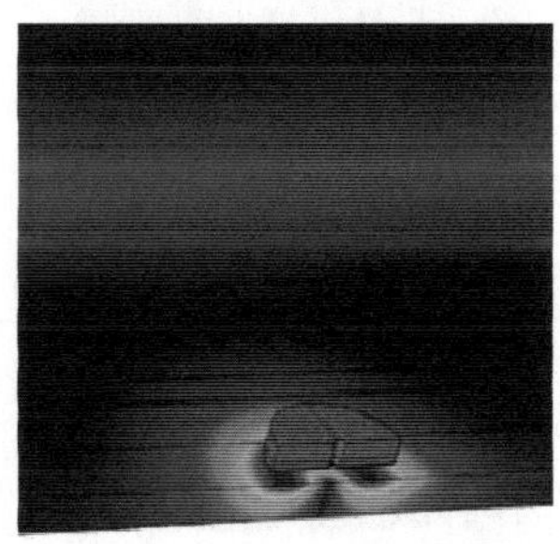

- Can measure a voltage in the receive coil
  - Voltage changes as the PT-CT gap changes
  - Can measure PT-CT gap

PT-CT Gap Eddy Current Coupling

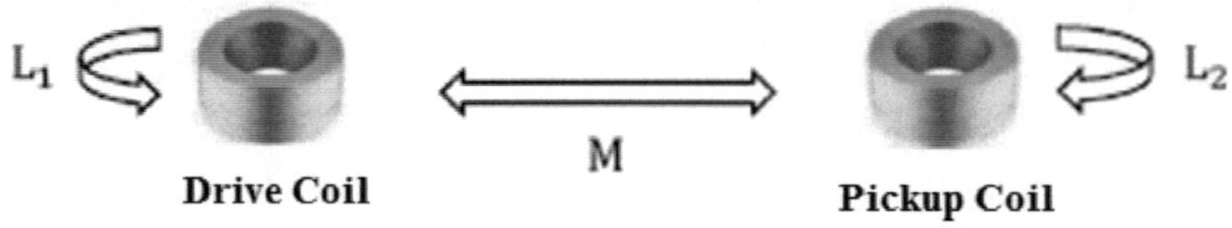

Figure 6: The coupling between the coils and the fuel channel for the drive coil and close receive coil (adapted from [4]).

[4] D.P.R. Desjardins, T.W. Krause, and L. Clapham, 'Transient Response of a Driver-Pick-up Coil Probe in Transient Eddy Current Testing', Nondestructive Testing and Evaluation International, 17, Oct. 2015, Pgs. 8-14.

PT-CT Gap Eddy Current Coupling

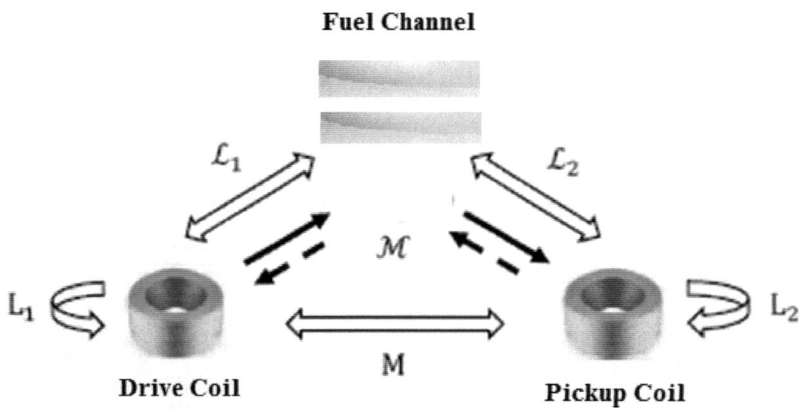

Figure 6: The coupling between the coils and the fuel channel for the drive coil and close receive coil (adapted from [4]).

[4] D.P.R. Desjardins, T.W. Krause, and L. Clapham, 'Transient Response of a Driver-Pick-up Coil Probe in Transient Eddy Current Testing', Nondestructive Testing and Evaluation International, 17, Oct. 2015, Pgs. 8-14.

COMSOL® Model – Real Geometry

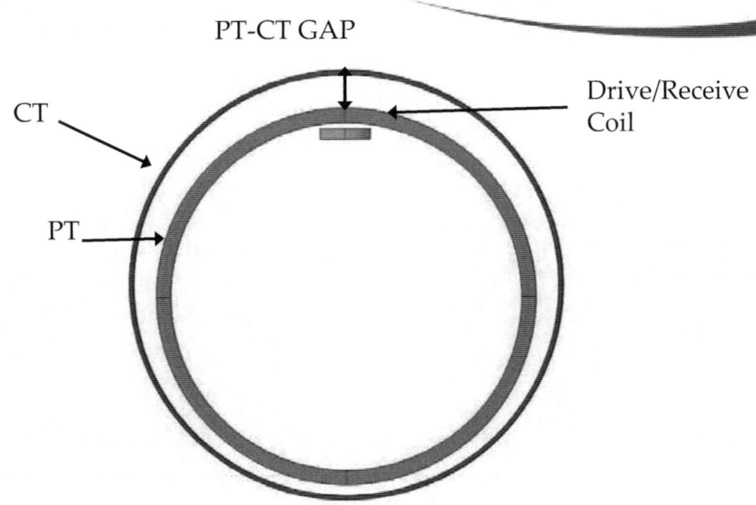

Figure 9: Simplified view of the curved model showing the cross section with the PT, CT and one coil.

COMSOL® Model

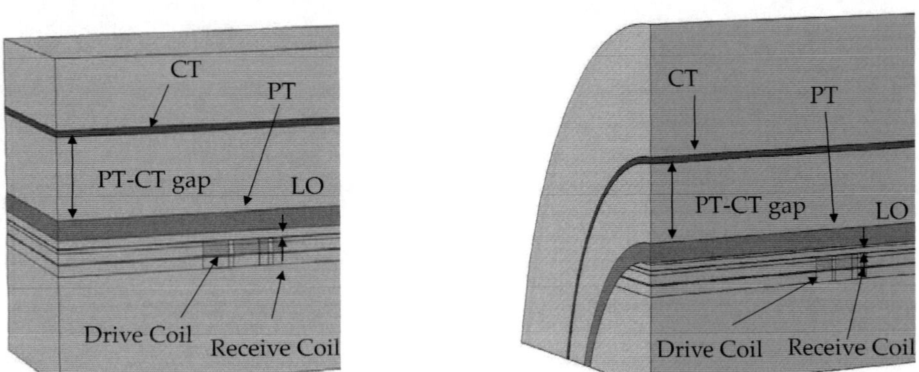

Figure 7: A screenshot of the 3D COMSOL® mode for the flat-plate geometry (left) and the curved geometry (right).

Mesh of the Models

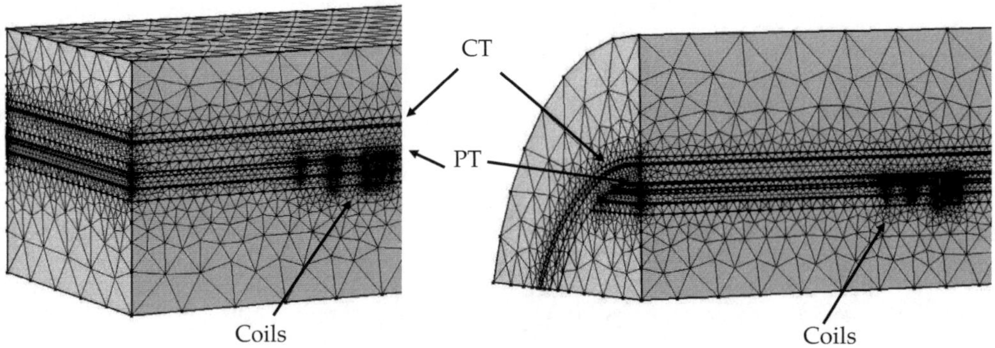

Figure 8: The left and right images show the mesh of the previous flat plate and curved COMSOL® models, respectively.

Flat-Plate vs Curved

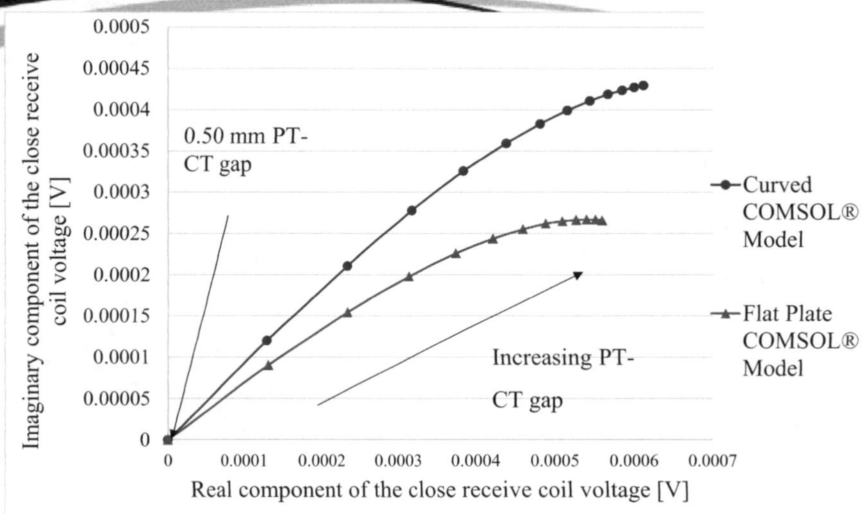

Figure 10: PT-CT gap response for the far pickup coil for the flat-plate and curved geometry models. PT-CT gap ranges from 0.5 mm (left) to 16.9 mm (right).

Flat-Plate vs Curved Cont.

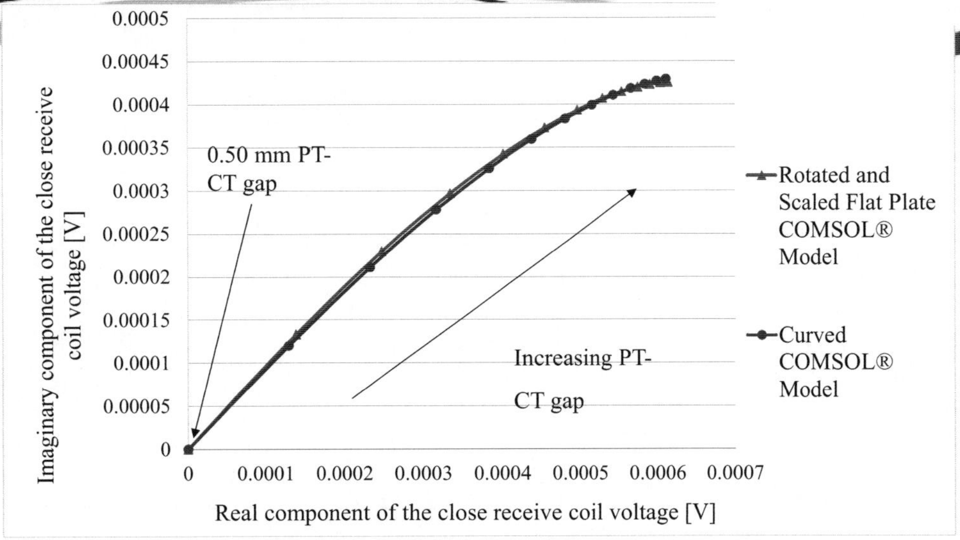

Figure 11: PT-CT gap response for the far pickup coil for the flat-plate and curved geometry models. PT-CT gap ranges from 0.5 mm (left) to 16.9 mm (right).

6th International CANDU ISI Workshop/NDT in Canada 2016 Conference 17

Flat-Plate vs Curved Cont.

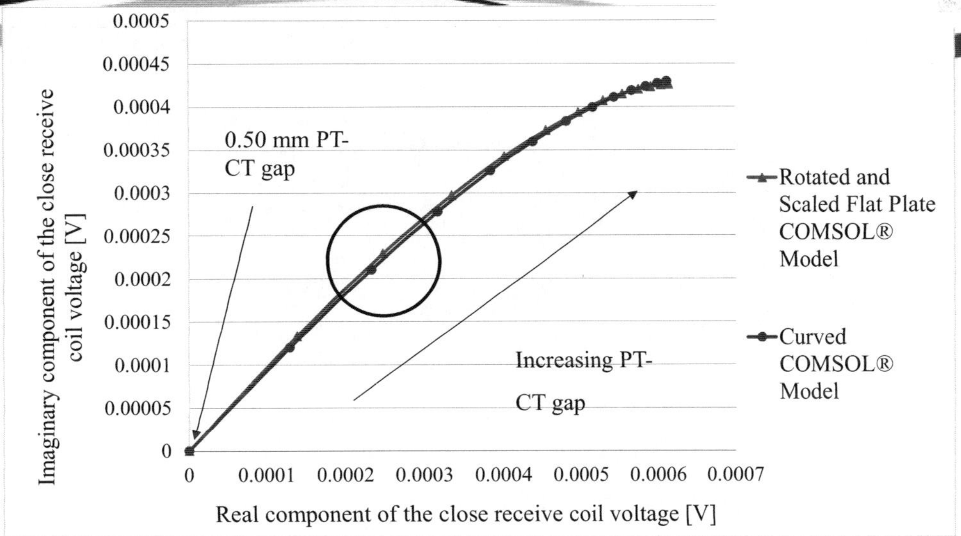

Figure 11: PT-CT gap response for the far pickup coil for the flat-plate and curved geometry models. PT-CT gap ranges from 0.5 mm (left) to 16.9 mm (right).

6th International CANDU ISI Workshop/NDT in Canada 2016 Conference 18

Flat-Plate vs Curved Cont.

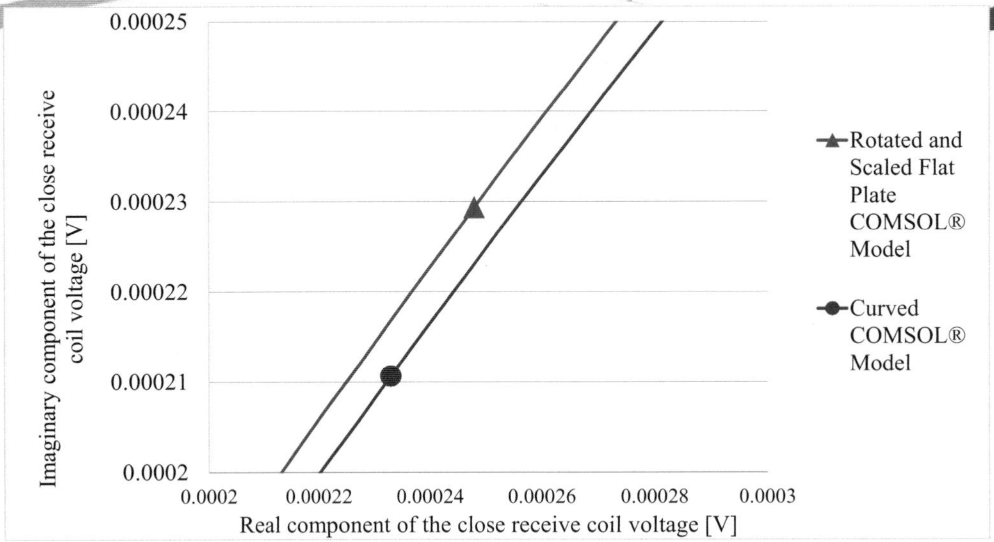

Figure 12: Focused in portion of Figure 6, at 3.2 mm.

Probe Design

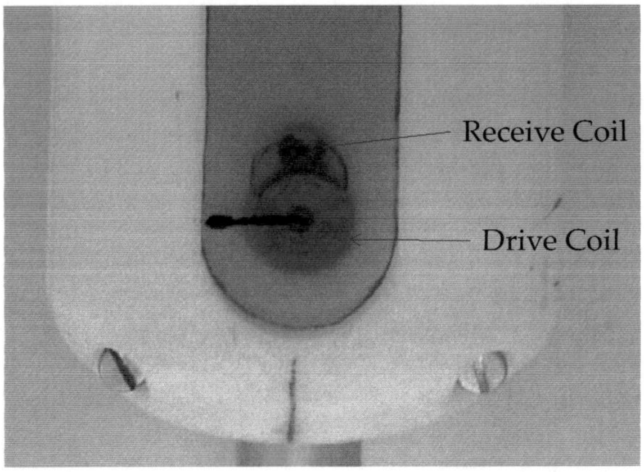

Figure 13: An image of the eddy current probe used.

COMSOL® WT Validation

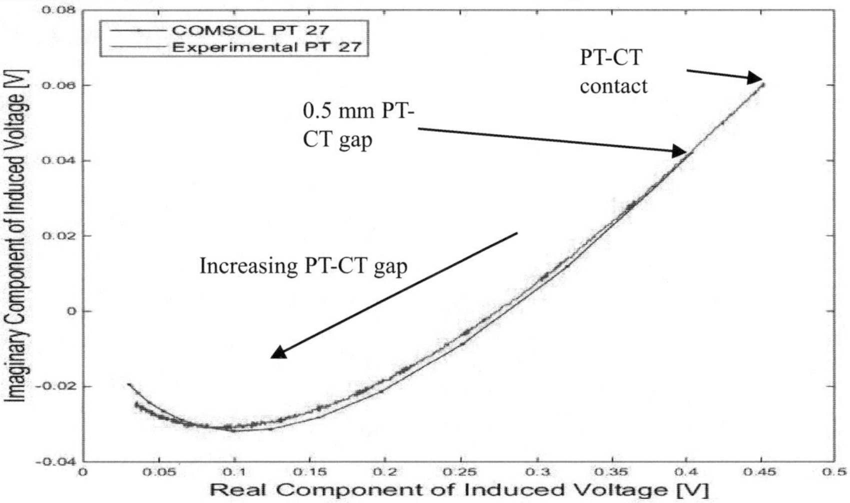

Figure 14: Comparison between COMSOL and experimental PT-CT gap measurements for PT sample with 4.38 mm PT wall thickness.

Error in PT-CT Gap Measurements

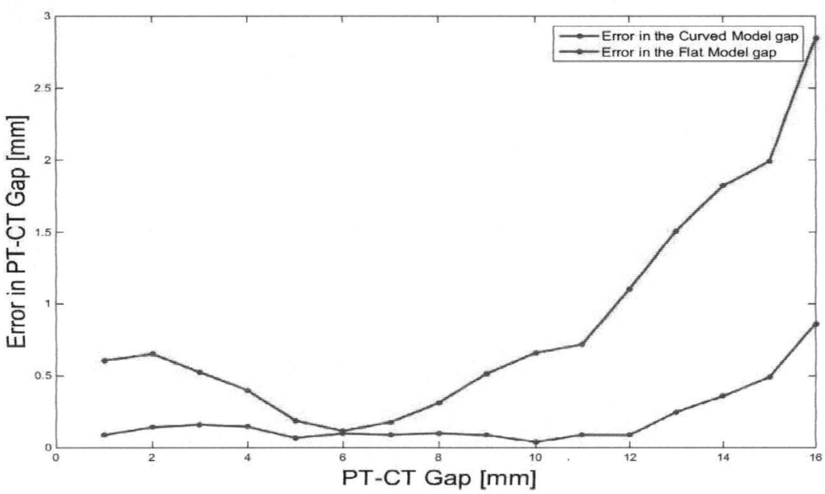

Figure 15: Errors in the modelled PT-CT gaps for the flat plate and curved FEM models.

Error in PT-CT Gap Measurement

Pressure Tube Wall Thickness [mm]	PT-CT Gap Error [mm]			
	1 mm Gap	5 mm Gap	10 mm Gap	16 mm Gap
4.38	0.09	0.07	0.04	0.9
4.36	0.06	0.12	0.05	1.4
4.34	0.07	0.25	0.15	1.3
4.28	0.18	0.22	0.17	0.9
4.26	0.26	0.18	0.14	1.4
Average	0.12	0.15	0.12	1.4

Table 1: Error in the modeled PT-CT gap for the curved model.

Summary

- Measurements of the PT-CT gap are obtained using eddy current based technology
  - Inspection qualification requires evidence of how essential parameter variations affect PT-CT gap measurements
- Models designed in COMSOL® to compare flat-plate and curved models
  - Curvature of the geometry is necessary to consider
  - Curved model has better estimate of the PT-CT gap
- Models designed to vary essential parameters of the PT-CT gap measurements
  - This was done using COMSOL®
    - Validated wall thickness variation
    - Comparison with experiment shows that curvature is necessary for PT-CT gap models to be accurate to within 0.2 mm for gaps < 10 mm

Future Work

- Continue validating FEM model on other experimental parameters:
  - LO variation;
  - PT resistivity variation;
  - Variations in curvature
- Observe which parameters have largest effect on measurements
  - Sensitivity analysis

6th International CANDU ISI Workshop/NDT in Canada 2016 Conference 25

Acknowledgements

- Thank you Ontario Power Generation for providing the pressure tubes that were used to take the experimental data.
- Funding was provided by University Network for Excellence in Nuclear Engineering (UNENE) and by the Natural Engineering and Research Council of Canada (NSERC) through a NSERC-CRD grant.

6th International CANDU ISI Workshop/NDT in Canada 2016 Conference 26

Questions?

Crack detection with self-powered wireless smart coating sensor

Nan WU [1], Samuel OSHO [1], Farjana FAISAl [1], Shengjie ZHAO [1]

[1] Department of Mechanical Engineering, University of Manitoba, Winnipeg, Manitoba, R3T 5V6, Canada
Phone: +01 204 474-7368, e-mail: nan.wu@umanitoba.ca, oshos@myumanitoba.ca, faisalf@myumanitoba.ca, zhaos348@myumanitoba.ca

Abstract
A low cost and efficient damage detection method is developed by introducing a new piezoelectric composite coating combined with a self-powered wireless signal transmission module. The new piezoelectric composite coating composed with a piezoelectric layer, an electrical conductive layer and an insulator, is designed as the sensor and to be easily applied on rough surfaces of engineering structures, such as concrete and the welding joint area. An energy harvesting circuit and an automatic electronic switch controlled by the piezoelectric coating sensor are designed to power and control the wireless signal transmitter sending warning signals of the damage occurrence. Methodology is explained in detail, and successful detection of the happening and occurrence of a crack propagation on a beam structure is realized through experimental testing.

Keywords: Non-destructive damage detection, piezoelectric composite coating, self-powered sensor, wireless sensor.

1. Introduction

Detection of structural damages especially at their earliest stage is a vital process in engineering to avoid calamitous and irreparable damages. Bearing this in mind for the last few decades, structural health monitoring (SHM) with different damage identification techniques have been a largely concerned field of study for numerous researchers from both industrial and academic communities.

Local damage detection methods which are mostly referred to as non-destructive tests (NDTs), such as guided ultrasonic waves, optical fibre sensors, X-ray test, ultrasonic, and magnetic particle test etc., have been well studied [1, 2 and 3]. However, these techniques have been proven to be unsuitable in particular cases due to long inspection times [3], expensiveness of some methods, inability to reach the damaged area and most importantly they are not capable of on-line real-time continuous detection of damages [4]. Vibration signals of a structure carries great amount of information about the healthiness of that particular structure. Numerous reviews [4 - 8] can be found in the literature on vibration based damage detection methodologies proposed in the past several decades.

In addition to traditional methods for detecting damage involve disassembling the engineering parts, there are new non-destructive damage detection techniques have been developed in the recent decades, including crack detection using the vibration power flow [9] lamb wave actuation by piezoelectric actuator [10], and wavelets transformation to extract information from different kinds of data [11]. The noise in the dynamic signal, which may reduce the accuracy and effectiveness of damage detection using classic time-frequency analysis, is usually inevitable. In addition to dynamic analysis for the health monitoring, space domain wavelet analysis was also proven to be efficient in the accurate detection and location of damage [12-15]. Previous studies required experiments to be conducted in lab or workshop environments with complex experimental set-ups and usually disassembled structures. New techniques that can realize the low cost on-board real time testing on various mechanical and civil structures are still desirable in industry and the SHM research field. An approach proposed in this research would help to evaluate the working condition of active structures and predict the possibility of future structural failures.

In this research, a new data acquisition system with smart material coatings that are easy to be applied on the different surface conditions of engineering structures will be developed in order to detect crack or fracture damages. The objective of this research is to develop a feasible on-board testing process to achieve efficient crack/fracture detection with self-powered wireless sensor made from piezoelectric composite coatings.

2. Methodology and design of the self-powered wireless crack sensor

In this section, the design of the new piezoelectric composite coating is explained in details fist to show the crack detection methodology. Then, the whole sensor system design is given with the description of the self-powering wireless signal transmitter module.

The piezoelectric composite coating can be applied easily on the different types of surfaces, such as concrete structures and welding joint area. The substrate structures are supposed to be electrically conductive grounded with zero electrical potential. If the substructure is not electrically conductive. A conductive layer/coating should be applied between the substructure and the piezoelectric composite coating. Figure 1 gives the design of the piezoelectric composite coating to detect the possible crack. First, a piezoelectric layer is attached to the substructure surface close to the tested position. The piezoelectric layer is then short connected by a thin insulator and conductive composite layer, which can be applied on the surface easily as a paint. During the operation of the substructure, the structure will be vibrating, and electrical charge and voltage will be generated on the piezoelectric layer surface. However, since the piezoelectric layer is short connected, without crack or fracture on the substructure, the voltage reading on the surface of the piezoelectric layer will be zero. When a crack or fracture occurs at the location covered by the piezoelectric composite coating, the conductor coating will be broken accordingly, and the short circuit on the piezoelectric layer will be cut off; even for a very short period, there will be a non-zero charge and voltage generated on the piezoelectric layer surface indicating when and where the crack or fracture occurs. The non-zero charge and voltage generated by the piezoelectric composite coating will be captured by a small capacitor and an electronic switch connected with it to turn on the wireless data transmitter sending the warning signals. The small capacitor, electronic switch and the self-powered wireless data transmitter are given and illustrated in figure 2.

Figure 2 illustrates the design of the whole self-powered wireless crack sensor system. The piezoelectric composite coating and the wireless signal transition module will be combined to realize the on-board real-time fracture/crack detection. During the operation of the host structure, the additional piezoelectric patch for energy harvesting shown in figure 1 will keep collecting kinetic energy from the host structure and generating electrical signal to charge the super capacitor through the energy harvesting circuit shown in figure 2. In this case, the super capacitor is assumed to be always fully charged before the damage happens. In figure 2, it is seen that an electronic switch, which can be a simple triode, is located between the wireless signal transmitter and its power source, the super capacitor. The default position of electronic switch is at 'off', when there is no damage at the position covered by the composite coating sensor. Once the fracture/crack happens, the piezo-layer in the composite coating will no longer be short connected. With the vibration during the operation of the host structure, none-zero voltage, V, will be generated on the piezo-layer for damage detection. However, it is noted that the voltage generated from the piezoelectric coating is dynamic AC voltage but not a constant value. This makes it hard to control/turn on the electronic switch giving continuous power supply to the wireless signal transmitter. Therefore, a diode bridge and small capacitor is designed to stabilize the voltage generated on the piezoelectric layer in the composite coating. When the small capacitor is fully charged, it turns on the electrical switch in a short period to activate/turn on the wireless signal transmitter. Then, the wireless signal transmitter, which is powered by the fully charged super capacitor, will send warning signals indicating the occurrence of the fracture/crack damage. It is noted that each piezoelectric composite coating is marked with a unique number indicating its position. When the warning signal is sent out by the wireless signal transmitter, the damage location information will be sent at the same time. The detailed realization process is given in the next section.

Photos of the piezoelectric composite coatings, wireless signal transmitter and energy harvesting circuit are given in the next section to describe the realization of the whole sensor system and the successful crack detection by experimental studies.

3. Experimental preparation and testing

In this section, the experimental realization of the damage detection with the self-powered wireless smart coating sensor is proposed. The preparation of the piezoelectric composite coating is described first. The whole experimental set up and testing process are then given and explained in details.

Figure 3 shows the piezoelectric composite coatings without and with the conductive layer. In figure 3 (a), two circular piezoelectric patches (0.012m in diameter, 0.002m in thickness) are bounded on the surface of an aluminium beam by 3M Epoxy metal weldment DP 100. The left piezoelectric patch (patch A) shown in figure 3 (a) is used as an energy harvester to charge the super capacitor powering the wireless signal transmitter, while the right one (patch B) is used for the crack detection. The piezoelectric patches are coated with silver electrodes on their top and bottom surfaces and connected (grounded) with aluminium substrate beam. It is noted the 3M epoxy metal weldment, which is used to bind the right piezoelectric patch, is extended to a certain area of the substrate beam acting as the isolation layer of the piezoelectric composite coating. In figure 3 (b), an electrical conductive silver epoxy coating is added on the top of the isolation layer made from the 3M epoxy metal weldment. Following the design given in figure 1, the electrical conductive layer covers little bit larger area of the beam than the isolation layer making sure the top and bottom electrodes of the piezoelectric patch is short connected. Figure 3 (b) shows a large notch that reaching close to the location covered by the composite coating. In this case, when vibration take place on the beam with the notch, crack will be initiated at the notch tip and propagate across the composite coating. Once the crack across the composite coating, the short connection circuit of the piezoelectric patch is cut off, and the non-zero voltage will be generated on the surface of the piezoelectric patch B and considered as the warning signal of the crack occurrence.

Corresponding to the sensor system diagram given in figure 2, figure 4 gives the whole experimental lay out of the damage detection testing with the piezoelectric composite coating combined with the self-powered wireless data transmission module. An aluminium cantilever beam (0.6m in length, 0.003m in thickness and 0.02m in width) with a large notch located at 0.13m from the fixed end of the beam is used for the testing. A random dynamic loading is applied at the left side of the notch by a modal shaker. Two super capacitors with 5F capacitance and 5V fully charged voltage are connected in series to power the wireless signal transmitter. It is noted that it takes quite long time (more than 12 hours) to fully charge the super capacitor with the small piezoelectric patch used in our experiment, but the charging time can be significantly reduced if piezoelectric patch with larger sizes is used as the energy harvester. The electronic switch and wireless transmitter modules are described as follows.

The diode bridge, small capacitor and the electronic switch controlling the wireless signal transmitter are integrated on a solderless breadboard, which is shown in figure 5. Triodes are employed to act as the electronic switch, which is controlled by the piezoelectric layer in the composite coating and the small capacitor.

An Arduino microcontroller board and a radio transmitter are used as the wireless signal transmitter module and shown in figure 6. The Arduino microcontroller board is a development chip board that can be programed and control external devices, such as the radio transmitter, Wi-Fi and Blue-tooth devices. During our experiment, once the damage occurs and electronic switch is tuned on by the piezoelectric composite coating, the super capacitor starts to provide power supply to the Arduino board and the radio transmitter to send out the digital warning signals. The digital signal sent out from the radio transmitter includes the piezoelectric composite coating number and position information to indicate the damage location. The piezoelectric composite coating number and position information can be obtained directly from the non-zero voltage on the piezoelectric layer as an input to the Arduino board. The Arduino board used in our experiment has 6 analog signal input channels and hence can be programed to recognize 6 damage locations.

At the signal receiver side shown in figure 7, another Arduino board and a radio receiver are employed to obtain the damage occurrence and location information. The Arduino board is connected with a personal computer to show and save this useful information. Again, the one Arduino board used in our experiment can receive different signals from 6 radio receivers.

Figure 8 shows the damage detection result, when a crack is initiated at the notch tip and propagate across the piezoelectric composite coating after around 30 minutes vibration of the beam. From the zoomed in screen shot of the personal computer, it is seen that the sensor system successfully detects the occurrence of the crack damage. Since there is only one piezoelectric composite coating and one damage simulated in our test, only the #1 joint is detected as broken.

Finally, the total cost of the whole system is estimated as 84USD. The rough costs for different parts and modules are given as below:

piezoelectric patches: 2USD*2; super capacitors: 4USD*2; Arduino board: 25USD*2; radio transmitter: 5USD*1; radio receiver: 7USD*1; diodes, triodes, small capacitor: 10USD.

4. Conclusions

A low cost on-board real time damage detection method is developed by introducing a new piezoelectric composite coating combined with a self-powered wireless signal transmission module. Experimental testing is conducted to realize the successful crack detection with the proposed technique and the design of the self-powered wireless sensor system. For the future works, different wireless signal transformation modules, such as blue-tooth and WiFi transmitters, will be tested to send more comprehensive damage data at multipole testing location. Further data analysis method will be developed to evaluate the whole structure stability.

References

1. Sylvester GJ. Electronic Theses and Dissertations Damage Detection in Plate Structures using Guided Ultrasonic Waves. 2013.
2. Kuang K. Distributed damage detection of offshore steel structurusing plastic optical fibre sensors. Sensors Actuators A Phys. 2015; 229: 59–67.
3. Yan G, De Stefano A, Matta E, et al. A novel approach to detecting breathing-fatigue cracks based on dynamic characteristics. J Sound Vib 2013; 332: 407–422.
4. Doebling SW, Farrar CR, Prime MB and Shevitz DW. Damage identification and health monitoring of structural and mechanical systems from changes in their vibration characteristics: A literature review. Report for Los Alamos National Lab Report no. LA-13070-MS, 1996.
5. Salawu OS. Detection of structural damage through changes in frequency: A review. Eng Struct 1997; 19: 718–723.
6. Doebling SW, Farrar CR and Prime MB. A summary review of vibration-based damage identification methods. The Shock and Vibration Digest 1998; 30: 91–105.
7. Montalvao D. A review of vibration-based structural health monitoring with special emphasis on composite materials. The Shock and Vibration Digest 2006; 38: 295–324.
8. Zhao S, Wu N and Wang Q. Damage detection of beams by a vibration characteristics tuning technique through an optimal design of piezoelectric. International Journal of Structural Stability and Dynamics 2016; 16:1550070.
9. Wang XQ, Wong WO, Cheng L. Modal power flow with application to damage detection. International Journal of Engineering Science 2009; 47:512-523.
10. Tua PS, Quek ST, Wang Q. Detection of cracks in cylindrical pipes and plates using piezo-actuated Lamb waves. Smart Mater. Struct. 2004; 14:1325–1342.
11. Newland DE. Wavelet analysis of vibration Part 1: theory. J. Vibr. Acoust 1994; 116:409–416.
12. Xu YL, Huang Q, and Zhan S et al. FRF-based structural damage detection of controlled buildings with podium structures: experimental investigation. Journal of Sound and Vibration 2014; 333:2762-2775.
13. Wang Q and Deng X. Damage detection with spatial wavelets. International Journal of Solids and Structures 1999; 36:3443-3468.
14. Wu N and Wang Q. Experimental study on damage detection of beam structures with wavelets transform. International Journal of Engineering Science 2011; 49:253-261.
15. Wang Q and Wu N. Detecting the delamination location of a beam with wavelet transform: An experimental study. Smart Materials and Structures 2011; 20:012002.

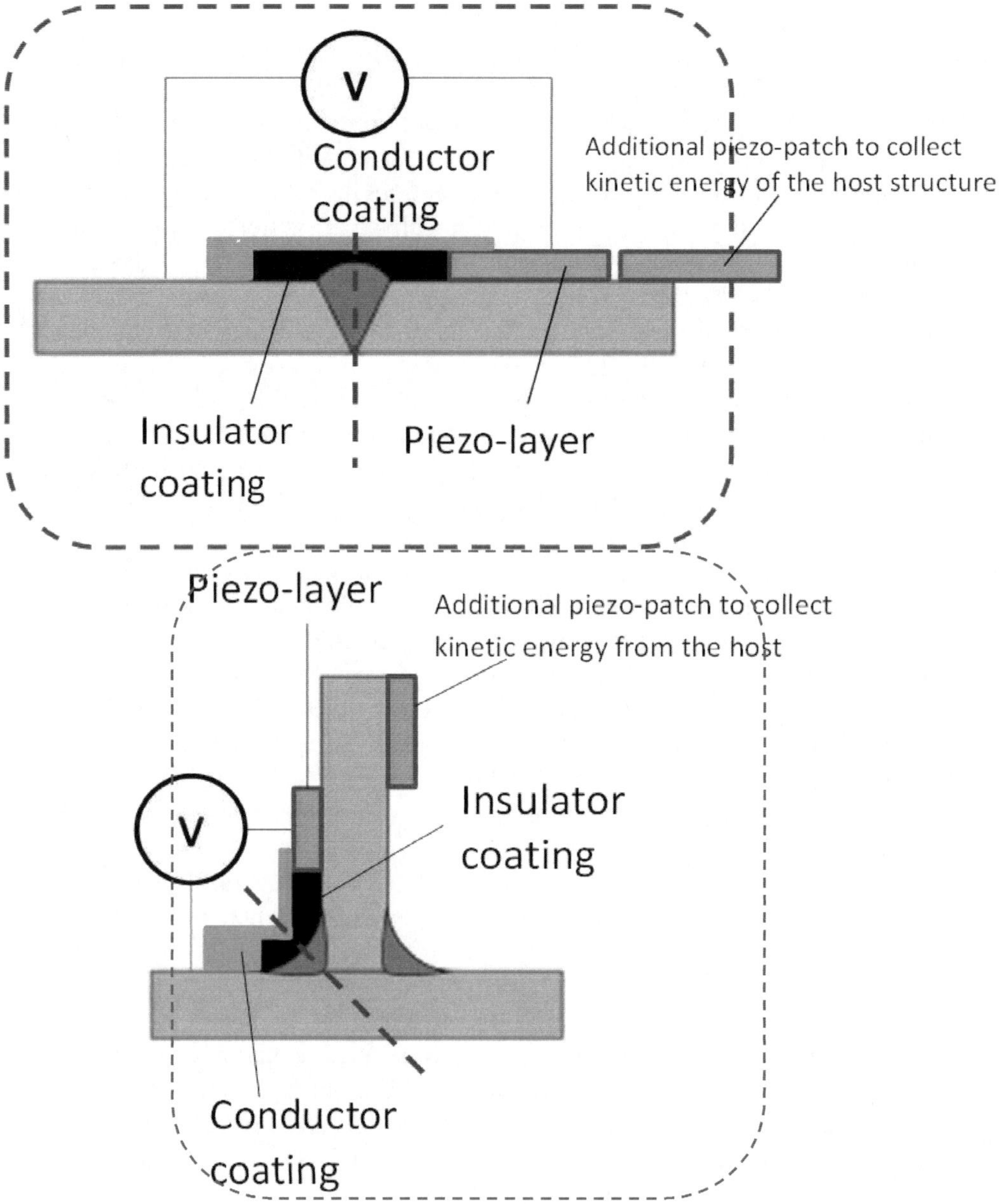

Figure 1 Damage detection module with energy harvesting piezo-patch.

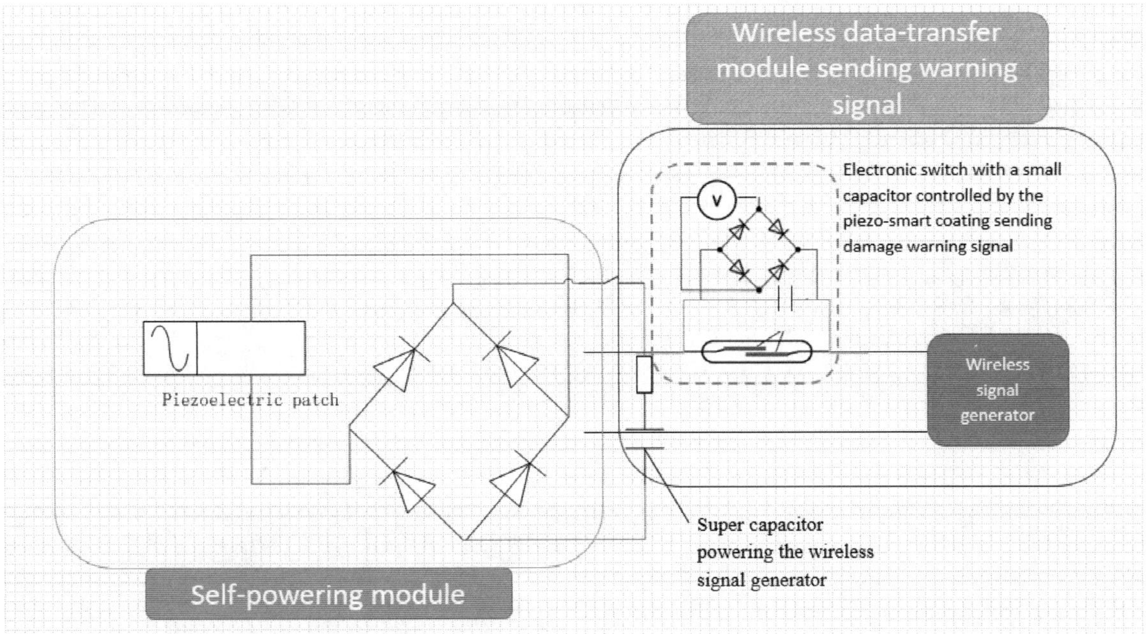

Figure 2 Wireless signal generation module with energy harvesting circuit to charge the super capacitor, which powers the wireless signal generator.

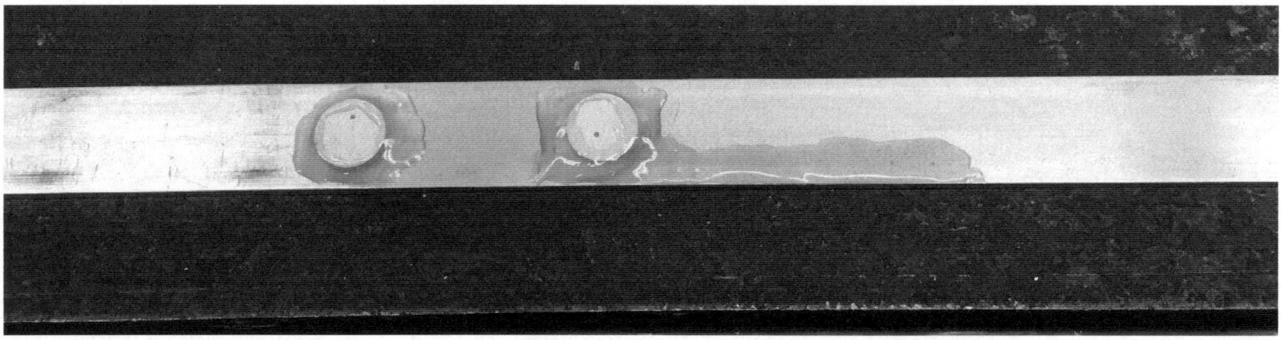

(a)

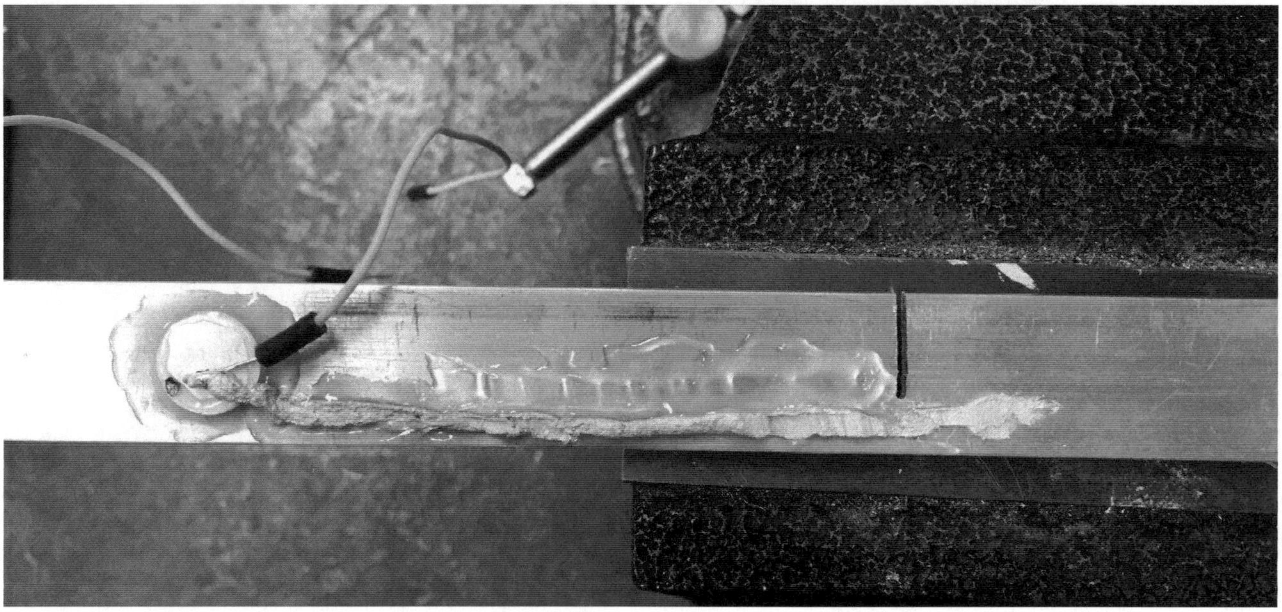

(b)

Figure 3 Preparation of the piezoelectric composite coating.

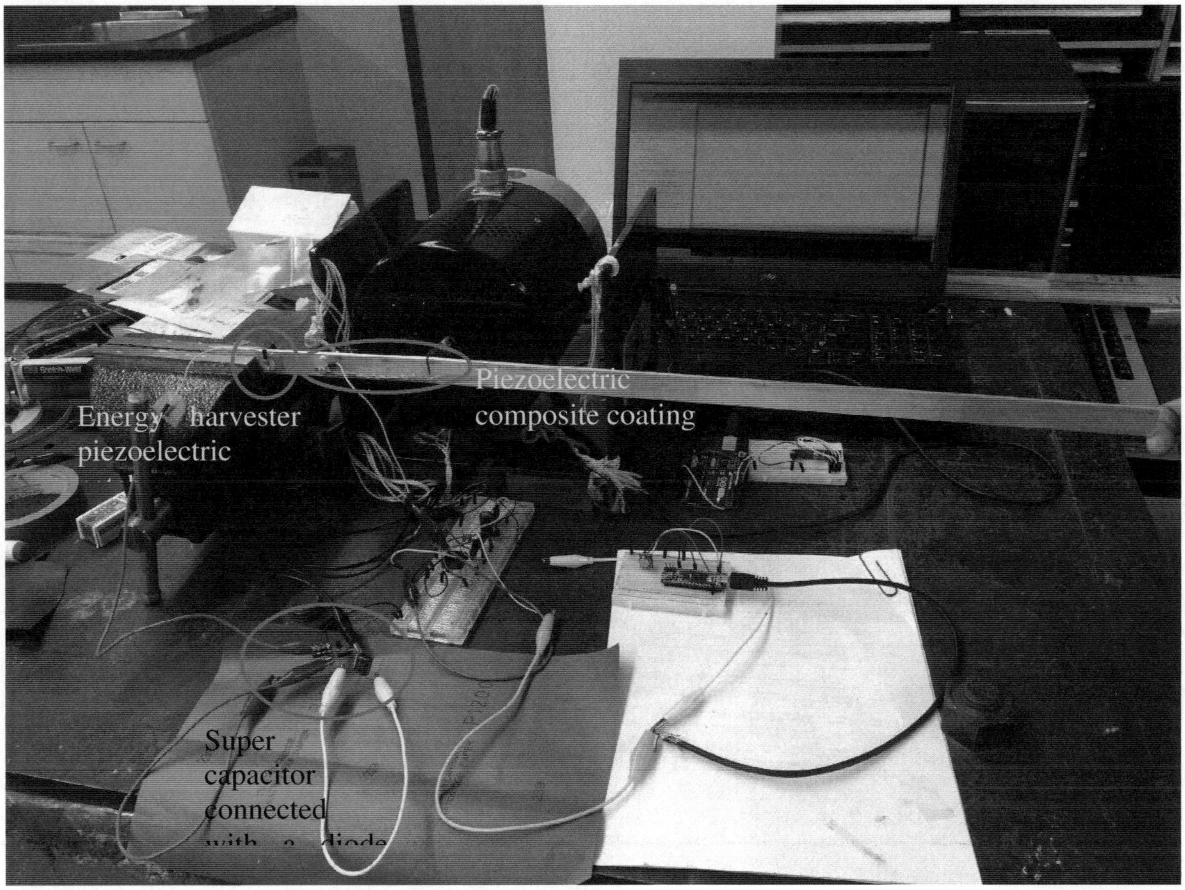

Figure 4 The whole lay out of the experimental set up.

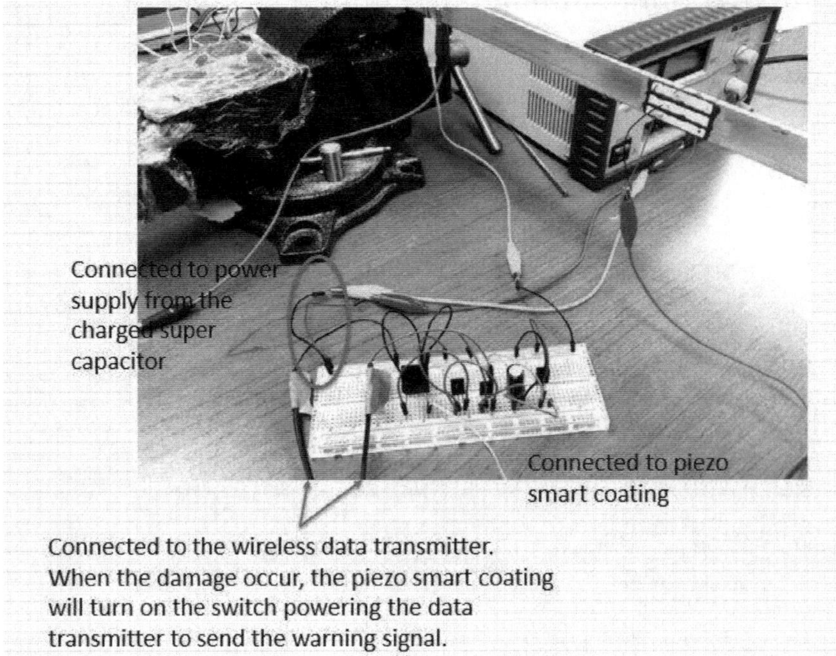

Figure 5 The triode electronic switch connected with a diode bridge.

Figure 6 The arduino board and radio transmitter acting as the wireless signal transmitter module.

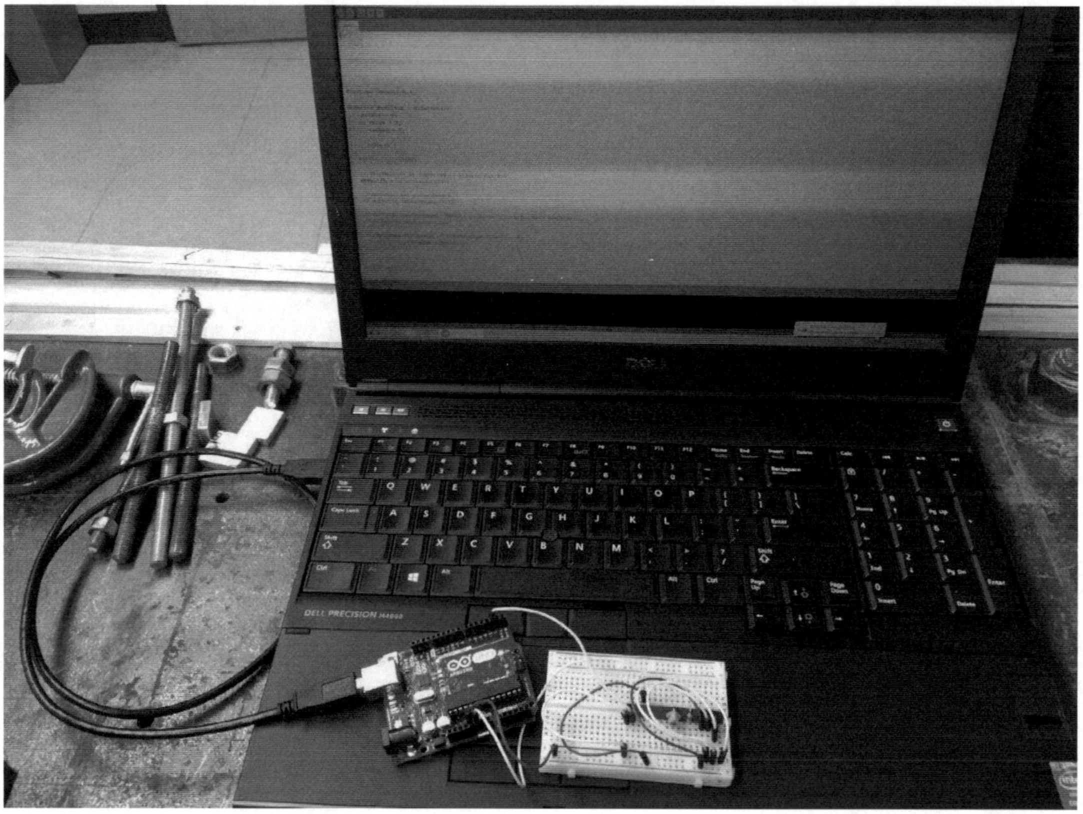

Figure 7 An arduino board and a radio receiver connected with a Dell mobile workstation.

Figure 8 Occurrence of the crack indicated by the message shown on the Dell mobile work station.

Steam Generator Tube Inspection and Foreign Object Retrieval in Wolsong NPP

Lee, Jeong Uk (jwlee357@khnp.co.kr)
Wolsong NPP1 Chemistry Dep't
KOREA HYDRO & NUCLEAR POWER CO., LTD.
South Korea

Diagnostic Program for S/G Reliability

- Recently, To enhance integrity of secondary side tube of steam generator(S/G) is strengthening the requirements of non-destructive inspection
  - ECT
  - FOSAR(Foreign Object Search and Retrieval)
- CANDU type steam generator tube is hardly defective
- But results of a diagnostic program may be affected by
  - Structural characteristics of S/G model
  - Retention of optimized equipment, Operation period
  - Inspection evaluation
  - Foreign substance characteristics (**Especially**)

Foreign Object Effect in S/G

- In steam generator with many years of operation,
  - Equipment access restriction due to hard sludge
  - Difficult to access for foreign objects inspection
  - ECT Detectivity effect : Co-existence of non-contact non-magnetic foreign objects and sedimentary sludge
- Effect of Foreign Object remaining between tubes
  - Increase the tube defects probability by flow, vibration
- Complementary relationship for Foreign Object detection
  - ECT
  - FOSAR

KHNP Wolsong NPP1

6th International CANDU In-Service Inspection Workshop NDT in Canada 2016 Conference November 15-17, 2016 Holiday Inn Burlington Hotel & Conference Centre Burlington, ON

Wolsong NPP Condition

Item	Unit 1 (1)	Unit 2, 3, 4
Manufacturer	Foster Wheeler	Bobcock & Wilcox
Quantity	4	4
Tube	3,558	3,530
Tube Material	Alloy 800	Alloy 800
2nd Side Handhole	3 (2" 1ea, 4" 2ea)	4 (2" 1ea, 4" 2ea)
2nd Side Manway	1	1
Tube Gap	4.74mm	5.02mm
Sludge Lancing	No (2)	Yes (4)
In-Bundle FOSAR	Yes (3)	Yes (4)
Tube ECT	Yes	Yes

(1) Continued Operation after license extension

(2) No sludge lancing after first by B&W in 2001

(3) First In-Bundle FOSAR in 2016(After Equipment Development)

(4) Sludge lancing & FOSAR(2 S/G) for Every outage since 2008

KHNP Wolsong NPP1

6th International CANDU In-Service Inspection Workshop NDT in Canada 2016 Conference November 15-17, 2016 Holiday Inn Burlington Hotel & Conference Centre Burlington, ON

NDT Requirement : Performance for S/G

- Unit 1 (24th Outage, '16.3)

Inspection Methods	Item	SGMP Requirement	Performance				Remarks
			S/G #1	#2	#3	#4	
Bobbin	Full Length	100% / 60EFPM	-	100%	-	100%	
MRPC	H/L	100% / 60EFPM	10.5%	100%	10.5%	100%	
	C/L	-	-	9.6%	-	9.6%	3 Col. of Annulus
Sludge Depth	H/L	Once / 5 period	-	100%	-	100%	
	C/L		-	20%	-	20%	
Sludge Lancing	H/L	Once / 4 period	-	-	-	-	No equipment
FOSAR	H/L	Once / 4 period	50%	50%	50%	50%	Hard sludge effect

- Unit 2 (15th Outage, '15.11)

Inspection Methods	Item	SGMP Requirement	Performance				Remarks
			S/G #1	#2	#3	#4	
Bobbin	Full Length	100% / 60EFPM	100%	-	100%	-	
MRPC	H/L	100% / 60EFPM	50.7%	-	50.3%	-	
	C/L	-	-	-	-	-	
Sludge Depth	H/L	Once / 5 period	100%	-	100%		
	C/L		20%	-	20%		
Sludge Lancing	H/L, C/L	Once / 4 period		100%	-	100%	Once / 2 period
FOSAR	H/L, C/L	Once / 4 period	100%	100%	100%	100%	Once / 2 period

Development of FOSAR equipment

- FOSAR Mock-Up Simulation and Training
  - Moving magnetized Rail assembly to the annulus area of S/G
  - 2.95mm Sword video probe(GE custom application) inspection

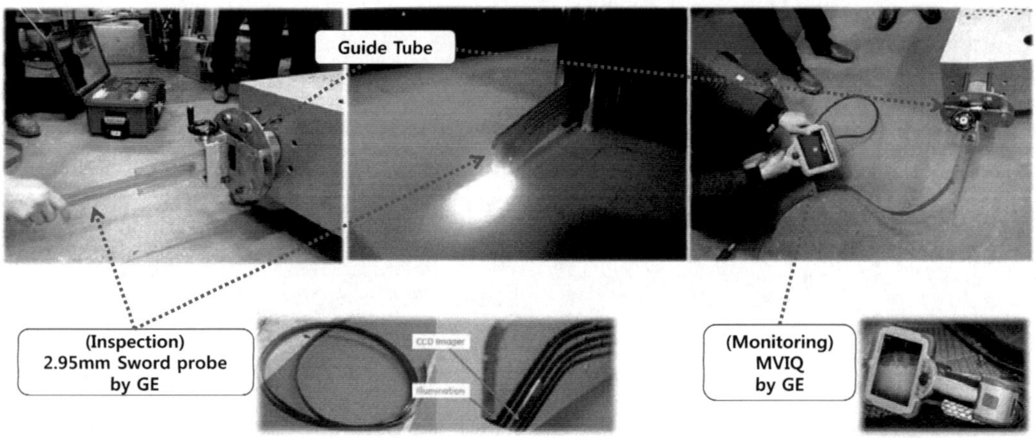

Development of FOSAR equipment

- S/G Structural characteristic and FOSAR Equipment

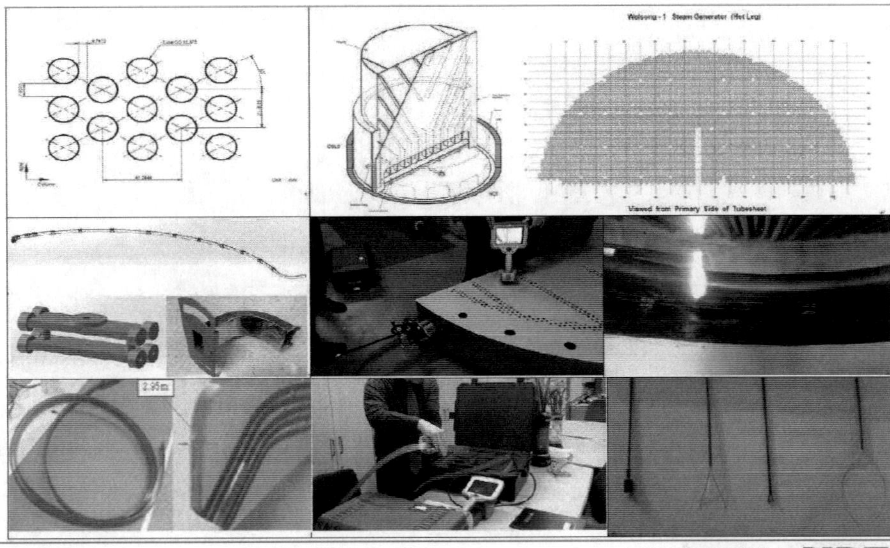

Inspection Results

- ECT & FOSAR
  - Unit 1&2 : No signal of degradation mechanism besides DSI(1)
  - Unit 1 : A number of ECT PLP signal
    - ✓Detection : ECT 81(404 signals) + FOSAR 146 = Total 227
    - ✓Removal : FOSAR 108
    - ✓Residue : 119
      - ❖ Access restriction area of FOSAR equipment because of sludge
      - ❖ Prevention Maintenance according to the evaluation results
        - ✓ Tube plugging : 1 tube (Remaining lifetime : 1.2 yr)
  - ECT PLP evaluation was proved as FOSAR performance
    - ✓In previous outages it was evaluated as deposit
    - ✓After development of FOSAR equipment, it was confirmed foreign objects
  - Unit 2 : ECT PLP (1) but FOSAR result was a hard sludge

Inspection Results

- ECT & FOSAR
  - Unit 1 : Foreign object 108(73.9%) of 146 were removed
    - ✓ Most foreign objects were thin plate type, Processing chip (Mn 0.38%, Si 0.60% : Corroded carbon steel)
    - ✓ Applying the development(R&D) of FOSAR Equipment

S / G	Location	Detection	Removal	Non Removal
#1	In bundle	30	25	5
	Annulus	0	0	0
	No tube lane	2	2	0
#2	In bundle	50	38	12
	Annulus	0	0	0
	No tube lane	0	0	0
#3	In bundle	48	34	14
	Annulus	1	1	0
	No tube lane	1	1	0
#4	In bundle	14	7	7
	Annulus	0	0	0
	No tube lane	0	0	0
소계	In bundle	142	104	38
	Annulus	1	1	0
	No tube lane	3	3	0

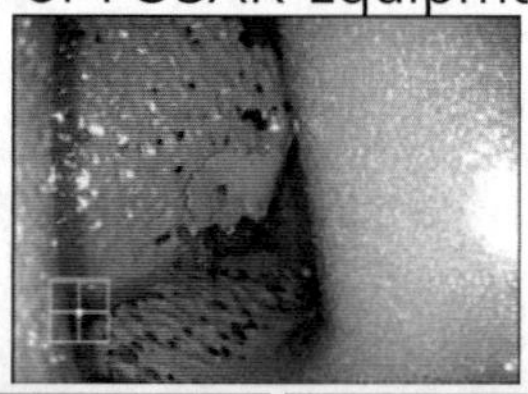

- ✓ Influx Route : S/G moisture separator remodeling work in 1984

Inspection Results

- ECT : DSI(Distorted Support Indication) of Unit 1
  - +Point MRPC performance for Anti-Vibration Bar 6th
  - Result : Wear defect 8% (Plugging Criteria : 40%)

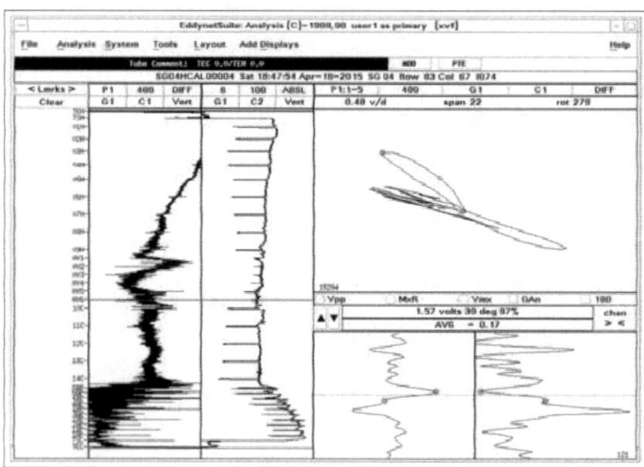

[23th Outage]

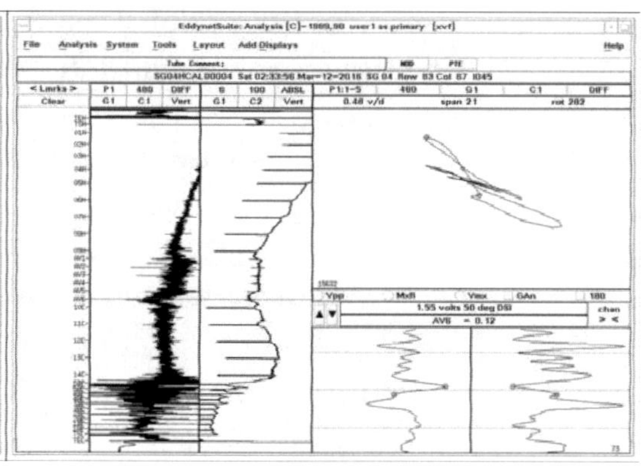

[24th Outage]

Inspection Results

- Sludge Depth & Sludge Lancing
  - Unit 1 : Steadily increase & distribution around hot leg center

Date	S/G	Sludge Depth(inch)		Sludge Lancing (kg)	Remarks
		Max	Avg		
15th('01.09)	#1			30	B&W Canada
	#2				
	#3				
	#4				
:	:	:	:	:	:
21th('10.06)	#1	6.25	2.46		Refurbishment ('09.4~'11.7)
	#2	6.73	3.20		
	#3	7.36	2.85		
	#4	7.96	3.00		
22th('12.07)	#2	7.49	3.13		
	#4	8.61	3.33		
23th('15.03)	#1	8.56	3.13		
	#3	9.93	3.51		
24th('16.03)	#2	8.33	3.00		
	#4	9.19	3.24		

KHNP Wolsong NPP1

6th International CANDU In-Service Inspection Workshop · NDT in Canada 2016 Conference · November 15-17, 2016 · Holiday Inn Burlington Hotel & Conference Centre · Burlington, ON

Inspection Results

- Sludge Depth & Sludge Lancing
  - Unit 2 : Slightly increase

Date	S/G	Sludge Depth(inch)		Sludge Lancing (kg)	Remarks
		Max	Avg		
12th('12.02)	#2	2.94	1.00	4.3	
	#4	2.02	1.08	1.7	
13th('13.05)	#1	1.29	0.68	2.6	
	#3	2.02	0.76	2.5	
14th('14.07)	#2	3.12	1.04	2.6	
	#4	2.27	0.99	2.4	
15th('15.10)	#1	1.72	0.87	3.0	
	#3	2.32	0.77	1.9	

KHNP Wolsong NPP1

6th International CANDU In-Service Inspection Workshop · NDT in Canada 2016 Conference · November 15-17, 2016 · Holiday Inn Burlington Hotel & Conference Centre · Burlington, ON

Conclusions

- In Unit 1, many foreign objects present inside S/G were removed through the FOSAR
  - S/G Reliability was proved by performing the residue evaluation
- Developed FOSAR system was identified the performance
  - Field test was successful and some improvement is needed
- ECT signal evaluation should be cautious if it's an old plant and FOSAR has not been performed for a long time
- For next, to overcome the access restriction area of FOSAR system
  - Development of Sludge Lancing system required(proceeding, ~'17.9)
  - (First) Sludge Lancing (later) FOSAR

Thank you

(jwlee357@khnp.co.kr)

KHNP Wolsong NPP1
6th International CANDU In-Service Inspection Workshop NDT in Canada 2016 Conference November 15-17, 2016 Holiday Inn Burlington Hotel & Conference Centre Burlington, ON

AUTHOR INDEX

Baron, J. 207
Bendada, A. 358
Buck, J.A. 47, 339
Carignan, J. 179
Cascante, G. 350
Chen, A. 369
Cyr, P. 233
Demers-Carpentier, V. 55
Dizeu, F.B.D. 358
Eaton, C. 39
Faisai, F. 409
Faurschou, K. 280
Franca, D.R. 189
Grenier, M. 55
Grotenhuis, R.T. 369
Hardy, F. 55
Hassan, M.U. 103
Hesabi, S. 358
Holloway, P. 179
Hong, A. 369
Horn, D. 310
Jang, K. 154, 207
Jayasinghe, V. 39
Johnston, D.P. 339
Karpelson, A. 241
Kenefick, S. 62
Khan, T.M.R. 103
Kleiman, J. 378
Klein, G. 395
Kramer, C. 47
Krause, T.W. 280, 318, 339, 395
Kudryavtsev, Y. 378
Latiolais, C. 62, 164, 173
Laurendeau, D. 358
Lee, J.U. 422
Lei, J. 47, 310
Lepine, B.A. 47, 310
Luloff, M. 395
Magny, J. 179
Marx, N. 293
Millejours, B. 39
Mirsadeghi, S.M. 112
Mohseni, E. 189
Morelli, J. 280, 339, 395
O'Regan, P. 199
Osho, S. 409
Ouellet, T. 179
Ouyan, H. 20
Pandey, M.D. 350
Rioux, P. 233
Rochette, M. 55
Rodriguez-Roblero, M.J. 350
Shang, J. 1, 11
Sisto, M.M. 55
Sullivan, S. 88, 124
Tallavo, F.J. 350
Tremblay, C. 55
Turcotte, J. 233
Turgeon, M. 55
Turnbow, M.L. 223
Underhill, P.R. 280, 318, 339
Vaidya, V. 179
Verma, Y. 369
Viens, M. 189
Warraich, S.A. 103
Weed, J. 154, 207
Wright, M. 146
Wu, N. 409
Yang, S. 207
Yee, R. 207
Young, N.K. 132
Zhao, S. 409